Problem Books in Mathematics

Series Editor:

Peter Winkler
Department of Mathematics
Dartmouth College
Hanover, NH
USA

More information about this series at http://www.springer.com/series/714

Antonio Caminha Muniz Neto

An Excursion through Elementary Mathematics, Volume II

Euclidean Geometry

 Springer

Antonio Caminha Muniz Neto
Universidade Federal do Ceará
Fortaleza, Ceará, Brazil

ISSN 0941-3502   ISSN 2197-8506   (electronic)
Problem Books in Mathematics
ISBN 978-3-030-08589-6   ISBN 978-3-319-77974-4   (eBook)
https://doi.org/10.1007/978-3-319-77974-4

Printed on acid-free paper

This Springer imprint is published by the registered company Springer International Publishing AG part
of Springer Nature.
The registered company address is: Gewerbestrasse 11, 6330 Cham, Switzerland

To my dearest wife Monica,
for all that goes without saying.

Preface

This is the second volume of a series of three volumes (the other ones being [5] and [6]) devoted to the mathematics of mathematical olympiads. Generally speaking, they are somewhat expanded versions of a collection of six volumes, first published in Portuguese by the Brazilian Mathematical Society in 2012 and currently in its second edition.

The material collected here and in the other two volumes is based on course notes that evolved over the years since 1991, when I first began coaching students of Fortaleza to the Brazilian Mathematical Olympiad and to the International Mathematical Olympiad. Some ten years ago, preliminary versions of the Portuguese texts also served as textbooks for several editions of summer courses delivered at UFC to math teachers of the Cape Verde Republic.

All volumes were carefully planned to be a balanced mixture of a smooth and self-contained introduction to the fascinating world of mathematical competitions, as well as to serve as textbooks for students and instructors involved with math clubs for gifted high school students.

Upon writing the books, I have stuck myself to an invaluable advice of the eminent Hungarian-American mathematician George Pólya, who used to say that one cannot learn mathematics without *getting one's hands dirty*. That's why, in several points throughout the text, I left to the reader the task of checking minor aspects of more general developments. These appear either as small omitted details in proofs or as subsidiary extensions of the theory. In this last case, I sometimes refer the reader to specific problems along the book, which are marked with an * and whose solutions are considered to be an essential part of the text. In general, in each section I collect a list of problems, carefully chosen in the direction of applying the material and ideas presented in the text. Dozens of them are taken from former editions of mathematical competitions and range from the almost immediate to real challenging ones. Regardless of their level of difficulty, generous hints, or even complete solutions, are provided to virtually all of them.

A quick glance through the Contents promptly shows that this second volume deals with plane and solid Euclidean geometry. Generally speaking, Chaps. 1 to 9 deal with plane geometry, whereas the remaining ones with solid geometry. We now describe the material covered a bit more specifically.

The text begins in a somewhat informal way, relying on the reader's previous knowledge of the basics of geometry and emphasizing simple geometric constructions. This is done purposefully, so that the axiomatic method does not engulf the exposition, from the start, with an amount of formalism unnecessary for our goals. Nevertheless, as the text evolves and deeper results are presented, the synthetic method of Euclid gains paramount importance, and from this time on several beautiful classical theorems, usually absent from high school textbooks, make their appearance.

After a quick review of the most elementary concepts and results, Chaps. 2 to 5 discuss the central ideas of congruence, locus, similarity, and area. Apart from what is usually expected, a number of additional topics and results are discussed, among which Ptolemy's problem on the locus of points with prescribed ratio of distances to two other given points, the collinearity and concurrence results of Menelaus and Ceva, some of Euler's classical results on the geometry of the triangle, the notion of power of a point with respect to a circle and Apollonius' tangency problems, and the isoperimetric problem for triangles. Also, from a theoretical standpoint, a careful development of the notion of area and circumference of a circle is presented in Chap. 5.

The last four chapters dealing with plane geometry present analytic geometry, trigonometry, vectors, and some projective geometry as distinct, though interrelated, tools for the study of plane Euclidean geometry. We do this without being too encyclopedic, so as not to overshadow the central ideas. On the one hand, we believe this way we make it easier for the reader to grasp the role of each such portion of knowledge amid the whole of geometry. On the other hand, such additional methods are applied both to expand the theory and to get further insight on previous results and examples. For instance, the text brings three sections on conics, two in Chap. 6 using analytic and synthetic tools, and a third one in Chap. 10, using simple solid geometry concepts to extend to conics a bunch of results of projective nature, discussed in Chap. 9.

Reflecting the current trend in mathematical competitions, our exposition of solid geometry is shorter than that of plane geometry. Nevertheless, it covers all of what is usually present in high school curricula, as well as some other more profound topics. Among these, we would like to mention the representation of conics as conic sections, the use of central projections to the study of some projective properties of conics, the discussion of some aspects of the interesting class of isosceles tetrahedra, the presentation of a complete—though simple—proof of Euler's theorem on convex polyhedra, the classification and construction of all regular polyhedra, as well as the computation of their volumes, and a glimpse on inversion in three-dimensional space. Since the reader is expected to reach the solid geometry chapters with a thorough grounding on plane geometry, some of these topics are partially covered amid the proposed problems. However, whenever we do so, we provide essentially full solutions to them.

Several people and institutions contributed throughout the years for my efforts of turning a bunch of handwritten notes into these books. The State of Ceará Mathematical Olympiad, created by the Mathematics Department of the Federal University of Ceará (UFC) back in 1980 and now in its 37th edition, has since then motivated hundreds of youngsters of Fortaleza to deepen their studies of mathematics. I was one such student in the late 1980s, and my involvement with this competition and with the Brazilian Mathematical Olympiad a few years later had a decisive influence on my choice of career. Throughout the 1990s, I had the honor of coaching several brilliant students of Fortaleza to the Brazilian Mathematical Olympiad. Some of them entered Brazilian teams to the IMO or other international competitions, and their doubts, comments, and criticisms were of great help in shaping my view on mathematical competitions. In this sense, sincere thanks go to João Luiz de A. A. Falcão, Roney Rodger S. de Castro, Marcelo M. de Oliveira, Marcondes C. França Jr., Marcelo C. de Souza, Eduardo C. Balreira, Breno de A. A. Falcão, Fabrício S. Benevides, Rui F. Vigelis, Daniel P. Sobreira, Samuel B. Feitosa, Davi Máximo A. Nogueira, and Yuri G. Lima.

Professor João Lucas Barbosa, upon inviting me to write the textbooks to the Amílcar Cabral Educational Cooperation Project with Cape Verde Republic, had unconsciously provided me with the motivation to complete the Portuguese version of these books. The continuous support of Professor Hilário Alencar, president of the Brazilian Mathematical Society when the Portuguese edition was first published, was also of great importance for me. Special thanks go to my colleagues— Professors Samuel B. Feitosa and Fernanda E. C. Camargo—who read the entire English version and helped me improve it in a number of ways. If it weren't for my editor at Springer-Verlag, Mr. Robinson dos Santos, I almost surely would not have had the courage to embrace the task of translating more than 1500 pages from Portuguese into English. I acknowledge all the staff of Springer involved with this project in his name.

Finally, and mostly, I would like to express my deepest gratitude to my parents Antonio and Rosemary, my wife Monica, and our kids Gabriel and Isabela. From early childhood, my parents have always called my attention to the importance of a solid education, having done all they could for me and my brothers to attend the best possible schools. My wife and kids fulfilled our home with the harmony and softness I needed to get to endure on several months of work while translating this book.

Fortaleza, Brazil   Antonio Caminha Muniz Neto
December 2017

Contents

Chapter 1
Basic Geometric Concepts

This book is devoted to the study of *Euclidean Geometry*, so named after the famous book *Elements* [11], of Euclid of Alexandria.[1]

We will guide our discussions, as much as possible, by the use of logical reasoning. However, we shall not be too seriously concerned with the problem of listing an exhaustive set of postulates from which one could construct Euclidean Geometry axiomatically.[2] To such a program, we refer the reader to [12] or [19]. For more than we shall present here, we suggest [1, 7, 14, 16, 24–26] or [27].

In this chapter, we present the most basic concepts and results involved in the construction of Euclidean Geometry *in a plane*.

1.1 Introduction

The reader certainly has a good idea, from both daily experience and previous studies, of what is a **point**, a **line** and a **plane**. Therefore, we shall assume these notions as **primitive concepts**, so that we shall not present formal definitions of them. We shall further assume that every line is a set of (at least two) points.[3]

---

[1]Euclid of Alexandria, Greek mathematician of the fourth and third centuries BC, and one of the most important mathematicians of classical antiquity. The greatest contribution of Euclid to Mathematics, and to science in general, was the treatise *Elements*, in which he systematically exposed all knowledge of his time in Geometry and Arithmetic. The importance of the *Elements* lies in the fact that it was the first book ever written in which a body of mathematical knowledge was presented in an axiomatic way, with all arguments relying solely on Logic.

[2]An *axiom* or *postulate* is a property imposed as true, without the need of a proof. The use of the *axiomatic method* is one of the most fundamental characteristics of Mathematics as a whole.

[3]For the time being, we also implicitly assume that all points under consideration are contained in a single plane, and that there exists at least three points not situated in the same line.

Figure 1.1 shows points A and B and lines r and s (we shall denote points and lines by upper case and lower case Latin letters, respectively). Roughly speaking, we could say that Plane Euclidean Geometry studies the properties of points and lines situated in a plane.

The following discussion will serve as a basis for all further developments of the theory. Throughout, all statements made without proofs should be taken as axioms.

Given a point P and a line r, there are only two possibilities: either the point P belongs to the line r or it doesn't; in the first case, we write $P \in r$ (P *belongs to* r) whereas, in the second, we write $P \notin r$ (P *does not belong to* r). In Fig. 1.2, we have $A \in r$ and $B \notin r$.

At this point, it is natural to ask how many lines join two different given points. We assume (as an axiom!) that there is *exactly one* such line. In short, *there is exactly one line passing through two distinct given points* (cf. Fig. 1.3). In this case, letting r be such a line, we denote $r = \overleftrightarrow{AB}$, whenever convenient.

A point A on a line r divides it into two *pieces*, namely, the **half-lines** of origin A. By choosing points B and C on r, each one situated in one of these two pieces, we can denote the half-lines of origin A by $\overrightarrow{AB}$ and $\overrightarrow{AC}$. In Fig. 1.4, we show the portion of line r corresponding to the half-line $\overrightarrow{AB}$ (the portion corresponding to the half-line $\overrightarrow{AC}$ has been erased).

Given two distinct points A and B on a line r, the **line segment** AB is the portion of r situated *from A to B*. We write $\overline{AB}$ to denote the length of the line segment AB (and, unless we explicitly state otherwise, such a length will be measured in centimeters). In order to decide whether two given line segments are *equal* (i.e., have equal lengths) or, on the contrary, each of which has greatest length, we can use a compass, *transporting* one of these line segments to the line determined by the other. This is done in the coming.

Fig. 1.1 Points and lines in a plane

Fig. 1.2 Relative positions of points and lines

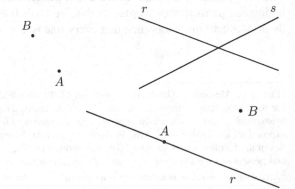

Fig. 1.3 Two distinct points
determine one single line

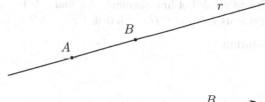

Fig. 1.4 Half-line $\overrightarrow{AB}$, of
origin A

Example 1.1 Transport line segment AB to line $\overleftrightarrow{CD}$ and decide whether $\overline{AB} > \overline{CD}$ or vice-versa.[4]

Solution

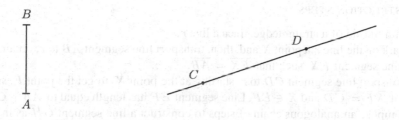

CONSTRUCTION STEPS

1. Place the nail end of the compass at A and open it until the other end is exactly over B.
2. By maintaining the opening of the compass, place its nail end at C and draw an arc that intersects half-line $\overrightarrow{CD}$ at a point E, so that $\overline{CE} = \overline{AB}$.
3. Compare the lengths of line segments $\overline{AB} = \overline{CE}$ and $\overline{CD}$.

□

We can also use straightedge and compass to *add* line segments and to *multiply* a line segment by a natural number, as the next example shows.

---

[4]This is the first one of a series of examples whose purpose is to develop, in the reader, a relative ability on the use of straightedge and compass. In all of them, we list a sequence of steps which, once followed, execute a specific geometric construction. After attentively reading each of these examples, we strongly urge the reader to reproduce the listed steps by him/herself to actually execute the geometric constructions under consideration. Finally, and for the sake of rigor, we observe that a geometric construction *does not* constitute a proof of a geometric property, for it necessarily involves precision errors and particular choices of positions. Its main purpose is to help developing geometric intuition.

Example 1.2 Let line segments AB and CD be given, as shown below. Draw line segments EF and GH, such that $\overline{EF} = \overline{AB} + \overline{CD}$ and $\overline{GH} = 3\,\overline{AB}$.

Solution

CONSTRUCTION STEPS

1. With the aid of a straightedge, trace a line r.
2. Mark on the line r a point X and, then, transport line segment AB to r, obtaining a line segment EX, such that $\overline{EX} = \overline{AB}$.
3. Transport line segment CD to r, starting at the point X, to get the point F, such that $\overline{XF} = \overline{CD}$ and $X \in EF$. Line segment EF has length equal to $\overline{AB} + \overline{CD}$.
4. Compose an analogous chain of steps to construct a line segment GH as in the statement of the example. To this end, note that $3\,\overline{AB} = \overline{AB} + \overline{AB} + \overline{AB}$.

$\square$

We shall make a last (though important) remark on line segments: given points A and B, we define the **distance** $d(A, B)$ between them as the length $\overline{AB}$ of the line segment AB:

$$d(A, B) = \overline{AB}.$$

Besides points, lines, half-lines and segments, *circles* will be of great importance in our study of Plane Euclidean Geometry. In order to define them in a precise way, let a point O and a real number $r > 0$ (that should be thought of as the length of a line segment) be given (Fig. 1.5). The **circle** of **center** O and **radius** r is the set of points P in the plane that are at a distance r from O, i.e., such that $\overline{OP} = r$.

In a more concrete way, the circle of center O and radius r is the *plane curve* we get when we fix the nail end of a compass at the point O, fix its opening as equal to the length r and let its pencil end complete a whole turn around O. The complement of a circle in the plane consists of two regions, a *limited* one, called its **interior**, and an *unlimited* one, called its **exterior**. More precisely, the interior of the circle of center O and radius r is the set of points P in the plane, whose distance from the center O is *less than* r, i.e., such that $\overline{OP} < r$ (cf. Fig. 1.6); analogously, the exterior of such a circle is the set of points P in the plane whose distance from the center O is *greater than* r, i.e., such that $\overline{OP} > r$. Sometimes, we shall refer to the set of points P in the plane for which $\overline{OP} \leq r$ as the (closed) **disk** of center O and radius r.

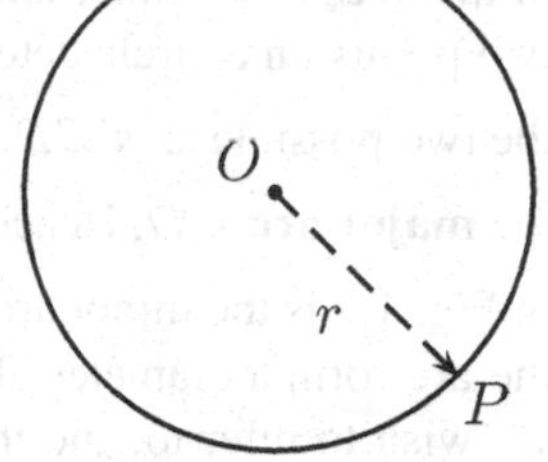

Fig. 1.5 The circle of center
O and radius r

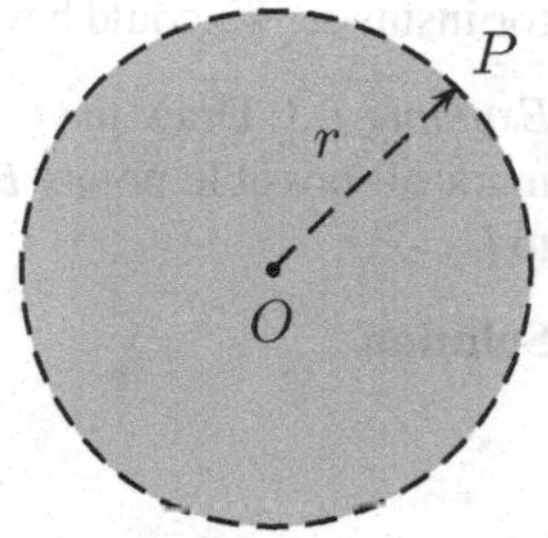

Fig. 1.6 Interior of the circle
of center O and radius r

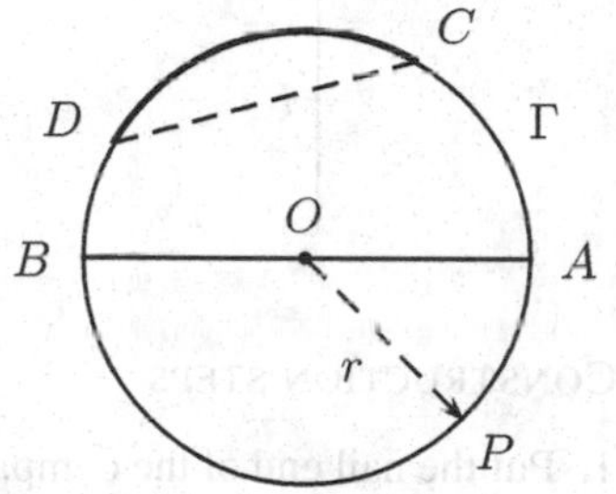

Fig. 1.7 Elements of a circle

As a general rule, we shall denote circles by upper case greek letters. For instance, the circle of Fig. 1.7 is denoted by Γ (one reads *gamma*), and we usually write $\Gamma(O; r)$ to stress that O is the center and r is the radius of Γ.

Given a circle Γ of center O and radius r, we also say that every line segment joining O to a point of Γ is a **radius** of it; for example, in Fig. 1.7, line segments OA, OB and OP are all radii of Γ. A **chord** of Γ is a line segment joining two of its points; a **diameter** of Γ is a chord that passes through O. In the notations of Fig. 1.7, AB and CD are chords of Γ, and AB is also a diameter. Every diameter of a circle divides it into two equal[5] parts, called two **semicircles** of the circle; conversely, if a chord of a circle divides it into two equal parts, then such a chord must necessarily be a diameter of it.

Concerning Fig. 1.7, the reader has surely noticed that a bold line traces out one of the portions of Γ limited by points C and D. Such a portion corresponds to one

---

[5]Of course, at this point we rely on the reader's intuition, or previous knowledge, for the meaning of *equal*.

of the **arcs** of Γ whose **endpoints** are C and D, and will be denoted by $\overset{\frown}{CD}$. Since two points on a circle determine two different arcs, in order to distinguish between the two possible arcs $\overset{\frown}{CD}$, we shall generally refer to the **minor arc** $\overset{\frown}{CD}$, or else to the **major arc** $\overset{\frown}{CD}$. Hence, we should more properly say that the bold portion of Γ in Fig. 1.7 is the minor arc $\overset{\frown}{CD}$. Another possibility (quite useful when the ends of the arc form a diameter of the circle) would be to choose another point on the arc we wish to refer to, and then denote it with the aid of this extra point; in Fig. 1.7, for instance, we could have written $\overset{\frown}{CPD}$ to denote the major arc $\overset{\frown}{CD}$.

Example 1.3 Draw the circle of center O and passing through the point A. Then, mark all possible points B on this circle, such that the chord AB has length equal to l.

Solution

CONSTRUCTION STEPS

1. Put the nail end of the compass at O, with opening equal to $\overline{OA}$. Then, draw the circle of center O and radius $\overline{OA}$.
2. As in item 1., draw the circle of center A and radius l.
3. The possible positions of point B are the intersection points of the circles drawn in items 1. and 2.

□

Problems: Sect. 1.1

1. Let A, B, C and D be distinct points on a line r. How many half-lines are contained in r and have one of these points as origin?
2. Points A, B and C are all situated on a line r, with $C \in AB$. If $\overline{AB} = 10$cm and $\overline{AC} = 4\,\overline{BC}$, compute $\overline{AC}$.
3. Let A, B, C and D be distinct points on a line r, such that $D \in \overrightarrow{AC}$, $B \in \overrightarrow{DC}$ and $\overline{AC} = \overline{BD}$. Prove that $\overline{AB} = \overline{CD}$.
4. On a line r points A, B and C are given, so that $B \in AC$, $\overline{AB} = 3$cm and $\overline{AC} = 5.5$cm. Use a compass to find, on the segment AB, a point D such that $\overline{AD} = \overline{BC}$.

5. With the aid of straightedge and compass, mark points A, B and C on the plane, such that $\overline{AB} = 5$cm, $\overline{AC} = 6$cm and $\overline{BC} = 4$cm.

1.2 Angles

The concepts of angle and angle measurement are absolutely fundamental to almost everything we will do in the rest of these notes. Prior to discussing them, we shall need to introduce another important notion, which is the object of our first formal

Definition 1.4 A region $\mathcal{R}$ in the plane is said to be **convex** if, for all points A, $B \in \mathcal{R}$, we have $AB \subset \mathcal{R}$. Otherwise, $\mathcal{R}$ is said to be a **non-convex** region.

According to the previous definition, for a region $\mathcal{R}$ in the plane to be non-convex, it suffices to find points A, $B \in \mathcal{R}$ such that at least one point of the line segment AB does not belong to $\mathcal{R}$ (Fig. 1.8).

A line r in the plane divides it into two convex regions, called the (**closed**) **half-planes** bounded by r. By definition, the line r is considered to be a part of each one of them. On the other hand, given points A and B, one in each of these two half-planes, we always have $AB \cap r \neq \emptyset$ (cf. Fig. 1.9).

Definition 1.5 Let $\overrightarrow{OA}$ and $\overrightarrow{OB}$ be two distinct half-lines in the plane. An **angle** or **angular region**) of **vertex** O and **sides** $\overrightarrow{OA}$ and $\overrightarrow{OB}$ is one of the regions in which the plane is divided by the half-lines $\overrightarrow{OA}$ and $\overrightarrow{OB}$.

An angle can be convex or non-convex; in Fig. 1.10, the angle at the left side is convex, whereas that at the right side is nonconvex. We shall denote an angle with sides $\overrightarrow{OA}$ and $\overrightarrow{OB}$ by writing $\angle AOB$; the context will make it clear whether we are referring to the convex angle or to the non-convex one.

Now, it is our purpose to associate to each angle a measure of the region of the plane it occupies. To this end (cf. Fig. 1.11), let us consider a circle Γ centered at O, and divide it into 360 equal arcs; then, take two of the 360 partitioning points, say X and Y, which are the ends of one of these 360 equal arcs. We then say that the **measure** of the angle $\angle XOY$ is one **degree**, and denote this by writing

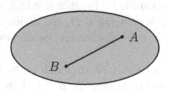
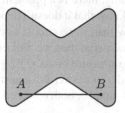

Fig. 1.8 A convex region (left) and a non-convex one (right)

Fig. 1.9 (Closed) half-planes
determined by the line r

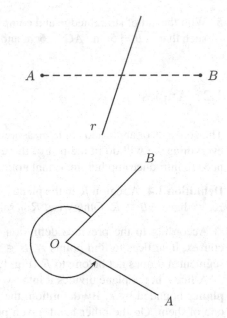

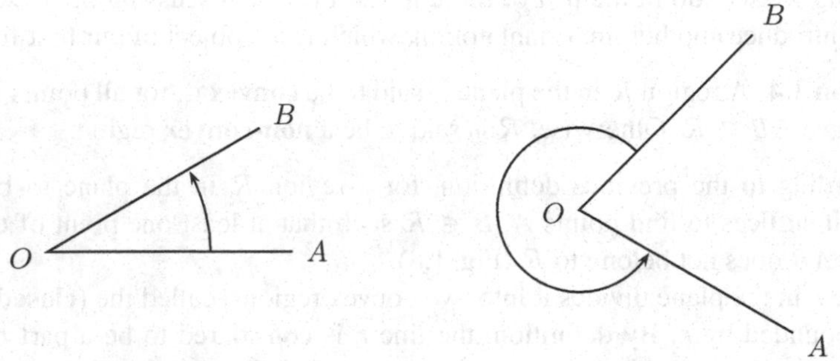

Fig. 1.10 Two angular regions in the plane

Fig. 1.11 Degree as unit of
measure of angles

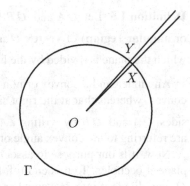

$$X\widehat{O}Y = 1°.$$

Actually, there is a problem with our definition of degree, as given above. How can we know that it doesn't depend on the chosen circle Γ? In other words, how can we know that, if we choose another circle Γ', also centered at O, and divide it into 360 equal parts, then we will obtain an angle $\angle X'OY'$ which could be said to have the same measure as $\angle XOY$?

In order to answer this question, let us consider Fig. 1.12, in which we have two circles Γ and Γ', both centered at the point O, and two points $A, B \in \Gamma$. Letting A' and B' be the points of intersection of the half-lines $\overrightarrow{OA}$ and $\overrightarrow{OB}$ with Γ', we shall assume as an *axiom* that the fraction of Γ represented by the bold arc $\widehat{AB}$ equals

Fig. 1.12 Well definiteness
of the notion of degree

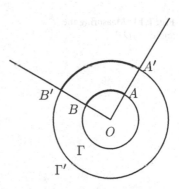

the fraction of Γ' represented by the bold arc $\overset{\frown}{A'B'}$. Therefore, if in the definition of degree we had taken a circle Γ', also centered at O but of radius different from that of Γ, then we would have got the same angle representing $1°$.

From the definition of degree, it is clear that a complete circle corresponds to $360°$. On the other hand, given an angle $\angle AOB$, there remains the question of how one can actually measure it. In order to answer such a question,[6] we start by making the following construction: we draw any circle Γ, centered at O, and mark the points A' and B' in which Γ intersects the sides $\overrightarrow{OA}$ and $\overrightarrow{OB}$ of $\angle AOB$ (cf. Fig. 1.13).

Then, we compute the fraction of the total length of Γ the arc $\overset{\frown}{A'B'}$ represents. The measure $A\widehat{O}B$ of the angle $\angle AOB$ will be equal to that fraction of $360°$. For instance, if the length of the arc $\overset{\frown}{A'B'}$ equals $\frac{1}{6}$ of the total length of Γ, then the measure of $\angle AOB$ will be

$$A\widehat{O}B = \frac{1}{6} \cdot 360° = 60°.$$

Remarks 1.6

  i. We shall say that two angles are **equal** if their measures are equal.
 ii. In order to avoid any danger of confusion, we shall always use different notations for an angle and its measure. More precisely, $\angle AOB$ will always refer to the angle with vertex O and sides $\overrightarrow{OA}$ and $\overrightarrow{OB}$ (which is a region in the plane), whereas $A\widehat{O}B$ will always refer to the measure, in degrees, of the angle $\angle AOB$.

---

[6]Strictly speaking, the following argument is fallacious, for, among other things, it invokes the notion of length of an arc of a circle, something which has not yet been defined. Nevertheless, it develops quite a useful intuition for the measurement of angles that is enough for our purposes along these notes. For a thorough discussion of the measurement of angles, we refer the reader to [19].

Fig. 1.13 Measuring the
angle $\angle AOB$

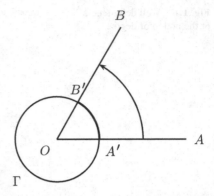

iii. In context, we will also frequently use lower case greek letters to denote
measures of angles.[7] For example, we shall write $A\widehat{O}B = \theta$ (one reads *teta*) to
mean that the measure of the angle $\angle AOB$ is θ degrees.

The next example teaches us how to use straightedge and compass to construct
an angle with a prescribed vertex, side and measure. The construction steps to be
listed will be justified when we study the SSS postulate of congruence of triangles,
in Sect. 2.1.

Example 1.7 Construct an angle with vertex O', such that one of its sides is
contained in the line r and its measure equals α.

Solution

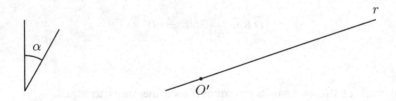

$\square$

Fig. 1.14 An angle of 180°

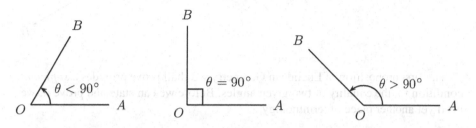

Fig. 1.15 Acute (left), right (center) and obtuse (right) angles

CONSTRUCTION STEPS

1. Draw a circle of some radius R, centered at the vertex O of the given angle, thus marking points X and Y along the sides of it.
2. Draw another circle of radius R, this time centered at O', and mark one of the points, say Y', in which it intersects line r.
3. Mark X' as one of the points of intersection of the circles $\Gamma(O'; R)$ and $\Sigma(Y'; \overline{XY})$.
4. The angle $\angle X'O'Y'$ has measure equal to α.

As we have pointed out before, each diameter of a circle divides it into two equal parts. Therefore, if $\angle AOB$ is such that $\overrightarrow{OA}$ and $\overrightarrow{OB}$ are *opposite* half-lines (i.e., such that A, O and B lie on the same line, with $O \in AB$—see Fig. 1.14), then $A\widehat{O}B = 180°$.

We shall rarely encounter angles whose measures are greater than 180°. Thus, in all that follows and unless stated otherwise, whenever we write $\angle AOB$ we will be referring to the *convex* angle $\angle AOB$ (i.e., to the angle $\angle AOB$ such that $0° < A\widehat{O}B \le 180°$. Accordingly (cf. Fig. 1.15) we shall say that an angle $\angle AOB$ is **acute** if $0° < A\widehat{O}B < 90°$, **right** if $A\widehat{O}B = 90°$ and **obtuse** if $90° < A\widehat{O}B < 180°$. We call the reader's attention (cf. Fig. 1.15), to the particular way of denoting a right angle.

It will be sometimes useful to have a special name attached to two angles whose sum of measures equals 90°; from now on, we shall say that two such angles are **complementary**. This way, if α and β are the measures of two complementary angles, then $\alpha + \beta = 90°$. Also, in such a case we shall say that the angle of measure α (or that α itself) is the **complement** of the angle of measure β (or of β itself), and vice-versa. For example, two angles whose measures are 25° and 65° are complementary, since $25° + 65° = 90°$; we can then say that one angle is the complement of the other, or that 25° is the complement of 65° (and vice-versa).

Fig. 1.16 Opposite angles
∠AOB and ∠COD

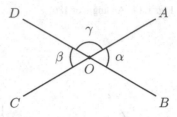

The first proposition of Euclidean Geometry we shall prove provides a *sufficient* condition for the equality of two given angles. Before we can state and prove it, we need yet another piece of terminology.

Definition 1.8 Angles ∠AOB and ∠COD (with a common vertex O) are said to be **opposite** (we abbreviate **OPP**) if their sides, in pairs, are opposite half-lines.

As indicated, angles ∠AOB and ∠COD in Fig. 1.16 are opposite, for $\overrightarrow{OA}$ and $\overrightarrow{OC}$, as well as $\overrightarrow{OB}$ and $\overrightarrow{OD}$, are opposite half-lines. Note that the same is true of angles ∠AOD and ∠BOC.

Proposition 1.9 *Two opposite angles have equal measures.*

Proof In the notations of Fig. 1.16, since $\overrightarrow{OB}$ and $\overrightarrow{OD}$ are opposite to each other, we have $\alpha + \gamma = 180°$. Similarly, $\beta + \gamma = 180°$, so that

$$\alpha = 180° - \gamma = \beta.$$

□

Problems: Sect. 1.2

1. Assume that the intersection of two convex regions of a plane is a nonempty set. Prove that it is also a convex region.
2. If we add the measure of an angle to the triple of the measure of its complement, we get $210°$. What is the measure of the angle?
3. Compute the measures of two complementary angles, knowing that the complement of the double of one of them equals one third of the measure of the other.
4. The measures α and β of two opposite angles are expressed, in degrees, by $9x - 2$ and $4x + 8$. Compute the measure, in degrees, of $\alpha + \beta$.
5. Compute the measure of an acute angle, knowing that it exceeds the measure of its complement in $76°$.
6. * If two lines intersect, prove that one of the angles formed by them equals $90°$ if and only if all four angles formed by them equal $90°$.

7. In the figure below, the angle α equals one sixth of the angle γ, plus half of the angle β. Compute α.

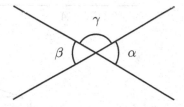

8. Five half-lines, of same origin O, form five angles that cover all of the plane and have measures, in degrees, respectively proportional to the numbers 2, 3, 4, 5 and 6. Compute the measure of the greatest of these five angles.

9. Construct, with straightedge and compass, an angle whose measure equals the sum of the measures of the angles $\angle AOB$ and $\angle A'O'B'$ of the figure below:

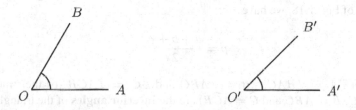

10. Points A, B and C are chosen on a circle of center O, such that the measures of the minor arcs $\overset{\frown}{AB}$, $\overset{\frown}{BC}$ and $\overset{\frown}{AC}$ are all equal. Compute the measures of the angles $\angle AOB$, $\angle BOC$ and $\angle AOC$.

11. Three half-lines of the same origin O form three angles that cover all of the plane. Show that at least one of these angles is greater than or equal to $120°$, and at least one of them is less than or equal to $120°$.

1.3 Convex Polygons

Let A, B and C be three given points in the plane. If C is on the line $\overleftrightarrow{AB}$, we say that A, B and C are **collinear**; otherwise, we say that A, B and C are **non collinear** (cf. Fig. 1.17).

Three non collinear points A, B and C form a **triangle** ABC. In this case, we say that A, B and C are the **vertices** of the triangle ABC, and that the line segments AB, AC and BC are the **sides** or **edges** of ABC. The **triangular region** corresponding to the triangle ABC (also denoted by ABC, whenever there is no danger of confusion) is the bounded portion of the plane whose boundary is the union of the sides of ABC (the dashed portion of the plane in Fig. 1.17).

Fig. 1.17 Three non collinear points

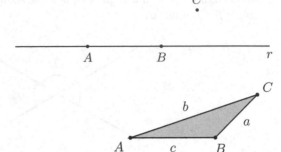

Fig. 1.18 Triangle and triangular region ABC

Given a triangle ABC, whenever there is no danger of confusion we shall also refer to the lengths of the line segments AB, BC and AC as the *sides* of the triangle. In this case, we usually write (cf. Fig. 1.18) $\overline{AB} = c$, $\overline{AC} = b$ and $\overline{BC} = a$. The sum of the lengths of the sides of a triangle is its **perimeter**, which will be denoted, from now on, by $2p$. Hence, p is the **semiperimeter** of the triangle and, in the notations of Fig. 1.18, we have

$$p = \frac{a+b+c}{2}. \tag{1.1}$$

Angles $\angle A = \angle BAC$, $\angle B = \angle ABC$ and $\angle C = \angle ACB$ (or their measures $\widehat{A} = B\widehat{A}C$, $\widehat{B} = A\widehat{B}C$ and $\widehat{C} = A\widehat{C}B$) are the **interior angles** of the triangle.

Triangles are usually classified in two different ways: with respect to the lengths of their sides or with respect to the measures of their interior angles. For the time being, let us see how to classify a triangle with respect to the lengths of its sides. Since each triangle has exactly three sides, the only possibilities for their lengths are that at least two of them are equal or that they are pairwise distinct. Thus, we have the following definition.

Definition 1.10 A triangle ABC is said to be:

(a) **Equilateral**, if $\overline{AB} = \overline{AC} = \overline{BC}$.
(b) **Isosceles**, if at least two of $\overline{AB}$, $\overline{AC}$ and $\overline{BC}$ are equal.
(c) **Scalene**, if $\overline{AB} \neq \overline{AC} \neq \overline{BC} \neq \overline{AB}$.

Note that, by the previous definition, every equilateral triangle is also isosceles. However, the converse statement is not true; for instance, for the triangle ABC at the center of Fig. 1.19, we clearly have $\overline{AB} = \overline{AC} \neq \overline{BC}$, so that it is isosceles but not equilateral.

If ABC is an isosceles triangle with $\overline{AB} = \overline{AC}$, we say that the third side BC is a **basis** for the triangle. Thus, each side of an equilateral triangle is a basis of it, but we rarely use this term in the context of equilateral triangles. In other words, we usually reserve the word *basis* for the unequal side of an isosceles triangle which is not equilateral.

A triangle is a particular type of *convex polygon*, according to the definition that follows.

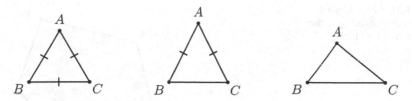

Fig. 1.19 Equilateral (left), isosceles (center) and scalene (right) triangles

Fig. 1.20 A convex polygon
of five vertices (and sides)

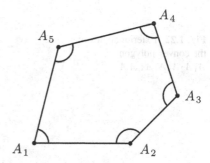

Definition 1.11 Let $n \geq 3$ be a natural number, and $A_1, A_2, \ldots, A_n$ be n distinct points in the plane. We say that $A_1 A_2 \ldots A_n$ is a **convex polygon** if, for $1 \leq i \leq n$, the two following conditions are satisfied (in both items below, we take $A_{n+1} = A_1$):

(a) $\overleftrightarrow{A_i A_{i+1}}$ does not contain any of the other points A_j.

(b) $\overleftrightarrow{A_i A_{i+1}}$ leaves all of the other points A_j in a single half-plane, of the two it determines.

The points $A_1, A_2, \ldots, A_n$ are the **vertices** of the convex polygon $A_1 A_2 \ldots A_n$; the line segments $A_1 A_2, A_2 A_3, \ldots, A_{n-1} A_n, A_n A_1$ (ou, whenever there is no danger of confusion and is convenient, their lengths) are the **sides** or **edges** of it. As with triangles, the sum of the lengths of the sides of a convex polygon is its **perimeter**. Also, the **polygonal region** corresponding to the convex polygon $A_1 A_2 \ldots A_n$ is the bounded region of the plane whose boundary is the union of the sides of the polygon. Figure 1.21 shows the polygonal region corresponding to the convex polygon of Fig. 1.20.

As in Figs. 1.18 and 1.21, it can be shown that the polygonal region corresponding to a convex polygon is a convex region. From now on, we shall assume this to be true, without further comments.

A **diagonal** of a convex polygon is any of the line segments $A_i A_j$ that is not a side of the polygon; for example, the convex polygon $A_1 A_2 \ldots A_5$ of Fig. 1.20 has exactly five diagonals: $A_1 A_3$, $A_1 A_4$, $A_2 A_4$, $A_2 A_5$ and $A_3 A_5$. In Proposition 1.12, we shall prove that every convex polygon of n sides has exactly $\frac{n(n-3)}{2}$ diagonals (see, also, Problem 1).

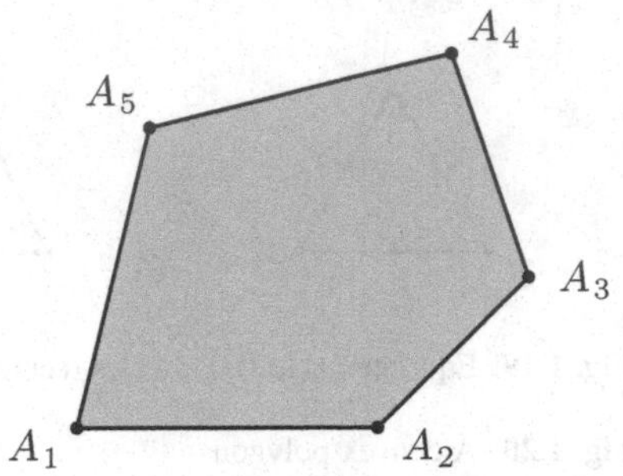

Fig. 1.21 The polygonal
region corresponding to the
polygon of Fig. 1.20

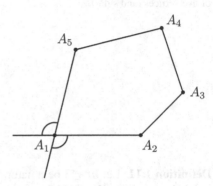

Fig. 1.22 Exterior angles of
the convex polygon
$A_1A_2A_3A_4A_5$ at A_1

The **interior angles** of the convex polygon $A_1A_2\ldots A_n$ are the convex angles
$\angle A_{i-1}A_iA_{i+1}$ (with the convention that $A_{-1} = A_n$ and $A_{n+1} = A_1$), or simply
$\angle A_i$, if there is no danger of confusion. In particular, $A_1A_2\ldots A_n$ has exactly n
interior angles. The **exterior angles** of $A_1A_2\ldots A_n$ at the vertex A_i are the opposite
angles that are the supplements of the interior angle of the polygon at A_i. Figure 1.20
shows the interior angles of the convex polygon $A_1A_2\ldots A_5$, and Fig. 1.22 shows
the exterior angles of $A_1A_2\ldots A_5$ at vertex A_1.

In general, we say that a convex polygon $A_1A_2\ldots A_n$ is a convex **n-gon**, in
reference to the fact that it has n sides (and n vertices). However, we use to say that
the polygon is a **quadrilateral** if $n = 4$, a **pentagon** if $n = 5$, an **hexagon** if $n = 6$,
heptagon if $n = 7$, **octagon** if $n = 8$ and a **decagon** if $n = 10$. Also concerning
specific numbers of sides, we shall sometimes label the vertices of a polygon with
upper case distinct Latin letters. For instance, we shall generally let $ABCD$ denote
a quadrilateral and, in this case, we will always assume, unless stated otherwise, that
their sides are AB, BC, CD and DA. Analogous remarks are valid for pentagons,
hexagons and so on.

The following proposition counts the number of diagonals of a convex n-gon.

Proposition 1.12 *Every convex n-gon has exactly $\frac{n(n-3)}{2}$ diagonals.*

Proof If $n = 3$ there is nothing to prove, for triangles have no diagonals and
$\frac{n(n-3)}{2} = 0$ for $n = 3$. Suppose, then, that $n \geq 4$. Joining vertex A_1 to the other
$n - 1$ vertices $A_2, \ldots, A_n$ we obtain $n - 1$ line segments, two of which are sides

Fig. 1.23 Diagonals of a
convex n-gon departing
from A_1

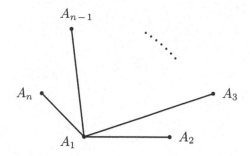

of the polygon (A_1A_2 and A_1A_n) and the remaining $n - 3$ ($A_1A_3, \ldots, A_1A_{n-1}$)
are diagonals (cf. Fig. 1.23). Since an analogous reasoning is valid for every other
vertex of the polygon, we conclude that exactly $n - 3$ diagonals depart from each
vertex.

Adding all of these diagonals would give us a total of $n(n - 3)$ diagonals (i.e.,
$n - 3$ diagonals for each of the n vertices). Nevertheless, this reasoning has counted
each diagonal A_iA_j exactly twice: one when we counted the diagonals departing
from A_i and the other when we counted those departing from A_j. Hence, in order
to obtain the correct number of diagonals, we have to divide $n(n - 3)$ by 2, thus
obtaining $\frac{n(n-3)}{2}$ diagonals.   □

Problems: Sect. 1.3

1. Use the principle of mathematical induction (cf. [5], for instance) to prove the
   formula for the number of diagonals of a convex polygon.
2. A convex polygon is such that, from one of its vertices, we can draw as many
   diagonals as those of a convex hexagon. How many sides does the polygon
   have?
3. Three convex polygons have numbers of sides equal to three consecutive natural
   numbers. Knowing that the sum of their numbers of diagonals is equal to
   133, compute the number of sides of the polygon with the largest number of
   diagonals.

Chapter 2
Congruence of Triangles

This chapter is devoted to the study of the usual sets of necessary and sufficient conditions for two triangles to be considered *the same*, in a sense it will soon be made precise. We also discuss here the important *fifth axiom* of Euclid (known as the axiom of parallels), as well as several interesting and important consequences of it, most notably the triangle inequality. Finally, in the last section of the chapter, several special types of quadrilaterals will make their first appearance.

2.1 The SAS, ASA and SSS Cases

Let's start by considering the following example.

Example 2.1 Use straightedge and compass to construct an equilateral triangle ABC whose sides have lengths equal to l.

Solution

l

CONSTRUCTION STEPS

1. Mark an arbitrary point A in the plane.
2. Use the compass to draw the circle of center A and radius l.
3. Mark an arbitrary point B on this circle.
4. Use the compass to draw the circle of center B and radius l.

© Springer International Publishing AG, part of Springer Nature 2018
A. Caminha Muniz Neto, *An Excursion through Elementary Mathematics, Volume II*,
Problem Books in Mathematics, https://doi.org/10.1007/978-3-319-77974-4_2

5. Letting C denote any of the two points of intersection of these two circles, we
   get the desired triangle ABC.

   $\square$

In the above example, we have constructed a triangle having certain pre-established properties (in this case, being equilateral). While solving the example, we implicitly assumed that there was *essentially one* triangle satisfying the desired properties; in other words, we tacitly assumed that any other equilateral triangle of side lengths equal to l we could have constructed should be regarded as *equal* to the one we actually constructed, for it would differ from this one only by its position in the plane.

The discussion of the previous paragraph motivates a notion of *equality* (rigorously speaking, *equivalence*) for triangles, which receives the special name of **congruence**: we say that two triangles are **congruent** if it is possible to move one of them in space, without deforming it, until we make it coincide with the other one.

Thus, if two triangles ABC and $A'B'C'$ are congruent, there must exist a *correspondence* between the vertices of one triangle and those of the other, in such a way that the interior angles at corresponding vertices are equal and the sides opposite to corresponding vertices are also equal. Figure 2.1 shows two congruent triangles ABC and $A'B'C'$, with the correspondence of vertices

$$A \longleftrightarrow A'; \quad B \longleftrightarrow B'; \quad C \longleftrightarrow C'.$$

For such triangles, we have, then

$$\begin{cases} \widehat{A} = \widehat{A'}; \ \widehat{B} = \widehat{B'}; \ \widehat{C} = \widehat{C'} \\ \overline{AB} = \overline{A'B'}; \ \overline{AC} = \overline{A'C'}; \ \overline{BC} = \overline{B'C'} \end{cases}.$$

It is immediate to see that the notion of congruence of triangles has the following properties[1]:

1. **Symmetry**: it does not matter if we say that a triangle ABC is congruent to a triangle DEF or that DEF is congruent to ABC, or even that ABC and DEF are congruent. Indeed, if we can move ABC is space (without deforming it) until we make it coincide with DEF, then we can certainly move DEF backwards, until superpose it with ABC.

Fig. 2.1 Two congruent
triangles

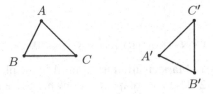

---

[1]The reader with some previous knowledge of Euclidean Geometry will promptly notice that we do not list below the *reflexive* property of congruence of triangles. In this sense, whenever we refer, in a certain context, to *two triangles*, we will implicitly assume that they are *distinct* ones.

2. **Transitivity**: if a triangle ABC is congruent to a triangle DEF and, in turn, triangle DEF is congruent to a third triangle GHI, then ABC will be also congruent to GHI. This is so because we can move ABC is space until make it coincide with GHI in two steps: first we move ABC until it coincides with DEF and, then, we continue the motion until it coincides with GHI.

From now on, we will write

$$ABC \equiv A'B'C'$$

to denote that triangles ABC and $A'B'C'$ are congruent, with the correspondence of vertices

$$A \longleftrightarrow A'; \quad B \longleftrightarrow B'; \quad C \longleftrightarrow C'.$$

It would be interesting (and useful) if we had at our disposal a *minimal* set of criteria to decide whether two given triangles are congruent or not. Moreover, they should be as simple as possible, in order to ease the checking of the desired congruence. Actually, more than one such set of criteria do exist, and these are usually referred to as the **cases of congruence of triangles**.

In what follows, we will study the usual cases of congruence of triangles under an informal point of view. Each such case is preceded by an example involving construction with straightedge and compass, whose solution motivates the case under study.

Example 2.2 Use straightedge and compass to construct a triangle ABC, knowing that $\overline{BC} = a$, $\overline{AC} = b$ and $\widehat{C} = \gamma$.

Solution

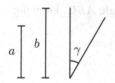

CONSTRUCTION STEPS

1. Mark a point C in the plane and draw a half-line $\overrightarrow{CX}$.

2. Construct an angle $X\widehat{C}Y = \gamma$, thus determining another half-line $\overrightarrow{CY}$.

3. On the half-lines $\overrightarrow{CX}$ and $\overrightarrow{CY}$ mark points B and A such that $\overline{AC} = b$ a $\overline{BC} = a$, respectively.

□

Upon executing the construction steps above several times, we become more and more confident that, if we choose any two distinct positions for the vertex C and two

Fig. 2.2 The SAS
congruence case

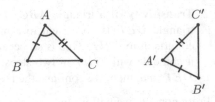

different directions for the side $\overrightarrow{CX}$ of angle $\angle XCY$, then we will get two triangles ABC which can be qualified as being congruent to one another. This discussion motivates our first case of congruence, known as the **SAS** case.

Axiom 2.3 (SAS Congruence Case) *If two sides of a triangle and the interior angle formed by them are respectively equal to two sides of another triangle and to the interior angle formed by them, then the two triangles are congruent.*

In symbols, for triangles ABC and $A'B'C'$ as in Fig. 2.2, the SAS congruence case assures that

$$\left.\begin{array}{c} \overline{AB} = \overline{A'B'} \\ \overline{AC} = \overline{A'C'} \\ \widehat{A} = \widehat{A'} \end{array}\right\} \overset{\text{SAS}}{\Longrightarrow} ABC \equiv A'B'C',$$

with the correspondence of vertices $A \leftrightarrow A'$, $B \leftrightarrow B'$, $C \leftrightarrow C'$. In particular, it follows that

$$\widehat{B} = \widehat{B'}, \quad \widehat{C} = \widehat{C'} \text{ and } \overline{BC} = \overline{B'C'}.$$

Let us now consider the following example.

Example 2.4 Use straightedge and compass to construct a triangle ABC, knowing that $\overline{BC} = a$, $\widehat{B} = \beta$ and $\widehat{C} = \gamma$.

Solution

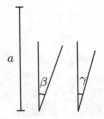

Fig. 2.3 The ASA
congruence case

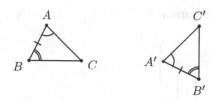

CONSTRUCTION STEPS

1. Draw a line r and mark points B and C on it, such that $\overline{BC} = a$.
2. Draw a half-line $\overrightarrow{BX}$ such that $C\widehat{B}X = \beta$.
3. Line r divides the plane in two half-planes. In the one that contains X, draw the half-line $\overrightarrow{CY}$ such that $B\widehat{C}Y = \gamma$.
4. Mark point A as the intersection of the half-lines $\overrightarrow{BX}$ and $\overrightarrow{CY}$.

□

As in the previous example, by repeating the construction steps above several times, choosing distinct positions for the side BC (subjected, of course, to the restriction $\overline{BC} = a$), the construction of triangle ABC rests *completely determined* by the measures angles $\angle B$ and $\angle C$ ought to have. In other words, in doing so we will obtain triangles in distinct positions, but will see no problem in qualifying all of them as being congruent to the one originally constructed. This motivates our second case of congruence, called the **ASA** case.

Axiom 2.5 (ASA Congruence Case) *If two angles and the included side of a triangle are respectively equal to two angles and the included side of another triangle, then the two triangles are congruent.*

In symbols, for triangles ABC and $A'B'C'$ as in Fig. 2.3, the ASA congruence case assures that

$$\left. \begin{array}{c} \widehat{A} = \widehat{A'} \\ \widehat{B} = \widehat{B'} \\ \overline{AB} = \overline{A'B'} \end{array} \right\} \overset{\text{ASA}}{\Longrightarrow} ABC \equiv A'B'C',$$

with the correspondence of vertices $A \leftrightarrow A'$, $B \leftrightarrow B'$, $C \leftrightarrow C'$. In particular, it follows that

$$\widehat{C} = \widehat{C'}, \quad \overline{AC} = \overline{A'C'} \text{ and } \overline{BC} = \overline{B'C'}.$$

Let's now turn to the example that motivates our third congruence case.

Example 2.6 Use straightedge and compass to construct a triangle ABC, given $\overline{AB} = c$, $\overline{AC} = b$ and $\overline{BC} = a$.

Solution

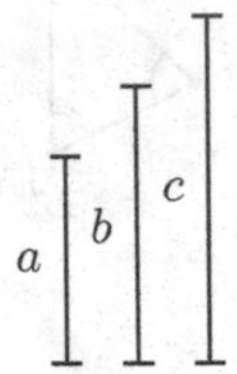

CONSTRUCTION STEPS

1. Draw a line segment BC such that $\overline{BC} = a$.
2. Draw the circles of center B and radius c, and of center C and radius b.
3. Mark point A as one of the intersection points of the circles constructed in item 2.

$\qquad\qquad\qquad\qquad\qquad\qquad\qquad\qquad\qquad\qquad\qquad\qquad\qquad\qquad\quad$ □

As in the two previous examples, by choosing a different position for the side BC (maintaining, of course, the condition $\overline{BC} = a$), and performing the subsequent constructions prescribed above, we would get another triangle, which would qualify as congruent to the originally constructed one. This motivates our third congruence case, the **SSS** case, stated below.

Axiom 2.7 (SSS Congruence Case) *If the three sides of a triangle are, in some order, congruent to the three sides of another triangle, then the two triangles are congruent.*

In symbols, for triangles ABC and $A'B'C'$ as in Fig. 2.4, the SSS congruence case assures that

$$\left.\begin{array}{c} \overline{AB} = \overline{A'B'} \\ \overline{BC} = \overline{B'C'} \\ \overline{CA} = \overline{C'A'} \end{array}\right\} \stackrel{\text{LLL}}{\Longrightarrow} ABC \equiv A'B'C',$$

with the correspondence of vertices $A \leftrightarrow A'$, $B \leftrightarrow B'$, $C \leftrightarrow C'$. In particular, this gives

$$\widehat{A} = \widehat{A'}, \quad \widehat{B} = \widehat{B'} \text{ and } \widehat{C} = \widehat{C'}.$$

It is worth remarking that the ASA and SSS congruence cases follow from the SAS case, in the following sense: given two triangles in the plane, it can be shown that the validity of a set of conditions ASA or SSS imply that the two triangles

Fig. 2.4 The SSS
congruence case

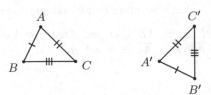

satisfy a set of conditions SAS. Nevertheless, for the purposes of these notes, such deductions would not give us any substantial gain, and hence we shall not discuss them here. (For a careful exposition, we refer the reader to [19].)

We shall present the last two cases of congruence of triangles in Corollary 2.22 and Problem 1, page 36, showing how these two additional cases follow from the ASA and SSS cases.

Finally, we observe that, once the congruence of two given triangles is established and whenever there is no danger of confusion, we shall omit the correspondence between their vertices. Since this usage will be followed several times along these notes, for the sake of understanding we urge the reader to check carefully the implied vertex correspondence every time we establish some congruence of triangles.

Problems: Sect. 2.1

1. (a) Give an example showing two congruent triangles for which it is not possible to *rigidly move* (i.e., to move without deforming) one of them *in the plane*, until make it coincide with the other. Nevertheless, show how one can rigidly move one of them in *space*, until make it coincide with the other.
   (b) In what sense do the two congruent triangles of the example given in item (a) differ from one another, so that such a difference explains the fact that we cannot perform the motion along the plane?

2.2 Applications of Congruence

In this section we collect some useful applications of the congruence cases studied in the previous one. Such applications will appear so frequently along the rest of the book that we advise the reader to memorize them as fast as possible.

We start with some terminology.

Definition 2.8 Given an angle $\angle AOB$, its **bisector** is the half-line $\overrightarrow{OC}$, contained in $\angle AOB$ and that splits it into two equal angles. In this case, we also say that $\overrightarrow{OC}$ bisects $\angle AOB$. Thus,

$$\overrightarrow{OC} \text{ bisects } \angle AOB \iff A\widehat{O}C = B\widehat{O}C.$$

We shall assume, without proof, that the bisector of an angle, if exists, is unique. The next example shows how to construct it.

Example 2.9 Use straightedge and compass to construct the bisector of the angle $\angle AOB$ given below.

Solution

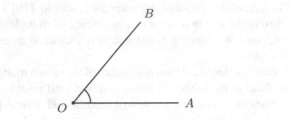

CONSTRUCTION STEPS

1. Draw the circle with center O and some radius r, and mark its intersection points X and Y, respectively wit $\overrightarrow{OA}$ and $\overrightarrow{OB}$.
2. Choose an opening $s > \frac{1}{2}\overline{XY}$, draw the circles centered at X and Y and with radii s, and let C be one of their intersection points ($C \neq O$). The half-line $\overrightarrow{OC}$ is the bisector of $\angle AOB$.

In order to justify why the construction steps described above really do the job, start by observing that, with respect to triangles XOC and YOC, we have $\overline{OX} = \overline{OY} = r$ and $\overline{XC} = \overline{YC} = s$. Since OC is a common side of these two triangles, it follows from the SSS congruence case that $XOC \equiv YOC$. Therefore, $X\widehat{O}C = Y\widehat{O}C$ or, which is the same, $A\widehat{O}C = B\widehat{O}C$.   □

In a triangle ABC, the **internal bisector** relative to the side BC (or to the vertex A) is the portion AP of the bisector of the internal angle $\angle A$ of the triangle, that goes from A to the point P in which it intersects side BC. Point $P \in BC$ is the **foot of the internal bisector** relative to BC. Analogously, we have in ABC the internal bisectors relative to the sides AC and AB (or, which is the same, to the vertices B and C, respectively), so that every triangle has exactly three internal bisectors. At this point, it is instructive for the reader to draw a triangle ABC, together with its internal bisector relative to the vertex A and the corresponding foot; in this respect, see also Problem 1.

Combining the SSS and SAS cases we can construct the **midpoint** of a line segment, i.e., the point that divides the segment into two equal parts. The next example shows how to do it.

Example 2.10 Use straightedge and compass to construct the midpoint of a line segment AB.

Solution

CONSTRUCTION STEPS

1. Fix an opening $r > \frac{1}{2}\overline{AB}$ and draw the circles centered at A and B and with radii r. Let X and Y be their intersection points.

2. The intersection point M of the line $\overleftrightarrow{XY}$ and the line segment AB is the midpoint of AB.

Indeed, with respect to triangles AXY and BXY, we have $\overline{AX} = \overline{BX} = r$ and $\overline{AY} = \overline{BY} = r$. Since XY is a common side of both of these triangles, it follows from the SSS congruence case that $AXY \equiv BXY$. Hence, $A\widehat{X}Y = B\widehat{X}Y$ or, which is the same, $A\widehat{X}M = B\widehat{X}M$. Now, with respect to triangles AXM and BXM, we have $\overline{AX} = \overline{BX}$ and $A\widehat{X}M = B\widehat{X}M$. However, since XM is a side of both these triangles, it follows from the SAS congruence case that $AXM \equiv BXM$. Therefore, $\overline{AM} = \overline{BM}$. □

In a triangle ABC, the **median** relative to the side BC (or to the vertex A) is the line segment that goes from the vertex A to the midpoint of side BC. Analogously, we have in ABC the medians relative to the sides AC and AB (or, which is the same, to the vertices B and C, respectively), so that every triangle has exactly three medians. At this point, we suggest the reader to draw a triangle ABC, together with its median relative to the vertex A and the midpoint of side BC; we also urge him/her to take a look at Problem 1.

Given lines r and s in the plane, we say that r is *perpendicular* to s, that s is perpendicular to r or simply that r and s are **perpendicular** if r and s have a point in common and form an angle of 90° at such a point (in this respect, see also Problem 6, page 12). We shall write $r \perp s$ to denote that r and s are perpendicular. The next example shows how to use the congruence cases studied before to construct the line perpendicular to another given line and passing through a given point.

Example 2.11 Given in the plane a line r and a point A, use straightedge and compass to construct a line s, such that $r \perp s$ and $A \in s$.

Solution There are two distinct cases to consider:

(a) $A \notin r$:

CONSTRUCTION STEPS

1. Center the compass at A and draw an arc of circle that intersects r at two distinct points, say B and C.

2. Construct the midpoint M of the line segment BC and let $s = \overleftrightarrow{AM}$.

To see that the construction above is correct, we start by establishing the congruence of triangles ABM and ACM. This follows from the SSS case, for $\overline{AB} = \overline{AC}$, $\overline{BM} = \overline{CM}$ and AM is a common side of these two triangles. Then, we get $A\widehat{M}B = A\widehat{M}C$. However, since $A\widehat{M}B + A\widehat{M}C = 180°$, we conclude that $A\widehat{M}B = A\widehat{M}C = 90°$ or, which is the same, that $\overleftrightarrow{AM}\perp r$.

(b) $A \in r$:

CONSTRUCTION STEPS

1. Center the compass at A and draw a semicircle that intersects r at points B and C.

2. Now, draw two circles of radii $R > \frac{1}{2}\overline{BC}$, centered at B and C, respectively; if A' one of the intersection points of these two circles, then $\overleftrightarrow{A'A}\perp r$.

Later, in Sect. 2.4, we shall show that the choice $R > \frac{1}{2}\overline{BC}$ really forces the two circles to have intersection points. Taking this for granted by now, we have $ABA' \equiv ACA'$ by the SSS case and, hence, $A'\widehat{A}B = A'\widehat{A}C$. On the other hand, since $A'\widehat{A}B + A'\widehat{A}C = 180°$, it follows that $A'\widehat{A}B = A'\widehat{A}C = 90°$.   □

In the notations of the previous example, if $A \notin r$ and s is the perpendicular to r passing through A, then the intersection point of lines r and s is called the **foot of the perpendicular** dropped from A to r.

Remark 2.12 Given in the plane a point A and a line r, it is possible to show that the line s, perpendicular to r and passing through A, in indeed *unique* (in this respect, see Problem 17, page 39).

We consider again a point A and a line r in the plane, with $A \notin r$. We then define the **distance** from A to r as the length of the line segment AP, where P is the foot of the perpendicular dropped from A to r (cf. Fig. 2.5). In other words, letting d denote the distance from A to r, we have $d = \overline{AP}$.

In Sect. 2.4 (cf. Corollary 2.24), we shall prove that the length of the segment AP is smaller than the length of any other line segment joining A to a point $P' \in r$ (i.e., with $P' \neq P$); in the notations of Fig. 2.5, this means that $d < \overline{AP'}$.

Given a triangle ABC, its **height** or **altitude** relative to the side BC (or to the vertex A) is the segment that joins A to the foot of the perpendicular dropped from

Fig. 2.5 Distance from point
A to line r

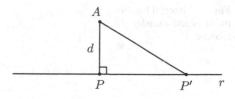

Fig. 2.6 ABC isosceles $\Rightarrow$
$\widehat{B} = \widehat{C}$

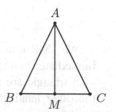

A to the *line* $\overleftrightarrow{BC}$. In this case, we usually say that the perpendicular foot is the **foot of the altitude** relative to BC. Analogously, we have in triangle ABC the altitudes relative to sides AC and AB (or to the vertices B and C, respectively), so that ABC has exactly three altitudes. At this point, we urge the reader to draw a triangle ABC, together with its altitude relative to the vertex A and the corresponding foot (in this respect, see also Problem 1).

We finish this section by studying a very important property of isosceles triangles, which will be used several times in the sequel.

Proposition 2.13 *If ABC is an isosceles triangle of base BC, then $\widehat{B} = \widehat{C}$.*

Proof The proof of this result is a direct consequence of the explanation we gave for the correctness of the construction presented for the midpoint of a line segment. Nevertheless, we repeat it here for the reader's convenience.

Let M be the midpoint of side BC (cf. Fig. 2.6). Since $\overline{BM} = \overline{CM}$, $\overline{AB} = \overline{AC}$ and AM is a common side of triangles AMB and AMC, the SSS congruence case assures that these two triangles are congruent. Therefore, $A\widehat{B}M = A\widehat{C}M$. □

Corollary 2.14 *The measures of the interior angles of an equilateral triangle are all equal.*

Proof It suffices to note that every side of an equilateral triangle can be seen as a basis for it, when looked at as an isosceles triangle. □

Problems: Sect. 2.2

1. Use straightedge and compass to construct the internal bisectors, the medians and the heights of the triangle of Fig. 2.7.

Fig. 2.7 Internal bisectors, medians and altitudes of a triangle

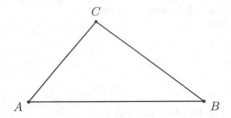

We shall now make some comments on the problem above. Firstly, it is immediate from the given definitions that the internal bisectors and the medians of a triangle are always contained in the corresponding triangular region; on the other hand, this is not necessarily the case for the altitudes, as you can easily see upon solving the last part of the problem. Moreover, after performing these constructions, one will probably notice that the internal bisectors of the given triangle all pass through a common point, the same happening with the medians and the altitudes. Such a property has nothing to do with the triangle we have; actually, to prove that the internal bisectors (respectively the medians, the altitudes) of every triangle pass through a single point will be the object of Sect. 3.2.

2. * Let a point A and a line r, with $A \notin r$, be given in the plane. We say that a point A' is the **symmetric** of A with respect to r if $\overleftrightarrow{AA'} \perp r$ and r passes through the midpoint of the line segment AA'. Show how to construct A' with straightedge and compass.[2]

3. Use straightedge and compass to construct a triangle ABC, given the lengths $\overline{AB} = c$ and $\overline{BC} = a$, as well as the length m_a of the median relative to A.

4. Use straightedge and compass to construct a triangle ABC, given the lengths $\overline{AB} = c$ and $\overline{AC} = b$, as well as the length m_a of the median relative to A.

5. Use straightedge and compass to construct a triangle ABC, given the length c of side $\overline{AB}$, the length β_a of the internal bisector relative to A and the measure α of $\angle BAC$.

6. * Let ABC be an isosceles triangle with basis BC. Prove that the internal bisector, the median and the altitude relative to BC coincide.

7. * Let ABC be a given triangle and P, M and H be the feet of the internal bisector, the median and the altitude relative to A, respectively. If P and H or M and H coincide,[3] prove that ABC is isosceles with basis BC.

---

[2] A *reflection along a line* is the *transformation* of the plane that associates to each point its symmetric with respect to a fixed line. So, this problem opens the way to the study of reflections along lines as *geometric transformations*. We shall not pursue such a viewpoint here; instead, we refer the reader to the superb book [24].

[3] The case $M = P$ will be dealt with in Problem 19, page 83.

8. * Let Γ be a circle of center O, and let AB be a chord of it. If M is a point on the line segment AB, prove that

$$\overleftrightarrow{OM} \perp \overleftrightarrow{AB} \Leftrightarrow \overline{AM} = \overline{BM}.$$

2.3 Parallelism

Given two distinct lines in the plane, there are only two possibilities for them: either they have or do not have a common point; in the first case, the lines are said to be **concurrent**; in the second, they are said to be **parallel** (cf. Fig. 2.8).

Given a line r and a point A, with $A \notin r$, we would like to study the problem of constructing (if it indeed exists) a line parallel to r and passing through A. To this end, we need the following auxiliary result, known as the **exterior angle inequality**.

Lemma 2.15 *In every triangle, the measure of each exterior angle is greater than the measures of the interior angles not adjacent to it.*

Proof Let ABC be a given triangle and M be the midpoint of side AC (cf. Fig. 2.9). Mark a point B' on the half-line $\overrightarrow{BM}$, such that M is the midpoint of BB'. Since $\overline{AM} = \overline{CM}$, $\overline{BM} = \overline{B'M}$ and $A\widehat{M}B = C\widehat{M}B'$ (OPP angles), the SAS congruence case gives $AMB \equiv CMB'$, and hence $B'\widehat{C}M = B\widehat{A}M$. Therefore,

$$X\widehat{C}A > B'\widehat{C}A = B'\widehat{C}M = B\widehat{A}M = B\widehat{A}C.$$

Analogously, one proves that $Y\widehat{C}B > A\widehat{B}C$ (and note that $X\widehat{C}A = Y\widehat{C}B$). □

The next example shows how to perform the most important of all elementary constructions with straightedge and compass, namely, that of a line parallel to another one and passing through a given point.

Fig. 2.8 Concurrent (left) and parallel (right) lines

Fig. 2.9 The exterior angle inequality

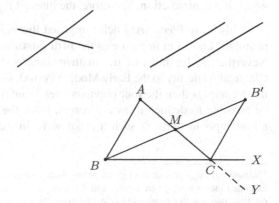

Example 2.16 Use straightedge and compass to construct a line s, parallel to a given line r and passing through the point A.

Solution

$$A$$
$$\bullet$$

$$\rule{10cm}{0.4pt}\ \ r$$

CONSTRUCTION STEPS

1. Take points C and X on r and join A to C.
2. Construct an angle $\angle CAY$ such that $C\widehat{A}Y = A\widehat{C}X$ and X and Y are in opposite half-planes with respect to $\overleftrightarrow{AC}$.
3. Line $s = \overleftrightarrow{AY}$ is parallel to r.

In order to prove that the construction delineated above actually gives a parallel to r, suppose, by the sake of contradiction, that $\overleftrightarrow{AY}$ intersects r at a point B (cf. Fig. 2.10). Let's look at the case in which $C \in BX$ (the cases $B \in CX$ and $X \in BC$ can be treated in totally analogous ways).

By construction, we have

$$B\widehat{A}C = Y\widehat{A}C = A\widehat{C}X;$$

on the other hand, since $\angle ACX$ is an external angle of triangle ABC, it follows from the previous lemma that

$$B\widehat{A}C < A\widehat{C}X,$$

which is a contradiction. Therefore, the lines $\overleftrightarrow{AY}$ and r are indeed parallel.   □

In his book *Elements*, Euclid imposed the uniqueness of the parallel line as a postulate, known in literature as the **fifth postulate**, or as the **parallels' postulate**. Nevertheless, for most of the mathematicians that studied Euclid's book, since Classical Antiquity to the Early Modern Period, such a postulate seemed to be much more complex than the four previous ones,[4] and this made them think that it would be possible to deduce it, as a theorem, from these four previous ones. However, all attempts to find out such a proof were in vain. It then happened that, in the

---

[4]Namely: through two distinct given points there passes only one line; every line segment can be extended into a line; given a point and a line segment having this point as an end, there exists a circle centered at that point and having the given line segment as a radius; all right angles are equal.

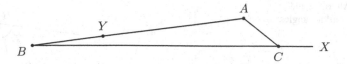

Fig. 2.10 Construction of a parallel to a given line through a given point

beginnings of the nineteenth century, the Hungarian mathematician János Bolyai and the Russian mathematician Nikolai Lobatchevsky, working independently, showed that it was really necessary *to assume* the uniqueness of the parallel as a postulate. More precisely, what they did was to construct *another type of geometry*, known today as *Hyperbolic Geometry*, in which the first four postulates of Euclid are still valid, albeit through a point not belonging to a given line there passes *infinitely many* lines parallel to the given one.[5]

All of the above being said, we now formally assume the uniqueness of the parallel as a postulate.

Postulate 2.17 (Fifth Postulate) *Given in the plane a line r and a point A, with A ∉ r, there exists only one line parallel to r and passing through A.*

In view of the above postulate, Example 2.16 showed how to use straightedge and compass to construct the parallel to a given line, passing through a point not belonging to it. Nevertheless, a much simpler construction will be seen in Sect. 2.5 (cf. Example 2.32).

If two lines r and s are parallel, we will write $r \parallel s$.

Now that we have the fifth postulate at our disposal, we can state and prove some of the most important results of Euclidean Geometry.

For the first of them, let be given, in the plane, lines r, s and t, such that t intersects r and s at the distinct points A and B, respectively. In the notations of Fig. 2.11, angles α and β form a pair of **alternate interior angles**, while angles α and γ form a pair of **consecutive interior angles** with respect to line t. The following criterion for the parallelism of two given lines will be proved to be quite useful.

Corollary 2.18 *In the notations of Fig. 2.11, we have*

$$r \parallel s \Leftrightarrow \alpha = \beta \Leftrightarrow \alpha + \gamma = 180°.$$

Proof Note firstly that, since $\beta + \gamma = 180°$, we have $\alpha = \beta \Leftrightarrow \alpha + \gamma = 180°$. Thus, it suffices to prove that $r \parallel s \Leftrightarrow \alpha = \beta$.

---

[5]For an elementary introduction to Hyperbolic Geometry, as well as for a discussion of the unsuccessful efforts to prove the fifth postulate, we recommend to the reader references [11] and [19].

Fig. 2.11 Alternate and
consecutive interior angles

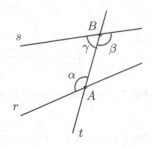

Fig. 2.12 Sum of the
measures of the interior
angles of a triangle

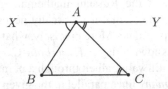

In Example 2.16, we have already seen that $\alpha = \beta \Rightarrow r \parallel s$. Thus, we are left
to establish the converse implication. To this end, let r and s be parallel lines. Then,
by the fifth postulate, s is *the only* line parallel to r and passing through B, so that it
can be constructed exactly as described in Example 2.16. Therefore, it follows from
the description of that construction that $\alpha = \beta$.   □

Yet another consequence of the above discussion, and an extremely relevant one,
is that of the following result.

Theorem 2.19 *The sum of the measures of the interior angles of any triangle equals*
180°.

Proof Let ABC be a triangle and $\overleftrightarrow{XY}$ be the parallel to $\overleftrightarrow{BC}$ passing through A (cf.
Fig. 2.12). By Corollary 2.18, we have $\widehat{B} = B\widehat{A}X$ and $\widehat{C} = C\widehat{A}Y$, so that

$$\widehat{A} + \widehat{B} + \widehat{C} = \widehat{A} + B\widehat{A}X + C\widehat{A}Y = 180°.$$

□

Corollary 2.20 *The interior angles of every equilateral triangle are equal to* 60°.

Proof By Corollary 2.14, every equilateral triangle has three equal angles. On the
other hand, the previous theorem assures that their sum is equal to 180°. Hence, they
must all be equal to $\frac{1}{3} \cdot 180° = 60°$.   □

The next result is known in mathematical literature as the **exterior angle
theorem**, and is a sharp refinement of Lemma 2.15.

Corollary 2.21 *In every triangle, the measure of each exterior angle is equal to the
sum of the measures of the two interior angles which are not adjacent to it.*

Fig. 2.13 The exterior angle
theorem

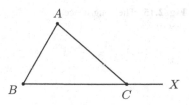

Fig. 2.14 A right triangle (at
left) and an obtuse triangle (at
right)

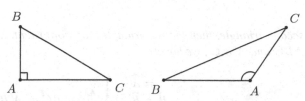

Proof In the notations of Fig. 2.13, it suffices to see that

$$A\widehat{C}X = 180° - \widehat{C} = \widehat{A} + \widehat{B},$$

where we used Theorem 2.19 in the last equality above. □

Theorem 2.19 also allows us to classify a triangle with respect to the measures of its interior angles. Indeed, it guarantees that every triangle has at most one interior angle greater than or equal to 90°, for if ABC was a triangle with $\widehat{A} \geq 90°$ and $\widehat{B} \geq 90°$, we would have

$$\widehat{A} + \widehat{B} + \widehat{C} > \widehat{A} + \widehat{B} \geq 90° + 90° = 180°,$$

which is an absurd.

In view of the above, an **acute triangle** has three acute interior angles, a **right triangle** has (exactly) one right interior angle and an **obtuse triangle** has (exactly) one obtuse interior angle (see Fig. 2.14). If a right (resp. obtuse) triangle ABC is such that $\widehat{A} = 90°$ (resp. $\widehat{A} > 90°$), we also say that ABC is **right at A** (resp. **obtuse at A**).

Also in the case of right triangles, the side opposite to the right angle is called the **hypotenuse** of the triangle, whereas the other two sides are its **legs**. In the notations of the Fig. 2.14 (left), BC is the hypotenuse, whereas AB and AC are the legs of the right triangle ABC. We will have more to say about right triangles in Sect. 4.2.

We finish this section studying yet another set of sufficient conditions for the congruence of two given triangles, known as the **AAS**. The last set of sufficient conditions for the congruence of two given triangles, which works specifically for right triangles, will be the object of Problem 1.

Corollary 2.22 (AAS Congruence Case) *Let two triangles ABC and $A'B'C'$ be given in the plane. If, in the first triangle, two angles and the side opposite to one of them are respectively equal to two angles and the side opposite to one of them in the*

Fig. 2.15 The congruence
case AAS

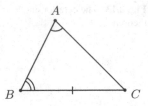

second triangle, then the two triangles are congruent. In symbols, given triangles
ABC and $A'B'C'$, we have:

$$\left.\begin{array}{r} \widehat{A} = \widehat{A'} \\ \widehat{B} = \widehat{B'} \\ \overline{BC} = \overline{B'C'} \end{array}\right\} \stackrel{AAS}{\Longrightarrow} ABC \equiv A'B'C',$$

with the correspondence of vertices $A \leftrightarrow A'$, $B \leftrightarrow B'$ and $C \leftrightarrow C'$. In particular,
we also have

$$\widehat{C} = \widehat{C'}, \quad \overline{AC} = \overline{A'C'} \ and \ \overline{AB} = \overline{A'B'}.$$

Proof It suffices to note that the conditions $\widehat{A} = \widehat{A'}$ and $\widehat{B} = \widehat{B'}$ imply

$$\widehat{C} = 180° - \widehat{A} - \widehat{B} = 180° - \widehat{A'} - \widehat{B'} = \widehat{C'}.$$

Therefore, for the two given triangles, we have

$$\widehat{B} = \widehat{B'}, \quad \overline{BC} = \overline{B'C'} \ and \ \widehat{C} = \widehat{C'},$$

so that, by the ASA case, they are indeed congruent.   □

  Also with respect to the AAS congruence case, suppose we know the length of
one side and the measures of two interior angles of a triangle, with one of these
angles being opposite to the given side (Fig. 2.15). The problem of using straightedge and compass to construct the triangle will be discussed in Example 2.34.

Problems: Sect. 2.3

1. * Let two right triangles be such that the hypotenuse and one of the legs of one
   of them are respectively congruent to the hypotenuse and one of the legs of the
   other one. Prove that the two triangles are congruent. This is the **hypotenuse-
   leg** (abbreviated **HL**) congruence case for right triangles.

2. * ABC is an isosceles triangle of basis BC, and $D \in AB$, $E \in AC$ are points such that $\overleftrightarrow{DE} \parallel \overleftrightarrow{BC}$. If F is the intersection point of the line segments CD and BE, prove that $\overline{BF} = \overline{CF}$.

3. Let ABC be an isosceles triangle of basis BC. Prove that the altitudes, medians and internal bisectors relative to the sides AB and AC have equal lengths.

4. In a triangle ABC, we have $\widehat{A} = 90°$. Letting $P \in AC$ be the foot of the internal bisector relative to B and knowing that the distance from P to the side BC is 2cm, compute the length of AP.

5. In the figure below, lines $\overleftrightarrow{AB}$ and $\overleftrightarrow{CD}$ are parallel. If the measures of the angles $\angle ABC$ and $\angle BCD$ are respectively equal to $3x - 20°$ and $x + 40°$, compute the value of x in degrees.

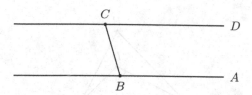

6. In the figure below, prove that $r \parallel s \Leftrightarrow \alpha = \beta$ (one says that α and β are **corresponding angles**).

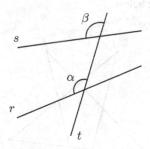

7. In the figure below, if $r \parallel s$, prove that $\alpha + \beta = \gamma$.

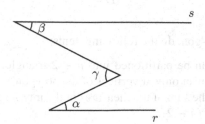

8. In the figure below, we have $A\widehat{B}C = 20°$, $B\widehat{C}D = 60°$ and $D\widehat{E}F = 25°$. If $\overleftrightarrow{AB}$ and $\overleftrightarrow{EF}$ are parallel, compute the measure of the angle $\angle CDE$.

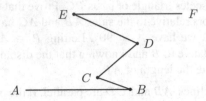

9. In the figure below, prove that $\alpha = D\widehat{A}B + A\widehat{B}C + B\widehat{C}D$.

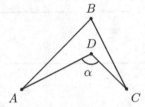

10. Compute the sum of the angles at the vertices A, B, C, D and E of the five-pointed star of the figure below.

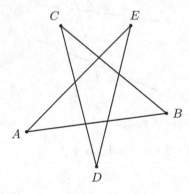

11. * Given a convex n-gon, do the following items:

   (a) Prove that it can be partitioned into $n - 2$ triangles, by means of $n - 3$ diagonals that meet only at vertices of the polygon.
   (b) Conclude that the sum of the measures of the interior angles of the polygon is equal to $180°(n - 2)$.
   (c) Conclude that the sum of the measures of the exterior angles of the polygon (one angle per vertex) is equal to $360°$.

12. * In a triangle ABC, let M be the midpoint of side BC. If $\overline{AM} = \frac{1}{2}\overline{BC}$, show that $B\widehat{A}C = 90°$.

13. * Let I be the intersection point of the internal bisectors of triangle ABC departing B and C. Prove that $B\widehat{I}C = 90° + \frac{1}{2}B\widehat{A}C$.

14. In a triangle ABC, we know that the measure of $\widehat{A}$ equals $\frac{1}{8}$ of the measure of the obtuse angle formed by the internal bisectors relative to the vertices B and C. Compute the measure of $\angle A$.

15. * In a triangle ABC, let I_a be the intersection point of the external bisectors relative to the vertices B and C. Prove that $B\widehat{I_a}C = 90° - \frac{1}{2}B\widehat{A}C$.

16. A triangle ABC is isosceles of basis BC. Points D on BC and E on AC are such that $\overline{AD} = \overline{AE}$ and $B\widehat{A}D = 48°$. Compute $C\widehat{D}E$.

17. * Given a line r and a point A in the plane, prove that there is exactly one line s such that $r \perp s$ and $A \in s$.

18. In triangle ABC, point $D \in BC$ is the foot of the internal bisector relative to A. Prove that $A\widehat{D}C - A\widehat{D}B = \widehat{B} - \widehat{C}$.

19. Triangle ABC is isosceles of basis BC. Points D and F on the side AB, and E on the side AC are chosen so that $\overline{BC} = \overline{CD} = \overline{DE} = \overline{EF} = \overline{FA}$. Compute the measure of $\angle BAC$.

20. (TT) $ABCDEF$ is a convex hexagon whose diagonals AD, BE and CF meet at a common point M. If M is the midpoint of each of these three diagonals, prove that $\widehat{A} + \widehat{B} + \widehat{C} = 180°$.

21. ABC is an isosceles triangle of basis BC, in which the altitudes relative to the equal sides measure 10cm each.

    (a) If P is a point on the basis BC, compute the sum of the distances from P to the sides AB and AC.

    (b) Let Q be a point on the line $\overleftrightarrow{BC}$, though not on the basis BC. Compute the absolute value of the difference of the distances from Q to the lines $\overleftrightarrow{AB}$ and $\overleftrightarrow{AC}$.

22. (Soviet Union) Triangle ABC is isosceles of basis BC and such that $B\widehat{A}C = 20°$. We mark points $D \in AC$ and $E \in AB$ such that $D\widehat{B}C = 60°$ and $E\widehat{C}B = 50°$. Compute $B\widehat{D}E$.

2.4 The Triangle Inequality

The main purpose of this brief section is to prove that, in every triangle, the length of each side is less than the sum of those of the other two. This will be the object of Theorem 2.26. For the time being, we start by relating the lengths of the sides of a triangle to the measures of their opposite angles.

Fig. 2.16 Comparing sides
and angles of a triangle

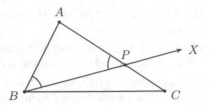

Proposition 2.23 *If ABC is a triangle such that $\widehat{B} > \widehat{C}$, then $\overline{AC} > \overline{AB}$.*

Proof Since $\widehat{B} > \widehat{C}$, we may draw (cf. Fig. 2.16) half-line $\overrightarrow{BX}$, which crosses the interior of ABC and satisfies $C\widehat{B}X = \frac{1}{2}(\widehat{B} - \widehat{C})$. Letting P be its intersection point with side AC, it follows from the exterior angle theorem that

$$A\widehat{P}B = C\widehat{B}P + B\widehat{C}P = \frac{1}{2}(\widehat{B} - \widehat{C}) + \widehat{C} = \frac{1}{2}(\widehat{B} + \widehat{C}).$$

However, since $A\widehat{B}P = \widehat{B} - \frac{1}{2}(\widehat{B} - \widehat{C}) = \frac{1}{2}(\widehat{B} + \widehat{C})$, it follows that ABP is an isosceles triangle of basis BP. Hence,

$$\overline{AB} = \overline{AP} < \overline{AC}.$$

$\square$

Corollary 2.24 *If a triangle ABC is such that $\widehat{A} \geq 90°$, then $\overline{BC}$ is its greatest side. In particular, in a right triangle the hypothenuse is the greatest side.*

Proof It suffices to note that, if $\widehat{A} \geq 90°$, then $\widehat{A}$ is the greatest angle of ABC, so that, by the previous proposition, $\overline{BC}$ is its greatest side. $\square$

Corollary 2.25 *Let ABC and $A'B'C'$ be two given triangles, such that $\overline{AB} = \overline{A'B'}$ and $\overline{AC} = \overline{A'C'}$. If $B\widehat{A}C < B'\widehat{A}'C'$, then $\overline{BC} < \overline{B'C'}$.*

Proof Out of the half-planes determined by straight line $\overleftrightarrow{AC}$, let α be the one containing B. Letting D be the point of α such that $\overline{AD} = \overline{A'B'}$ and $D\widehat{A}C = B'\widehat{A}'C'$, we have $DAC \equiv B'A'C'$ by the SAS congruence case, so that $\overline{DC} = \overline{B'C'}$. It then suffices to show that $\overline{DC} > \overline{BC}$ or, by Proposition 2.23, that $D\widehat{B}C > B\widehat{D}C$. There are two different cases to look at:

(i) A and D lie in a single half-plane, from those determined by $\overleftrightarrow{BC}$ (cf. Fig. 2.17, left): since ABD is isosceles of basis BD, we get

$$D\widehat{B}C > D\widehat{B}A = B\widehat{D}A > B\widehat{D}C.$$

(ii) A and D lie in opposite half-planes with respect to $\overleftrightarrow{BC}$ (cf. Fig. 2.17, right): since ABD is isosceles of basis BD, we have $A\widehat{B}D < 90°$. Hence, letting E be the intersection point of $\overrightarrow{AB}$ and CD, we get

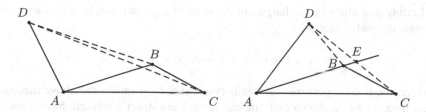

Fig. 2.17 Triangles with two pairs of equal sides and unequal angles

Fig. 2.18 The triangle
inequality

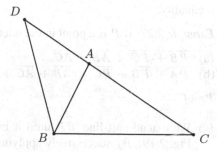

$$D\widehat{B}C = D\widehat{B}E + E\widehat{B}C > D\widehat{B}E = 180° - D\widehat{B}A > 90°.$$

However, since a triangle has at most one obtuse angle, it comes that $D\widehat{B}C > B\widehat{C}D$.   □

As anticipated at the beginning of this section, the following is our main result, and is known as the **triangle inequality**.

Theorem 2.26 *In every triangle, the length of each side is less than the sum of the lengths of the other two sides.*

Proof Let ABC be such that $\overline{AB} = c$, $\overline{AC} = b$ and $\overline{BC} = a$. We shall show that $a < b + c$, the proof of the other two inequalities being completely analogous. Mark (cf. Fig. 2.18) point D on $\overrightarrow{CA}$ such that $A \in CD$ and $\overline{AD} = \overline{AB}$.

Since

$$\overline{CD} = \overline{AC} + \overline{AD} = \overline{AC} + \overline{AB} = b + c,$$

Proposition 2.23 guarantees that it suffices for us to show that $B\widehat{D}C < D\widehat{B}C$. However, since $B\widehat{D}A = D\widehat{B}A$, we just have to note that

$$B\widehat{D}C = B\widehat{D}A = D\widehat{B}A < D\widehat{B}A + A\widehat{B}C = D\widehat{B}C.$$

□

Letting a, b and c be the lengths of the sides of a given triangle, it follows from triangle inequality that

$$a < b + c, \quad b < a + c, \quad c < a + b.$$

Conversely, if three positive real numbers a, b and c do satisfy the above inequalities, then it can be shown that one can always construct a triangle having these numbers as lengths of its sides.

We end this section by presenting two interesting consequences of the triangle inequality.

Example 2.27 If P is a point in the interior of a triangle ABC, then:

(a) $\overline{PB} + \overline{PC} < \overline{AB} + \overline{AC}$.
(b) $\overline{PA} + \overline{PB} + \overline{PC} < \overline{AB} + \overline{AC} + \overline{BC}$.

Proof

(a) We extend half-line $\overrightarrow{BP}$ until it intersects side AC at the point Q, say (cf. Fig. 2.19). By successively applying the triangle inequality to triangles CPQ and ABQ, we obtain

$$\overline{PB} + \overline{PC} < \overline{PB} + (\overline{PQ} + \overline{CQ}) = \overline{BQ} + \overline{CQ}$$

$$< (\overline{AB} + \overline{AQ}) + \overline{CQ} = \overline{AB} + \overline{AC}.$$

(b) Arguing analogously to the proof of item (a), we have $\overline{PA} + \overline{PB} < \overline{AC} + \overline{BC}$ and $\overline{PA} + \overline{PC} < \overline{AB} + \overline{BC}$. Termwise adding these two inequalities to that of item (a), we arrive at

$$2(\overline{PA} + \overline{PB} + \overline{PC}) < 2(\overline{AB} + \overline{AC} + \overline{BC}),$$

as desired.   □

Example 2.28 In Fig. 2.20, use straightedge and compass to construct the point $P \in r$ for which $\overline{PA} + \overline{PB}$ is as small as possible.

Fig. 2.19 Consequences of the triangle inequality

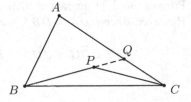

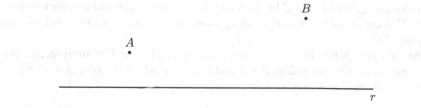

Fig. 2.20 Least path touching a line

Solution Letting A' be the symmetric to A with respect to r (cf. Problem 2, page 30), we claim that there is just one possible location for P, and that is the intersection point of $A'B$ and r. In order to prove this fact, let Q be another point on r (sketch a figure to follow the reasoning). The fact that A' is the symmetric of A with respect to r guarantees that $\overline{AQ} = \overline{A'Q}$ and, analogously, $\overline{AP} = \overline{A'P}$ (prove that!) Such equalities, together with the triangle inequality, successively give

$$\overline{AP} + \overline{BP} = \overline{A'P} + \overline{BP} = \overline{A'B}$$
$$< \overline{A'Q} + \overline{BQ} = \overline{AQ} + \overline{BQ}.$$

$\square$

Problems: Sect. 2.4

1. The lengths of two sides of an isosceles triangle are 38cm and 14cm. Compute its perimeter.
2. Find the range of $x \in \mathbb{R}$, knowing that the lengths of the sides of a certain triangle are $x + 10$, $2x + 4$ and $20 - 2x$.
3. The length (in centimeters) of side AB of triangle ABC is an integer. Compute the largest possible value for such a length, knowing that $\overline{AC} = 27$cm, $\overline{BC} = 16$cm and $\widehat{C} < \widehat{A} < \widehat{B}$.
4. In a triangle ABC, we choose at random points $P \in BC$, $Q \in AC$ and $R \in AB$, all distinct from the vertices of ABC. Prove that the perimeter of triangle PQR is less than that of ABC.
5. If a, b and c are the lengths of the sides of a triangle, prove that $|b - c| < a$.
6. (TT) Let a, b and c be the lengths of the sides of a triangle. Prove that $a^3 + b^3 + 3abc > c^3$.
7. Given a convex quadrilateral $ABCD$, prove that the point P in the plane for which the sum $\overline{PA} + \overline{PB} + \overline{PC} + \overline{PD}$ is minimum is the point of intersection of the diagonals of $ABCD$.
8. Let $n \geq 3$ be a given integer. Prove that, in every convex n-gon, the length of each side is less than the sum of the lengths of the $n - 1$ remaining ones.

9. Let $m, n \geq 3$ and $\mathcal{P}$ and $\mathcal{Q}$ be respectively a convex n-gon and a convex m-gon. If $\mathcal{P}$ is contained in the interior of $\mathcal{Q}$, show that the perimeter of $\mathcal{P}$ is less than that of $\mathcal{Q}$.
10. In the figure below, half-lines r and s are perpendicular. Use straightedge and compass to construct points $B \in r$ and $C \in s$ for which the sum $\overline{AB} + \overline{BC} + \overline{CD}$ is as small as possible.

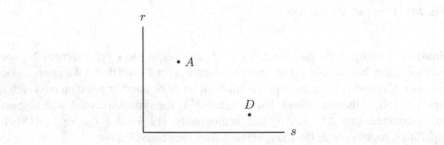

11. * Let ABC be an equilateral triangle of side length l. If P and Q are points on the sides AB and AC, respectively, both distinct from the vertices of ABC, prove that $\overline{BQ} + \overline{PQ} + \overline{CP} > 2l$.
12. Let ABC be a triangle with $\overline{AB} \geq \overline{AC} \geq \overline{BC}$. Given a point P in the interior of ABC, prove that $\overline{PA} + \overline{PB} + \overline{PC} < \overline{AB} + \overline{AC}$.
13. (Soviet Union) In a certain country, the distances between the cities are pairwise distinct. If a plane departs from each city to the nearest one, prove that no more than five planes landed in any city.
14. $A_1 A_2 \ldots A_n$ is a convex polygon a P is a point inside it. Prove that there exists $1 \leq i \leq n$ such that the foot of the perpendicular from P to $\overleftrightarrow{A_i A_{i+1}}$ (with $A_{n+1} = A_1$) lies inside the line segment $A_i A_{i+1}$.

2.5 Special Quadrilaterals

In this section we begin the systematic study of the geometry of convex quadrilaterals. Among the various particular types of them, the main ones are surely *parallelograms*, which are qualified in the coming

Definition 2.29 A convex quadrilateral is said to be a **parallelogram** if its opposite sides are pairwise parallel.

In the sequel, we shall discuss some equivalent ways of defining parallelograms. The reader must keep those results as notable properties of this class of quadrilaterals, for we shall use them over and over.

Fig. 2.21 $ABCD$ parallelogram $\Leftrightarrow$ $\overleftrightarrow{AB} \parallel \overleftrightarrow{CD}$ and $\overleftrightarrow{AD} \parallel \overleftrightarrow{BC}$

Fig. 2.22 $\widehat{A} = \widehat{C}$ and $\widehat{B} = \widehat{D} \Rightarrow$ $ABCD$ parallelogram

Fig. 2.23 $ABCD$ parallelogram $\Rightarrow$ $\overline{AB} = \overline{CD}$ and $\overline{AD} = \overline{BC}$

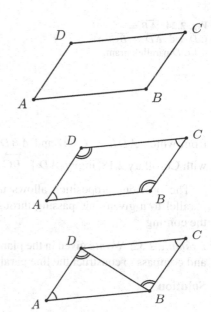

Proposition 2.30 *A convex quadrilateral is a parallelogram if and only if its opposite angles are equal.*

Proof Suppose first that the convex quadrilateral $ABCD$ is a parallelogram (cf. Fig. 2.21). Then, $\overleftrightarrow{AD} \parallel \overleftrightarrow{BC}$ and, since angles $\angle A$ and $\angle B$ of the parallelogram are consecutive with respect to $\overleftrightarrow{AB}$, we have $\widehat{A} + \widehat{B} = 180°$. Analogously, $\widehat{B} + \widehat{C} = 180°$, so that $\widehat{A} = 180° - \widehat{B} = \widehat{C}$. Likewise, $\widehat{B} = \widehat{D}$.

Conversely, let $ABCD$ be a convex quadrilateral in which $\widehat{A} = \widehat{C}$ and $\widehat{B} = \widehat{D}$ (cf. Fig. 2.22). Then, $\widehat{A} + \widehat{B} = \widehat{C} + \widehat{D}$ and, since $\widehat{A} + \widehat{B} + \widehat{C} + \widehat{D} = 360°$ (cf. Problem 11, page 38), we have $\widehat{A} + \widehat{B} = \widehat{C} + \widehat{D} = 180°$. Analogously, $\widehat{A} + \widehat{D} = \widehat{B} + \widehat{C} = 180°$. Now, since $\widehat{A} + \widehat{B} = 180°$, Corollary 2.18 guarantees that $\overleftrightarrow{AD} \parallel \overleftrightarrow{BC}$. Accordingly, $\widehat{B} + \widehat{C} = 180°$ gives us $\overleftrightarrow{AB} \parallel \overleftrightarrow{CD}$, so that $ABCD$ has parallel opposite sides, i.e., is a parallelogram. □

Proposition 2.31 *A convex quadrilateral is a parallelogram if and only if its pairs of opposite sides have equal lengths.*

Proof Suppose first that the convex quadrilateral $ABCD$ is a parallelogram (cf. Fig. 2.23). Then, we already know that $\widehat{A} = \widehat{C}$. On the other hand, since $\overleftrightarrow{AD} \parallel \overleftrightarrow{BC}$, we have $A\widehat{D}B = C\widehat{B}D$. Hence, triangles ABD and CDB are congruent by AAS, and it follows that $\overline{AB} = \overline{CD}$ and $\overline{AD} = \overline{BC}$.

Conversely, let $ABCD$ be a convex quadrilateral such that $\overline{AB} = \overline{CD}$ and $\overline{AD} = \overline{BC}$ (cf. Fig. 2.24). Then, triangles ABD and CDB are congruent by SSS,

Fig. 2.24 $\overline{AB} =$
$\overline{CD}$ and $\overline{AD} = \overline{BC} \Rightarrow$
$ABCD$ parallelogram

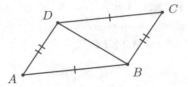

from which $A\widehat{D}B = C\widehat{B}D$ and $A\widehat{B}D = C\widehat{D}B$. In turn, such equalities, together with Corollary 2.18, imply $\overleftrightarrow{AD} \parallel \overleftrightarrow{BC}$ and $\overleftrightarrow{AB} \parallel \overleftrightarrow{CD}$.   □

The previous proposition allows us to present a simple construction for the parallel to a given line, passing through a point not belonging to it, as shown by the coming

Example 2.32 We are given in the plane a line r and a point $A \notin r$. Use straightedge and compass to construct the line parallel to r and passing through A.

Solution

$A_{\bullet}$

_____
                                                                            r

CONSTRUCTION STEPS

1. Draw a circle α, centered at A and intersecting r in two distinct points, B and C.
2. Draw the circle β of center A and radius equal to $\overline{BC}$.
3. Draw the circle γ, centered at C and with radius equal to that of α.
4. Mark the intersection point D of β and γ, such that A and D are in a single half-plane with respect to r.
5. By the previous proposition, $ABCD$ is a parallelogram; in particular $\overleftrightarrow{AD}$ is parallel to r.

                                                                            □

The next two examples bring useful applications of the construction delineated above.

Example 2.33 Use straightedge and compass to construct a parallelogram, knowing the lengths a and b of its sides, as well as the acute angle α formed by them.

Solution

CONSTRUCTION STEPS

1. Draw a line r and mark on it a line segment AB of length b (or a; the choice is immaterial).
2. Construct an angle $\angle BAX$ of measure α.
3. Plot, on half-line $\overrightarrow{AX}$, the point D such that $\overline{AD} = a$.
4. Trace through D the parallel to $\overleftrightarrow{AB}$ and through B the parallel to $\overleftrightarrow{AD}$; then, mark the intersection point C of these two lines.
5. $ABCD$ is a parallelogram satisfying the stated conditions.

□

Example 2.34 Use straightedge and compass to construct triangle ABC, given the length a of side BC and the measures α and β of the internal angles at A and B, respectively.

Solution

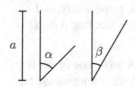

CONSTRUCTION STEPS

1. Draw a line r and mark on it a line segment BC of length a.
2. Construct an angle $\angle CBX$ of measure β.
3. Construct an angle $\angle BXY$, of measure α, with Y lying in the same half-plane as C with respect to $\overleftrightarrow{BX}$.
4. Draw, through C, the parallel to $\overleftrightarrow{XY}$; then, plot A as the intersection point of this parallel with line $\overleftrightarrow{BX}$.

□

Back to the general discussion of parallelograms, the next result brings yet another useful characterization of them.

Proposition 2.35 *A convex quadrilateral is a parallelogram if and only if its diagonals intersect each other in the respective midpoints.*

Fig. 2.25 $ABCD$ parallelogram $\Rightarrow$ $\overline{AM} = \overline{CM}$ and $\overline{BM} = \overline{DM}$

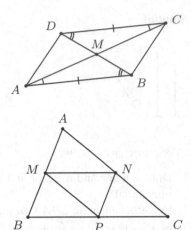

Fig. 2.26 The midsegments of a triangle

Proof Firstly, let $ABCD$ be a parallelogram and M the point of intersection of its diagonals (cf. Fig. 2.25). From $\overleftrightarrow{AB} \parallel \overleftrightarrow{CD}$, it comes that $B\widehat{A}M = D\widehat{C}M$ and $A\widehat{B}M = C\widehat{D}M$. Since we already know that $\overline{AB} = \overline{CD}$, we conclude that triangles ABM and CDM are congruent, by ASA. Therefore, $\overline{AM} = \overline{CM}$ and $\overline{BM} = \overline{DM}$.

Conversely (see also Fig. 2.25), let $ABCD$ be a quadrilateral whose diagonals AC and BD intersect at M, midpoint of both of them. Then, $\overline{MA} = \overline{MC}$, $\overline{MB} = \overline{MD}$ and $A\widehat{M}B = C\widehat{M}D$ (OPP angles), so that triangles ABM and CDM are congruent by SAS. Analogously, BCM and DAM are also congruent by SAS. In turn, such congruences give us $\overline{AB} = \overline{CD}$ and $\overline{BC} = \overline{AD}$, respectively, and we already know that these last two equalities are equivalent to the fact that $ABCD$ is a parallelogram. $\qquad\qquad\square$

For what comes next, we define a **midsegment** of a triangle to be one of the line segments joining the midpoints of two sides of the triangle. Thus, each triangle has *exactly three* midsegments. In the notations of Fig. 2.26, the midsegments of triangle ABC are MN, NP and MP. We also say that MN is the midsegment relative to vertex A (or to side BC); analogously, NP and MP are the midsegments of ABC relative to the vertices B and C (or to the sides AB and AC), respectively. Finally, triangle MNP (i.e., the triangle having the midsegments of ABC as its sides) is the **medial triangle** of ABC.

The properties of parallelograms obtained up to this point allows us to prove, in the coming proposition, an important result on the midsegments of a triangle, which is known as the **midsegment theorem**.

Proposition 2.36 *Let ABC be any triangle. If MN is the midsegment relative to BC, then $\overleftrightarrow{MN} \parallel \overleftrightarrow{BC}$. Conversely, if through the midpoint M of side AB we draw the line parallel to $\overleftrightarrow{BC}$, then it intersects side AC in its midpoint N. Moreover, if this is so, then*

Fig. 2.27 Length of a
midsegment of a triangle

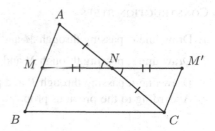

$$\overline{MN} = \frac{1}{2}\,\overline{BC}.$$

Proof For the first part, in the notations of Fig. 2.27, take M' on $\overrightarrow{MN}$ such that
$\overline{MN} = \overline{NM'}$. Since N is the midpoint of AC and $A\widehat{N}M = C\widehat{N}M'$ (OPP angles),
triangles AMN and $CM'N$ are congruent by SAS. Hence, $\overline{M'C} = \overline{MA}$ and
$M'\widehat{C}N = M\widehat{A}N$, whence (by Corollary 2.18) $\overleftrightarrow{M'C} \parallel \overleftrightarrow{AM}$. Thus,

$$\overline{BM} = \overline{AM} = \overline{M'C} \text{ and } \overleftrightarrow{BM} = \overleftrightarrow{AM} \parallel \overleftrightarrow{M'C}.$$

Having a pair of opposite sides equal and parallel, Problem 1, page 58 guarantees
that the convex quadrilateral $MBCM'$ is a parallelogram. However, since in all
parallelograms both pairs of opposite sides are equal and parallel, we conclude that

$$\overleftrightarrow{BC} \parallel \overleftrightarrow{MM'} = \overrightarrow{MN} \text{ and } \overline{BC} = \overline{MM'} = 2\,\overline{MN}.$$

Conversely, let r be the line passing through the midpoint M of side AB and
parallel to $\overleftrightarrow{BC}$. Since $\overleftrightarrow{MN}$ also passes through M and is parallel to $\overleftrightarrow{BC}$, the fifth
postulate 2.17 of Euclid assures that r coincides with $\overleftrightarrow{MN}$; in particular, $N \in r$. $\square$

The coming example brings a first application of the midsegment theorem.

Example 2.37 Construct triangle ABC, knowing the positions of the midpoints M,
N and P of the sides BC, CA and AB, respectively.

Solution

$P\;\bullet$ $\overset{\displaystyle N}{\bullet}$

$\underset{\displaystyle M}{\bullet}$

CONSTRUCTION STEPS

1. Draw line r, passing through M and parallel to $\overleftrightarrow{NP}$.
2. Draw line s, passing through N and parallel to $\overleftrightarrow{MP}$.
3. Draw line t, passing through P and parallel to $\overleftrightarrow{MN}$.
4. According to the previous proposition, we have $s \cap t = \{A\}$, $r \cap t = \{B\}$ and $r \cap s = \{C\}$.

$\square$

For what follows, recall that a *median* of a triangle is the line segment joining a vertex of the triangle to the midpoint of the opposite side. Obviously, every triangle has exactly three medians. On the other hand, and as an application of the midsegment theorem together with the properties of parallelograms, we shall show in the coming proposition that the medians of each triangle intersect at a single point, which is called its *barycenter*.

Proposition 2.38 *In every triangle, the three medians pass through a single point, the **barycenter** of the triangle. Moreover, the barycenter divides each median, from the corresponding vertex, in the ratio* 2 : 1.

Proof Let N and P be the midpoints of the sides AC and AB, respectively, and let $BN \cap CP = \{G_1\}$ (cf. Fig. 2.28). Let also S and T be the midpoints of the line segments BG_1 and CG_1, respectively. Note that NP is the midsegment of ABC relative to BC, whereas ST is the midsegment of BCG_1 relative to BC; therefore, the midsegment theorem assures that both NP and ST are parallel to BC and have length equal to half of that of $\overline{BC}$. Hence, $\overline{NP} = \overline{ST}$ and $\overleftrightarrow{NP} \| \overleftrightarrow{ST}$, so that Problem 1, page 58, guarantees that $NPST$ is a parallelogram. It then follows from Proposition 2.35 that $\overline{PG_1} = \overline{G_1T}$ and $\overline{NG_1} = \overline{G_1S}$. However, since $\overline{BS} = \overline{SG_1}$ and $\overline{CT} = \overline{TG_1}$, we conclude that $\overline{BS} = \overline{SG_1} = \overline{G_1N}$ and $\overline{CT} = \overline{TG_1} = \overline{G_1P}$. In turn, such equalities readily furnish $\overline{BG_1} = 2\,\overline{G_1N}$ and $\overline{CG_1} = 2\,\overline{G_1P}$.

Now, if M is the midpoint of BC and G_2 is the point of intersection of the medians AM and BN, we conclude in an analogous way that G_2 divides AM and BN in the ratio 2 : 1 from each vertex. But then, G_1 and G_2 are such that $\overline{BG_1} = 2\,\overline{G_1N}$ and $\overline{BG_2} = 2\,\overline{G_2N}$, and this clearly implies $G_1 \equiv G_2$.

Fig. 2.28 The medians and
the barycenter of a triangle

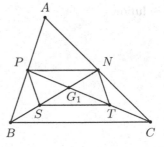

Fig. 2.29 A trapezoid
$ABCD$, with $\overleftrightarrow{AB} \parallel \overleftrightarrow{CD}$

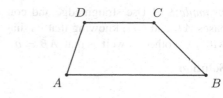

Fig. 2.30 Parallelograms
associated to the trapezoid
$ABCD$

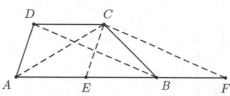

Finally, letting G denote the point $G_1 \equiv G_2$, we have proved that AM, BN and CP concur at G, which divides each median in the ratio $2 : 1$, from the corresponding vertex to the opposite side. $\square$

Hereafter, unless explicitly stated otherwise, we shall denote the barycenter of a triangle ABC by G. O The barycenter of a triangle is one of its *notable points*; others (the *circumcenter*, the *incenter* and the *orthocenter*, to name just the more popular ones) will be studied in Sect. 3.2.

As we have previously pointed out, Problem 1, page 58, guarantees that a quadrilateral having a pair of equal and parallel opposite sides is a parallelogram. Nevertheless, it may happen that two opposite sides of a quadrilateral are parallel, being equal or not. In such a case, we say that the quadrilateral is a **trapezoid** or **trapezium** (cf. Fig. 2.29). Thus, every parallelogram is in particular a trapezoid, albeit the converse is obviously not true.

In every trapezoid, the two sides which are known to be parallel are its **bases**; and one uses to refer to them as the **larger** and **smaller** bases of the trapezoid. The other two sides (about which we at first know nothing, but which can also be parallel—provided the trapezoid is in particular a parallelogram) are the **legs** of the trapezoid. In the notations of Fig. 2.29, AB and CD are the bases, and BC and AD are the legs of trapezoid $ABCD$.

Whenever we deal with problems involving geometric constructions in a trapezoid $ABCD$ as that of Fig. 2.29, it is frequently useful to observe (cf. Fig. 2.30) that if E and F are points on the line $\overleftrightarrow{AB}$ for which $ADCE$ and $BDCF$ are parallelograms, then:

i. Triangle BCE is such that $\overline{BE} = \overline{AB} - \overline{CD}$, $\overline{CE} = \overline{AD}$ and $B\widehat{C}E$ = measure of the angle formed by lines $\overleftrightarrow{AD}$ and $\overleftrightarrow{BC}$.
ii. Triangle ACF is such that $\overline{AF} = \overline{AB} + \overline{CD}$, $\overline{CF} = \overline{BD}$ and $A\widehat{C}F$ = measure of the angle formed by the diagonals AC and BD.

We shall use the above discussion in the coming example.

Example 2.39 Use straightedge and compass to construct a trapezoid $ABCD$ of bases AB and CD, knowing that its diagonals AC and BD form an angle of $45°$ with each other, as well as that $\overline{AB} = a$, $\overline{AC} = d_1$ and $\overline{BD} = d_2$.

Solution

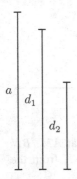

CONSTRUCTION STEPS

1. Inspired by Fig. 2.30, construct a triangle ACF such that $\overline{AC} = d_1$, $\overline{CF} = d_2$ and $A\widehat{C}F = 45°$ or $180° - 45° = 135°$ (there will be two distinct solutions).
2. Mark point B on $\overrightarrow{AE}$ such that $\overline{AB} = a$.
3. Draw the line r, parallel to $\overleftrightarrow{CF}$ and passing through B, and line s, parallel to $\overleftrightarrow{AF}$ and passing through C.
4. Mark D as the intersection point of r and s.   □

Before we proceed, we need to set a few more conventions on trapezoids, namely: the line segment joining the midpoints of the legs is the **midsegment** of the trapezoid, whereas the line segment joining the midpoints of its diagonals is the **Euler median**.[6] The coming proposition teaches us how to compute the lengths of such segments in terms of the lengths of the basis of the trapezoid.

Proposition 2.40 *Let $ABCD$ be a trapezoid of basis AB and CD, and legs AD and BC. Let also M and N be the midpoints of the legs AD and BC, respectively, and P and Q be the midpoints of the diagonals AC and BD, also respectively (cf. Fig. 2.31). Then:*

(a) M, N, P and Q are collinear and $\overleftrightarrow{MN} \parallel \overleftrightarrow{AB}, \overleftrightarrow{CD}$.
(b) $\overline{MN} = \frac{1}{2}(\overline{AB} + \overline{CD})$ and $\overline{PQ} = \frac{1}{2}|\overline{AB} - \overline{CD}|$.

Fig. 2.31 Midsegment and
Euler median of a trapezoid

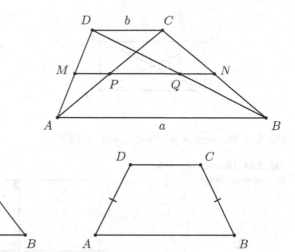

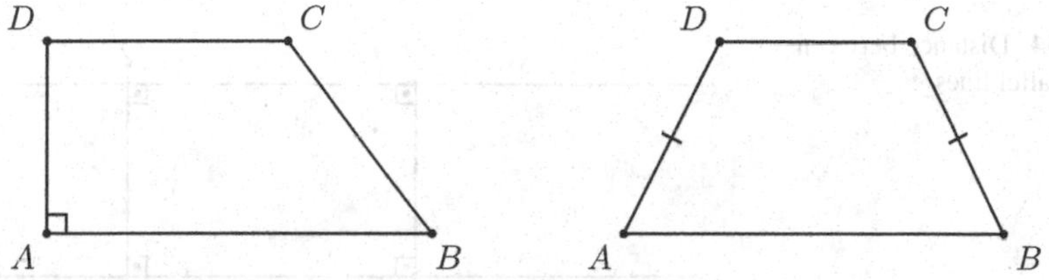

Fig. 2.32 A right (left) and isosceles (right) trapezoid $ABCD$

Proof In the notations of Fig. 2.31, since MP is a midsegment of triangle DAC, Proposition 2.36 assures that $\overleftrightarrow{MP} \parallel \overleftrightarrow{CD}$ and $\overline{MP} = \frac{b}{2}$. On the other hand, since MQ is a midsegment of triangle ADB, Proposition 2.36 also gives $\overleftrightarrow{MQ} \parallel \overleftrightarrow{AB}$ and $\overline{MQ} = \frac{a}{2}$. However, since $\overleftrightarrow{AB} \parallel \overleftrightarrow{CD}$, the fifth postulate of Euclid guarantees that $\overleftrightarrow{MP} = \overleftrightarrow{MQ}$, and hence M, P and Q are collinear. Moreover,

$$\overline{PQ} = \overline{MQ} - \overline{MP} = \frac{a}{2} - \frac{b}{2} = \frac{a-b}{2}.$$

Now, arguing analogously with the midsegments NQ and NP of triangles CBD and ABC, respectively, we conclude that P, Q and N are collinear and $\overline{NQ} = \frac{b}{2}$. Hence, it follows from what we did above that

$$\overline{MN} = \overline{MQ} + \overline{NQ} = \frac{a}{2} + \frac{b}{2} = \frac{a+b}{2}.$$

□

We finish this initial study of trapezoids by fixing some terminology, for future reference: given a trapezoid $ABCD$ of bases AB and CD and legs AD and BC, we say that $ABCD$ is **right** at A if $D\widehat{A}B = 90°$; also $ABCD$ is **isosceles** if $\overline{AD} = \overline{BC}$ (see Fig. 2.32).

In order to complete our study of the more elementary particular types of quadrilaterals, let us now look at *rectangles* and *rhombuses*. A convex quadrilateral is a **rectangle** if all of its interior angles have equal measures. Since Problem 11, page 38, showed that the sum of the measures of the interior angles of a convex

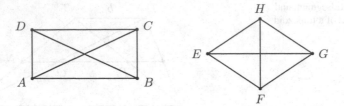

Fig. 2.33 Rectangle $ABCD$ and rhombus $EFGH$

Fig. 2.34 Distance between
two parallel lines

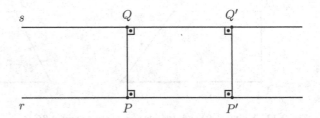

quadrilateral is always equal to 360°, we conclude that a convex quadrilateral is a
rectangle if and only if the measures of its four interior angles are equal to 90°.
A convex quadrilateral is a **rhombus** if the lengths of its four sides are all equal.
Figure 2.33 shows examples of a rectangle and a rhombus.

Since the opposite sides of a rectangle are always parallel (for they are both
perpendicular to the other two sides), we conclude that *every rectangle is a
parallelogram*. On the other hand, Proposition 2.31 guarantees that *every rhombus
is a parallelogram* too.

The discussion above allows us to define the distance between two parallel lines.
To this end, note first that if r and s are parallel lines, then (cf. Corollary 2.18) a line
t is perpendicular to r if and only if it is also perpendicular to s.

Definition 2.41 If r and s are parallel lines, the **distance** between them is the length
of any line segment PQ such that $P \in r$, $Q \in s$ and $\overleftrightarrow{PQ} \perp r, s$.

In order to see that the previous definition makes sense, take $P, P' \in r$ and let
$Q, Q' \in s$ be such that $\overleftrightarrow{PQ}, \overleftrightarrow{P'Q'} \perp r$ (cf. Fig. 2.34). Then, $PQQ'P'$ has its four
interior angles equal to 90°, hence is a rectangle. In particular, $\overline{PQ} = \overline{P'Q'}$.

Also with respect to the previous definition, the coming example shows how to
construct the parallels to a given line, situated at a given distance of it.

Example 2.42 Use straightedge and compass to construct the parallels to line r,
situated at a distance d of r.

Solution

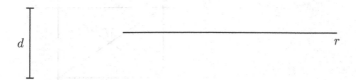

CONSTRUCTION STEPS

1. Mark a point A on r and draw line t, perpendicular to r and passing through A.
2. Mark on t points B and B', such that $\overline{AB} = \overline{AB'} = d$.
3. Draw, through B and B', respectively, lines s and s' parallel to r.

□

Back to the general discussion of rectangles and rhombuses, propositions 2.43 and 2.46 collect useful characterizations of these quadrilaterals.

Proposition 2.43 *A parallelogram is a rectangle if and only if its diagonals have equal lengths.*

Proof If $ABCD$ is a rectangle of diagonals AC and BD (cf. Fig. 2.33), then $D\widehat{A}B = A\widehat{D}C = 90°$ and (since $ABCD$ is also a parallelogram) $\overline{AB} = \overline{DC}$. However, since triangles DAB and ADC share the common side AD, they are congruent by SAS; in particular, $\overline{AC} = \overline{BD}$.

Conversely, assume that $ABCD$ is a parallelogram in which $\overline{AC} = \overline{BD}$ (cf. Fig. 2.35). Since we also have $\overline{AB} = \overline{DC}$, triangles DAB and ADC (which share side AD) are again congruent, this time by SSS LLL. Therefore, $D\widehat{A}B = A\widehat{D}C$. However, since $ABCD$ is a parallelogram, we have $D\widehat{A}B + A\widehat{D}C = 180°$, so that $D\widehat{A}B = A\widehat{D}C = 90°$. Analogously, $A\widehat{B}C = D\widehat{C}B = 90°$, and $ABCD$ is a rectangle.

□

The coming corollary brings an extremely useful consequence of the former proposition.

Corollary 2.44 *The length of the median of a right triangle relative to its hypotenuse is half the length of the hypotenuse itself.*

Proof Let ABC be a right triangle, with $\widehat{A} = 90°$ (cf. Fig. 2.36). Draw through B the parallel to AC, through C the parallel to AB and let D be their point of intersection. Since $B\widehat{A}C + A\widehat{B}D = 180°$ and $B\widehat{A}C = 90°$, it comes that $A\widehat{B}D = 90°$. Analogously, $A\widehat{C}D = 90°$ and, since the sum of the angles of $ABDC$ is 360°,

Fig. 2.35 $ABCD$ is a parallelogram such that $\overline{AC} = \overline{BD}$

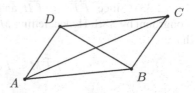

Fig. 2.36 The median
relative to the hypotenuse of a
right triangle

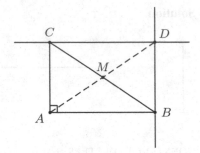

we conclude that $B\widehat{D}C = 90°$. Hence, quadrilateral $ABDC$ is a rectangle, whence $\overline{AD} = \overline{BC}$ and the point M of intersection of AD and BC is the midpoint of both such segments. Therefore, $\overline{BC} = \overline{AD} = 2\,\overline{AM}$.   □

Example 2.45 Construct a right triangle ABC, knowing the lengths m and h of the median and the height relative to hypotenuse BC, respectively.

Solution

m $\left|\rule{0pt}{1.5em}\right.$ h $\left|\rule{0pt}{1.5em}\right.$

CONSTRUCTION STEPS

1. According with the previous corollary, we have $\overline{BC} = 2m$. Construct such a segment BC, together with its midpoint M.
2. Draw (cf. Example 2.42) a line r parallel to $\overleftrightarrow{BC}$ and at a distance h of r.
3. Obtain the possible positions of vertex A as one of the intersections of line r and the circle centered at M and with radius m.

□

We now turn to the promised characterization of rhombuses.

Proposition 2.46 *A parallelogram is a rhombus if and only if it has perpendicular diagonals.*

Proof Suppose first that $EFGH$ is a rhombus of diagonals EG and FH (cf. Fig. 2.33). Since $\overline{EF} = \overline{EH}$ and $\overline{GF} = \overline{GH}$, triangles EFG and EHG are congruent by SSS. Hence, letting M be the point of intersection of EG and FH, we have

$$F\widehat{E}M = F\widehat{E}G = H\widehat{E}G = H\widehat{E}M.$$

Fig. 2.37
$EG \perp FH \Rightarrow EFGH$
rhombus

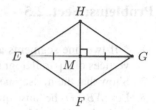

Fig. 2.38 Square $ABCD$

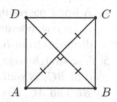

Thus, EM is the interior bisector of angle $\angle FEH$ of triangle EFH, which is isosceles of basis FH; Problem 6, page 30, assures that EM is also the height of EFH relative to FH. Therefore, $\overleftrightarrow{FH} \perp \overleftrightarrow{EM} = \overleftrightarrow{EG}$.

Conversely, let $EFGH$ be a parallelogram with perpendicular diagonals EG and FH (cf. Fig. 2.37). Since EG and FH intersect at the midpoint M of both of them (for $EFGH$ is a parallelogram), it comes that in triangle EHG segment HM is at the same time median and height relative to EG. Hence, Problem 7, page 30 gives $\overline{EH} = \overline{GH}$. However, since $\overline{EH} = \overline{FG}$ and $\overline{EF} = \overline{GH}$, there is nothing left to do. □

The last type of convex quadrilateral we wish to study at this point is the **square**, namely, a quadrilateral which is simultaneously a rectangle and a rhombus (cf. Fig. 2.38). Thus, a square has equal sides and interior angles; moreover, its diagonals are also equal and perpendicular, intersect at their midpoints and form angles of 45° with the sides of the square. (Prove this last claim!)

Remark 2.47 Letting $\mathcal{T}$ denote the set of trapezoids, $\mathcal{P}$ the set of parallelograms, $\mathcal{R}$ the set of rectangles, $\mathcal{L}$ the set of rhombuses and $\mathcal{Q}$ that of the squares, it follows from the material of this section that

$$\begin{cases} \mathcal{R} \cup \mathcal{L} \subset \mathcal{P} \subset \mathcal{T} \\ \mathcal{R} \cap \mathcal{L} = \mathcal{Q} \end{cases},$$

with all inclusions being strict ones.

Problems: Sect. 2.5

1. * If two line segments are equal and parallel, prove that its endpoints are the vertices of a parallelogram.
2. Show that the midsegments of a triangle divide it into four congruent triangles.
3. Let $ABCD$ be any quadrilateral (not necessarily convex). Show that the midpoints of its sides are the vertices of a parallelogram.
4. A line r passes through the barycenter G of a triangle ABC and leaves vertex A on one side and vertices B and C on the other. Prove that the sum of the distances from B and C to the line r is equal to the distance from A to r.
5. Use straightedge and compass to construct triangle ABC, knowing the length a of BC, as well as the lengths m_a and m_b of the medians relative to the sides BC and AC, respectively.
6. Prove that, in every triangle, the sum of the lengths of the medians is less that $\frac{3}{2}$ of the perimeter and greater than $\frac{3}{4}$ of the perimeter of the triangle.
7. (England) Consider a circle of center O and diameter AB. Extend a chord AP up to the point Q such that P is the midpoint of AQ. If $OQ \cap BP = \{R\}$, compute the ratio of the lengths of the segments $\overline{RQ}$ and $\overline{RO}$.
8. Let $ABCD$ be a trapezoid of bases $\overline{AB} = 7$cm and $\overline{CD} = 3$cm and legs AD and BC. If $\widehat{A} = 43°$ and $\widehat{B} = 47°$, compute the distance between the midpoints of the bases of the trapezoid.
9. We are given in the plane a parallelogram $ABCD$ of diagonals AC and BD, and a line r which does not intersect $ABCD$. Knowing that the distances of the points A, B and C to the line r are respectively equal to 2, 3 and 6 centimeters, compute the distance from D to r.
10. The bases AB and CD of a trapezoid have lengths a and b, respectively, with $a > b$. If the legs of the trapezoid are AD and BC, and $\angle BCD = 2\angle DAB$, prove that $\overline{BC} = a - b$.
11. Let $ABCD$ be a trapezoid in which the length of the larger basis AB is equal to the sum of the length of the smaller basis CD with that of the leg BC. If $\widehat{A} = 70°$, compute $\widehat{C}$.
12. Use straightedge and compass to construct a trapezoid, knowing the lengths a and b of its bases, and c and d of its legs.
13. * (OCM) A triangle ABC is rectangle at A and such that $\overline{BC} = 2\overline{AB}$. Compute the measures, in degrees, of its interior angles.
14. In a triangle ABC, let M be the midpoint of side BC and H_b and H_c be the feet of the heights relative to AC and AB, respectively. Prove that triangle MH_bH_c is isosceles.
15. Let $ABCD$ be a square of diagonals AC and BD, and E be a point on side CD, such that $\overline{AE} = \overline{AB} + \overline{CE}$. Letting F be the midpoint of side CD, prove that $E\widehat{A}B = 2 \cdot F\widehat{A}D$.

16. (TT) Let $ABCD$ be a rectangle of diagonals AC and BD, and M, N, P and Q be points situated on the sides AB, BC, CD and AD, respectively, all distinct from the vertices of the rectangle. Show that the perimeter of the quadrilateral $MNPQ$ is greater than or equal to twice that of the rectangle $ABCD$. When does equality take place?

17. * (Hungary) In a triangle ABC, let M be the midpoint of side BC and P be the foot of the perpendicular dropped from B to the internal bisector relative to vertex A. Prove that

$$\overline{PM} = \frac{1}{2}|\overline{AB} - \overline{AC}|.$$

Chapter 3
Loci in the Plane

The concept of locus, developed in this chapter, turns out to be essential for a deeper understanding of the *synthetic method* in Euclidean Geometry. After we have mastered that notion, we will be able to discuss various additional interesting properties of triangles and quadrilaterals, among which we highlight the problem of inscribing such polygons in a circle.

3.1 Basic Loci

We start this section by presenting the general concept of *locus*.

Definition 3.1 Given a property $\mathcal{P}$ relative to points in the plane, the **locus** of the point which possess property $\mathcal{P}$ is the subset $\mathcal{L}$ of the plane satisfying the two following conditions:

(a) Every point of $\mathcal{L}$ has the property $\mathcal{P}$.
(b) Every point of the plane having the property $\mathcal{P}$ belongs to $\mathcal{L}$.

Rewording, $\mathcal{L}$ is the locus of the property $\mathcal{P}$ if $\mathcal{L}$ if formed *precisely* by the points of the plane satisfying property $\mathcal{P}$, no more, no less. In what comes next, we shall study a few elementary loci, as well as some applications of them (Fig. 3.1).

Example 3.2 Given positive real number r and a point O in the plane, the locus of the points of the plane situated at a distance r from O is the circle of center O and radius r:

$$\overline{AO} = r \Longleftrightarrow A \in \Gamma(O; r).$$

Fig. 3.1 Circle as locus

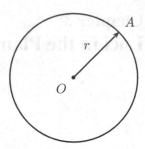

Example 3.3 As we saw in Example 2.42, the locus of the points in the plane situated at a distance d from a given line r is the union of lines s and s', parallel to r and each of which situated at a distance d from r.

For the coming example, given distinct points A and B in the plane, we define the **perpendicular bisector** of AB as the line perpendicular to AB and passing through its midpoint.

Example 3.4 Use straightedge and compasso to construct the perpendicular bisector of the line segment AB given below.

Solution

CONSTRUCTION STEPS

1. With a single opening $r > \frac{1}{2}\overline{AB}$, draw the circles of radius r and centered at A and B; if X and Y are their points of intersection, then $\overleftrightarrow{XY}$ is the perpendicular bisector of AB.

Indeed, letting M be the point of intersection of segments XY and AB, we saw in Example 2.10 that M is the midpoint of AB. On the other hand, since triangle XAB is isosceles of basis AB and XM is its median relative to the basis, the result of Problem 6, page 30, guarantees that XM is also a height of XAB. Therefore, $\overleftrightarrow{XY}$ passes through the midpoint of AB and is perpendicular to AB, thus coincides with the perpendicular bisector of this line segment.   □

The coming result characterizes the perpendicular bisector of a segment as a locus.

Proposition 3.5 *Given distinct points A and B in the plane, the perpendicular bisector of AB is the locus of the points in the plane which are at the same distance from A and B.*

Fig. 3.2 $P \in$
(perpendicular bisector of AB) $\Rightarrow$
$\overline{PA} = \overline{PB}$

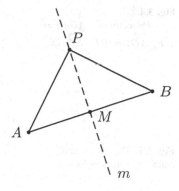

Fig. 3.3 $\overline{PA} = \overline{PB} \Rightarrow P \in$
(perpendicular bisector of
AB)

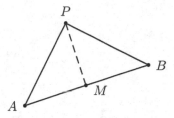

Proof Let M be the midpoint and m the perpendicular bisector of AB (cf. Fig. 3.2).
If $P \in m$, then, in triangle PAB, PM is median and height, so that the result of
Problem 7, page 30, gives PAB isosceles of basis AB. Therefore, $\overline{PA} = \overline{PB}$.

Conversely, let P be a point in the plane for which $\overline{PA} = \overline{PB}$ (see Fig. 3.3).
Then, triangle PAB is isosceles of basis AB, whence the median and height of PAB
with respect to AB coincide. However, since the median of PAB relative to AB is
the line segment PM, we conclude that $PM \perp AB$, which is the same as saying that
$\overleftrightarrow{PM}$ is the perpendicular bisector of AB.   □

The role of the bisector of an angle as a locus is essentially contained in the
coming result.

Proposition 3.6 *Let $\angle AOB$ be a given angle. If P is a point of it, then*

$$d(P, \overrightarrow{AO}) = d(P, \overrightarrow{BO}) \iff P \in (bisector \ of \ \angle AOB).$$

Proof Suppose first that P belongs to the bisector of $\angle AOB$ (cf. Fig. 3.4), and
let M and N be the feet of the perpendiculars dropped from P to the lines $\overleftrightarrow{AO}$
and $\overleftrightarrow{BO}$, respectively. Triangles OMP and ONP are congruent by AAS, since
$M\widehat{O}P = N\widehat{O}P$, $O\widehat{M}P = O\widehat{N}P = 90°$ and OP is a common side of both these
triangles. Hence, $\overline{PM} = \overline{PN}$, which is the same as $d(P, \overleftrightarrow{AO}) = d(P, \overleftrightarrow{BO})$.

Conversely, let P be a point in the interior of angle $\angle AOB$, such that $\overline{PM} =$
$\overline{PN}$, with M and N being the feet of the perpendiculars dropped from P to the lines

Fig. 3.4
$P \in$ (bisector of $\angle AOB$) $\Rightarrow$
$d(P, \overrightarrow{AO}) = d(P, \overrightarrow{BO})$

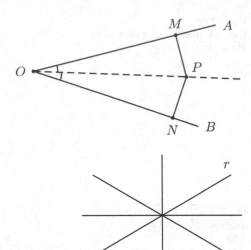

Fig. 3.5 The bisectors of two
concurrent lines as locus

$\overleftrightarrow{AO}$ and $\overleftrightarrow{BO}$, respectively. Then, triangles MOP and NOP are again congruent, this time by the HL congruence case (for OP is the common hypotenuse and $\overline{PM} = \overline{PN}$—see Problem 1, page 36). But then, $M\widehat{O}P = N\widehat{O}P$, so that P belongs to the bisector of $\angle AOB$.   □

Example 3.7 Given in the plane lines r and s, concurrent at O, we saw in the previous proposition that a point P of the plane is at the same distance from r and s if and only if P lies on one of the lines bisecting the angles formed by r and s (the bold lines in Fig. 3.5). Thus, the locus of points in the plane equidistant from two concurrent lines is the union of the bisectors of the angles formed by them.

Having studied some basic loci, it is worth talking a bit on the general problem of construction, with straightedge and compass, of a figure satisfying given geometric conditions. Yet in another way, the standard approach to such a problem basically consists of performing the two following steps:

1. *Assume that the problem has been solved*: we sketch the desired figure, identifying the problem data as well as the geometric elements that can lead us to the actual construction.
2. *Construct the key points for the solution*: a **key point** is any point that, once constructed, lets the necessary subsequent constructions immediate or almost immediate, thus allowing us to solve the problem. In order to construct the key point(s) corresponding to a certain problem, we should carefully examine the geometric properties involved in the situation under scrutiny in order to identify, in each case, two different loci to which the desired point belongs to. This way, the key point(s) is (are) determined as the intersection points of those loci.

Let us examine, at a simple example, how the actual como funciona a execução do programa acima.

Example 3.8 Construct, with straightedge and compass, a circle passing through two given points A and B and having its center on the line r.

Proof

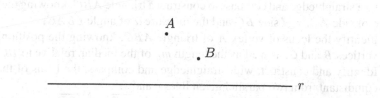

CONSTRUCTION STEPS

1. Assuming that the problem has been solved, we want to construct a circle as the one in the figure below:

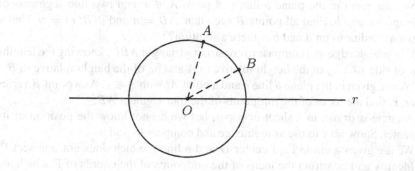

2. The key point will be the center O of the circle, for once we find O, it will suffice to put the nail end of the compass on it and, with opening $\overline{OA}$, to construct the desired circle. In order to construct the point O, we need to have two distinct loci in which O must lie. One of them is the line r itself, for the problem asks O to lie on it. On the other hand, since OA and OB are radii, we have $\overline{OA} = \overline{OB}$, so that O must also lie on the perpendicular bisector of the line segment AB; therefore, this is our second locus.

Having executed the analysis above, we are left to constructing the perpendicular bisector of the line segment AB, thus obtaining its intersection O with the line r and, then, drawing the desired circle (i.e., that of center O and radius $\overline{OA} = \overline{OB}$).

□

Problems: Sect. 3.1

1. Construct a circle of given radius r and passing through two given points A and B. Under what conditions is there a solution?
2. Identify and construct, with straightedge and compass, the locus of the endpoint A of a line segment AB, knowing the position of B and the length c of AB.
3. Use straightedge and compass to construct a triangle ABC, knowing the lengths c of side AB, a of side BC and the measure α of angle $\angle BAC$.
4. Identify the locus of vertex A of triangle ABC, knowing the positions of the vertices B and C, as well as the length m_a of the median relative to BC.
5. Identify and construct, with straightedge and compass, the locus of the points equidistant from two parallel given lines r and s.
6. Construct triangle ABC, knowing the lengths $\overline{AB} = c$, $\overline{BC} = a$ and h_a of the height relative to A.
7. Construct triangle ABC, given the concurrent lines r and s that contain sides AB and AC, respectively, as well as the lengths h_b and h_c of the heights relative to B and C, also respectively.
8. We are given in the plane a line r, a point $A \notin r$ and two line segments of lengths a and b. Find all points B such that $\overline{AB} = a$ and $d(B; r) = b$. Under what conditions on a and b is there a solution?
9. Use straightedge and compass to construct a triangle ABC, knowing the lengths a of side BC, h_a of the height relative to A and h_b of the height relative to B.
10. We are given in the plane a line r and a point A, with $A \notin r$. As a point B varies on r, find the locus of the midpoints of the line segment AB.
11. A circle is drawn in a sheet of paper, but we do not know the position of its center. Show how to use straightedge and compass to find it.
12. We are given a circle Γ of center O and a line r which does not intersect Γ. Identify and construct the locus of the midpoints of the chords of Γ which are parallel to r.
13. Construct triangle ABC, knowing its semiperimeter p and the measures β and γ of its interior angles at vertices B and C, respectively.
14. (Netherlands) We are given a line segment AB and a point P on it. At the same side of $\overleftrightarrow{AB}$ we construct the isosceles right triangles APQ and BPR, of hypotenuses AP and BP, respectively; then, we mark the midpoint M of QR. Find the locus described by M as P varies on AB.

3.2 Notable Points of a Triangle

In this section we apply the concept of locus to study a few more *notable points* of a triangle, namely, the *circumcenter*, the *orthocenter* and the *incenter*. Recall that we have already defined and studied the main property of the barycenter in Proposition 2.38.

Fig. 3.6 The circumcenter of a triangle

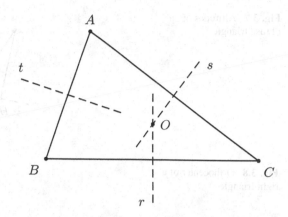

Proposition 3.9 *In every triangle, the perpendicular bisectors of its sides all pass through a single point, called the* **circumcenter** *of the triangle.*

Proof Let ABC be a triangle and r, s and t be the perpendicular bisectors of sides BC, CA and AB, respectively. Let also O be the point of intersection of r and s (see Fig. 3.6).

The characterization of the perpendicular bisector of a line segment as a locus gives $\overline{OB} = \overline{OC}$ (since $O \in r$) and $\overline{OC} = \overline{OA}$ (since $O \in s$). Hence, $\overline{OB} = \overline{OA}$ and, once more from the characterization of the perpendicular bisector as locus, we conclude that $O \in t$. ☐

Example 3.10 Construct the circumcenter of the triangle ABC given below.

Solution

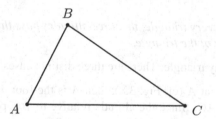

CONSTRUCTION STEPS

1. Draw the perpendicular bisectors of sides AB and AC.
2. The circumcenter of ABC is their intersection point.

☐

As a corollary to the above discussion, we can study the problem of concurrence of the altitudes of a triangle. Note first that, if the triangle is obtuse (Fig. 3.7), then the altitudes departing from the acute angled vertices are exterior to the triangle.

Fig. 3.7 Altitudes of an
obtuse triangle

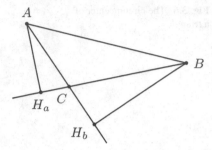

Fig. 3.8 Orthocenter of a
right triangle

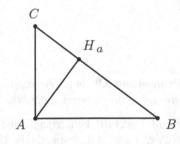

Fig. 3.9 Orthocenter of an
acute triangle

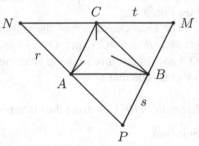

Proposition 3.11 *In every triangle, the three altitudes pass through a single point,*
*called the **orthocenter** of the triangle.*

Proof Let ABC be any triangle. There are three distinct cases to consider:

(a) ABC is right, say at A (cf. Fig. 3.8): then A is the foot of the heights relative
to sides AB and AC. Since the altitude relative to BC passes (by definition)
through A, we conclude that the altitudes of ABC concur at A.

(b) ABC is acute (cf. Fig. 3.9): draw, through A, B and C lines r, s and t,
respectively, parallel to BC, CA and AB (also respectively), and let $r \cap s = \{P\}$,
$s \cap t = \{M\}$ and $t \cap r = \{N\}$. Quadrilaterals $ABCN$ and $ABMC$ are
parallelograms, so that $\overline{CN} = \overline{AB} = \overline{CM}$, and hence C is the midpoint of
MN. Analogously, B is the midpoint of MP and A is that of NP.

On the other hand, the altitude of ABC relative to BC is also perpendicular
to $\overleftrightarrow{NP}$, for $\overleftrightarrow{BC}$ and $\overleftrightarrow{NP}$ are parallel. Likewise, the altitudes relative to AC and
AB are perpendicular to MP and MN, respectively. It follows that the altitudes

of triangle ABC are the perpendicular bisectors of the sides of triangle MNP. However, we have already proved that the perpendicular bisectors of the sides of any triangle are concurrent, so that the altitudes of ABC must pass through a single point.

(c) ABC is obtuse: the proof is entirely analogous to that of (b).

□

In the coming corollary we collect an interesting consequence of the above proof. For the statement of it, we recall (cf. discussion at page 48) that the *medial triangle* of a triangle ABC is the one whose vertices are the midpoints of the sides of ABC.

Corollary 3.12 *The circumcenter of a triangle is the orthocenter of its medial triangle.*

Proof In the notations of item (b) of the proof of the last proposition, ABC is the medial triangle of MNP and the perpendicular bisectors of the sides of MNP are the heights of ABC; hence, the circumcenter of MNP coincides with the orthocenter of ABC. The remaining cases are entirely analogous. □

Example 3.13 Use straightedge and compass to construct the orthocenter of triangle ABC given below.

Solution

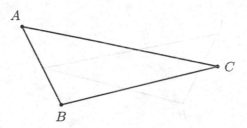

CONSTRUCTION STEPS

1. Draw line r, perpendicular to $\overleftrightarrow{BC}$ and passing through A.
2. Draw line s, perpendicular to $\overleftrightarrow{AC}$ and passing through B.
3. The orthocenter of ABC is the intersection point of r and s.

□

Let us finally examine the point of intersection of the internal bisectors of a triangle.

Proposition 3.14 *The internal bisectors of every triangle concur at a point, called the **incenter** of the triangle.*

Proof Let r, s and t denote the internal bisectors of the angles $\angle A$, $\angle B$ and $\angle C$ of triangle ABC (cf. Fig. 3.10), and I be the intersection point of lines r and s.

Fig. 3.10 Incenter of a
triangle

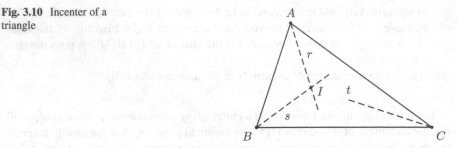

Since $I \in r$, it follows from the characterization of bisectors as locus, given in
Proposition 3.6, that I is at the same distance from the sides AB and AC of ABC.
Analogously, $I \in s$ guarantees that I is at the same distance from the sides AB and
BC. Hence, I is also at the same distance from AC and BC, so that, by invoking
again the characterization of bisectors as locus, we conclude that I belongs to the
internal bisector of angle $\angle C$, i.e., to the line t. But this is the same as saying that r,
s and t concur at I.   □

Example 3.15 Construct the incenter of the triangle ABC given below.

Solution

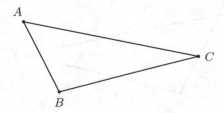

CONSTRUCTION STEPS

1. Draw half-line $\overrightarrow{AX}$, internal bisector of ABC relative to A.
2. Draw half-line $\overrightarrow{BY}$, internal bisector of ABC relative to B.
3. The incenter of ABC is the intersection point of the half-lines $\overrightarrow{AX}$ and $\overrightarrow{BY}$.

                                                                                                □

We finish this section with an important notational remark: most of the times,
whenever we are dealing with the geometry of some given triangle ABC, unless
stated otherwise we shall denote its barycenter by G, its orthocenter by H, its
incenter by I and its circumcenter by O.

Problems: Sect. 3.2

1. From a triangle ABC we know the positions of vertices B and C and of the circumcenter O. Explain why these data do not suffice to determine the position of vertex A.
2. From a triangle ABC we know the positions of vertices B and C and of the incenter I. Construct vertex A.
3. From a triangle ABC we know the positions of vertices B and C and of the orthocenter H. Construct vertex A.
4. Two concurrent lines r and s are drawn in a piece of paper. It happens that, due to the size of the paper, the intersection point A of r and s does not lie in it. Let P be a point in the paper such that the feet of the perpendiculars dropped from P to r and s lie in the paper. Show how to construct a line t, passing through P and passing through A.
5. Let ABC be a triangle of orthocenter H, incenter I and circumcenter O. Show that ABC is equilateral if and only if any two of H, I and O coincide.
6. (OIM) Use compass and straightedge to construct triangle ABC, given the positions of the midpoints of sides AB and AC, as well as that of its orthocenter.

3.3 Tangency and Angles in the Circle

We start this section by studying one of the most important notions of elementary Euclidean Geometry, namely, that of a line and a circle *tangent* to each other.

We say that a circle Γ and a line r are **tangent**, or also that the line r is tangent to the circle Γ, if r and Γ have *exactly one* common point P. In this case, P is said to be the **tangency point** of r and Γ.

The coming result teaches us how to construct a line tangent to a circle and passing through a point of it.

Proposition 3.16 *Let Γ be a circle of center O and P be a point of Γ. If t is the line passing through P and perpendicular to $\overleftrightarrow{OP}$, then t is tangent to Γ.*

Proof Let R be the radius of Γ. If $Q \neq P$ is another point on t (cf. Fig. 3.11), we have $\overline{QO} > \overline{PO} = R$, for $Q\widehat{P}O = 90°$ is the largest angle of triangle OPQ. Hence, $Q \notin \Gamma$, so that P is the only common point of t and Γ. $\qquad\square$

The next example exercises the construction delineated at the proof of the previous proposition.

Example 3.17 In the notations of the figure below, construct a line r, tangent to Γ at P.

Fig. 3.11 Tangent to a circle
through a point of it

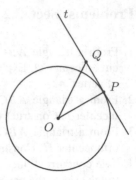

Solution

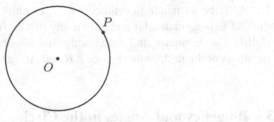

CONSTRUCTION STEPS

1. Draw line $\overleftrightarrow{OP}$.
2. Construct, through P, the line r perpendicular to $\overleftrightarrow{OP}$.

□

It is not difficult to show (cf. Problem 1, page 81) that the line tangent to a circle
Γ through a point P of it is unique. On the other hand, if P lies in the interior of the
circle, we shall see in Proposition 3.26 that there are *exactly two* lines tangent to Γ
and passing through P.

We now turn to the study of certain angles in a circle. Given in the plane a circle
Γ of center O, a **central angle** in Γ is an angle of vertex O, having two radii
OA and OB of Γ as sides. In general, such a central angle will be denoted by
$\angle AOB$, and the context will make it clear to which of the two angles $\angle AOB$ we are
referring. By definition, the *measure* of the central angle $\angle AOB$ is the measure of
the corresponding arc $\overset{\frown}{AB}$. The coming example will show that equal central angles
correspond to equal chords.

Example 3.18 If A, B, C and D are points on a circle Γ, such that the central angles
$\angle AOB$ and $\angle COD$ are equal, then $\overline{AB} = \overline{CD}$.

Proof Suppose (cf. Fig. 3.12) that $A\widehat{O}B = C\widehat{O}D < 180°$ (the case $A\widehat{O}B = C\widehat{O}D > 180°$ can be dealt with in an analogous way). Since $\overline{AO} = \overline{CO}$, $\overline{BO} =$

Fig. 3.12 Chords of equal central angles are equal

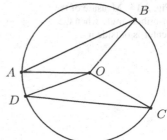

Fig. 3.13 Measure of the inscribed angle when the center is inside it

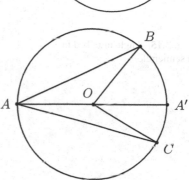

$\overline{DO}$ and $A\widehat{O}B = C\widehat{O}D$, triangles AOB and COD are congruent by SAS, so that $\overline{AB} = \overline{CD}$. □

Another important class of angles in a circle is that formed by *inscribed* angles. By definition, an **inscribed angle** in a circle is an angle whose vertex is a point of the circle and whose sides are two chords of it. The next result, known as the **inscribed angle theorem**, teaches us how to compute the measure of such an angle.

Proposition 3.19 *If AB and AC are chords of a circle of center O, then the measure of the inscribed angle $\angle BAC$ is equal to half of the measure of the corresponding central angle $\angle BOC$.*

Proof We look at three separate cases:

(a) Angle $\angle BAC$ contains the center O in its interior (see Fig. 3.13): since triangles OAC and OAB are isosceles of bases AC and AB, respectively, we have $O\widehat{A}C = O\widehat{C}A = \alpha$ and $O\widehat{A}B = O\widehat{B}A = \beta$, say. It then follows that $B\widehat{A}C = \alpha + \beta$ and, by the exterior angle theorem (cf. Corollary 2.21), we obtain $C\widehat{O}A' = 2\alpha$ and $B\widehat{O}A' = 2\beta$. Hence,

$$B\widehat{O}C = B\widehat{O}A' + C\widehat{O}A' = 2(\alpha + \beta) = 2B\widehat{A}C.$$

(b) Angle $\angle BAC$ does not contain the center O (see Fig. 3.14): once again, we have OAC and OAB isosceles of bases AC and AB. Moreover, letting $O\widehat{A}C = O\widehat{C}A = \alpha$ and $O\widehat{A}B = O\widehat{B}A = \beta$, we have $B\widehat{A}C = \beta - \alpha$ and, once more from the exterior angle theorem, $C\widehat{O}A' = 2\alpha$ and $B\widehat{O}A' = 2\beta$. Therefore,

Fig. 3.14 Measure of the
inscribed angle when the
center is outside it

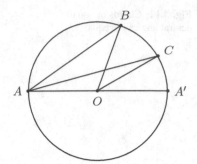

Fig. 3.15 Angle inscribed in
a semicircle

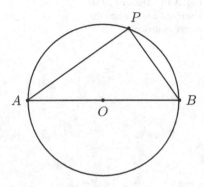

$$B\widehat{O}C = B\widehat{O}A' - C\widehat{O}A' = 2(\beta - \alpha) = 2B\widehat{A}C.$$

(c) The center O lies on one of the sides of $\angle BAC$: the analysis of this case is
analogous to the previous ones and will be left as an exercise for the reader.   □

Given a circle Γ of center O and a chord AB of Γ, an important particular case of
the former proposition is that in which AB is a diameter of Γ (cf. Fig. 3.15). Letting
P be a point of Γ distinct from A and B, the referred proposition gives

$$A\widehat{P}B = \frac{1}{2} \cdot 180° = 90°.$$

The *limit case* of an inscribed angle is that of a **tangent-chord angle** (cf.
Fig. 3.16): its vertex is a point of the circle and its sides are a chord of and a tangent
to the circle. The coming proposition shows that we can compute the measure of
tangent-chord angles in pretty much the same way as we compute the measure of
inscribed ones.

Fig. 3.16 Measure of a
tangent-chord angle

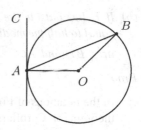

Fig. 3.17 Measure of an
interior angle in a circle

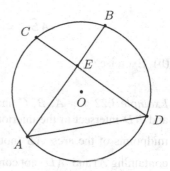

Proposition 3.20 *In the notations of Fig. 3.16, the measure of the tangent-chord angle* $\angle BAC$ *is equal to half the measure of the corresponding central angle* $\angle AOB$.

Proof Let $B\widehat{A}C = \alpha$. Since $\overleftrightarrow{AC} \perp \overleftrightarrow{AO}$, we have $A\widehat{B}O = B\widehat{A}O = 90° - \alpha$ and hence

$$B\widehat{O}A = 180° - 2(90° - \alpha) = 2\alpha = 2B\widehat{A}C.$$

$\square$

Yet another useful way of generalizing inscribed angles is to consider *interior* and *exterior* angles in a circle. An **interior angle** in a circle (cf. Fig. 3.17) is an angle whose sides are two chords of the circle which intersect inside it; an **exterior angle** in a circle is an angle whose sides are two chords of the circle which intersect outside it (make a drawing!).

We now learn how to compute the measures of such angles in terms of the measures of the arcs into which the involved chords partition the circle. In this respect, see also Problem 16, page 83.

Proposition 3.21 *Let AB and CD be two chord of a circle, such that the corresponding lines intersect at a point E.*

(a) *If E lies inside the circle, then the measure of the interior angle* $\angle AEC$ *is equal to the arithmetic mean of the measures of the corresponding arcs* $\widehat{AC}$ *and* $\widehat{BD}$.

(b) *If E lies outside the circle, then the measure of the exterior angle* ∠*AEC is equal to half the modulus of the difference of the measures of the corresponding arcs* $\overset{\frown}{BD}$ *and* $\overset{\frown}{AC}$.

Proof

(a) In the notations of Fig. 3.17, it suffices to successively apply the exterior angle theorem (cf. Corollary 2.21) and the result of Proposition 3.19:

$$A\widehat{E}C = A\widehat{D}C + B\widehat{A}D = \frac{1}{2}\overset{\frown}{AC} + \frac{1}{2}\overset{\frown}{BD}.$$

(b) Exercise.

□

Example 3.22 Let A, B, C and D be points on a circle Γ, such that chords AC and BD intersect in the interior of Γ. If M, N, P and Q, respectively denote the midpoints of the arcs $\overset{\frown}{AB}$ (not containing C), $\overset{\frown}{BC}$ (not containing D), $\overset{\frown}{CD}$ (not containing A) and $\overset{\frown}{AD}$ (not containing B), prove that $\overleftrightarrow{MP} \perp \overleftrightarrow{NQ}$.

Proof In accordance with the statement above and Fig. 3.18, let $\overset{\frown}{AB} = 2\alpha$, $\overset{\frown}{BC} = 2\beta$, $\overset{\frown}{CD} = 2\gamma$ and $\overset{\frown}{AD} = 2\delta$. Then, $\alpha + \beta + \gamma + \delta = 180°$, $\overset{\frown}{MN} = \alpha + \beta$ and $\overset{\frown}{PQ} = \gamma + \delta$. Therefore, letting E denote the intersection point of MP and NQ, we have

$$M\widehat{E}N = \frac{1}{2}(\overset{\frown}{MN} + \overset{\frown}{PQ}) = \frac{1}{2}((\alpha + \beta) + (\gamma + \delta)) = 90°.$$

□

We now turn to the analysis of another important locus, the *arc capable* of a given angle on a given segment.

Fig. 3.18 An interior angle of 90°

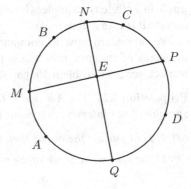

Fig. 3.19 $A\widehat{P}B = A\widehat{P'}B$

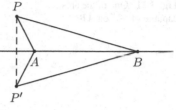

Fig. 3.20 One of the arcs capable of α on AB

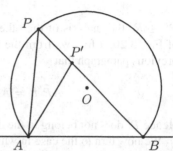

Proposition 3.23 *Given a line segment AB and an angle α, with $0° < \alpha < 180°$, the locus of the points P in the plane for which $A\widehat{P}B = \alpha$ is the union of two arcs of circle, symmetric with respect to line $\overleftrightarrow{AB}$ and having points A and B as their endpoints. Such arcs are said to be **capable** of α on AB.*

Proof Let us first look at the case $0° < \alpha < 90°$. To this end (see Fig. 3.19), start by choosing a point P such that $A\widehat{P}B = \alpha$. If P' is the symmetric of P with respect to $\overleftrightarrow{AB}$ (cf. Problem 2, page 30), then $\overleftrightarrow{AB}$ is the perpendicular bisector of PP', so that $\overline{AP} = \overline{AP'}$ and $\overline{BP} = \overline{BP'}$. Hence, triangles ABP and ABP' are congruent by SSS, which gives $A\widehat{P'}B = A\widehat{P}B = \alpha$. Analogously, $A\widehat{P'}B = \alpha$ implies $A\widehat{P}B = \alpha$, so that it suffices to restrict ourselves to the points P located at one of the half-planes determined by $\overleftrightarrow{AB}$. From now one, we shall assume that such a half-plane is the one lying *above* $\overleftrightarrow{AB}$ (cf. Fig. 3.20), which we shall refer to as the *upper* half-plane.

In the upper half-plane, let O be such that AOB is isosceles of basis AB, with $A\widehat{O}B = 2\alpha$ (note that $0° < 2\alpha < 180°$ in the present case), and denote $\overline{OA} = \overline{OB} = R$. Let Γ denote the arc of the circle of center O and radius R contained in the upper half-plane. If P is any point on Γ, the inscribed angle theorem gives

$$A\widehat{P}B = \frac{1}{2}A\widehat{O}B = \alpha,$$

so that P belongs to the desired locus. Conversely, let P' be a point of the upper half-plane, with $P' \notin \Gamma$; we shall show that P' does not belong to the locus. To the end, letting $\mathcal{R}$ denote the region of the upper half-plane with boundary $\Gamma \cup AB$, there are two possibilities: $P' \in \mathcal{R}$ or $P \notin \mathcal{R} \cup \Gamma$. Let us look at the case in which

Fig. 3.21 One of the arcs
capable of 90° on AB

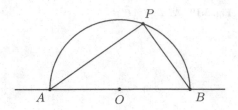

$P' \in \mathcal{R}$, the analysis of the other case being entirely analogous. In the notations
of Fig. 3.20, it follows from the exterior angle theorem and the discussion of the
previous paragraph that

$$A\widehat{P'}B = A\widehat{P}B + P\widehat{A}P' > A\widehat{P}B = \alpha.$$

Hence P' does not belong to the desired locus.

We now turn to the case in which $\alpha = 90°$, noticing first that, as in the previous
case, a symmetry argument reduces the problem to the points of the upper half-
plane. The discussion at the paragraph following the proof of Proposition 3.19
guarantees that every point of the semicircle of diameter AB lying in the upper half-
plane belongs to the locus (see Fig. 3.21). Conversely, if P is a point of the upper
half-plane for which $A\widehat{P}B = 90°$ and O is the midpoint of AB, then Corollary 2.44
guarantees that $\overline{PO} = \frac{1}{2}\overline{AB} = \overline{AO}$. Thus, P belongs to the semicircle of center
O and diameter AB.

Finally, the case $90° < \alpha < 180°$ is left as an exercise, for which we refer the
reader to Problem 17, page 83.   □

The proof of the previous proposition also teaches us how to construct the arcs
capable of an angle α on a segment AB, when $0° < \alpha \le 90°$. Indeed, if $\alpha = 90°$ we
only have to construct the circle of diameter AB. Suppose, then, that $0° < \alpha < 90°$.
In the notations of the proof of just presented, since $O\widehat{A}B = O\widehat{B}A$, we have

$$O\widehat{A}B = O\widehat{B}A = \frac{1}{2}(180° - A\widehat{O}B) = \frac{1}{2}(180° - 2\alpha) = 90° - \alpha;$$

thus, we obtain the center O of the upper arc capable of α on AB as the intersection
point of the half-lines departing from A and B, lying in the upper half-plane and
forming angles of $90° - \alpha$ with AB. The case $90° < \alpha < 180°$ can be treated
analogously (cf. Problem 17, page 83).

Example 3.24 Use straightedge and compass to construct the upper arc capable of
α on AB.

Proof

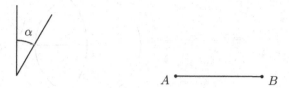

CONSTRUCTION STEPS

1. In accordance with the discussion above, construct half-lines $\overrightarrow{AX}$ and $\overrightarrow{BY}$ in the upper half-plane such that $B\widehat{A}X = A\widehat{B}Y = 90° - \alpha$.
2. The center O of the desired arc is the intersection point of $\overrightarrow{AX}$ and $\overrightarrow{BY}$.

□

The next example will show that there is a simple (and, as we shall see in the problems of this section, quite useful) relation between the arcs capable of angles α and $\frac{1}{2}\alpha$.

Example 3.25 The figure below shows the upper arc capable of an angle α on the line segment AB. Construct the upper arc capable of $\frac{1}{2}\alpha$ on AB.

Solution

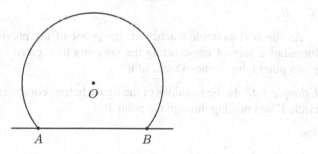

CONSTRUCTION STEPS

1. Draw the perpendicular bisector of AB and mark its intersection point O' with the given arc.
2. Use the inscribed angle theorem to show that the arc we are looking for is the arc of the circle of center O' and radius $\overline{OA} = \overline{OB}$, situated in the upper half-plane.

□

Among other interesting problems, one can use the notion of arc capable to address the problem of constructing the tangents to a given circle passing through an exterior point. We do this now.

Fig. 3.22 Tangents to a circle drawn through an exterior point

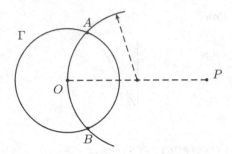

Proposition 3.26 *Given a circle Γ and a point P exterior to it, there are exactly two tangents to Γ passing through P.*

Proof Let O be the center of the given circle and A and B be its intersection points with that of diameter OP (cf. Fig. 3.22). What we have done so far allows us to look at the upper and lower semicircles of the circle of diameter OP as the arcs capable of $90°$ on OP. Hence, $O\widehat{A}P = O\widehat{B}P = 90°$, which is the same as $OA, OB \perp \overleftrightarrow{AP}$. Therefore, Proposition 3.16 assures that $\overleftrightarrow{AP}$ and $\overleftrightarrow{BP}$ are tangent to Γ.

Conversely, let r be a line passing through P and tangent to Γ, say at X. Then, $OX \perp \overleftrightarrow{XP}$, which is the same as $O\widehat{X}P = 90°$. Therefore, X lies in one of the arcs capable of $90°$ on OP, i.e., X belongs to the circle of diameter OP. This way, X is one of the intersection points of Γ and the circle of diameter OP, so that $X = A$ or $X = B$. □

As the next example teaches us, the proof of the previous proposition readily furnishes a way of constructing the tangents to a given circle passing through a given point lying in the exterior of it.

Example 3.27 In the notations of the figure below, construct the lines tangent to the circle Γ and passing through the point P.

Proof

CONSTRUCTION STEPS

1. Mark the midpoint M of the line segment OP.
2. Draw the circle γ of center M and radius $\overline{OM} = \overline{MP}$.
3. Mark A and B as the intersection points of γ and Γ; the searched tangents are lines $\overleftrightarrow{AP}$ and $\overleftrightarrow{BP}$.

□

Fig. 3.23 Properties of the tangents through an exterior point

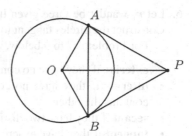

We now establish two quite useful properties of the tangents to a circle passing through an exterior point.

Proposition 3.28 *Let Γ be a circle of center O and P be an exterior point. If $A, B \in \Gamma$ are such that $\overleftrightarrow{PA}$ and $\overleftrightarrow{PB}$ are tangent to Γ (cf. Fig. 3.23), then:*

(a) $\overline{PA} = \overline{PB}$.

(b) $\overleftrightarrow{PO}$ is the perpendicular bisector of AB.

(c) $\overleftrightarrow{PO}$ is the bisector of both $\angle AOB$ and $\angle APB$.

(d) $\overleftrightarrow{PO} \perp \overleftrightarrow{AB}$.

Proof Since $\overline{OA} = \overline{OB}$ and $P\hat{A}O = P\hat{B}O = 90°$, triangles POA and POB are congruent by the special case HL of congruence of right triangles (cf. Problem 1, page 36); in particular, $\overline{PA} = \overline{PB}$, $A\hat{P}O = B\hat{P}O$ and $A\hat{O}P = B\hat{O}P$.

Now, since P and O are at equal distances from A and B, Proposition 3.5 assures that $\overleftrightarrow{PO}$ is the perpendicular bisector of AB. Thus, $\overleftrightarrow{PO} \perp \overleftrightarrow{AB}$. □

Problems: Sect. 3.3

1. * Given in the plane a circle Γ and a point P on it, show that the tangent to Γ at P is *unique*.
2. We are given in the plane a line r and a point $A \in r$. Identify and construct, with straightedge and compass, the locus of the points in the plane which are centers of the circles tangents to r at A.
3. In the plane, to concurrent lines r and s and a point $P \in r$ are given. Construct the circles tangent to r and s, if P is the tangency point with r.
4. We have a line segment of length R and a line r. Identify and construct the locus of the points in the plane which are centers of circles of radius R and tangent to r.
5. Two concurrent lines r and s, as well as a line segment of length R, are given in the plane. Construct all circles of radius R and tangent r and s.

6. Let r, s and t be three given lines, with $r \parallel s$ and t concurrent with r and s. Construct the circles tangent to r, s and t.

   For Problems 7 to 9 below, we say that two circles are :

   - **exterior** if they have no common points and have disjoint interiors;
   - **interior** if they have no common points but the interior of one of them contains the other;
   - **secant** if they have two distinct points in common;
   - **tangent** if they have exactly one common point; if this is the case, they are said to be **exteriorly tangent** if they have disjoint interiors and **internally tangent** otherwise.

7. * Given circles $\Gamma_1(O_1; R_1)$ and $\Gamma_2(O_2; R_2)$, prove that Γ_1 and Γ_2 are :

   (a) exterior if and only if $\overline{O_1 O_2} > R_1 + R_2$.
   (b) externally tangent if and only if $\overline{O_1 O_2} = R_1 + R_2$.
   (c) secant if and only if $|R_1 - R_2| < \overline{O_1 O_2} < R_1 + R_2$.
   (d) internally tangent if and only if $\overline{O_1 O_2} = |R_1 - R_2|$.
   (e) interior if and only if $\overline{O_1 O_2} < |R_1 - R_2|$.

8. Let a circle Γ of center O and a point $A \in \Gamma$ be given. Identify and construct the locus of the centers of circles tangent to Γ at A.

9. We have a circle Γ of center O and radius R, and a line segment of length r. Identify and construct, with straightedge and compass, the locus of the centers of the circles of radius r, tangent to Γ. To what extent does the locus depend on the values of R and r?

10. Let Γ be a given circle. Points A, P and Q are such that AP and AQ are tangent to Γ at P and Q, respectively. Points $B \in AP$ and $C \in AQ$ are chosen in such a way that BC is also tangent to Γ. If $\overrightarrow{AP} = 5$cm, compute the possible values of the perimeter of triangle ABC.

11. (TT) Let $ABCD$ be a square of side length a and Γ be the circle of center A and radius a. Points M and N are marked on the sides BC and CD, , respectively, such that MN is tangent to Γ. What are the possible values of the measure of angle $\angle MAN$?

12. Lines r and s pass through A and are tangent to a circle Γ of center O. Points $P \in r$ and $Q \in s$ are such that $\overleftrightarrow{PQ}$ is tangent to Γ and leaves A and O in opposite half-planes. If $P\widehat{A}Q = 30°$, compute $P\widehat{O}Q$.

13. Two circles Γ and Σ intersect each other at points A and B. We choose $X \in \Gamma$ and $Y \in \Sigma$ such that $A \in XY$. Prove that the measure of $\angle XBY$ does not depend on the direction of line $\overleftrightarrow{XY}$.

14. Chords AB and CD of a circle Γ are perpendicular at a point E of the interior of Γ. The perpendicular to AC passing through E intersects segment BD at F. Show that F is the midpoint of BD.

15. Let A, B and C be points on a circle, such that the smaller arcs $\overset{\frown}{AB}$, $\overset{\frown}{AC}$ and $\overset{\frown}{BC}$ are all equal to 120°. If P is a point on the smaller arc $\overset{\frown}{BC}$, prove that $\overline{PA} = \overline{PB} + \overline{PC}$.

16. Prove item (b) of Proposition 3.21. Also, check that the formulas for the measures of interior and exterior angles in the circle remain valid when one side of the angle contains a chord and the other is tangent to the circle.

17. * Analyse the construction of the arcs capable of an angle α on a segment AB when $90° < \alpha < 180°$.

18. Construct triangle ABC, knowing the lengths a of BC and h_a of the altitude relative to BC, as well as the measure α of $\angle A$.

19. * Let ABC be a given triangle, and P and M be the feet of the internal bisector and the median relative to BC, respectively. If P and M coincide, prove that ABC is isosceles of basis BC.

20. * From a square $ABCD$ we know the length l of its sides and the positions of points M, N and P, lying respectively on the sides AB, AD and on the diagonal AC. Show how to find the positions of its vertices.

21. From a triangle ABC we know the positions of vertices B and C and the measure α of $\angle BAC$. As A varies on the arcs capable of α on BC, find the locus described by the incenter I of ABC.

22. * We are given in the plane two exterior circles Γ and Γ'. Show how to construct the **common internal** and **external tangents** to Γ and Γ', i.e., the straight lines simultaneously tangent to both of these circles.

23. * The secant circles $\Gamma_1(O_1; R_1)$ and $\Gamma_2(O_2; R_2)$ intersect at points A and B. Given a line segment of length l, explain how to construct a straight line passing through A (said to be a *secant* to Γ_1 and Γ_2), intersecting Γ_1 and Γ_2 at points X and Y, respectively (with $X, Y \neq A$), and such that $\overline{XY} = l$. Under which conditions is there a solution?

24. Two secant circles $\Gamma_1(O_1; R_1)$, $\Gamma_2(O_2; R_2)$ intersect at points A and B. Explain how to construct a secant to Γ_1 and Γ_2, passing through A and having the greatest possible length.

25. We are given a triangle ABC and a line segment of length a. Show how to use straightedge and compass to construct an equilateral triangle MNP, of side length a and such that $A \in NP$, $B \in MP$ and $C \in MN$.

26. We are given a triangle ABC in the plane. Show how to use straightedge and compass to construct an equilateral triangle MNP, of the greatest possible side length and such that $A \in NP$, $B \in MP$ and $C \in MN$.

27. From a triangle ABC we know the positions of vertices B and C and the measure α of $\angle A$. Knowing the sum l of the lengths of sides AB and AC, show how to find the position of vertex A.

The result of the coming problem is known as the **broken chord theorem**, and is due to Archimedes.[1]

---

[1] Archimedes of Syracuse, one of the greatest scientists of all time. Apart from other genial discoveries, like Archimedes' law in Hydrostatics and the lever principle in Statics, Archimedes anticipated the central ideas of the Integral Calculus in some $2,000$ years!

28. We are given a circle Γ and points A, B and C on it, such that $\overline{AB} > \overline{AC}$.
    We mark the midpoint M of the arc $\overset{\frown}{BC}$ that contains A, as well as the foot
    N of the perpendicular dropped from M to the line segment AB. Prove that
    $\overline{BN} = \overline{AN} + \overline{AC}$.

3.4   Circles Associated to a Triangle

With the concepts of arc capable of an angle and of tangency of lines and circles, we
now go through the study of some notable circles associated to a triangle. We start
by with the following proposition, which shows that every triangle admits a unique
circle passing through its vertices.

Proposition 3.29 *The circumcenter of a triangle is the center of the only circle
passing through the vertices of the triangle.*

Proof Let ABC be a triangle of circumcenter O (cf. Fig. 3.24). Since O is the
intersection point of the perpendicular bisectors of the sides of the triangle, we have
$\overline{OA} = \overline{OB} = \overline{OC}$. Letting R denote this common distance, we conclude that the
circle of center O and radius R passes through A, B and C. Thus, there exists a
circle passing through the vertices of ABC.

  Conversely, the center of a circle that passes through the vertices of ABC must be
at the same distance from them. Hence, it must belong to the perpendicular bisectors
of the sides of ABC, so that it coincides with the point de intersection of them, i.e.,
with the circumcenter O of ABC.

  Finally, the radius of the circle, being the distance from O to the vertices of ABC,
is equal to R.   □

  In the notations of the above proposition, the circle just constructed is called the
circumcircle of the triangle. It is also said to be **circumscribed** to the triangle, and
its radius to be the **circumradius** of the triangle. Whenever there is no danger of
confusion, we shall write R to denote the circumradius of a triangle.

Fig. 3.24 Circumcenter and
the circle through the vertices
of a triangle

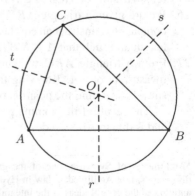

Fig. 3.25 O is in the interior
of ABC

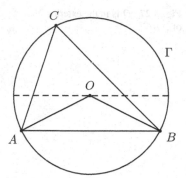

Fig. 3.26 O lies in one side
of ABC

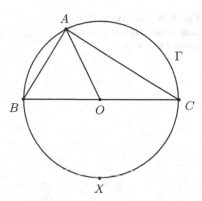

We now relate the location of the circumcenter to the shape of the corresponding triangle.

Proposition 3.30 *If ABC is a triangle of circumcenter O, then O lies in the interior (resp. on one side, in the exterior) of ABC if and only if ABC is an acute(resp. right, obtuse) triangle.*

Proof Let Γ be the circle circumscribed to ABC. There are three cases to look at:

(i) O lies in the interior of ABC (cf. Fig. 3.25): in triangle AOB, we have $A\widehat{O}B = 2A\widehat{C}B$. On the other hand, $0° < A\widehat{O}B < 180°$, so that $2A\widehat{C}B < 180°$ or, which is the same, $A\widehat{C}B < 90°$. Analogously, $A\widehat{B}C < 90°$ and $B\widehat{A}C < 90°$, so that ABC is an acute triangle.

(ii) O lies in one side of ABC (cf. Fig. 3.26): suppose, without loss of generality, that $O \in BC$. In this case, BC is a diameter of Γ and O is the midpoint of BC.

$$B\widehat{A}C = \frac{1}{2} B\widehat{X}C = \frac{1}{2}180° = 90°.$$

(iii) O is in the exterior of ABC (cf. Fig. 3.27): assume, without loss of generality, that O and A are in opposite half-planes with respect to $\overleftrightarrow{BC}$. Since the measure

Fig. 3.27 O is in the exterior
of ABC

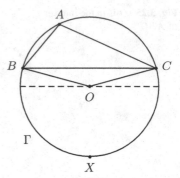

Fig. 3.28 Comparing the
central and internal angles
relative to a vertex

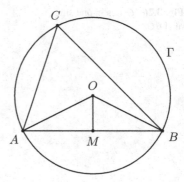

of arc $B\widehat{X}C$ is clearly greater than $180°$, we have

$$B\widehat{A}C = \frac{1}{2} B\widehat{X}C > \frac{1}{2}180° = 90°,$$

and ABC is obtuse at A.

$\square$

We now collect a useful elaboration of the discussion of item (a) of the previous proposition.

Corollary 3.31 *Let ABC be an acute triangle of circumcenter O. If M is the midpoint of AB, then $A\widehat{O}M = B\widehat{O}M = A\widehat{C}B$.*

Proof In the notations of Fig. 3.28, OM is the median relative to the basis AB of the isosceles triangle AOB. Hence, Problem 6, page 30, guarantees that OM is also the internal bisector of $\angle AOB$. Thus,

$$A\widehat{O}M = B\widehat{O}M = \frac{1}{2}A\widehat{O}B = A\widehat{C}B,$$

where we have applied the inscribed angle theorem in the last equality above. $\square$

Fig. 3.29 Circle inscribed in a triangle

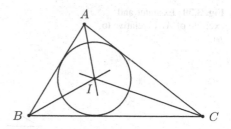

Another fact worth noticing is that every triangle admits four circles tangent to its sides. The next result presents the most important of such circles.

Proposition 3.32 *The incenter of a triangle is the center of the unique circle tangent to the sides and contained in the triangle.*

Proof Let *I* be the incenter of a triangle *ABC* (see Fig. 3.29). Since *I* is the intersection point of the internal bisectors of *ABC*, we know that *I* is at the same distance from the sides of *ABC*. Letting *r* denote such common distance, we conclude that the circle of center *I* and radius *r* is contained in *ABC* and is also tangent to its sides.

For the uniqueness, any circle lying inside the triangle and tangent to its sides must be such that its center is at the same distance from the three sides. Therefore, the characterization of angle bisectors as locus assures that the center must be the intersection point of the internal bisectors of *ABC*, thus coinciding with the incenter of it. □

In the notations of the previous proposition, we shall call the circle just constructed the **incircle** of the triangle. We shall also say that it is **inscribed** in the triangle, and that its radius is the **inradius** of the triangle. Whenever there is no danger of confusion, we shall write *r* to denote the inradius of a triangle.

Example 3.33 Use straightedge and compass to construct the circumcircle and incircle of the given triangle *ABC*.

Solution

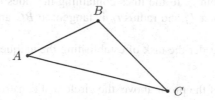

Fig. 3.30 Excenter and
excircle of ABC relative to
BC

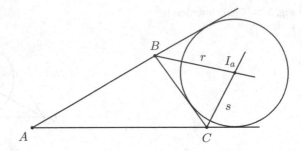

CONSTRUCTION STEPS

1. For the incircle, start by finding the incenter I of ABC.
2. Then, draw the line r, perpendicular to BC and passing through I.
3. If M is the intersection point of r and BC, then the incircle has radius IM.
4. For the circumcircle, first construct the circumcenter O of ABC. Its radius will
   be OA.

$\square$

As we have anticipated a few moments ago, there are three other notable
circles associated to a triangle, its *excircles*. The next proposition establishes their
existence.

Proposition 3.34 *In every triangle, there exists a unique circle tangent to one side
and to the extensions of the other two sides.*

Proof Let (see Fig. 3.30) r and s be the external bisectors of a triangle ABC relative
to the vertices B and C and I_a be its intersection point (the reader can easily check
that the half-lines of r and s situated within $\angle BAC$ form acute angles with the side
BC, so that r and s intersect inside such an angle). Since $I_a \in r$ and r is the bisector
of the external angle of ABC at B, we get

$$d(I_a, \overleftrightarrow{BC}) = d(I_a, \overleftrightarrow{AB}).$$

Likewise, since $I_a \in s$, we obtain $d(I_a, \overleftrightarrow{BC}) = d(I_a, \overleftrightarrow{AC})$. Letting r_a denote the
common distance from I_a to the lines containing the sides of ABC, we conclude
that the circle of center I_a and radius r_a is tangent to BC and to the extensions of
AB and AC.

We leave to the reader the task of establishing the uniqueness of the circle with
the stated properties. $\square$

In the notations of the proof above, the circle just constructed is the **excircle** of
ABC relative to BC (or to A); its center is the **excenter** and its radius is the **exradius**
of ABC relative to BC (or to A). Every triangle has exactly three excenters and three
excircles. For a triangle ABC, we shall usually write I_a, I_b and I_c to denote the
excenters and r_a, r_b and r_c to denote the exradii relative to A, B and C, respectively.

A useful way of looking at the proof of the previous result is as set in the coming.

Corollary 3.35 *In every triangle, the bisector of an internal angle and the external bisectors of the external angles at the other two vertices are concurrent at an excenter of the triangle.*

Proof In the notations of the proof of the proposition, we showed that

$$d(I_a, \overleftrightarrow{AB}) = d(I_a, \overleftrightarrow{BC}) \text{ and } d(I_a, \overleftrightarrow{AC}) = d(I_a, \overleftrightarrow{BC}).$$

Hence, $d(I_a, \overleftrightarrow{AB}) = d(I_a, \overleftrightarrow{AC})$ and Proposition 3.6 assures that I_a belongs to the bisector of $\angle BAC$. $\qquad\qquad\square$

We now compute, among others, the lengths of the line segments determined on the sides of a triangle by the tangency points of its incircle and excircles.

Proposition 3.36 *Let ABC be a triangle of sides $\overline{AB} = c$, $\overline{BC} = a$, $\overline{CA} = b$ and semiperimeter p (see Fig. 3.31). Let D, E and F be the points in which the incircle touches the sides BC, CA and AB, respectively, and M, N and P be the points in which the excircle relative to A touches the side BC and the extensions of sides AC and AB, also respectively. Then:*

(a) *$\overline{BD} = \overline{BF} = p - b$, $\overline{CD} = \overline{CE} = p - c$, $\overline{AF} = \overline{AE} = p - a$.*
(b) *$\overline{AN} = \overline{AP} = p$.*
(c) *$\overline{BM} = \overline{BP} = p - c$, $\overline{CM} = \overline{CN} = p - b$.*
(d) *$\overline{EN} = \overline{FP} = a$.*
(e) *The midpoint of BC is also the midpoint of DM.*

Proof

(a) With the aid of item (a) of Proposition 3.28, denote $\overline{AE} = \overline{AF} = x$, $\overline{BD} = \overline{BF} = y$ and $\overline{CD} = \overline{CE} = z$. This way, we obtain the linear system of equations

$$\begin{cases} x + y = c \\ y + z = a \; . \\ z + x = b \end{cases}$$

By termwise addition of its equations, we obtain $x + y + z = 2p$, and hence

$$x = (x + y + z) - (y + z) = p - a.$$

Analogously, $y = p - b$ and $z = p - c$.
(b) Letting $\overline{AN} = \overline{AP} = u$ and invoking again item (a) of Proposition 3.28, we get

Fig. 3.31 Computing some
notable segments of triangle
ABC

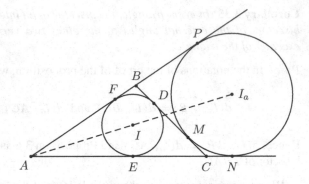

$$2u = \overline{AN} + \overline{AP} = (\overline{AC} + \overline{CN}) + (\overline{AB} + \overline{BP})$$

$$= (\overline{AC} + \overline{AB}) + (\overline{CN} + \overline{BP})$$

$$= (b + c) + (\overline{CM} + \overline{BM})$$

$$= b + c + \overline{BC} = a + b + c = 2p,$$

so that $u = p$.

(c) As in (b), we have $\overline{BM} = \overline{BP}$ and $\overline{CM} = \overline{CN}$. On the other hand,

$$\overline{BP} = \overline{AP} - \overline{AB} = p - c \quad \text{and} \quad \overline{CN} = \overline{AN} - \overline{AC} = p - b.$$

(d) Let us prove that $\overline{EN} = a$, the proof of $\overline{FP} = a$ being totally analogous:

$$\overline{EN} = \overline{AN} - \overline{AE} = p - (p - a) = a.$$

(e) It suffices to show that $\overline{CM} = \overline{BD}$, and we have already done this.

$$\square$$

As the reader will notice, the computations in the above proposition will be quite useful for the solution of many problems, so much that it is worth trying to memorize the results of items (a), (b), (d) and (e) as soon as possible. Note also that items (c), (d) and (e) are almost immediate consequences of items (a) and (b).

We finish this section by presenting a result that furnishes an important relation between the incenter and excenters of a triangle.

Proposition 3.37 *Let ABC be a triangle of incenter I and excenter I_a relative to BC. If M is the point in which the circle circumscribed to ABC intersects segment II_a (cf. Fig. 3.32), then M is the midpoint of the arc $\overset{\frown}{BC}$ not containing A, and*

$$\overline{MB} = \overline{MC} = \overline{MI} = \overline{MI_a}.$$

Fig. 3.32 Incenter, excenter
and midpoint of the arc $\overset{\frown}{BC}$

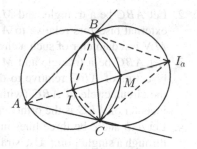

Proof Since $M\widehat{A}B = M\widehat{A}C = \frac{1}{2}\widehat{A}$, it follows from the inscribed angle theorem
that the arcs $\overset{\frown}{MB}$ and $\overset{\frown}{MC}$ not containing A are equal; hence, M is the midpoint of
the arc $\overset{\frown}{BC}$ not containing A.

Since equal arcs correspond to equal chords, we have $\overline{MB} = \overline{MC}$. Now, again
from the inscribed angle theorem, we have $B\widehat{M}I = B\widehat{M}A = B\widehat{C}A = \widehat{C}$ and

$$I\widehat{B}M = I\widehat{B}C + C\widehat{B}M = \frac{1}{2}\widehat{B} + C\widehat{A}M$$

$$= \frac{1}{2}\widehat{B} + \frac{1}{2}\widehat{A}.$$

Hence,

$$B\widehat{I}M = 180° - I\widehat{B}M - B\widehat{M}I$$

$$= 180° - \frac{1}{2}\widehat{B} - \frac{1}{2}\widehat{A} - \widehat{C}$$

$$= \widehat{A} + \widehat{B} + \widehat{C} - \frac{1}{2}\widehat{B} - \frac{1}{2}\widehat{A} - \widehat{C}$$

$$= \frac{1}{2}\widehat{B} + \frac{1}{2}\widehat{A} = I\widehat{B}M,$$

so that triangle IBM is isosceles of basis IB. Thus, $\overline{IM} = \overline{BM} = \overline{CM}$.

We leave to the reader the task of proving the equality $\overline{BM} = \overline{MI_a}$; the argument
is quite similar to the one executed above.   □

Problems: Sect. 3.4

1. Construct a triangle ABC, knowing the lengths of the circumradius and of the
   sides BC and AC.

2. Let ABC be a triangle, and M and N be the points in which the internal and external bisectors relative to A intersect the circumcircle of ABC. Show that MN is a diameter of such a circle.

3. Let ABC be a triangle and M, N and P be the points in which the internal bisectors of ABC relative to the vertices A, B and C, respectively, intersect the circumcircle of ABC (with $M \neq A$, $N \neq B$ and $P \neq C$). Prove that the incenter of ABC is the orthocenter of MNP.

4. Let a, b and c be three lines in the plane, pairwise concurrent but not passing through a single point. Use straightedge and compass to construct the points in the plane lying at the same distance from a, b and c.

5. * Let ABC be a triangle of orthocenter H and circumcenter O. Prove that the internal bisector relative to BC also bisects the angle $\angle HAO$.

6. * Prove that, in every triangle, the symmetrics of the orthocenter with respect to the straightlines containing the sides of the triangle lie on the circumscribed circle.

7. From a triangle ABC we know the positions of vertices B and C, the measure α of $\angle BAC$ and the half-plane β, of those determined by $\overleftrightarrow{BC}$, to which vertex A belongs. As A varies on the arc capable of α on BC situated in β, find the locus described by the orthocenter H of ABC.

8. From a triangle ABC we know the positions of vertices B and C, the measure α of $\angle BAC$ and the half-plane β, of those determined by $\overleftrightarrow{BC}$, to which vertex A belongs. As A varies on the arc capable of α on BC situated in β, find the locus described by the excenter I_a relative to BC?

9. Let ABC be a right triangle and H be the foot of the altitude relative to the hypotenuse BC. Let also I_1 and I_2 be the incenters of triangles ABH and ACH. Prove that A is the excenter of triangle $I_1 H I_2$ relative to $I_1 I_2$.

10. Construct the square $ABCD$, knowing the positions of four points M, N, P and Q, respectively situated on the sides AB, BC, CD and DA.

11. (OIM) In a triangle ABC, let Q and R be the in which the incircle and the excircle relative to A touch side BC. If P is the foot of the perpendicular dropped from B to the internal bisector of ABC relative to A, show that $Q\widehat{P}R = 90°$.

12. Use compass and straightedge to construct triangle ABC, knowing the lengths p of the semiperimeter, a of the side BC and r_a of the excircle relative to BC.

13. (China) In a triangle ABC, the smallest angle is $\widehat{C} = 30°$ and we choose points $D \in BC$ and $E \in AC$ such that $\overline{AB} = \overline{AE} = \overline{BD}$. If I and O stand for the incenter and the circumcenter of ABC, respectively, prove that $\overline{DE} = \overline{OI}$ and $\overleftrightarrow{DE} \perp \overleftrightarrow{OI}$.

3.5 Cyclic and Tangential Quadrilaterals

Although every triangle admits a circumcircle, this is not always the case with convex quadrilaterals. To see this, it suffices to take a triangle ABD and a point C not belonging to its circumcircle (cf. Fig. 3.33). On the other hand, we say that a convex quadrilateral is **cyclic** or **inscribed** if there exists a circle passing through its vertices. In this case, we also say that the vertices of the quadrilateral form a set of **concyclic** points.

The uniqueness of the circumcircle of a triangle obviously assures that the circle passing through the vertices of a cyclic quadrilateral is unique; from now on, such a circle will be called the **circle circumscribed** to the quadrilateral.

One can show (cf. Problem 7, page 99) that a quadrilateral is cyclic if and only if the perpendicular bisectors to its sides intersect at a single point, the **circumcenter**. Nevertheless, for the applications we have in mind, the following characterization of cyclic quadrilaterals is much more useful.

Proposition 3.38 *A convex quadrilateral $ABCD$, of sides AB, BC, CD and DA, is cyclic if and only if any of the following conditions is satisfied:*

(a) $D\widehat{A}B + B\widehat{C}D = 180°$.
(b) $B\widehat{A}C = B\widehat{D}C$.

Proof Assume first that $ABCD$ is cyclic (cf. Fig. 3.34). Then, the inscribed angle theorem gives $B\widehat{A}C = B\widehat{D}C$ and

$$D\widehat{A}B + B\widehat{C}D = \frac{1}{2} B\widehat{C}D + \frac{1}{2} B\widehat{A}D = 180°.$$

Fig. 3.33 A non-cyclic
quadrilateral

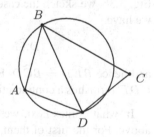

Fig. 3.34 $ABCD$ cyclic
$\Rightarrow D\widehat{A}B + B\widehat{C}D = 180°$
and $B\widehat{A}C = B\widehat{D}C$

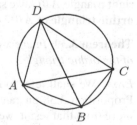

Fig. 3.35
$B\widehat{A}C = B\widehat{D}C \Rightarrow ABCD$
inscribed

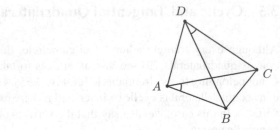

Fig. 3.36 $B\widehat{A}C + B\widehat{D}C =$
$180° \Rightarrow ABCD$ cyclic

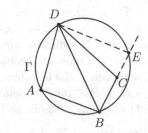

Conversely (cf. Fig. 3.35), suppose first that $B\widehat{A}C = B\widehat{D}C$. Since $ABCD$ is convex and its vertices are labelled consecutively, it follows that A and D are in a single half-plane, of those determined by $\overleftrightarrow{BC}$. Letting θ be the common value of the measures of $\angle BAC$ and $\angle BDC$, we conclude that A and D lie in the arc capable of θ on BC and situated in such half-plane. Therefore, the circle containing such an arc circumscribes $ABCD$.

We now suppose that $D\widehat{A}B + B\widehat{C}D = 180°$ (cf. Fig. 3.36), and consider the circumcircle Γ of triangle BAD. If $C \notin \Gamma$, let $\overleftrightarrow{BC} \cap \Gamma = \{E\}$, with $E \neq B, C$ (in Fig. 3.36, we sketch the case in which C belongs to the interior of Γ). By item (a), we have

$$D\widehat{A}B + B\widehat{E}D = 180° = D\widehat{A}B + B\widehat{C}D,$$

and hence $B\widehat{E}D = B\widehat{C}D$. However, the exterior angle theorem applied to triangle CDE furnishes a contradiction.   □

In what comes next, we present two important applications of the proposition above. For the first of them, we shall need the following convention: given a non-right triangle ABC, we say that the triangle formed by the feet of its altitudes is the **orthic triangle** of ABC (cf. Fig. 3.37).

Theorem 3.39 *In every acute triangle, the orthocenter coincides with the incenter of the orthic triangle.*

Proof We shall refer to Fig. 3.37. Since $H\widehat{H_a}B + H\widehat{H_c}B = 90° + 90° = 180°$, Proposition 3.38 guarantees that HH_aBH_c is a cyclic quadrilateral. Hence, once more from that result, we obtain

$$H\widehat{H_a}H_c = H\widehat{B}H_c = H_b\widehat{B}A = 90° - \widehat{A}.$$

Fig. 3.37 The orthic triangle $H_a H_b H_c$ of ABC

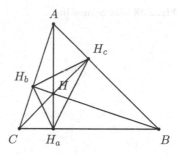

Reasoning in a similar way we get $H\widehat{H_a}C + H\widehat{H_b}C = 180°$, so that $H H_a C H_b$ is also cyclic. Thus, we conclude that

$$H\widehat{H_a}H_b = H\widehat{C}H_b = H_c\widehat{C}A = 90° - \widehat{A}.$$

The argument above has proved that $H\widehat{H_a}H_c = H\widehat{H_a}H_b$, i.e., that the line segment $H H_a$ is the internal bisector of $\angle H_c H_a H_b$. Analogously, $H H_b$ and $H H_c$ are the internal bisectors of the other two angles of the orthic triangle, so that H (the orthocenter of ABC) is also the incenter of $H_a H_b H_c$. □

For our second application, we say that four points are in **general position** if any three of them are non collinear. We shall be concerned with the following situation: given in the plane a triangle ABC and a point P such that A, B, C, P are in general position, we mark points D, E and F, feet of the perpendiculars dropped from P to $\overleftrightarrow{BC}$, $\overleftrightarrow{CA}$ and $\overleftrightarrow{AB}$, respectively. Triangle DEF thus obtained is said to be the **pedal triangle** of P with respect to ABC. For instance, the orthic triangle of a non-right triangle ABC (cf. Fig. 3.37) is the pedal triangle of the orthocenter of ABC with respect to itself.

The result of the proposition below, known as the **Simson-Wallace theorem**,[2] explains when the pedal triangle of a point with respect to a given triangle *degenerates* (i.e., in the notations of the previous paragraph, when D, E and F are collinear). Such a result will be of fundamental importance for the material of Sect. 7.5 (cf. Theorem 7.37).

Theorem 3.40 (Simson-Wallace) *We are given a triangle ABC and a point P such that A, B, C and P are in general position. The pedal triangle of P with respect to ABC is degenerated if and only if P lies in the circumcircle of ABC.*

Proof For P to lie in the circumcircle of ABC, the only possibility is that P lies outside AC and in one of the angular regions $\angle BAC$, $\angle ABC$ or $\angle BCA$. Analogously, a simple checking shows that, for the pedal triangle of P with respect

---

[2]After the scottish mathematicians of eighteenth and nineteenth centuries, Robert Simson and William Wallace.

Fig. 3.38 The Simson line

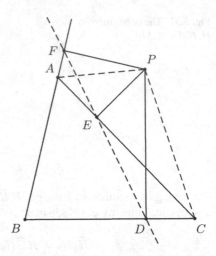

to ABC to be degenerated, P must lie outside ABC and in one of those angular regions. We can thus assume, without any loss of generality, that P is exterior to ABC and in the angular region $\angle ABC$ (cf. Fig. 3.38).

Let D, E and F be the feet of the perpendiculars dropped from P to the lines $\overleftrightarrow{BC}$, $\overleftrightarrow{AC}$ and $\overleftrightarrow{AB}$, respectively. We can also suppose, with no loss of generality, that D and E lie on the sides BC and AC, respectively, whereas F lies on the extension of side AB. Since $P\widehat{F}A = P\widehat{E}A = 90°$, quadrilateral $PFAE$ is cyclic. Analogously, quadrilateral $PEDC$ is also cyclic. Therefore,

$$A\widehat{P}C - D\widehat{P}F = D\widehat{P}C - F\widehat{P}A = D\widehat{E}C - F\widehat{E}A,$$

so that

$$A\widehat{P}C = D\widehat{P}F \Leftrightarrow D\widehat{E}C = F\widehat{E}A \Leftrightarrow D, E \text{ and } F \text{ are collinear.}$$

Finally, computing the sum of the measures of the internal angles of quadrilateral $BCPF$, we obtain $D\widehat{P}F = 180° - A\widehat{B}C$, and hence

$$A\widehat{P}C = D\widehat{P}F \Leftrightarrow A\widehat{P}C + A\widehat{B}C = 180° \Leftrightarrow ABCP \text{ is cyclic.}$$

□

In the notations of the discussion above, whenever P lies in the circumcircle of ABC we shall say that the line passing through D, E and F is the **Simson line**[3] of P relative to ABC.

---

[3] Although Theorem 3.40 is usually attributed to Simson and Wallace, we shall follow tradition, referring to the line through D, E and F as the Simson line, instead of the Simson-Wallace line.

Fig. 3.39 $ABCD$ tangential
$\Rightarrow$ equal sums of lengths of
opposite sides

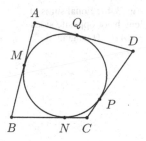

Back to the discussion of the first paragraph of this section, we now observe that
for not every convex quadrilateral one can find a circle tangent to all of its sides
(the reader can easily construct a counterexample). When such a circle does exist,
we shall say that the quadrilateral is **tangential**, and that the corresponding circle
is **inscribed** in the quadrilateral. The coming result, known as **Pitot's theorem**,[4]
provides a useful characterization of tangential quadrilaterals.

Theorem 3.41 (Pitot) *A convex quadrilateral $ABCD$, of sides AB, BC, CD and
DA, is tangential if and only if*

$$\overline{AB} + \overline{CD} = \overline{AD} + \overline{BC}.$$

Proof Suppose first that $ABCD$ is tangential, and let M, N, P and Q be the
tangency points of the sides AB, BC, CD and DA with the inscribed circle,
respectively (cf. Fig. 3.39).

By applying item (a) of Proposition 3.28 several times, we obtain

$$\overline{AB} + \overline{CD} = (\overline{AM} + \overline{MB}) + (\overline{CP} + \overline{PD})$$

$$= \overline{AQ} + \overline{BN} + \overline{CN} + \overline{DQ}$$

$$= (\overline{AQ} + \overline{DQ}) + (\overline{BN} + \overline{CN})$$

$$= \overline{AD} + \overline{BC}.$$

Conversely, arguing by contradiction, assume that $\overline{AB} + \overline{CD} = \overline{AD} + \overline{BC}$
but $ABCD$ is not tangential. If O is the intersection point of the bisectors of the
internal angles $\angle DAB$ and $\angle ABC$ of $ABCD$, then Proposition 3.6 assures that O
is the center of a circle tangent to sides AD, AB and BC of $ABCD$ (cf. Fig. 3.40).
Since we are assuming that $ABCD$ is not tangential, we conclude that such a circle
is not tangent to the side CD of $ABCD$.

Let E be the point on the half-line $\overrightarrow{AD}$ such that CE is tangent to the circle of
Fig. 3.40 (there, we are looking at the case of E lying on AD; the other cases are
entirely analogous). By the first part above, it follows that

---

[4]Henri Pitot, French engineer of the seventeenth century.

Fig. 3.40 Equal sums of
lengths of opposite sides
$\Rightarrow ABCD$ tangential

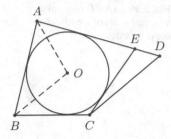

$$\overline{AB} + \overline{CE} = \overline{AE} + \overline{BC}.$$

However, since $\overline{AB} + \overline{CD} = \overline{AD} + \overline{BC}$ by hypothesis, we thus get

$$\overline{CD} - \overline{CE} = \overline{AD} - \overline{AE} = \overline{DE},$$

so that $\overline{CD} = \overline{CE} + \overline{ED}$. Finally, such an equality contradicts the triangle
inequality on triangle CDE.   □

Problems: Sect. 3.5

1. On each side of an acute triangle ABC we construct a circle having that side
   as a diameter. Prove that the three circles thus obtained pairwise intersect in six
   points, three of which are the feet of the altitudes of ABC.
2. * Let ABC be an acute triangle of circumcenter O, and H_a, H_b and H_c be the
   feet of the altitudes relative to BC, CA and AB, respectively. Prove that:
   (a) $A\widehat{H_b}H_c = A\widehat{B}C$ and $A\widehat{H_c}H_b = A\widehat{C}B$.
   (b) $\overleftrightarrow{OA} \perp \overleftrightarrow{H_bH_c}$.
3. We are given in the plane four straightlines, no two of which parallel and no
   three of which passing through a single point. Prove that the circumcircles of
   the four triangles they determine all pass through a single point.
4. Quadrilateral $ABCD$ is inscribed in a circle Γ of diameter BD. Let $M \in \Gamma$
   be such that $M \neq A$ and $AM \perp BD$, and N be the foot of the perpendicular
   dropped from A to BD. If the parallel to $\overleftrightarrow{AC}$ drawn by N intersects $\overleftrightarrow{CD}$ at P
   and $\overleftrightarrow{BC}$ at Q, prove that $CPMQ$ is a rectangle.
5. Given a triangle ABC with circumcircle Γ, let P be a point situated on the
   arco $\widehat{AC}$ of Γ not containing vertex B, and D be the foot of the perpendicular
   dropped from P to the line $\overleftrightarrow{BC}$. If $Q \neq P$ is the other point of intersection

of $\overleftrightarrow{DP}$ with Γ and r denotes the Simson line of P with respect to ABC, prove that $r \parallel \overleftrightarrow{AQ}$.

6. Let ABC be a triangle of circumcircle Γ, and P and P' points lying on the arc $\overset{\frown}{AC}$ of Γ not passing through B. If r and r' denote the Simson lines of P and P' with respect to ABC, respectively, prove that the angle between r and r' is equal to half of the measure of the arc $\overset{\frown}{PP'}$ of Γ not passing through A.

7. * A convex polygon is **cyclic** if there exists a circle passing through all of its vertices, which is then said to be **circumscribed** to the polygon.[5] Prove that a convex polygon is cyclic if and only if the perpendicular bisectors of their sides pass through a single point.

8. Let $ABCD$ be a tangential quadrilateral. Show that the incircles of triangles ABC and ACD touch diagonal AC at the same point.

9. * A convex polygon is **tangential** if there exists a circle tangent to all of its sides, which is then said to be **inscribed** in the polygon. Prove that a convex polygon is tangential if and only if the bisectors of the internal angles of the polygon all pass through a single point. Moreover, if this is so, show that the inscribed circle is unique.

10. Let $ABCD$ be a cyclic quadrilateral and E be the point of intersection of its diagonals. Also, let M, N, P and Q be the feet of the perpendiculars dropped from E to the sides AB, BC, CD and DA, respectively. Prove that quadrilateral $MNPQ$ is tangential.

11. If a convex hexagon $A_1A_2A_3 \ldots A_6$ is tangential (cf. Problem 9), prove that

$$\overline{A_1A_2} + \overline{A_3A_4} + \overline{A_5A_6} = \overline{A_2A_3} + \overline{A_4A_5} + \overline{A_6A_1}.$$

12. (IMO) On a circle Γ we are given three distinct points A, B and C. Show how to use straightedge and compass to construct a fourth point D on Γ, such that $ABCD$ is a tangential (convex) quadrilateral.

---

[5]If it does exist, such a circle is clearly unique, for it also circumscribes every triangle formed by three of the vertices of the polygon.

Chapter 4
Proportionality and Similarity

This chapter develops a set of tools which will allow us to start a systematic study of the *metric* aspects of Plane Euclidean Geometry; generally speaking, the central problem with which we shall be concerned here is that of comparing ratios of lengths of line segments. Among several interesting and important applications, the most prominent ones are the theorems of Thales and Pythagoras, which will reveal themselves to be almost indispensable hereafter. We also present a series of classical results, among which we highlight the study of the Apollonius circle and the solution of the Apollonius tangency problem, the collinearity and concurrence theorems of Ceva and Menelao, and some of the many theorems of Euler on the geometry of the triangle.

4.1 Thales' Theorem

We consider the following situation: we have in the plane parallel lines r, s and t (cf. Fig. 4.1). Then, we draw two transversals lines u and u', wit u (resp. u') intersecting r, s and t at points A, B and C (resp. A', B' and C'), respectively.

Trace line u'', parallel to u' and passing through C, and let A'' and B'' be its intersections with r and s, respectively. If we had $\overline{AB} = \overline{BC}$ (which does not seem to be the case in Fig. 4.1), then the midsegment theorem (cf. Proposition 2.40) applied to trapezoid $AA'C'C$ would give us

$$\overline{A'B'} = \overline{B'C'}.$$

To put it in another way,

$$\frac{\overline{AB}}{\overline{BC}} = 1 \Rightarrow \frac{\overline{A'B'}}{\overline{B'C'}} = 1.$$

Fig. 4.1 Transversals u and
u' to the parallels r, s and t

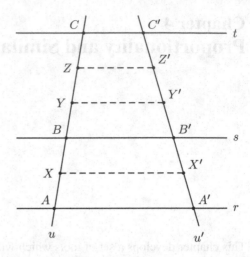

Now, assume that $\frac{\overline{AB}}{\overline{BC}}$ is equal to some rational number, say $\frac{2}{3}$. Divide segments
AB and BC in two and three equal parts, respectively, thus getting points X, Y and
Z in u, such that

$$\overline{AX} = \overline{XB} = \overline{BY} = \overline{YZ} = \overline{ZC}$$

(look again at Fig. 4.1). Draw through X, Y and Z the parallels to r, s and t, which
intersect u' respectively at X', Y' and Z'; three more applications of the midsegment
theorem for trapezoids give

$$\overline{A'X'} = \overline{X'B'} = \overline{B'Y'} = \overline{Y'Z'} = \overline{Z'C'}$$

so that

$$\frac{\overline{AB}}{\overline{BC}} = \frac{2}{3} \Rightarrow \frac{\overline{A'B'}}{\overline{B'C'}} = \frac{2}{3}.$$

By continuing with such a reasoning, suppose that we had $\frac{\overline{AB}}{\overline{BC}} = \frac{m}{n}$, with $m, n \in$
$\mathbb{N}$. A slight modification of the above argument (this time by first dividing AB and
BC into m and n equal parts, respectively) would give $\frac{\overline{A'B'}}{\overline{B'C'}} = \frac{m}{n}$, so that

$$\frac{\overline{AB}}{\overline{BC}} = \frac{m}{n} \Rightarrow \frac{\overline{A'B'}}{\overline{B'C'}} = \frac{m}{n}.$$

What we have done up to this point assures that the equality

$$\frac{\overline{AB}}{\overline{BC}} = \frac{\overline{A'B'}}{\overline{B'C'}}$$

holds whenever the left (or right) hand side is equal to a rational number. The natural question at this moment is the following: does the equality above remains true if one of its sides is equal to an irrational number? The answer is yes, and to understand why we shall use the following standard fact on real numbers[1]:

Lemma 4.1 *Every positive irrational number is the limit of a sequence of positive rational numbers.*

Now, suppose that

$$\frac{\overline{AB}}{\overline{BC}} = x,$$

with x being an irrational number. Choose (by the above lemma) a sequence $(a_n)_{n \geq 1}$ of positive rationals such that

$$x < a_n < x + \frac{1}{n}$$

for every $n \in \mathbb{N}$. Then, mark (cf. Fig. 4.2) point $C_n \in u$ with

$$\frac{\overline{AB}}{\overline{BC_n}} = a_n.$$

Let t_n be the line parallel to r, s and t through C_n, and C'_n be the point in which t_n intersects u'. Since $a_n \in \mathbb{Q}$, an argument analogous to what we did before show that

$$\frac{\overline{A'B'}}{\overline{B'C'_n}} = a_n.$$

Fig. 4.2 The case of an irrational ratio $\frac{\overline{AB}}{\overline{BC}}$

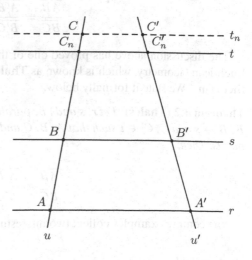

---

[1]For the necessary background, see sections 7.1 and 7.2 of [5], for instance.

Yet in another way, we have proved that

$$x < \frac{\overline{AB}}{\overline{BC_n}} < x + \frac{1}{n} \Rightarrow x < \frac{\overline{A'B'}}{\overline{B'C'_n}} < x + \frac{1}{n}$$

or, which is the same, that

$$\frac{\overline{AB}}{\overline{BC}} < \frac{\overline{AB}}{\overline{BC_n}} < \frac{\overline{AB}}{\overline{BC}} + \frac{1}{n} \Rightarrow \frac{\overline{AB}}{\overline{BC}} < \frac{\overline{A'B'}}{\overline{B'C'_n}} < \frac{\overline{AB}}{\overline{BC}} + \frac{1}{n}. \qquad (4.1)$$

Now, note that the inequalities of the left hand side above guarantee that, as n increases without bound, points C_n come closer and closer to C. However, since $t_n \parallel t$, it follows that points C'_n come closer and closer to C', so that the ratio $\frac{\overline{A'B'}}{\overline{B'C'_n}}$ converges to $\frac{\overline{A'B'}}{\overline{B'C'}}$. We abbreviate this by writing

$$\frac{\overline{A'B'}}{\overline{B'C'_n}} \longrightarrow \frac{\overline{A'B'}}{\overline{B'C'}} \quad \text{as } n \to +\infty.$$

On the other hand, we can clearly infer from the inequalities at the right hand side of (4.1) that

$$\frac{\overline{A'B'}}{\overline{B'C'_n}} \longrightarrow \frac{\overline{AB}}{\overline{BC}} \quad \text{as } n \to +\infty.$$

Since the limit of a convergent sequence of real numbers is unique, we are then forced to conclude that

$$\frac{\overline{AB}}{\overline{BC}} = \frac{\overline{A'B'}}{\overline{B'C'}}.$$

The discussion above has proved one of the most fundamental results of Plane Euclidean Geometry, which is known as **Thales' theorem**, or also as the **intercept theorem**.[2] We state it formally below.

Theorem 4.2 (Thales) *Let r, s and t be parallel lines, and choose points A, $A' \in r$, B, $B' \in s$ and C, $C' \in t$ such that A, B, C and A', B', C' are two triples of collinear points. Then,*

$$\frac{\overline{AB}}{\overline{BC}} = \frac{\overline{A'B'}}{\overline{B'C'}}.$$

The coming examples collect two interesting applications of Thales' theorem.

---

[2]After Thales of Miletus, Greek mathematician and philosopher of seventh century BC, the first of Classical Antiquity.

Example 4.3 Use straightedge and compass to divide the line segment AB below into five equal parts.

Solution

CONSTRUCTION STEPS.

1. Draw through A an arbitrary line r, different from $\overleftrightarrow{AB}$.
2. Mark on r points $C_0 = A$, C_1, C_2, C_3, C_4 and C_5 such that, for $0 \leq i \leq 4$, the lengths $\overline{C_i C_{i+1}}$ are all equal.
3. For $1 \leq i \leq 4$, trace the parallel to $\overleftrightarrow{C_5 B}$ passing through C_i.
4. If D_i is the intersection of such a parallel with the line segment AB, then Thales's theorem assures that points D_1, D_2, D_3 and D_4 divide AB into five equal parts.

For the next example, given positive real numbers a, b and c, we say that the positive real x is the **fourth proportional** of a, b and c (in this order) if

$$\frac{a}{b} = \frac{c}{x}.$$

If a, b and c are the lengths of three line segments, we shall also say that a line segment of length x, given as above, is the fourth proportional of the line segments of lengths a, b and c (in this order).

Example 4.4 Use straightedge and compass to construct the fourth proportional of the line segments given below.

Solution

CONSTRUCTION STEPS

1. Draw two lines r and s, concurrent at A.
2. Mark on r line segments AB and BC, with $\overline{AB} = a$ and $\overline{BC} = c$; mark on s the line segment AD, such that $\overline{AD} = b$.
3. Trace through C the parallel to $\overleftrightarrow{BD}$, which intersect s at E. By Thales' theorem, we have $\overline{DE} = \frac{bc}{a}$, as wished.

As important as Thales' theorem is its *partial converse* below, also due to Thales.

Fig. 4.3 Partial converse of
Thales' theorem

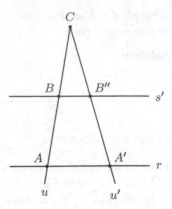

Corollary 4.5 *We are given in the plane lines r and s and points* $A, A' \in r$ *and* $B, B' \in s$, *with* $\overleftrightarrow{AB} \cap \overleftrightarrow{A'B'} = \{C\}$. *If* $\frac{\overline{AB}}{\overline{BC}} = \frac{\overline{A'B'}}{\overline{B'C}}$, *then* $r \parallel s$.

Proof Suppose that $B \in AC$ (cf. Fig. 4.3—the other possible cases can be dealt with analogously). Draw through B line $s' \parallel r$ and mark point B'' in which it intersects segment $A'C$. Thales' theorem gives $\frac{\overline{AB}}{\overline{BC}} = \frac{\overline{A'B''}}{\overline{B''C}}$, so that our hypotheses furnish

$$\frac{\overline{A'B'}}{\overline{B'C}} = \frac{\overline{A'B''}}{\overline{B''C}}.$$

Now, it follows from Problem 2, page 108, that $B' = B''$ or, which is the same, $s = s'$. Thus, $s \parallel r$.   □

The coming result is an important application of Thales' theorem, which is known in mathematical literature as the **angle bisector theorem** (Fig. 4.4).

Theorem 4.6 *Let ABC be a triangle in which* $\overline{AB} \neq \overline{AC}$.

(a) *If P is the foot of the internal bisector and Q is the foot of the external bisector relative to BC, then*

$$\frac{\overline{BP}}{\overline{PC}} = \frac{\overline{BQ}}{\overline{QC}} = \frac{\overline{BA}}{\overline{AC}}.$$

(b) *If $\overline{AB} = c$, $\overline{AC} = b$ and $\overline{BC} = a$, we have*

$$\begin{cases} \overline{BP} = \frac{ac}{b+c} \\ \overline{PC} = \frac{ab}{b+c}. \end{cases} \quad and \quad \begin{cases} \overline{BQ} = \frac{ac}{|b-c|} \\ \overline{QC} = \frac{ab}{|b-c|}. \end{cases}$$

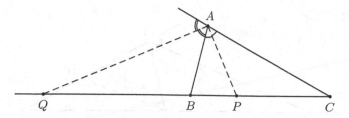

Fig. 4.4 The angle bisector theorem

Proof Item (b) follows immediately from (a): letting $\overline{BP} = x$ and $\overline{PC} = y$, we have $x + y = a$ and, by the result of (a), $\frac{x}{y} = \frac{c}{b}$. Solving the system of equations

$$\begin{cases} x + y = a \\ \frac{x}{y} = \frac{c}{b} \end{cases},$$

we obtain $x = \frac{ac}{b+c}$ and $y = \frac{ab}{b+c}$. The remaining formulas of item (b) are proved likewise.

In what concerns (a), let us show that $\frac{\overline{BQ}}{\overline{QC}} = \frac{\overline{BA}}{\overline{AC}}$, leaving the (analogous) proof of equality $\frac{\overline{BP}}{\overline{PC}} = \frac{\overline{BA}}{\overline{AC}}$ to the reader (cf. Problem 4, page 108).

Suppose (cf. Fig. 4.5) that $\overline{AB} < \overline{AC}$ (the case $\overline{AB} > \overline{AC}$ can be dealt with in the same way). Trace through B the parallel to $\overleftrightarrow{AQ}$ and mark its intersection point B' with side AC of ABC. Since $\overleftrightarrow{QA} \parallel \overleftrightarrow{BB'}$ and $\overrightarrow{AQ}$ is the angle bisector of $\angle BAX$, we obtain

$$A\widehat{B}B' = B\widehat{A}Q = Q\widehat{A}X = B\widehat{B'}A.$$

Therefore, triangle ABB' is isosceles of basis BB', so that $\overline{B'A} = \overline{BA}$. If we now apply Thales' theorem to the parallels $\overleftrightarrow{QA}$ and $\overleftrightarrow{BB'}$ with respect to the transversals $\overleftrightarrow{QC}$ and $\overleftrightarrow{AC}$, we get

$$\frac{\overline{BQ}}{\overline{QC}} = \frac{\overline{AB'}}{\overline{AC}} = \frac{\overline{BA}}{\overline{AC}}.$$

$\square$

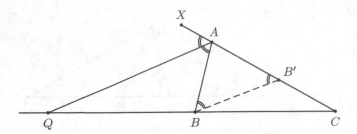

Fig. 4.5 Proof of the angle bisector theorem

Problems: Sect. 4.1

1. Lines r, s and t are parallel, with s lying between r and t. Transversals u and v determine, on r, s and t, points A, B, C and A', B', C', respectively, such that $\overline{AB} = x + 2$, $\overline{BC} = 2y$, $\overline{A'B'} = y$ and $\overline{B'C'} = (x - 10)/2$. Knowing that $x + y = 18$, compute the length of the line segment AB.
2. * Let P_1 and P_2 be points of the line segment AB, such that

$$\frac{\overline{AP_1}}{\overline{P_1B}} = \frac{\overline{AP_2}}{\overline{P_2B}}.$$

   Prove that P_1 and P_2 coincide.
3. Given line segments of lengths a and b, we say that a line segmento of length x is the **third proportional** of a and b (in this order) if

$$\frac{a}{b} = \frac{b}{x}.$$

   Show how to use straightedge and compass to construct it.
4. * Complete the proof of the angle bisector theorem.
5. * Let ABC be a triangle and P and M be the feet of the internal angle bisector and of the median relative to BC, respectively. If P and M coincide, prove that ABC is isosceles of basis BC.
6. In a triangle ABC, let P be the foot of the internal angle bisector relative to BC. Construct ABC, knowing the lengths $\overline{PB}$, $\overline{PC}$ and $\overline{AB}$.
7. In a triangle ABC, let P be the foot of the internal angle bisector relative to A. We mark respectively on AB and AC points M and N for which $\overline{BM} = \overline{BP}$ and $\overline{CN} = \overline{CP}$. Prove that $MN \parallel BC$.
8. Construct triangle ABC with straightedge and compass, knowing the lengths m_a, m_b and m_c of the medians relative to BC, AC and AB, respectively.

Fig. 4.6 Two similar triangles

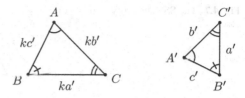

4.2 Similar Triangles

In this section we study the concept of *similarity* of triangles, which generalizes the notion of congruence of triangles and will be of paramount importance for all that follows.

Two triangles are said to be **similar** if there exists a one-to-one onto correspondence between the vertices of one and the other triangles such that the internal angles at corresponding vertices are equal and the ratios between the lengths of pairs of corresponding sides are always the same (cf. Fig. 4.6). Physically, two triangles are similar if we can *dilate* and/or *rotate* and/or *reflect* and/or *translate* one of them, obtaining the other one after performing such a set of operations.[3]

In Fig. 4.6, triangles ABC and $A'B'C'$ are similar, with correspondence of vertices $A \leftrightarrow A'$, $B \leftrightarrow B'$, $C \leftrightarrow C'$. Thus, $\widehat{A} = \widehat{A'}$, $\widehat{B} = \widehat{B'}$, $\widehat{C} = \widehat{C'}$ and there exists a real number $k > 0$ with

$$\frac{\overline{AB}}{\overline{A'B'}} = \frac{\overline{BC}}{\overline{B'C'}} = \frac{\overline{AC}}{\overline{A'C'}} = k.$$

Such a positive real k is called the **ratio of similitude** between ABC and $A'B'C'$, in this order (note that the ratio of similitude between $A'B'C'$ and ABC, in this order, is $\frac{1}{k}$).

We write $ABC \sim A'B'C'$ to indicate that triangles ABC and $A'B'C'$ are similar under the correspondence of vertices $A \leftrightarrow A'$, $B \leftrightarrow B'$, $C \leftrightarrow C'$.

If $ABC \sim A'B'C'$ with ratio of similitude k, it is possible to prove that k is equal to the ratio of the lengths of any two corresponding segments in triangles ABC and $A'B'C'$ (in this order); for instance, in the notations of Fig. 4.6, letting M be the midpoint of BC and M' that of $B'C'$, we have

$$\frac{\overline{MA}}{\overline{M'A'}} = \frac{a/2}{a'/2} = \frac{a}{a'} = k$$

(in this respect, see also Problem 3, page 115).

---

[3] Although such a point of view can be made precise by means of the study of *geometric transformations*, we shall not pursue it here. Nevertheless, we refer the interested reader to the marvelous books [24–26] of professor Yaglom, and [27] of professors Yaglom and Shenitzer.

Fig. 4.7 The SSS similarity case

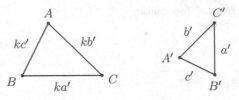

Fig. 4.8 Proof of the SSS similarity case

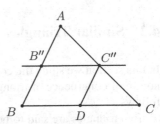

The three coming results establish the usual sets of necessary and sufficient conditions for two given triangles to be similar. For this reason, they are commonly referred to as the **cases of similarity of triangles**. Since their proofs are relatively simple consequences of the partial converse of Thales' theorem, we shall do the first of them, leaving the other two as exercises for the reader (cf. Problem 1, page 115).

The coming proposition is known as the **SSS** similarity case for triangles (Fig. 4.7).

Proposition 4.7 *Let ABC and $A'B'C'$ be two given triangles, with*

$$\frac{\overline{AB}}{\overline{A'B'}} = \frac{\overline{BC}}{\overline{B'C'}} = \frac{\overline{AC}}{\overline{A'C'}}.$$

Then, $ABC \sim A'B'C'$ under the correspondence of vertices $A \leftrightarrow A'$, $B \leftrightarrow B'$, $C \leftrightarrow C'$. In particular, $\widehat{A} = \widehat{A'}$, $\widehat{B} = \widehat{B'}$ and $\widehat{C} = \widehat{C'}$.

Proof Letting k denote the common value of the ratios in the statement of the proposition, we have $\overline{AB} = k \cdot \overline{A'B'}$, $\overline{BC} = k \cdot \overline{B'C'}$ and $\overline{AC} = k \cdot \overline{A'C'}$. Assume, without loss of generality, that $k > 1$ and mark (cf. Fig. 4.8) point $B'' \in AB$ such that $\overline{AB''} = \overline{A'B'}$.

Let C'' denote the intersection, with side AC, of the parallel to BC passing through B''. Thales' theorem gives

$$\frac{\overline{AC''}}{\overline{AC}} = \frac{\overline{AB''}}{\overline{AB}} = \frac{1}{k},$$

so that $\overline{AC''} = \frac{1}{k} \cdot \overline{AC} = \overline{A'C'}$.

Now, draw the parallel to AB passing through C'', which intersects BC at D. Then, since the quadrilateral $B''C''DB$ is a parallelogram, again from Thales'

Fig. 4.9 The SAS similarity case

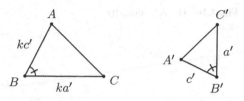

theorem we obtain

$$\frac{\overline{B''C''}}{\overline{BC}} = \frac{\overline{BD}}{\overline{BC}} = \frac{\overline{AC''}}{\overline{AC}} = \frac{1}{k}.$$

Therefore, $\overline{B''C''} = \frac{1}{k} \cdot \overline{BC} = \overline{B'C'}$.

The above discussion has shown that

$$\overline{AB''} = \overline{A'B'}, \quad \overline{AC''} = \overline{A'C'} \text{ and } \overline{B''C''} = \overline{B'C'},$$

i.e., that triangles $AB''C''$ and $A'B'C'$ are congruent, by the SSS congruence case. Hence, we get

$$\widehat{B} = A\widehat{B}C = A\widehat{B''}C'' = A'\widehat{B'}C' = \widehat{B'},$$

and likewise $\widehat{A} = \widehat{A'}$ and $\widehat{C} = \widehat{C'}$. □

We now turn to the statement of the **SAS** similarity case (Fig. 4.9).

Proposition 4.8 *Let ABC and $A'B'C'$ be given triangles, with*

$$\frac{\overline{AB}}{\overline{A'B'}} = \frac{\overline{BC}}{\overline{B'C'}} = k \text{ and } \widehat{B} = \widehat{B'}.$$

Then, $ABC \sim A'B'C'$, with the correspondence of vertices $A \leftrightarrow A'$, $B \leftrightarrow B'$, $C \leftrightarrow C'$. In particular, $\widehat{A} = \widehat{A'}$, $\widehat{C} = \widehat{C'}$ and $\frac{\overline{AC}}{\overline{A'C'}} = k$.

Finally, the proposition below presents the **AA** similarity case for triangles (Fig. 4.10).

Proposition 4.9 *Let ABC and $A'B'C'$ be given triangles, with*

$$\widehat{A} = \widehat{A'} \text{ and } \widehat{B} = \widehat{B'}.$$

Then, $ABC \sim A'B'C'$, with the correspondence of vertices $A \leftrightarrow A'$, $B \leftrightarrow B'$, $C \leftrightarrow C'$. In particular,

$$\frac{\overline{AB}}{\overline{A'B'}} = \frac{\overline{BC}}{\overline{B'C'}} = \frac{\overline{AC}}{\overline{A'C'}}.$$

Fig. 4.10 The AA similarity case

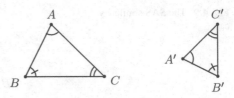

As an important consequence of the similarity cases studied above, we now derive the usual **metric relations in right triangles**. The content of item (c) is the famous and ubiquitous **Pythagoras' theorem**[4]; for another proof of it, see Example 5.8.

Theorem 4.10 *Let ABC be a right triangle at A, with legs $\overline{AB} = c$, $\overline{AC} = b$ and hypotenuse $\overline{BC} = a$. If H is the foot of the altitude relative to the hypotenuse, $\overline{CH} = x$, $\overline{BH} = y$ and $\overline{AH} = h$, then:*

(a) $ah = bc$.
(b) $ax = b^2$ and $ay = c^2$.
(c) $a^2 = b^2 + c^2$.
(d) $xy = h^2$.

Proof (a) and (b): since $A\widehat{H}B = C\widehat{A}B$ and $A\widehat{B}H = C\widehat{B}A$ (cf. Fig. 4.11), triangles BAH and BCA are similar by AA, with correspondence of vertices $A \leftrightarrow C$, $H \leftrightarrow A$, $B \leftrightarrow B$. Thus,

$$\frac{\overline{BH}}{\overline{AB}} = \frac{\overline{AB}}{\overline{BC}} \text{ and } \frac{\overline{AH}}{\overline{AC}} = \frac{\overline{AB}}{\overline{BC}}$$

or, which is the same,

$$\frac{y}{c} = \frac{c}{a} \text{ and } \frac{h}{b} = \frac{c}{a}.$$

Relation $ax = b^2$ is proved in pretty much the same way.

(c) Termwise addition of the relations of item (b) furnish the equality $a(x + y) = b^2 + c^2$. Now, since $x + y = a$, there is nothing left to do.

(d) Termwise multiplication of the relations of item (b) give $a^2 \cdot xy = (bc)^2$. By invoking the result of (a), we can write

$$xy = \left(\frac{bc}{a}\right)^2 = h^2,$$

as wished.   □

---

[4]Pythagoras of Samos was one of the greatest mathematicians of Classical Antiquity. The theorem that bears his name was already known to babylonians, at least two thousand years before he was born; nevertheless, Pythagoras was the first one to prove it. It is also attributed to him the first proof of the irrationality of $\sqrt{2}$.

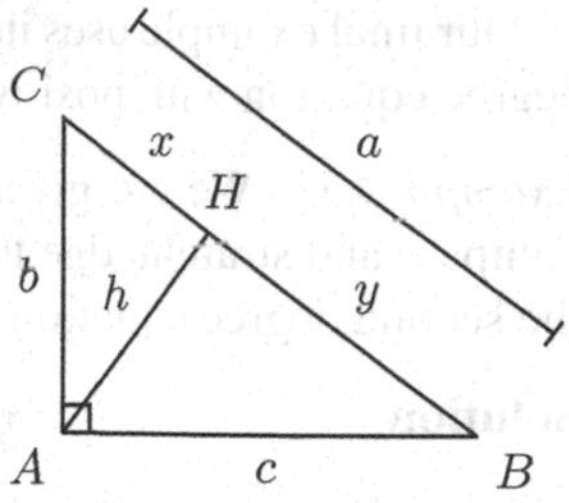

Fig. 4.11 Metric relations on
a right triangle

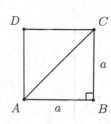

Fig. 4.12 Diagonal of a
square in terms of the length
of its sides

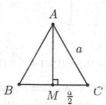

Fig. 4.13 Altitudes of an
equilateral triangle

The three corollaries below collect some important applications of Pythagoras' theorem.

Corollary 4.11 *The diagonals of a square of side length a are equal to $a\sqrt{2}$.*

Proof If $ABCD$ is a square of side length a and diagonals AC and BD (cf. Fig. 4.12), then ABC is a right isosceles triangle. Hence, Pythagoras' theorem gives

$$\overline{AC} = \sqrt{\overline{AB}^2 + \overline{BC}^2} = \sqrt{a^2 + a^2} = a\sqrt{2}.$$

□

Corollary 4.12 *The altitudes of an equilateral triangle of side length a are equal to $\frac{a\sqrt{3}}{2}$.*

Proof Let ABC be an equilateral triangle of side length a, and M be the midpoint of BC (cf. Fig. 4.13). Problem 6, page 30, guarantees that $AM \perp BC$. Hence, by applying Pythagoras' theorem to triangle ACM, we obtain

$$\overline{AM}^2 = \overline{AC}^2 - \overline{CM}^2 = a^2 - \left(\frac{a}{2}\right)^2 = \frac{3a^2}{4},$$

and the result follows.

□

Our final example uses item (d) of Theorem 4.10 to solve *geometrically* a second degree equation with positive roots.

Example 4.13 We are given line segments of lengths s and p, with $s > 2p$. Use compass and straightedge to construct line segments whose lengths are the roots of the second degree equation $x^2 - sx + p^2 = 0$.

Solution

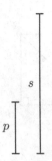

CONSTRUCTION STEPS.

1. Draw a line r and mark on it points B and C such that $\overline{BC} = s$. Then, construct a semicircle Γ of diameter BC.
2. Draw line r', parallel to r and at distance p from r, which intersects Γ at (distinct) points A and A' (since $p < \frac{s}{2}$).
3. The discussion following the proof of Proposition 3.19 assures that triangle ABC is right at A.
4. If H is the foot of the perpendicular dropped from A to BC, then $\overline{BH} + \overline{CH} = s$ and item (d) of Theorem 4.10 assures that $\overline{BH} \cdot \overline{CH} = p^2$. Therefore, basic algebra (cf. Section 2.3 of [5], for instance) assures that $\overline{BH}$ and $\overline{CH}$ are the roots of the second degree equation given in the statement of the example.   □

We finish this section by establishing the converse of Pythagoras' theorem (in this respect, see also Theorem 7.25).

Proposition 4.14 *Let ABC be a triangle with $\overline{AB} = c$, $\overline{BC} = a$ and $\overline{AC} = b$. If $a^2 = b^2 + c^2$, then ABC is right at A.*

Proof Firstly, note that $a^2 = b^2 + c^2$ implies $a > b, c$. Hence, Proposition 2.23 guarantees that $\angle A > \angle B, \angle C$, so that $\angle B$ and $\angle C$ are acute. Thus, if H stands for the foot of the altitude relative to BC, we conclude that $H \in BC$ (cf. Fig. 4.14). Let $\overline{AH} = h$, M be the midpoint of BC and $\overline{BH} = x$. We can assume, without loss of generality, that $H \in BM$. By applying Pythagoras' theorem to triangles AHC and AHB, we obtain

$$a^2 = b^2 + c^2 = (\overline{AH}^2 + \overline{CH}^2) + (\overline{AH}^2 + \overline{BH}^2) = 2h^2 + (a - x)^2 + x^2,$$

Fig. 4.14 Converse of
Pythagoras' theorem

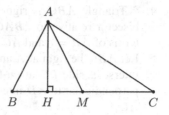

whence $h^2 = ax - x^2$. But then, applying Pythagoras' theorem to triangle AHM, we arrive at

$$\overline{AM}^2 = \overline{AH}^2 + \overline{HM}^2 = h^2 + (\overline{BM} - \overline{BH})^2$$

$$= (ax - x^2) + \left(\frac{a}{2} - x\right)^2 = \frac{a^2}{4},$$

so that $\overline{AM} = \frac{a}{2} = \frac{1}{2}\overline{BC}$. Therefore, M lies at equal distances from the vertices of ABC, and Proposition 3.30 assures that ABC is right at A. □

Problems: Sect. 4.2

1. * Prove that the sets of conditions listed in each of the Proposition 4.8 and 4.9 indeed suffice to assure the similarity of triangles ABC and $A'B'C'$.
2. In the figure below, the three quadrilaterals are squares and points X, Y and Z are collinear. Compute, in, em centimeters, the length of the sides of the largest square, knowing that the lengths of the other two are equal to 4cm and 6cm.

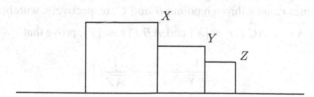

3. * ABC and $A'B'C'$ are similar triangles, with similitude ratio k. We let m_a and m'_a, h_a and h'_a, β_a and β'_a denote the lengths of the medians, altitudes and internal angle bisectors relative to A and A', respectively. Prove that

$$\frac{m_a}{m'_a} = \frac{h_a}{h'_a} = \frac{\beta_a}{\beta'_a} = k.$$

4. * Triangle ABC is right at A, and point $P \in BC$ is the foot of the angle bisector relative to $\angle BAC$. Compute the distance from P to the side AC in terms of $\overline{AB} = c$ and $\overline{AC} = b$.

5. Let ABC be right at A and such that $\overline{AB} = 1$. The internal bisector of $\angle BAC$ intersects side BC at D. Knowing that the line r that passes through D and is perpendicular to AD intersects side AC at its midpoint, compute the length of AC.

6. Let $ABCD$ be a parallelogram with diagonals AC and BD, and sides $\overline{AB} = 10$, $\overline{AD} = 24$. Let also E and F stand for the feet of the perpendiculars dropped from A to the sides BC and CD, respectively. If $\overline{AF} = 20$, compute the length of AE.

7. Two circles, of radii $r < R$, are externally tangent at P. Knowing that such circles are also tangent to the sides of an acute angle of vertex A, compute $\overline{AP}$ in terms of r and R.

8. Let ABC be a triangle such that $\overline{BC} = a$, $\overline{AC} = b$ and $\overline{AB} = c$, and let M, N and P be points respectively situated on AB, BC and AC, such that $AMNP$ is a rhombus.

   (a) Compute the length of the sides of such a rhombus in terms of a, b and c.
   (b) Show how to use compass and straightedge to find the position of point M.

9. ABC is an equilateral triangle of side length a, and M is the midpoint of AB. We choose a point D along $\overleftrightarrow{BC}$, with C between B and D, in such a way that $\overline{CD} = \frac{a}{2}$. If $AC \cap DM = \{E\}$, compute $\overline{AE}$ as a function of a.

10. $ABCD$ is a trapezoid of bases $\overline{AB} = a$, $\overline{CD} = b$, in which the legs are AD and BC. Through the intersection point P of the diagonals of $ABCD$ we draw the line segment MN, parallel to the bases of $ABCD$ and such that $M \in AD$ and $N \in BC$. Prove that P is the midpoint of MN and that $\overline{MN}$ is equal to the *harmonic mean* of a and b, i.e., prove that $\overline{MN} = \frac{2ab}{a+b}$.

11. * On the side BC of a triangle ABC we choose an arbitrary point Z. Then, we draw lines r and s through points B and C, respectively, with both of them parallel to $\overleftrightarrow{AZ}$. If $\overleftrightarrow{AC} \cap r = \{X\}$ and $\overleftrightarrow{AB} \cap s = \{Y\}$, prove that

$$\frac{1}{\overline{BX}} + \frac{1}{\overline{CY}} = \frac{1}{\overline{AZ}}.$$

12. $ABCD$ is a trapezoid of bases AB and CD and legs AD and BC. M is the midpoint of CD, and the line segment AM intersects the diagonal BD at F. We draw through F line r, parallel to the basis. If r intersects AD, AC and BC respectively at E, G and H, prove that $\overline{EF} = \overline{FG} = \overline{GH}$.

13. * (OCM) Let ABC be triangle with $\overline{AB} = c$, $\overline{AC} = b$ and $\overline{BC} = a$. If $A\widehat{B}C = 2A\widehat{C}B$, show that $b^2 = c(a + c)$.

14. (OCM) Triangle ABC is such that $A\widehat{C}B = 2B\widehat{A}C$ and $\overline{AC} = 2\overline{BC}$. Show that it is right at B.

15. * (OCS - adapted) Let $\Gamma(O; R)$ be the circumcircle of triangle ABC and H_a be the foot of its altitude relative to A. If A' stands for the symmetric of A with respect to O, prove that $AA'C \sim ABH_a$. Subsequently, conclude that if $\overline{AB} = c$, $\overline{AC} = b$ and $\overline{AH_a} = h_a$, then

$$h_a = \frac{bc}{2R}.$$

16. Lines r and s are tangent to the circumcircle of an acute triangle ABC respectively at B and C. Letting D, E and F be the feet of the perpendiculars dropped from A respectively to BC, r and s, prove that

$$\overline{AD}^2 = \overline{AE} \cdot \overline{AF}.$$

17. In a rectangle $ABCD$ of sides $\overline{AB} = 4$ and $\overline{BC} = 3$, we mark on diagonal AC the point M such that $DM \perp AC$. Compute the length of segment AM

18. Let ABC be a right triangle of legs b and c and height h relative to the hypotenuse. Prove that

$$\frac{1}{h^2} = \frac{1}{b^2} + \frac{1}{c^2}.$$

19. * Given positive reals a and b, let AB be a line segment of length $a + b$, and H be a point on it with $\overline{AH} = a$ and $\overline{BH} = b$. Draw a semicircle of diameter AB and mark point C as its intersection with the perpendicular to AB through H.

   (a) Show that $\overline{CH} = \sqrt{ab}$.
   (b) Conclude that the arithmetic-geometric means inequality for two positive real numbers (cf. Chapter 5 of [5], for instance), is essentially equivalent to the triangle inequality in the (possibly degenerate) triangle CHO, where O is the midpoint of AB.

20. * Given in the plane line segments of lengths a and b, use compass and straightedge to construct a line segment of length $\sqrt{ab}$.

21. Let M, N and P be points respectively on the sides BC, AC and AB of an equilateral triangle ABC of side length a, such that $\overline{BM} = \overline{CN} = \overline{AP} = \frac{a}{3}$. Show that triangle MNP is also equilateral and that its sides are perpendicular to those of ABC.

22. Given line segments of lengths a, b and c, use compass and straightedge to construct a segment of length comprimento $\sqrt{a^2 + b^2 - c^2}$, admitting that the expression under the square root sign is positive.

23. Hypotenuse BC of right triangle ABC is divided into four equal segments by points D, E and F. If $\overline{BC} = 20$, compute the value of $\overline{AD}^2 + \overline{AE}^2 + \overline{AF}^2$.

24. Identify and construct, with compass and straightedge, the locus of the midpoints of the chords of length l of a given circle $\Gamma(O; R)$.

15. * (OCS - adapted) Let $\Gamma(O; R)$ be the circumcircle of triangle ABC and H_a be the foot of its altitude relative to A. If A' stands for the symmetric of A with respect to O, prove that $AA'C \sim ABH_a$. Subsequently, conclude that if $\overline{AB} = c$, $\overline{AC} = b$ and $\overline{AH_a} = h_a$, then

$$h_a = \frac{bc}{2R}.$$

16. Lines r and s are tangent to the circumcircle of an acute triangle ABC respectively at B and C. Letting D, E and F be the feet of the perpendiculars dropped from A respectively to BC, r and s, prove that

$$\overline{AD}^2 = \overline{AE} \cdot \overline{AF}.$$

17. In a rectangle $ABCD$ of sides $\overline{AB} = 4$ and $\overline{BC} = 3$, we mark on diagonal AC the point M such that $DM \perp AC$. Compute the length of segment AM

18. Let ABC be a right triangle of legs b and c and height h relative to the hypotenuse. Prove that

$$\frac{1}{h^2} = \frac{1}{b^2} + \frac{1}{c^2}.$$

19. * Given positive reals a and b, let AB be a line segment of length $a + b$, and H be a point on it with $\overline{AH} = a$ and $\overline{BH} = b$. Draw a semicircle of diameter AB and mark point C as its intersection with the perpendicular to AB through H.

    (a) Show that $\overline{CH} = \sqrt{ab}$.
    (b) Conclude that the arithmetic-geometric means inequality for two positive real numbers (cf. Chapter 5 of [5], for instance), is essentially equivalent to the triangle inequality in the (possibly degenerate) triangle CHO, where O is the midpoint of AB.

20. * Given in the plane line segments of lengths a and b, use compass and straightedge to construct a line segment of length $\sqrt{ab}$.

21. Let M, N and P be points respectively on the sides BC, AC and AB of an equilateral triangle ABC of side length a, such that $\overline{BM} = \overline{CN} = \overline{AP} = \frac{a}{3}$. Show that triangle MNP is also equilateral and that its sides are perpendicular to those of ABC.

22. Given line segments of lengths a, b and c, use compass and straightedge to construct a segment of length comprimento $\sqrt{a^2 + b^2 - c^2}$, admitting that the expression under the square root sign is positive.

23. Hypotenuse BC of right triangle ABC is divided into four equal segments by points D, E and F. If $\overline{BC} = 20$, compute the value of $\overline{AD}^2 + \overline{AE}^2 + \overline{AF}^2$.

24. Identify and construct, with compass and straightedge, the locus of the midpoints of the chords of length l of a given circle $\Gamma(O; R)$.

25. Let Γ be a circle of center O and radius R. Prove that the locus of the points of the plano from which we can draw tangents of length l to Γ is the circle of center O and radius $\sqrt{R^2 + l^2}$.

26. (OCM) A pedestrian, standing 25m from the entrance of a building, visualizes it, from top to bottom, under a certain angle. He/she then walks away from the building more 50m and notes that, upon doing so, the new angle of visualization is exactly half of the former. Compute the height of the building.

27. (OCM) Lines r, s and t are parallel, with s lying between r and t, such that the distance from r to s is of 3m, whereas that from s to t is of 1m. Triangle ABC is equilateral and has a vertex on each of the lines r, s and t. Compute the lengths of its sides.

28. (OCM) Two towers, one of which 30m high and the other 40m high, lie at 50m from one another. Somewhere along the line segment connecting their ground floors lies a small fountain. At some moment, two birds started flying, from the top of each tower and with equal velocities, directly towards the fountain. If the birds arrived in the fountain at the same time, compute the distances from it to the bases of the towers.

29. The lengths of two sides of a triangle are 7 and $5\sqrt{2}$, and the enclosed angle is of 135°. Compute the length of the third side.

30. We are given a point P inside a rectangle $ABCD$ of diagonals AC and BD. Prove that $\overline{AP}^2 + \overline{CP}^2 = \overline{BP}^2 + \overline{DP}^2$.

31. $ABCD$ is a square of side length 10 and P is a point on its circumcircle. Compute the possible values of the sum $\overline{PA}^2 + \overline{PB}^2 + \overline{PC}^2 + \overline{PD}^2$.

32. If the diagonals of a convex quadrilateral are perpendicular to each other, prove that the sum of the squares of the lengths of the opposite sides of the quadrilateral are equal.

33. Let $ABCD$ be a trapezoid of bases AB and CD and legs AD and BC, right em A. If $\overline{AB} = 12$, $\overline{CD} = 4$ and $ABCD$ is tangential, compute the distances of B and C from the center of the inscribed circle.

34. Consider a line s and two circles of radii R and r, tangent to s at distinct points and externally to each other. A third circle, also tangent to s, is externally tangent to the two other circles. If x stands for the radius of this third circle, prove that

$$\frac{1}{\sqrt{x}} = \frac{1}{\sqrt{R}} + \frac{1}{\sqrt{r}}.$$

35. Let s stand for the real line and α for one of the half-planes it determines. For each $n \in \mathbb{Z}$, draw the circle of radius $\frac{1}{2}$, contained in α and tangent to s at n. Then, if Γ_1 and Γ_2 are two of these adjacent ones among these circles (so that they are externally tangent), draw the circle tangent to s and externally tangent to Γ_1 and Γ_2. Upon repeating this process recursively, prove that the set of

points of tangency with s of the circles thus obtained is contained in the set $\mathbb{Q}$ of rational numbers.[5]

36. We are given n points in the plane, not all of which being collinear. Prove the **Sylvester-Gallai theorem**[6]: there exists at least one line that passes through exactly two of these n points.

4.3 Some Applications

This section collects some more elaborate applications of similarity of triangles, paying a tribute to Apollonius of Perga, Claudius Ptolemy and Leonhard Euler.

Our first result establishes a converse of the angle bisector theorem. For the statement of it, recall that in every triangle the internal and external angle bisectors relative to a vertex are always perpendicular to each other.

Proposition 4.15 *Let ABC be a given triangle, and P and Q be points on $\overleftrightarrow{BC}$, with $P \in BC$ and $Q \notin BC$. If*

$$P\widehat{A}Q = 90° \ \text{ and } \ \frac{\overline{BP}}{\overline{PC}} = \frac{\overline{BQ}}{\overline{QC}},$$

then AP is the internal angle bisector and AQ is the external angle bisector of $\angle BAC$.

Proof Draw, through point P, the parallel to $\overleftrightarrow{AQ}$, and let U and V be its points of intersection with $\overleftrightarrow{AB}$ and $\overleftrightarrow{AC}$, respectively (cf. Fig. 4.15).

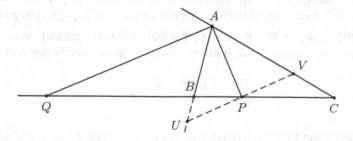

Fig. 4.15 Establishing the converse of the angle bisector theorem

---

[5]It can be shown that the set of points of tangency is actually equal to $\mathbb{Q}$. For a proof, we refer the reader to [21].

[6]James Sylvester, English mathematician, and Tibor Gallai, Hungarian mathematician, both of the twentieth century.

Corollary 2.18 guarantees that $B\widehat{U}P = B\widehat{A}Q$ and $P\widehat{V}C = Q\widehat{A}C$; on the other hand, $U\widehat{B}P = A\widehat{B}Q$ (for the corresponding angles are opposite) and $P\widehat{C}V = Q\widehat{C}A$. Hence, $BUP \sim BAQ$ and $PCV \sim QCA$ by AA, so that

$$\frac{\overline{BP}}{\overline{BQ}} = \frac{\overline{PU}}{\overline{QA}} \quad \text{and} \quad \frac{\overline{PC}}{\overline{CQ}} = \frac{\overline{PV}}{\overline{AQ}}.$$

However, since $\frac{\overline{BP}}{\overline{BQ}} = \frac{\overline{PC}}{\overline{QC}}$ by hypothesis, the similarity relations above give

$$\frac{\overline{PU}}{\overline{QA}} = \frac{\overline{PV}}{\overline{AQ}},$$

whence $\overline{PU} = \overline{PV}$. Now, from $\overleftrightarrow{AP} \perp \overleftrightarrow{AQ}$ and $\overleftrightarrow{UV} \| \overleftrightarrow{AQ}$ we obtain $\overleftrightarrow{UV} \perp \overleftrightarrow{AP}$, so that AP is median and altitude of AUV. Therefore, the result of Problem 6, page 30, assures that AP is also the bisector of $\angle UAV$, and thus of $\angle BAC$.

Finally, since the external bisector of $\angle BAC$ is perpendicular to the internal one, we conclude that AQ is the external bisector of $\angle BAC$.   □

We are now in position to state and prove the following theorem of Apollonius.[7]

Theorem 4.16 (Apollonius) *Given a positive real number $k \neq 1$ and distinct points B and C in the plane, the locus of points A in the plane for which $\overline{AB} = k \cdot \overline{AC}$ is the circle of diameter PQ, with $P \in BC$ and $Q \in \overleftrightarrow{BC} \setminus BC$ being such that*

$$\frac{\overline{BP}}{\overline{PC}} = \frac{\overline{BQ}}{\overline{QC}} = k.$$

Proof Let P and Q be as in the statement of the theorem. If $A \neq P, Q$ is such that $\overline{AB} = k \cdot \overline{AC}$, then a straightforward variant of problem 2, page 108, guarantees that $A \notin \overleftrightarrow{BC}$. Now, let $P', Q' \in \overleftrightarrow{BC}$ be the feet of the internal and external angle bisectors of ABC, relative to A. It follows from the angle bisector theorem that

$$\frac{\overline{BP'}}{\overline{P'C}} = \frac{\overline{BQ'}}{\overline{Q'C}} = \frac{\overline{BA}}{\overline{AC}} = k,$$

and invoking again Problem 2, page 108, we conclude that $P' = P$ and $Q' = Q$. However, since $P'\widehat{A}Q' = 90°$, it comes that A lies in the circle of diameter PQ.

---

[7]Apollonius of Perga, Greek mathematician of the third century BC Apollonius gave great contributions to Euclidean Geometry, notably to the study of *conics*. We shall have more to say on these curves in Sects. 6.3 and 10.3.

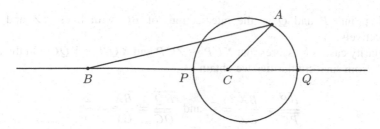

Fig. 4.16 Apollonius' circle on (B, C) in the ratio $k \neq 1$

Conversely, let $A \neq P$, Q be a point on the circle of diameter PQ. Then $A \notin \overleftrightarrow{BC}$ and $P\widehat{A}Q = 90°$. Since $\frac{BP}{PC} = \frac{BQ}{QC} = k$, the previous proposition assures that, in triangle ABC, segment AP is the internal angle bisector relative to A, whereas segment AQ is the corresponding external angle bisector. Hence, the angle bisector theorem gives

$$\frac{\overline{BA}}{\overline{AC}} = \frac{\overline{BP}}{\overline{PC}} = k,$$

so that A belongs to the desired locus. $\square$

As already quoted in Fig. 4.16, the locus described in the previous result in known as **Apollonius' circle** relative to (B, C), in the ratio $k \neq 1$. In particular, note that the Apollonius' circles relative to (B, C) and (C, B), both in the ratio $k \neq 1$, are distinct. For another proof of Apollonius' theorem, we refer the reader to Problem 15, page 202.

The coming example presents a straightforward way of constructing Apollonius' circle relative to a given ordered pair of points, in a given ratio $k \neq 1$.

Example 4.17 In the notations of the figure below, construct Apollonius circle relative to (B, C), in the ratio $\frac{2}{3}$.

Solution

CONSTRUCTION STEPS

1. Draw parallel lines r and s, distinct from $\overleftrightarrow{BC}$ and such that $B \in r$, $C \in s$.
2. Fix an arbitrary segment of length u, and mark on r a point X such that $\overline{BX} = 2u$ and on s distinct points Y and Z such that $\overline{CY} = \overline{CZ} = 3u$, with X and Y lying in the same half-plane with respect to $\overleftrightarrow{BC}$.

3. Mark points P and Q as the intersections of $\overleftrightarrow{BC}$ with lines $\overleftrightarrow{XZ}$ and $\overleftrightarrow{XY}$, respectively.
4. Similarity case AA assures that $XBP \sim ZCP$ and $XQB \sim YQC$. On the other hand, from such similarities we obtain

$$\frac{\overline{BP}}{\overline{PC}} = \frac{\overline{BX}}{\overline{CZ}} = \frac{2}{3} \text{ and } \frac{\overline{BQ}}{\overline{QC}} = \frac{\overline{BX}}{\overline{CY}} = \frac{2}{3},$$

so that the desired locus is the circle of diameter PQ.   □

The next application of the notion of similarity of triangles ia a famous theorem on cyclic quadrilaterals, due to Claudius Ptolemy.[8]

Theorem 4.18 (Ptolemy)   *If $ABCD$ is a cyclic quadrilateral with diagonals AC and BD, then*

$$\overline{AB} \cdot \overline{CD} + \overline{AD} \cdot \overline{BC} = \overline{AC} \cdot \overline{BD}.$$

Proof Mark point P on diagonal BD, such that $P\widehat{C}D = A\widehat{C}B$ (cf. Fig. 4.17). Since $B\widehat{A}C = B\widehat{D}C = \frac{1}{2} \overset{\frown}{BC}$, triangles ABC and DPC are similar by AA, so that

$$\frac{\overline{AB}}{\overline{AC}} = \frac{\overline{DP}}{\overline{CD}}.$$

Fig. 4.17 Proof of Ptolemy's theorem

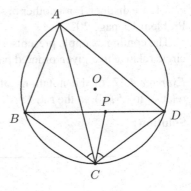

[8]Claudius Ptolemy, Greek astronomer and mathematician of the second century AD, gave great contributions to Euclidean Geometry. Ptolemy is mostly known for his work as an astronomer, especially for having proposed the (wrong) Geocentric Theory, according to which the Earth occupied the center of the Universe. Such a theory was accepted as a dogma by the Catholic Church for some 1400 years, and was the ultimate responsible for the judgement of Galileo Galilei by the Saint Inquisition.

Analogously, triangles ADC and BPC are also similar, so that

$$\frac{\overline{BP}}{\overline{BC}} = \frac{\overline{AD}}{\overline{AC}}.$$

Now, the two relations obtained above give

$$\overline{BD} = \overline{BP} + \overline{PD} = \frac{\overline{AD} \cdot \overline{BC}}{\overline{AC}} + \frac{\overline{AB} \cdot \overline{CD}}{\overline{AC}},$$

as wished.   □

The coming corollary isolates an interesting and important consequence of Ptolemy's theorem. In this respect, see also Problem 15, page 82.

Corollary 4.19 *If ABC is an equilateral triangle and P is a point on the smaller arc $\overset{\frown}{BC}$ of the circumcircle of ABC, then $\overline{PB} + \overline{PC} = \overline{PA}$.*

Proof By applying Ptolemy's theorem to the quadrilateral $PBAC$ of Fig. 4.18, we obtain

$$\overline{AB} \cdot \overline{PC} + \overline{AC} \cdot \overline{PB} = \overline{AP} \cdot \overline{BC}.$$

Therefore, by canceling out the equal lengths $\overline{AB} = \overline{AC} = \overline{BC}$, we obtain the desired result.   □

We shall present another interesting (and deeper) application of Ptolemy's theorem in Sect. 5.2 (cf. Theorem 5.14). On the other hand, a version of it for general convex quadrilaterals will be discussed in Sect. 7.5 (cf. Theorem 7.37).

Fig. 4.18 A useful consequence of Ptolemy's theorem

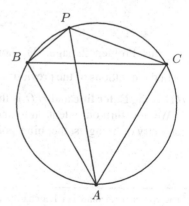

Continuing, we now pay a small tribute to the geniality of Leonhard Euler,[9] gathering two beautiful theorems of his (for a third one, see Theorem 4.32, in the coming section; others appear in [6]).

Theorem 4.20 (Euler) *If O, G and H respectively stand for the circumcenter, barycenter and orthocenter of a triangle ABC, then:*

(a) $\overline{AH} = 2\overline{OM}$, where M is the midpoint of BC.

(b) H, G and O are collinear, with $G \in HO$ and $\overline{HG} = 2\overline{GO}$.

Proof

(a) Letting N be the midpoint of AC (cf. Fig. 4.19), the midsegment theorem (for triangles) gives $MN \parallel AB$ and $\overline{MN} = \frac{1}{2}\overline{AB}$. Hence, $O\widehat{M}N = H\widehat{A}B$ and $O\widehat{N}M = H\widehat{B}A$, so that triangles OMN and HAB are similar, and then

$$\frac{\overline{OM}}{\overline{AH}} = \frac{\overline{MN}}{\overline{AB}} = \frac{1}{2}.$$

(b) If G' denotes the intersection point of line segments AM and HO (it is not difficult to see that these two segments always intersect), then $O\widehat{G'}M = H\widehat{G'}A$. Also, since $\overleftrightarrow{OM} \parallel \overleftrightarrow{AH}$, it follows that $O\widehat{M}G' = H\widehat{A}G'$. Therefore, $MOG' \sim AHG'$ by AA, and this, together with the result of item (a), furnishes

$$\frac{\overline{OG'}}{\overline{HG'}} = \frac{\overline{MG'}}{\overline{AG'}} = \frac{\overline{OM}}{\overline{AH}} = \frac{1}{2}.$$

It comes from the above and Proposition 2.38 that G and G' are points on the line segment AM for which

$$\frac{\overline{AG'}}{\overline{MG'}} = \frac{\overline{AG}}{\overline{MG}};$$

thus, Problem 2, page 108, shows that $G = G'$.   □

In the notations of the previous result, and as anticipated in Fig. 4.19, we say that $\overleftrightarrow{HO}$ is the **Euler line** and HO is the **Euler median** of triangle ABC.

We now turn our attention to another one of Euler's amazing discoveries on the geometry of triangles, the **nine-point circle** or **Euler's circle** of a triangle.

---

[9]The Swiss mathematician Leonhard Euler, who lived in the eighteenth century, is generally accepted to be one of the mathematicians who most published relevant works. His contributions vary, impressively, from Geometry to Combinatorics (in which he created Graph Theory), passing through Number Theory and Physics. In each one of these areas of Mathematics there is at least one celebrated *Euler's theorem*.

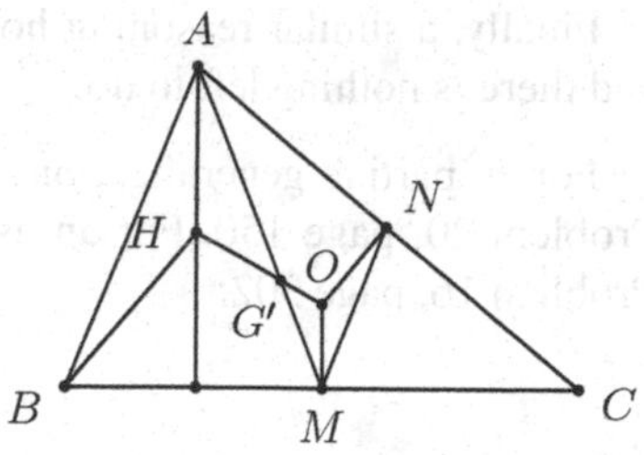

Fig. 4.19 The Euler line $\overleftrightarrow{HO}$ of ABC

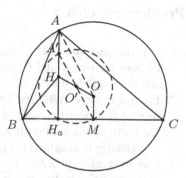

Fig. 4.20 The nine-point circle of triangle ABC

Theorem 4.21 (Euler) *In every triangle, the circumcenter of the orthic triangle coincides with the midpoint of the Euler median. The circumcircle of the orthic triangle is also the circumcircle of the medial triangle and passes through the midpoints of the line segments that join the vertices of the triangle to the orthocenter. Moreover, its radius is equal to half of the circumradius of the given triangle.*

Proof Let ABC be a triangle of circumcenter O and orthocenter H, let M be the midpoint of BC, A' that of AH and R the circumradius of ABC (cf. Fig. 4.20). Since $AA' \parallel OM$ and (by the previous theorem) $\overline{AA'} = \overline{OM}$, Problem 1, page 58, guarantees that $AA'MO$ is a parallelogram. Hence, $\overline{A'M} = \overline{AO} = R$.

On the other hand, since $A'H \parallel OM$ and (also from the previous result) $\overline{A'H} = \overline{OM}$, it follows again from Problem 1, page 58, that $A'HMO$ is a parallelogram. Therefore, its, diagonals cut in half, so that, letting O' be the midpoint of HO, we conclude that O' is also the midpoint of $A'M$. Thus,

$$\overline{O'A'} = \overline{O'M} = \frac{1}{2}\overline{A'M} = \frac{R}{2}.$$

Now, let H_a be the foot of the altitude relative to BC. Since in every right triangle the median relative to the hypotenuse is equal to half of it, in triangle $A'H_aM$ we have

$$\overline{O'H_a} = \frac{1}{2}\overline{A'M} = \frac{R}{2}.$$

In particular, the circle of center O' and radius $\frac{R}{2}$ passes through A', M and H_a.

Finally, a similar reasoning holds for the six remaining points of the statement, and there is nothing left to do.   □

For a partial generalization of the previous result, we refer the reader to Problem 20, page 150. For an asthonishing property of the nine-point circle, see Problem 16, page 302.

Problems: Sect. 4.3

1. Use compass and straightedge to construct triangle ABC, knowing the positions of vertices B and C, the foot of the internal angle bisector relative to A and the length b of side AC.

2. From a triangle ABC we know the positions of vertices B and C and the foot of the external angle bisector relative to A, as well as the length l of the corresponding internal angle bisector. Explain how to use compass and straightedge to construct the position of A.

3. Construct triangle ABC, knowing the lengths a of side BC and m_a of the median relative to BC, if $\overline{AB} = \frac{3}{4}\overline{AC}$.

4. * Let $k \neq 1$ be a positive real. Prove that the Apollonius' circle relative to (B, C) in the ratio k has radius equal to $\frac{k}{|k^2-1|} \cdot \overline{BC}$.

5. (IMO shortlist) $ABCD$ is a convex quadrilateral of diagonals AC and BD, which is inscribed in a circle Γ. Show that there exists $P \in \Gamma$ such that $\overline{PA} + \overline{PC} = \overline{PB} + \overline{PD}$.

6. * Let ABC be a triangle whose internal angles are less than $120°$. Externally to ABC, construct equilateral triangles BCD, ACE and ABF (which are known as **Napoleonic Triangles**[10] of ABC). Prove that:

    (a) The circumcircles of triangles BCD, ACE and ABF all pass through a single point P, which is called the **Fermat point**[11] of ABC.
    (b) $A\widehat{P}F = F\widehat{P}B = B\widehat{P}D = D\widehat{P}C = C\widehat{P}E = E\widehat{P}A = 60°$.
    (c) $\overline{AD} = \overline{BE} = \overline{CF} = \overline{PA} + \overline{PB} + \overline{PC}$.

7. We are given a natural number n and a circle Γ of radius 1. If AB is a diameter of Γ, prove that there exist points $C_1, \ldots, C_n \in \Gamma$ such that $\overline{AC_i}, \overline{BC_i} \in \mathbb{Q}$ for $1 \leq i \leq n$.

8. (IMO - adapted) Given $n \in \mathbb{N}$, show that there exists a circle Γ and points $A_1, \ldots, A_n \in \Gamma$ such that $\overline{A_iA_j} \in \mathbb{N}$ for $1 \leq i < j \leq n$.

---

[10] Some sources attribute the discovery of the facts listed in items (a) and (b) to the French emperor Napoleon Bonaparte, whereas others suggest that this is apocryphal.

[11] Pierre Simon de Fermat, French mathematician of the seventeenth century.

9. (Brazil) In a parallelogram $ABCD$ of diagonals AC and BD, we let H denote the orthocenter of ABD and O the circumcenter of BCD. Prove that points H, O and C are collinear.

10. Use compass and straightedge to construct triangle ABC, given the positions of its circumcenter O, its orthocenter H and of the midpoint M of side BC.

11. In a non equilateral triangle ABC of orthocenter H and circumcenter O, prove that the perpendicular bisector of line segment HO and the internal angle bisector relative to vertex A intersect each other on the circumcircle of triangle AHO.

12. In a cyclic quadrilateral $ABCD$ of diagonals AC and BD, let H_1 and H_2 be the orthocenters of triangles ACD and BCD, respectively. Prove that $H_1 H_2 BA$ is a parallelogram.

13. (IMO shortlist) In a triangle ABC of orthocenter H and circumcenter O, we have $\overline{AO} = \overline{AH}$. Find all possible values for the measure of $\angle BAC$.

14. Use compass and straightedge to construct an acute triangle ABC, given the length a of side BC, the circumradius R of ABC and the inradius r of the orthic triangle of ABC.

4.4 Collinearity and Concurrence

In this section we present two classical results on collinearity and concurrence, as well as some important applications of them. Later, in Chap. 9, we shall see that they are natural departing points for the beginnings of *Projective Geometry*.

Henceforth, we shall adopt the following conventions:

i. Given distinct points X and Y in the plane, XY denotes the ordinary line segment joining X and Y, **oriented** from X to Y. In particular, we shall write $XY = -YX$ as a reminder of the fact that the oriented line segments XY and YX have *distinct*, or *opposite*, *orientations*.

ii. Given collinear points X, Y and Z, we shall denote

$$\frac{XY}{YZ} = \begin{cases} \frac{\overline{XY}}{\overline{YZ}}, & \text{if } XY \text{ and } YZ \text{ have equal orientations} \\ -\frac{\overline{XY}}{\overline{YZ}}, & \text{if } XY \text{ and } YZ \text{ have distinct orientations} \end{cases}.$$

With the notations above at our disposal, we can state and prove the first of the aforementioned results, due to the Greek astronomer and mathematician of the first and second centuries AD Menelaus of Alexandria. It is known in mathematical literature as **Menelaus' theorem** of collinearity.

Theorem 4.22 (Menelaus) *Let ABC be a triangle and A', B' and C' be points on the lines $\overleftrightarrow{BC}$, $\overleftrightarrow{AC}$ and $\overleftrightarrow{AB}$, respectively, all distinct from the vertices of ABC. Then:*

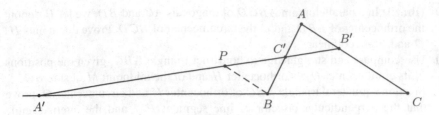

Fig. 4.21 Menelaus' theorem

$$\frac{BA'}{A'C} \cdot \frac{CB'}{B'A} \cdot \frac{AC'}{C'B} = -1 \qquad (4.2)$$

if and only if A', B' and C' are collinear.

Proof Initially assume (cf. Fig. 4.21) that $A' \in \overrightarrow{CB} \setminus BC$, $B' \in AC$, $C' \in AB$ and that A', B' and C' are collinear (the remaining cases are entirely analogous). Mark point $P \in \overleftrightarrow{A'B'}$ such that $\overleftrightarrow{BP} \parallel \overleftrightarrow{AC}$. Then, $A'BP \sim A'CB'$ and $PBC' \sim B'AC'$, so that

$$\frac{BA'}{A'C} = -\frac{\overline{BP}}{\overline{CB'}} \quad \text{and} \quad \frac{AC'}{C'B} = \frac{\overline{AB'}}{\overline{BP}}.$$

Therefore,

$$\frac{BA'}{A'C} \cdot \frac{CB'}{B'A} \cdot \frac{AC'}{C'B} = -\frac{\overline{BP}}{\overline{CB'}} \cdot \frac{\overline{CB'}}{\overline{B'A}} \cdot \frac{\overline{AB'}}{\overline{BP}} = -1.$$

Conversely, let A', B' and C' be points situated along lines $\overleftrightarrow{BC}$, $\overleftrightarrow{CA}$ and $\overleftrightarrow{AB}$, respectively, all distinct from the vertices of ABC and such that relation (4.2) holds. Mark the intersection point B'' of lines $\overleftrightarrow{A'C'}$ and $\overleftrightarrow{AC}$. Since A', B'' and C' are collinear, the first part of the proof guarantees that

$$\frac{BA'}{A'C} \cdot \frac{CB''}{B''A} \cdot \frac{AC'}{C'B} = -1.$$

By comparing such a relation with (4.2), we conclude that $\frac{CB'}{B'A} = \frac{CB''}{B''A}$; from this, and with the aid of the result of Problem 2, page 108, it is easy to conclude that $B' = B''$. Thus, A', B' and C' are collinear.   □

The coming example brings a typical application of Menelaus' theorem. For a wide generalization, see items (c) and (d) of Problem 5.

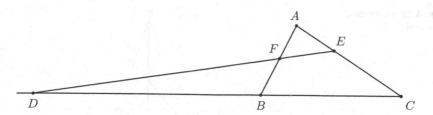

Fig. 4.22 Collinearity of the feet of certain angle bisectors

Example 4.23 Let ABC be a triangle in which $\overline{AB} \neq \overline{AC}$. If D is the foot of the external angle bisector relative to A, and E and F are the feet of the internal angle bisectors relative to B and C, prove that points D, E and F are collinear.

Proof Suppose, without loss of generality, that $\overline{AB} < \overline{AC}$, and hence that vertex B lies inside segment CD (cf. Fig. 4.22).

By Menelaus' theorem, it suffices to show that

$$\frac{AF}{FB} \cdot \frac{BD}{DC} \cdot \frac{CE}{EA} = -1.$$

To this end, the bisector theorem gives

$$\frac{AF}{FB} = \frac{\overline{AC}}{\overline{CB}}, \quad \frac{BD}{DC} = -\frac{\overline{AB}}{\overline{AC}}, \quad \text{and} \quad \frac{CE}{EA} = \frac{\overline{BC}}{\overline{AB}}.$$

Therefore, termwise multiplication of the three expressions above gives the desired equality. □

As an additional application of Menelaus' theorem we prove, next, a famous result of Girard Desargues, French mathematician of the seventeenth century, known as the founder of modern Projective Geometry.

Theorem 4.24 (Desargues) *If ABC and $A'B'C'$ are triangles such that $\overleftrightarrow{AB}$ $\cap$ $\overleftrightarrow{A'B'} = \{Z\}$, $\overleftrightarrow{BC} \cap \overleftrightarrow{B'C'} = \{X\}$ and $\overleftrightarrow{AC} \cap \overleftrightarrow{A'C'} = \{Y\}$, then points X, Y and Z are collinear if and only if lines $\overleftrightarrow{AA'}$, $\overleftrightarrow{BB'}$ are $\overleftrightarrow{CC'}$ either concurrent or parallel.*

Proof Suppose first that X, Y and Z are collinear (cf. Fig. 4.23). In order to show that lines $\overleftrightarrow{AA'}$, $\overleftrightarrow{BB'}$ and $\overleftrightarrow{CC'}$ are either concurrent or parallel, it suffices to assume that $\overleftrightarrow{AA'}$ and $\overleftrightarrow{BB'}$ concur at a point O, then showing that $O \in \overleftrightarrow{CC'}$, i.e., that O, C and C' are collinear. This will follow from Menelaus' theorem applied to triangle $AA'Y$, once we show that

$$\frac{YC'}{C'A'} \cdot \frac{A'O}{OA} \cdot \frac{AC}{CY} = -1. \tag{4.3}$$

Fig. 4.23 Desargues'
theorem

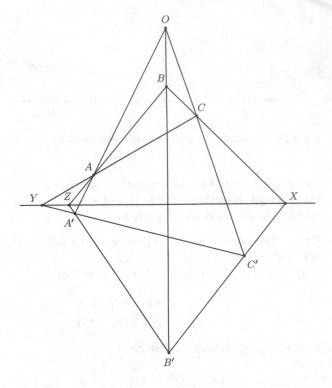

For what is left to do, apply Menelaus' theorem to triangles ZAA', YZA and
YZA', thus obtaining

$$\frac{ZB'}{B'A'} \cdot \frac{A'O}{OA} \cdot \frac{AB}{BZ} = -1, \quad \frac{YZ}{XZ} \cdot \frac{ZB}{BA} \cdot \frac{AC}{CY} = -1$$

and

$$\frac{YC'}{C'A'} \cdot \frac{A'B'}{B'Z} \cdot \frac{ZX}{XY} = -1.$$

By termwise multiplying the three relations above and taking into account that

$$\frac{YX}{XZ} \cdot \frac{ZX}{XY} = \frac{AB}{BZ} \cdot \frac{ZB}{BA} = \frac{ZB'}{B'A'} \cdot \frac{A'B'}{B'Z} = 1,$$

we obtain (4.3).

The proof of the converse is analogous and will be left as an exercise to the reader
(cf. Problem 1). □

In the language of visual arts, if X, Y and Z are collinear one says that triangles
ABC and $A'B'C'$ are **in perspective from a line** (the line of X, Y and Z), which

is then called the **horizon**; on the other hand, if lines $\overleftrightarrow{AA'}$, $\overleftrightarrow{BB'}$ and $\overleftrightarrow{CC'}$ are either concurrent or parallel, one says that ABC and $A'B'C'$ are **in perspective from a point**, namely, the intersection point of lines $\overleftrightarrow{AA'}$, $\overleftrightarrow{BB'}$ and $\overleftrightarrow{CC'}$ (which is then called the **vanishing point**), in case they concur, or the **point at infinity** in the direction of $\overleftrightarrow{AA'}$, $\overleftrightarrow{BB'}$ and $\overleftrightarrow{CC'}$, in case they are parallel. This being said, we can state Desargues' theorem in words by saying that, in the plane *two triangles are in perspective from a line if and only if they are in perspective from a point*. Such a point of view will be taken up again and deepened in Sect. 9.4.

Continuing, we now present a theorem analogous to that of Menelaus' but this time dealing with the concurrence of lines joining each vertex of a given triangle to a point lying on the line containing the opposite side. Such result is due to the Italian mathematician of seventeenth and eighteenth centuries Giovanni Ceva, and is known as **Ceva's theorem**.

Theorem 4.25 (Ceva) *Given a triangle ABC and points A', B', C', respectively situated on lines $\overleftrightarrow{BC}$, $\overleftrightarrow{CA}$ and $\overleftrightarrow{AB}$, we have*

$$\frac{BA'}{A'C} \cdot \frac{CB'}{B'A} \cdot \frac{AC'}{C'B} = 1 \tag{4.4}$$

if and only if $\overleftrightarrow{AA'}$, $\overleftrightarrow{BB'}$, $\overleftrightarrow{CC'}$ are either concurrent or parallel.

Proof Suppose first (cf. Fig. 4.24), that points $A' \in \overrightarrow{BC} \setminus BC$, $B' \in AC$ and $C' \in \overrightarrow{BA} \setminus AB$ are so situated that $\overleftrightarrow{AA'}$, $\overleftrightarrow{BB'}$ and $\overleftrightarrow{CC'}$ concur at a point P (the remaining cases are entirely analogous). Mark points $Q \in \overleftrightarrow{AA'}$ and $R \in \overleftrightarrow{CC'}$ such that $\overleftrightarrow{CQ}$, $\overleftrightarrow{AR} \| BP$. Then, $BPA' \sim CQA'$, $ARC' \sim BPC'$ and $CB'P \sim CAR$, so that

$$\frac{BA'}{A'C} = -\frac{\overline{BP}}{\overline{CQ}}, \quad \frac{AC'}{C'B} = -\frac{\overline{AR}}{\overline{BP}}$$

Fig. 4.24 Ceva's theorem

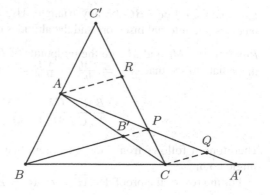

and

$$\frac{CB'}{B'A} = \frac{\overline{CB'}}{\overline{B'A}} = \frac{\overline{CA} - \overline{B'A}}{\overline{B'A}} = \frac{\overline{CA}}{\overline{B'A}} - 1 = \frac{\overline{CQ}}{\overline{B'P}} - 1.$$

Hence,

$$\frac{BA'}{A'C} \cdot \frac{CB'}{B'A} \cdot \frac{AC'}{C'B} = \frac{\overline{BP}}{\overline{CQ}}\left(\frac{\overline{CQ}}{\overline{B'P}} - 1\right)\frac{\overline{AR}}{\overline{BP}} = \overline{AR}\left(\frac{1}{\overline{B'P}} - \frac{1}{\overline{CQ}}\right),$$

and it suffices to show that the last expression above is equal to 1, i.e., that

$$\frac{1}{\overline{B'P}} = \frac{1}{\overline{AR}} + \frac{1}{\overline{CQ}}.$$

But this is exactly what says Problem 11, page 116.

Finally, we leave as an exercise for the reader (cf. Problem 2) the task of showing that if $\overleftrightarrow{AA'}$, $\overleftrightarrow{BB'}$ and $\overleftrightarrow{CC'}$ are parallel, then (4.4) also holds.

Conversely, suppose that (4.4) holds true and mark point $B'' \in \overleftrightarrow{AC}$ such that $\overleftrightarrow{AA'}$, $\overleftrightarrow{BB''}$ and $\overleftrightarrow{CC'}$ are either concurrent or parallel. The first part above gives

$$\frac{BA'}{A'C} \cdot \frac{CB'}{B'A} \cdot \frac{AC'}{C'B} = 1,$$

and comparing such a relation with 4.4 we obtain $\frac{CB'}{B'A} = \frac{CB''}{B''A}$. From here, and invoking once more the result of Problem 2, page 108, we immediately conclude that $B' = B''$. Therefore, $\overleftrightarrow{AA'}$, $\overleftrightarrow{BB'}$ and $\overleftrightarrow{CC'}$ are either concurrent or parallel.   □

For another proof of Ceva's theorem in the case in which $\overleftrightarrow{AA'}$, $\overleftrightarrow{BB'}$ and $\overleftrightarrow{CC'}$ concur, see Problem 16, page 157. For the time being, we shall show how it allows us to give alternative proofs for the concurrence of the medians, internal bisectors and altitudes of a triangle.

Example 4.26 Let ABC be any triangle. Use Ceva's theorem to prove that the medians, the internal bisectors and the altitudes of ABC concur.

Proof If M_a, M_b and M_c are the midpoints of BC, AC and AB, respectively, it is immediate to see that $\frac{BM_a}{M_aC} = \frac{CM_b}{M_bA} = \frac{AM_c}{M_cB} = 1$, whence

$$\frac{BM_a}{M_aC} \cdot \frac{CM_b}{M_bA} \cdot \frac{AM_c}{M_cB} = 1.$$

Therefore, it follows from Ceva's theorem that AM_a, BM_b and CM_c concur at a single point.

For the rest of the proof, let $\overline{AB} = c$, $\overline{AC} = b$ and $\overline{BC} = a$.

Let P_a, P_b and P_c denote the feet of the internal angle bisectors relative to A, B and C, respectively. The bisector theorem (cf. Theorem 4.6) gives $\frac{AP_c}{P_cB} = \frac{b}{a}$, $\frac{BP_a}{P_aC} = \frac{c}{b}$ and $\frac{CP_b}{P_bA} = \frac{a}{c}$, so that

$$\frac{AP_c}{P_cB} \cdot \frac{BP_a}{P_aC} \cdot \frac{CP_b}{P_bA} = \frac{b}{a} \cdot \frac{c}{b} \cdot \frac{a}{c} = 1.$$

Therefore, Ceva's theorem assures that AP_a, BP_b and CP_c all pass through a single point.

For the concurrence of the altitudes of ABC, we shall only consider the case in which ABC is acute, leaving the others as exercises to the reader (cf. Problem 3). This being said, let H_a, H_b and H_c stand for the feet of the altitudes relative to A, B and C, respectively. Problem 2, page 98, gives $AH_bH_c \sim ABC$, so that $\frac{\overline{AH_c}}{\overline{H_bA}} = \frac{\overline{AC}}{\overline{AB}} = \frac{b}{c}$; analogously, $\frac{\overline{BH_a}}{\overline{H_cB}} = \frac{c}{a}$ and $\frac{\overline{CH_b}}{\overline{H_aC}} = \frac{a}{b}$. Thus,

$$\frac{AH_c}{H_cB} \cdot \frac{BH_a}{H_aC} \cdot \frac{CH_b}{H_bA} = \frac{\overline{AH_c}}{\overline{H_bA}} \cdot \frac{\overline{BH_a}}{\overline{H_cB}} \cdot \frac{\overline{CH_b}}{\overline{H_aC}} = \frac{b}{c} \cdot \frac{c}{a} \cdot \frac{a}{b} = 1,$$

and it suffices to invoke Ceva's theorem once more. $\qquad\square$

Theorem 4.28 will present a beautiful application of Ceva's theorem. In order to properly state it, we define a **cevian** of a triangle as any line segment (or the corresponding line or half-line) joining a vertex of the triangle to a point on the line containing the opposite side; for instance, medians, angle bisectors (internal or external) and altitudes of a triangle are all examples of cevians of it.

Cevians AP and AP' of triangle ABC are said to be **isogonal** (with respect to A) if lines $\overleftrightarrow{AP}$ and $\overleftrightarrow{AP'}$ are symmetric with respect to the internal angle bisector relative to A. In the notations of Fig. 4.25, AP and AP' are isogonal if and only if $P\widehat{A}Q = P'\widehat{A}Q$.

It is immediate to see that for every cevian AP of a triangle ABC there exists a single cevian AP' of ABC such that AP and AP' are isogonal. In this case, we shall also say that AP' (resp. AP) is the cevian **isogonal** to AP (resp. to AP'). The coming lemma establishes a useful characterization of the isogonality of two cevians (departing from the same vertex, of course).

Fig. 4.25 Isogonal cevians

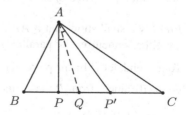

Fig. 4.26 A useful criterion
for isogonality

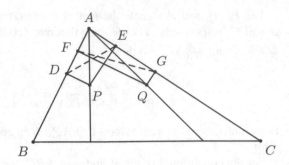

Lemma 4.27 *In the notations of Fig. 4.26, if D and E (resp. F and G) are the feet of the perpendiculars dropped from P (resp. Q) to the lines $\overleftrightarrow{AB}$ and $\overleftrightarrow{AC}$, then*

$$\overrightarrow{AP} \quad and \quad \overrightarrow{AQ} \quad are\ isogonal \;\Leftrightarrow\; \frac{\overline{PD}}{\overline{PE}} = \frac{\overline{QG}}{\overline{QF}}.$$

Proof In the notations of Fig. 4.26, start by noticing that quadrilaterals $ADPE$ and $AFQG$ are cyclic (for each one of them has two opposite angles equal to 90°). Therefore, $D\widehat{P}E = G\widehat{Q}F = 180° - B\widehat{A}C$, so that successively applying the SAS and AA similarity cases we conclude that

$$\frac{\overline{PD}}{\overline{PE}} = \frac{\overline{QG}}{\overline{QF}} \;\Leftrightarrow\; DPE \sim GQF \;\Leftrightarrow\; D\widehat{E}P = G\widehat{F}Q.$$

However, by using again the fact that the quadrilaterals $ADPE$ and $AFQG$ are cyclic, we conclude that the last equality above holds if and only if $D\widehat{A}P = G\widehat{A}Q$; in turn, this last equality is equivalent to the fact that $\overrightarrow{AP}$ and $\overrightarrow{AQ}$ are isogonal with respect to A.   □

We can finally state and prove the desired consequence of Ceva's theorem, which establishes the concurrence of the cevians isogonal to three concurrent given cevians.

Theorem 4.28 *Let AA', BB' and CC' be cevians of a triangle ABC, with isogonals AA'', BB'' and CC'', respectively. Lines $\overleftrightarrow{AA'}$, $\overleftrightarrow{BB'}$, $\overleftrightarrow{CC'}$ are concurrent or parallel if and only if so are lines $\overleftrightarrow{AA''}$, $\overleftrightarrow{BB''}$, $\overleftrightarrow{CC''}$.*

Proof We shall show that if $\overleftrightarrow{AA'}$, $\overleftrightarrow{BB'}$ and $\overleftrightarrow{CC'}$ concur, then $\overleftrightarrow{AA''}$, $\overleftrightarrow{BB''}$ and $\overleftrightarrow{CC''}$ are either concurrent or parallel (for the remaining case, see Problem 8, page 137. To this end, let P be the point of concurrence of $\overleftrightarrow{AA'}$, $\overleftrightarrow{BB'}$ and $\overleftrightarrow{CC'}$, and x, y and z be the distances from P to the lines $\overleftrightarrow{BC}$, $\overleftrightarrow{AC}$ and $\overleftrightarrow{AB}$, respectively (cf. Fig. 4.27).

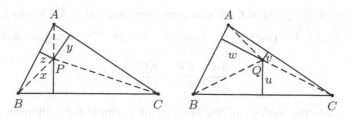

Fig. 4.27 Concurrence of isogonal cevians

It suffices to show that if $\overleftrightarrow{BB''}$ and $\overleftrightarrow{CC''}$ intersect at a point Q, then $\overleftrightarrow{AA''}$ also passes through Q. To this end, let u, v and w be the distances from Q to the lines $\overleftrightarrow{BC}$, $\overleftrightarrow{AC}$ and $\overleftrightarrow{AB}$, respectively. Since $\overleftrightarrow{BB'}$, $\overleftrightarrow{BB''}$ and $\overleftrightarrow{CC'}$, $\overleftrightarrow{CC''}$ are pairs of isogonal cevians, the previous lemma gives

$$\frac{x}{y} = \frac{v}{u} \quad \text{and} \quad \frac{z}{x} = \frac{u}{w}.$$

Hence,

$$\frac{z}{y} = \frac{z}{x} \cdot \frac{x}{y} = \frac{u}{w} \cdot \frac{v}{u} = \frac{v}{w},$$

so that, again from the previous lemma, points P and Q lie on cevians isogonal with respect to A. However, since $P \in \overleftrightarrow{AA'}$, we conclude that $Q \in \overleftrightarrow{AA''}$. $\qquad\square$

In the notations of the statement of the theorem above, if P (resp. Q) is the point of concurrence of cevians AA', BB' and CC' (resp. AA'', BB'' and CC''), we say that P and Q are **isogonal conjugates**, or that P (resp. Q) is the **isogonal conjugate** of Q (resp. P) with respect to ABC. For a relevant example, see Problem 9; another interesting example will be discussed by the end of Sect. 5.2. For another proof of Theorem 4.28, see Problem 21, page 263.

Problems: Sect. 4.4

1. * Complete the proof of Desargues' theorem. More precisely, show that if ABC and $A'B'C'$ are two triangles such that $\overleftrightarrow{AB} \cap \overleftrightarrow{A'B'} = \{Z\}$, $\overleftrightarrow{BC} \cap \overleftrightarrow{B'C'} = \{X\}$ and $\overleftrightarrow{AC} \cap \overleftrightarrow{A'C'} = \{Y\}$, and if AA', BB' and CC' are either concurrent or parallel, then X, Y and Z are collinear.

2. * Complete the proof of Ceva's theorem, showing that if A', B' and C' lie on $\overleftrightarrow{BC}$, $\overleftrightarrow{AC}$ and $\overleftrightarrow{AB}$, respectively, with $\overleftrightarrow{AA'}$, $\overleftrightarrow{BB'}$ and $\overleftrightarrow{CC'}$ being parallel, then

$$\frac{BA'}{A'C} \cdot \frac{CB'}{B'A} \cdot \frac{AC'}{C'B} = 1.$$

3. * Complete the analysis of the last part of Example 4.26, showing that the altitudes of an obtuse triangle pass through a single point.

4. * Let ABC be a scalene triangle. Use Ceva's theorem to show that the external angle bisectors relative to the vertices A and B concur with the internal angle bisector relative to C.

5. * We are given points P and Q of line $\overleftrightarrow{BC}$, such that $\{P, Q\} \cap \{B, C\} = \emptyset$. We say that P and Q (in this order) are **harmonic conjugates**[12] with respect to B and C (also in this order) if

$$\frac{BP}{PC} = -\frac{BQ}{QC}.$$

Concerning this notion, do the following items:

(a) If P and Q are harmonic conjugates with respect to B and C, then exactly one of the points P and Q lies in the segment BC, and both P and Q are distinct from the midpoint M of BC.

(b) For every point $P \in \overleftrightarrow{BC}$, with $P \neq B, C, M$ there exists a single point $Q \in \overleftrightarrow{BC}$ such that $Q \neq B, C$ and P and Q are harmonic conjugates with respect to B and C (thanks to this item, we say that Q (resp. P) is the **harmonic conjugate** of P (resp. of Q) with respect to B and C).

(c) In a triangle ABC with $\overline{AB} \neq \overline{AC}$, let P be the foot of the internal angle bisector and Q that of the external angle bisector relative to A. Show that P and Q are harmonic conjugates with respect to B and C.

(d) In a triangle ABC, let P, Q and R be points on $\overleftrightarrow{BC}$, $\overleftrightarrow{CA}$ and $\overleftrightarrow{AB}$, respectively, all distinct from the vertices of ABC. If P' is the harmonic conjugate of P with respect to B and C, prove that $\overleftrightarrow{AP}$, $\overleftrightarrow{BQ}$ and $\overleftrightarrow{CR}$ concur if and only if P', Q and R are collinear.

6. Prove that, in every triangle, the cevians joining each vertex to the point of tangency of the incircle with the opposite side are concurrent. Their common point is called the **Gergonne point**[13] of triangle ABC.

---

[12] Such a concept will be taken up again, in a thorough way, in Sect. 9.2.

[13] Joseph Gergonne, French mathematician of the nineteenth century.

7. Prove that, in every triangle, the cevians joining each vertex to the point of tangency of the excircle relative to that vertex with the opposite side are concurrent. Their common point is called the **Nagel point**[14] of triangle ABC.

8. * Complete the proof of Theorem 4.28. More precisely, show that if AA', BB' and CC' are parallel cevians of a triangle ABC, with isogonals AA'', BB'' and CC'', respectively, then $\overleftrightarrow{AA''}$, $\overleftrightarrow{BB''}$ and $\overleftrightarrow{CC''}$ are either concurrent or parallel.

9. * Prove that, in every triangle, the orthocenter and the circumcenter are isogonal conjugates.

10. * Generalize the previous problem, showing the following result: given a triangle ABC and a point P inside it, let Q and R be the feet of the perpendiculars dropped from P to the sides AB and AC; show that the isogonal conjugate to $\overrightarrow{AP}$ is the half-line departing from A and perpendicular to $\overleftrightarrow{QR}$.

11. Let ABC be a triangle and A', $A'' \in \overleftrightarrow{BC}$, B', $B'' \in \overleftrightarrow{AC}$, C', $C'' \in \overleftrightarrow{AB}$ be such that the midpoints of $A'A''$, $B'B''$, $C'C''$ coincide with the midpoints of BC, AC, AB, respectively. If AA', BB' and CC' are concurrent (resp. A', B' and C' are collinear), prove that AA'', BB'' and CC'' are also concurrent (resp. A'', B'' and C'' are collinear).

12. Prove **Pappus' theorem**[15]: we are given the triples of collinear points A, B, C and A', B', C'. If $\overleftrightarrow{AB'} \cap \overleftrightarrow{A'B} = \{F\}$, $\overleftrightarrow{AC'} \cap \overleftrightarrow{A'C} = \{E\}$ and $\overleftrightarrow{BC'} \cap \overleftrightarrow{B'C} = \{D\}$, then the points D, E and F are collinear.

13. (USA) Let ABC be a scalene triangle and r, s and t be the tangents to its circumcircle at vertices A, B and C, respectively. If P, Q and R denote the intersection points of r, s and t with $\overleftrightarrow{BC}$, $\overleftrightarrow{AC}$ and $\overleftrightarrow{AB}$, respectively, prove that P, Q and R are collinear.

14. In a scalene triangle ABC we draw altitude AH_a and, then, drop perpendiculars H_aD and H_aE to $\overleftrightarrow{AB}$ and $\overleftrightarrow{AC}$, respectively (with $D \in AB$ and $E \in \overleftrightarrow{AC}$); we then mark point P as the intersection point of $\overleftrightarrow{DE}$ and $\overleftrightarrow{BC}$. From the altitudes relative to the vertices B and C we construct, in a similar way, points $Q \in \overleftrightarrow{AC}$ and $R \in \overleftrightarrow{AB}$. Show that points P, Q and R are collinear.

15. Prove the following theorem of Monge[16]: we are given pairwise exterior circles Γ_1, Γ_2, Γ_3, of pairwise distinct radii and noncollinear centers. If X_i is the intersection point of the common external tangents to Γ_j and Γ_k, for $\{i, j, k\} = \{1, 2, 3\}$, then X_1, X_2, X_3 are collinear points.

---

[14]Christian Heinrich von Nagel, German mathematician of the nineteenth century.

[15]Pappus of Alexandria, Greek mathematician of the fourth century. For another proof of Pappus' theorem, see Problem 8, page 329.

[16]Gaspard Monge, French mathematician of the eighteenth and nineteenth century, gave several important contributions to Geometry, particularly to Differential Geometry.

4.5   The Theorem of Intersecting Chords

In this section, we continue to explore the elementary consequences of the notion of similarity of triangles. We begin by presenting the **theorem of intersecting chords**, which, in turn, will allow us to get a number of interesting and important results.

Proposition 4.29 *Given pairwise distinct points A, B, C, D and P, with* $\overleftrightarrow{AB}$ $\cap \overleftrightarrow{CD} = \{P\}$*, one has*

$$\overline{PA} \cdot \overline{PB} = \overline{PC} \cdot \overline{PD} \Leftrightarrow \text{ the convex quadrilateral of vertices}$$

$$A, B, C, D \text{ is cyclic.}$$

Proof Firstly, suppose that the convex quadrilateral of vertices A, B, C and D is cyclic, with circumscribed circle Γ. In principle, we have to separately consider the cases in which P lies inside or outside the disk bounded by Γ; nevertheless, the analysis of the second case is entirely analogous to that of the first, so we shall consider only this one (cf. Fig. 4.28).

Draw line segments AD and BC. The inscribed angle theorem gives $A\widehat{B}C = A\widehat{D}C$, or also $P\widehat{B}C = A\widehat{D}P$. Since $B\widehat{P}C = A\widehat{P}D$ (for they are OPP angles), the AA similarity case gives $PBC \sim PDA$. Therefore, we have $\frac{PB}{PC} = \frac{PD}{PA}$, and hence $\overline{PA} \cdot \overline{PB} = \overline{PC} \cdot \overline{PD}$.

Conversely, if $\overline{PA} \cdot \overline{PB} = \overline{PC} \cdot \overline{PD}$, then $\frac{PB}{PC} = \frac{PD}{PA}$. However, since $B\widehat{P}C = A\widehat{P}D$, the SAS similarity case assures that $PBC \sim PDA$, whence $P\widehat{B}C = A\widehat{D}P$. But this is the same as $A\widehat{B}C = A\widehat{D}C$, and Proposition 3.38 guarantees that $ABCD$ is cyclic.   □

The coming result can be seen as a limit situation of the previous one, when the point P lies outside the disk bounded by Γ. Therefore, its proof will be left as an exercise for the reader (cf. Problem 1).

Fig. 4.28 The theorem of intersecting chords

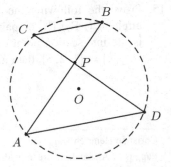

Fig. 4.29 Limit case of the intersecting chords theorem

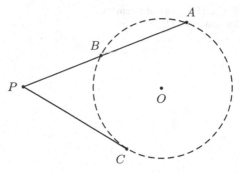

Fig. 4.30 Computing $\overline{PA} \cdot \overline{PB}$ in terms of $\overline{PO}$ and R

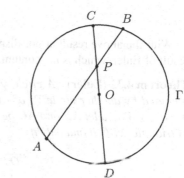

Proposition 4.30 *Let A, B, C and P be pairwise distinct points in the plane, with $B \in AP$ and $C \notin \overleftrightarrow{AB}$. Then, $\overline{PA} \cdot \overline{PB} = \overline{PC}^2$ if and only if the circumcircle of triangle ABC is tangent to $\overleftrightarrow{PC}$ at C (cf. Fig. 4.29).*

For future use, we now collect an important consequence of the intersecting chords theorem.

Corollary 4.31 *We are given in the plane a circle $\Gamma(O; R)$ and a point $P \notin \Gamma$. If a line passing through P intersects Γ at points A and B (possibly with $A = B$, if P lies outside the disk bounded by Γ and the line is tangent to Γ at A), then*

$$\overline{PA} \cdot \overline{PB} = |R^2 - \overline{OP}^2|. \tag{4.5}$$

Proof We only consider the case in which P lies inside the disk bounded by Γ (cf. Fig. 4.30); the case of P lying outside such a disk is completely analogous. Draw through P the diameter CD of Γ, with $P \in OC$. Then, $\overline{PC} = R - \overline{OP}$ and $\overline{PD} = R + \overline{OP}$, so that the intersecting chords theorem gives

$$\overline{PA} \cdot \overline{PB} = \overline{PC} \cdot \overline{PD} = (R - \overline{OP})(R + \overline{OP}) = R^2 - \overline{OP}^2.$$

$\square$

Fig. 4.31 Euler's theorem on
the distance $\overline{OI}$

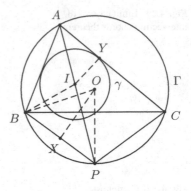

With the above result at our disposal, we can now present yet another beautiful result of Euler, which is the content of the coming

Theorem 4.32 (Euler) *A circle γ, of radius r and center I, lies inside the disk bounded by another circle Γ, of radius R and center O. We choose an arbitrary point $A \in \Gamma$ and let AB and AC be chords of Γ tangent to γ. Then, γ is the incircle of triangle ABC if and only if*

$$\overline{OI}^2 = R(R - 2r).$$

Proof Let P be the other point of intersection (i.e., distinct from A) of the angle bisector $\overrightarrow{AI}$ of $\angle BAC$ with Γ (cf. Fig. 4.31). The previous corollary gives

$$\overline{AI} \cdot \overline{IP} = R^2 - \overline{OI}^2. \tag{4.6}$$

Now, letting X and Y be the feet of the perpendiculars dropped from O and I to BP and AC, respectively, the fact that BOP is isosceles, together with the inscribed angle theorem, gives

$$B\widehat{O}X = \frac{1}{2}B\widehat{O}P = B\widehat{A}P = P\widehat{A}C = I\widehat{A}Y.$$

Since both triangles BOX and IAY have a 90° angle, the AA similarity case assures that $BOX \sim IAY$. Hence, $\frac{BX}{IY} = \frac{BO}{AI}$, or

$$\overline{BX} \cdot \overline{AI} = \overline{BO} \cdot \overline{IY}. \tag{4.7}$$

However, since $\overline{BO} = R$, $\overline{IY} = r$ and $\overline{BX} = \frac{1}{2}\overline{BP}$, relations (4.6) and (4.7) give

$$R^2 - \overline{OI}^2 = \overline{AI} \cdot \overline{IP} = 2Rr \cdot \frac{\overline{IP}}{\overline{BP}},$$

so that

$$\overline{OI}^2 = R^2 - 2Rr \Leftrightarrow \overline{BP} = \overline{IP}.$$

Finally, Proposition 3.37 guarantees that the last equality above holds if and only if I is the incenter of ABC.   □

We list two important corollaries of the above result, the first of which is immediate.

Corollary 4.33 *If r and R denote the inradius and the circumradius of a triangle ABC, then $R \geq 2r$, with equality if and only if ABC is equilateral.*

The second consequence is a particular case, also due to Euler, of a famous result of Poncelet on conics.[17]

Corollary 4.34 (Euler) *Let γ and Γ be respectively the incircle and the circumcircle of a triangle ABC. If $A' \neq A, B, C$ is another point of Γ, and $A'B'$ and $A'C'$ are chords of Γ tangent to γ, then γ is also the incircle of triangle $A'B'C'$ (cf. Fig. 4.32).*

Proof If $\gamma(I; r)$ and $\Gamma(O; R)$, the fact that γ is the incircle of ABC guarantees, via Euler's theorem 4.32, that $\overline{OI}^2 = R^2 - 2Rr$. In view of this equality, the referred theorem applied this time to triangle $A'B'C'$ assures that $B'C'$ is tangent to γ, as we wished to show.   □

Back to the general development of the theory and motivated by Corollary 4.31, we define the **power** of point P with respect to circle $\Gamma(O; R)$ as the real number

$$\mathrm{Pwr}_\Gamma(P) = \overline{OP}^2 - R^2. \qquad (4.8)$$

Fig. 4.32 Yet another one of
Euler's theorem

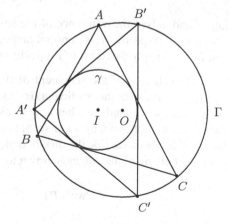

---

[17]Jean Victor Poncelet, French mathematician of the nineteenth century. For a proof of this general version of Poncelet, see Chapter 4 of the beautiful book [23].

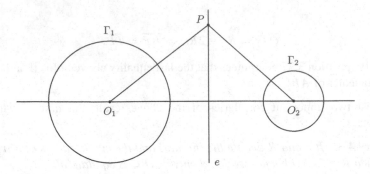

Fig. 4.33 The radical axis of two non concentric circles

Thus, $\mathrm{Pwr}_\Gamma(P) = 0$ if and only if $P \in \Gamma$, $\mathrm{Pwr}_\Gamma(P) > 0$ if and only if P lies outside the disk bounded by Γ and $\mathrm{Pwr}_\Gamma(P) < 0$ if and only if P lies inside such a disk. Note also that $\mathrm{Pwr}_\Gamma(P) \geq -R^2$, with equality holding if and only if $P = O$.

The importance of the concept of power of a point with respect to a circle lies in the coming theorem, whose proof will be completed in Sect. 6.2 (cf. Proposition 6.8).

Theorem 4.35 *If Γ_1 and Γ_2 are two non concentric circles, then the locus of the points P in the plane for which $\mathrm{Pwr}_{\Gamma_1}(P) = \mathrm{Pwr}_{\Gamma_2}(P)$ is a line perpendicular to the line joining the centers of Γ_1 and Γ_2 (line e, in Fig. 4.33).*

Proof If $\Gamma_1(O_1; R_1)$ and $\Gamma_2(O_2; R_2)$, then

$$\mathrm{Pwr}_{\Gamma_1}(P) = \mathrm{Pwr}_{\Gamma_2}(P) \Leftrightarrow \overline{PO_1}^2 - R_1^2 = \overline{PO_2}^2 - R_2^2$$
$$\Leftrightarrow \overline{PO_1}^2 - \overline{PO_2}^2 = R_1^2 - R_2^2, \tag{4.9}$$

i.e., if and only if the difference of the squares of the distances of P to the points O_1 and O_2, respectively, is constant and equal to $R_1^2 - R_2^2$. To conclude the proof, it suffices to apply the result of Proposition 6.8.   □

In the notations of the statement of the previous result, the described locus is usually referred to as the **radical axis**[18] of Γ_1 and Γ_2.

The coming example teaches how to use compass and straightedge to construct the radical axis of two tangent or secant circles. It is based on the following simple fact: if P is a point outside the disk bounded by a circle $\Gamma(O; R)$, and T is the point of contact of one of the tangents drawn to Γ through P, then Corollary 4.31 gives

$$\mathrm{Pwr}_\Gamma(P) = \overline{PO}^2 - R^2 = \overline{PT}^2. \tag{4.10}$$

---

[18]Such a name is due to the fact that, in the notations of the proof of Theorem 4.35, $P \in e$ if and only if $\sqrt{\overline{PO_1}^2 + R_2^2} = \sqrt{\overline{PO_2}^2 + R_1^2}$, so that we have *equal radicals*.

Fig. 4.34 Radical axis e of two externally tangent circles

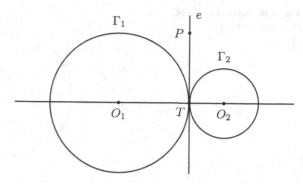

Fig. 4.35 Radical axis e of two externally tangent circles

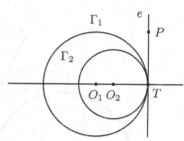

Example 4.36 Use compass and straightedge to construct the radical axis of two circles which are either tangent or secant.

Solution There are three different cases to consider:

(i) Γ_1 and Γ_2 are externally tangent: their radical axis is the inner common tangent e, shown in Fig. 4.34. Indeed, for every point $P \in e\backslash\{T\}$, it follows from (4.10) that

$$\mathrm{Pwr}_{\Gamma_1}(P) = \overline{PT}^2 = \mathrm{Pwr}_{\Gamma_2}(P).$$

(ii) Γ_1 and Γ_2 are internally tangent: their radical axis is their common tangent e, shown in Fig. 4.35. As in case (i), for every point $P \in e \setminus \{T\}$, (4.10) gives

$$\mathrm{Pwr}_{\Gamma_1}(P) = \overline{PT}^2 = \mathrm{Pwr}_{\Gamma_2}(P).$$

(iii) Γ_1 and Γ_2 are secant circles, intersecting at points A and B: their radical axis is the line $e = \overleftrightarrow{AB}$ of Fig. 4.36, since for every point $P \in e \setminus AB$ we have

$$\mathrm{Pwr}_{\Gamma_1}(P) = \overline{PA} \cdot \overline{PB} = \mathrm{Pwr}_{\Gamma_2}(P).$$

In order to show how to use compass and straightedge to construct the radical axis of two non concentric circles which are either interior or exterior, we shall first need to establish the following consequence of Theorem 4.35.

Fig. 4.36 Radical axis e of two secant circles

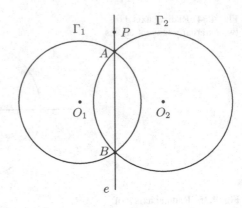

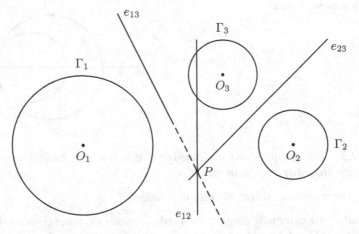

Fig. 4.37 The radical center of three circles with non collinear centers

Corollary 4.37 *If Γ_1, Γ_2 and Γ_3 are three pairwise non concentric circles with non collinear centers, then there exists a unique point P in the plane such that*

$$\mathrm{Pwr}_{\Gamma_1}(P) = \mathrm{Pwr}_{\Gamma_2}(P) = \mathrm{Pwr}_{\Gamma_3}(P).$$

Proof For $1 \leq i < j \leq 3$, let e_{ij} denote the radical axis of Γ_i and Γ_j (cf. Fig. 4.37, where we show the case of three pairwise exterior circles). Since the centers of the circles are non collinear, points, lines e_{12} and e_{23} are non parallel. Letting P stand for its intersection point, it follows from $P \in e_{12}$ that $\mathrm{Pwr}_{\Gamma_1}(P) = \mathrm{Pwr}_{\Gamma_2}(P)$; also, from $P \in e_{23}$ we obtain $\mathrm{Pwr}_{\Gamma_2}(P) = \mathrm{Pwr}_{\Gamma_3}(P)$. Therefore, by comparing these two relations, we get $\mathrm{Pwr}_{\Gamma_1}(P) = \mathrm{Pwr}_{\Gamma_3}(P)$, and hence $P \in e_{13}$.

The (easy) argument for uniqueness is left to the reader. $\square$

In the notations of the above corollary, and as anticipated in Fig. 4.37, the point P whose existence and uniqueness we have just proved is called the **radical center** of Γ_1, Γ_2 and Γ_3. For future use, it is worth keeping the fact that it lies in all of the radical axes e_{12}, e_{13} and e_{23}.

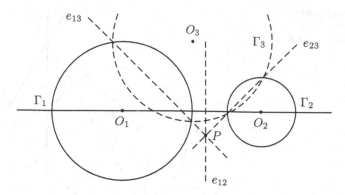

Fig. 4.38 Constructing the radical axis of two exterior circles

As we have said before, the notion of radical center allows us to present a construction, with compass and straightedge, of the radical axis of two interior or exterior circles. This is done in the coming

Example 4.38 Use straightedge and compass to construct the radical axis of the circles Γ_1 and Γ_2 shown in Fig. 4.38.

Solution Draw an auxiliary circle Γ_3 of center O_3, secant to both Γ_1 and Γ_2 and such that $O_3 \notin \overleftrightarrow{O_1 O_2}$ (to see that such a choice is always possible, just choose O_3 in the perpendicular bisector of segment $O_1 O_2$). For $i = 1, 2$, draw the radical axis e_{i3} of Γ_i and Γ_3, thus getting the radical center P of Γ_1, Γ_2 and Γ_3 as the intersection point of e_{13} and e_{23}. Finally, since P must also lie in the radical axis of Γ_1 and Γ_2, such an axis is the line passing through P and perpendicular to $\overleftrightarrow{O_1 O_2}$.

We leave to the reader the task of verifying that the construction described in the above example works equally well for two interior and non concentric circles. For another way of constructing the radical axis of two exterior circles, see Problem 9, page 149.

As a second application of the notion of radical center, we present, in the next two examples, two particular cases of **Apollonius' problem** on tangency of circles.

It is worth observing that the solution of the first example below reduces to a slight modification of the argument presented in the solution of the previous example.

Example 4.39 We are give in the plane a circle Γ and distinct points A and B, both of which lying outside the disk bounded by Γ. Use straightedge and compass to construct all circles α, tangent to Γ and passing through A and B.

Solution In the notations of Fig. 4.39, draw an auxiliary circle β, passing through A and B and secant to Γ. Then, draw the radical axis e of Γ and β, and mark the intersection point P of e and $\overleftrightarrow{AB}$.

Fig. 4.39 Constructing a
circle tangent to Γ and
passing through A and B

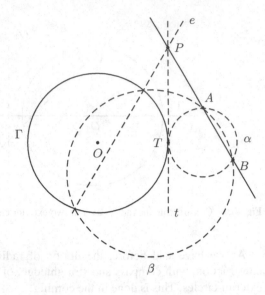

It is pretty clear that P is the radical center of α, β and Γ, so that P belongs to
the radical axis t of Γ and α. However, since Γ and α must be tangent, we know
from Example 4.36 that t is a common tangent of both such circles; therefore, by
invoking the construction delineated in Proposition 3.26, we can actually construct
t as being one of the tangent to Γ passing through P (in general, there are two
possible positions for line t, one of which is shown in Fig. 4.39). Finally, letting T
be the point of contact between t and Γ, we are left to constructing α as the circle
passing through points A, B and T.   □

For another approach to the previous example, see Problem 3, page 301.

We now examine Apollonius' problem for constructing a circle tangent to two
other given circles and passing through a given point.

Example 4.40 We are given exterior circles Γ_1 and Γ_2 and a point A, lying outside
the disks bounded by both Γ_1 and Γ_2. Use compass and straightedge to construct all
circles Γ, passing through A and simultaneously tangent to Γ_1 and Γ_2.

Solution Assuming that the problem has been solved, let P and Q be the points
of tangency of Γ with Γ_1 and Γ_2, respectively (cf. Fig. 4.40). Let also C be the
intersection point of lines $\overleftrightarrow{O_1 O_2}$ and $\overleftrightarrow{P Q}$, and B be the intersection of Γ and $\overleftrightarrow{AC}$.
Letting O denote the center of Γ, we have

$$O_2\widehat{R}Q = O_2\widehat{Q}R = O\widehat{Q}P = O\widehat{P}Q,$$

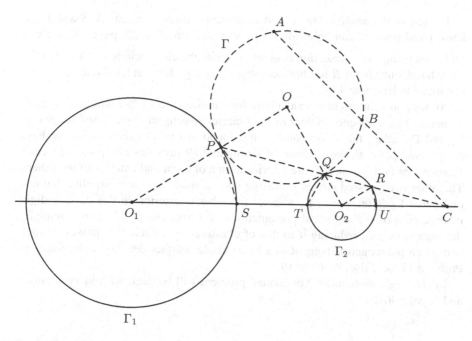

Fig. 4.40 Constructing a circle tangent to Γ_1 and Γ_2 and passing through A

so that $\overleftrightarrow{OP} \| \overleftrightarrow{O_2R}$. In particular, it is immediate to verify from this fact that C coincides with the intersection of the external tangents to Γ_1 and Γ_2 with line $\overleftrightarrow{O_1O_2}$ and, hence, that $\overleftrightarrow{RU} \| \overleftrightarrow{PS}$.

We claim that $PSTQ$ is a cyclic quadrilateral. Indeed, the parallelism of $\overleftrightarrow{RU}$ and $\overleftrightarrow{PS}$, together with the fact that $TQRU$ is cyclic, furnish

$$S\widehat{P}Q = U\widehat{R}C = 180° - P\widehat{R}U = 180° - Q\widehat{R}U = Q\widehat{T}U = 180° - Q\widehat{T}S,$$

so that $S\widehat{P}Q + Q\widehat{T}S = 180°$.

Now, by applying the intersecting chords theorem to the cyclic quadrilaterals $PSTQ$ and $PQBA$, we successively obtain

$$\overline{CS} \cdot \overline{CT} = \overline{CP} \cdot \overline{CQ} \text{ and } \overline{CP} \cdot \overline{CQ} = \overline{CA} \cdot \overline{CB};$$

therefore,

$$\overline{CS} \cdot \overline{CT} = \overline{CA} \cdot \overline{CB}.$$

With such a relation at our disposal, one further application of the intersecting chords theorem assures that $STBA$ is also cyclic.

In view of the analysis above, and since the positions of points A, S and T are known and point C can be easily constructed (cf. Problem 22, page 83), we can find B as being the intersection point of $\overleftrightarrow{AC}$ with the circumcircle of STA. On the other hand, once point B has been constructed, the problem at hand reduces to that discussed in Example 4.39.

At this point, the attentive reader must have noticed that his/her intuition suggests (generally) the existence of four distinct circles passing through A and tangent to Γ_1 and Γ_2, albeit the above solution has found just two of them (after we have constructed point B, the solution of Example 4.39 provides two possible circles Γ, such that both Γ_1 and Γ_2 are interior to one of them and exterior to the other). The other two guessed circles Γ emerge upon considering the possibility that one of Γ_1 and Γ_2 is interior to Γ, whereas the other is exterior; in this case, a slight modification of the argument presented above solves the problem in essentially the same way, i.e., reducing it to that of constructing a circle that passes through two given points and is tangent to a given circle. Further details can be found in Problem 17; see, also, Problem 19.

Interesting variations of Apollonius' problem will be dealt with in Problems 4 and 5, page 301.

Problems: Sect. 4.5

1. * Prove Proposition 4.30.
2. AB is a chord of a circle Γ of center O, of length 8cm. We mark on AB a point C, situated at 3cm from B. The radius of Γ passing through O and C intersects Γ at D, with $\overline{CD} = 1$cm. Find the length of the radius of Γ.
3. In a triangle ABC, $\overline{AB} = 8$cm. If M is the midpoint of AB, compute all possible lengths of side BC, so that the circumcircle of AMC is tangent to it.
4. Chords AB and CD of a circle are perpendicular and intersect at point E, situated inside the disk bounded by Γ and such that $\overline{AE} = 2$, $\overline{EB} = 6$ and $\overline{DE} = 3$. Compute the radius of the circle.
5. ABC is an isosceles triangle of basis $\overline{BC} = a$, and h_a is the length of its altitude relative to the basis. If R is the circumradius of ABC, show that

$$R = \frac{a^2 + 4h^2}{8h}.$$

6. Use the intersecting chords theorem to give another proof of Pythagoras' theorem.
7. We are given a line r and points A, B, $P \in r$, with $P \notin AB$. If a varying circle Γ passes through A and B, use compass and straightedge to construct the locus of the points of contact, with Γ, of the tangents drawn through P.

8. We are given a line r and points A and B is a single half-plane of those determined by r. Use compass and straightedge to construct all circles passing through A and B and tangent to r.

9. Let Γ_1 and Γ_2 be two exterior circles, and r and s be two common tangents to them. Let A_1 and A_2 (resp. B_1 and B_2) be the points of tangency of r (resp. s) with Γ_1 and Γ_2, respectively. If P and Q are the midpoints of line segments $A_1 A_2$ and $B_1 B_2$, also respectively, show that $\overleftrightarrow{PQ}$ is the radical axis of Γ_1 and Γ_2.

10. In a circle Γ we are given chords AB and CD, such that CD passes through the midpoint M of AB. Let Σ be the circle of diameter CD, and $E \in \Sigma$ be such that $\overleftrightarrow{ME} \perp \overleftrightarrow{CD}$. Prove that $A\widehat{E}B = 90°$.

11. (USA) In an acute triangle ABC, the altitude relative to AC intersects the circle of diameter AC at M and N, and the altitude relative to AB intersects the circle of diameter AB at P and Q. Prove that points M, N, P and Q are concyclic.

12. Let a real number $k > 1$ and distinct points B and C be given. Let also Γ be the Apollonius' circle relative to (B, C) in the ratio k. If O is the center of Γ and $X, Y \in \Gamma$ are such that B, X and Y are collinear, prove that points C, O, X and Y are concyclic.

13. Let Γ_1 and Γ_2 be two non concentric circles in the plane. Prove that the locus of points which are centers of circles that intersect Γ_1 and Γ_2 along diameters is the line symmetric to the radical axis of Γ_1 and Γ_2 with respect to the midpoint of the segment formed by their centers.

14. (USA) In a triangle ABC, let D, E and F be the feet of the internal angle bisectors relative to A, B and C, respectively. If $E\widehat{D}F = 90°$, find all possible values of $B\widehat{A}C$.

15. (BMO) A line passing through the incenter I of triangle ABC intersects its circumcircle at F and G, and its incircle at D and E, with $D \in EF$. If r stands for the inradius of ABC, prove that $\overline{DF} \cdot \overline{EG} \geq r^2$ and find out when the equality holds.

16. (Poland) In a triangle ABC, the internal angle bisectors relative to vertices A, B and C, respectively, intersect the opposite sides at points D, E and F, respectively, and the circumcircle of ABC again at points K, L and M, also respectively. Prove that

$$\frac{\overline{AD}}{\overline{DK}} + \frac{\overline{BE}}{\overline{EL}} + \frac{\overline{CF}}{\overline{FM}} \geq 9,$$

with equality holding if and only if ABC is equilateral.

17. * Complete the discussion of Example 4.40, constructing all circles α tangent to Γ_1 and Γ_2 and passing through point A, such that exactly one of Γ_1 and Γ_2 is interior to α.

18. Imitate the discussion of Example 4.40 and of the previous problem to construct all circles passing through a given point A and tangent to two given circles Γ_1 and Γ_2, in the case in which A and Γ_2 are both interior to Γ_1.

19. We are given three circles of noncollinear centers. Use straightedge and compass to construct all circles simultaneously tangent to all of these three circles.

    For the next problem, recall (according to the discussion preceding Theorem 3.40) that, given a triangle ABC and a point $P \neq A, B, C$, the *pedal triangle* of P with respect to ABC is the (possibly degenerate) triangle formed by the feet of the perpendiculars dropped from P to the lines $\overleftrightarrow{AB}, \overleftrightarrow{AC}, \overleftrightarrow{BC}$.

20. We are given a triangle ABC and two points P and Q lying in its interior. If P and Q are isogonal conjugates, prove that the vertices of the corresponding pedal triangles form a cyclic hexagon, such that the center of the circumscribed circle is the midpoint of segment PQ. Such a circle is called the **pedal circle** of $\{P, Q\}$ with respect to ABC.[19]

21. (China) $ABCD$ is a parallelogram and E and F are points on the diagonal BD for which there exists a circle α passing through E and F and tangent to lines $\overleftrightarrow{BC}$ and $\overleftrightarrow{CD}$. Prove that there exists a circle passing through E and F and tangent to lines $\overleftrightarrow{AB}$ and $\overleftrightarrow{AD}$.

22. Let α be the circumcircle of triangle ABC and M be the midpoint of arc $\overset{\frown}{BC}$ of α not containing vertex A. A circle β, passing through A and M, intersects AC at E and the extension of AB at F. If $AM \cap BC = \{D\}$, prove that D, E and F are collinear if and only if the incenters of ABC and AEF coincide.

---

[19]The attentive reader has certainly noticed that, according to Problem 9, page 137, if H and O are the orthocenter and the circumcenter of a triangle ABC, respectively, then the pedal circle of $\{H, O\}$ with respect to ABC is precisely the nine-point circle of ABC.

Chapter 5
Area of Plane Figures

Intuitively, the *area* of a plane region should be a positive number that we associate to the region and that serves to quantify the space it occupies. We refer the interested reader to the excellent book of E. Moise [19] for a proof that it is indeed possible to associate to each convex polygon in the plane a notion of area satisfying postulates 1. to 5. below. Our purpose in this chapter is mainly to *operationalize* the computation of areas, extracting from it some interesting applications.

5.1 Areas of Convex Polygons

For any concept of *area* for convex polygons to be useful, we *postulate* it ought to have the following (intuitively desirable) properties:

1. The area of a convex polygon is a positive real number.
2. Congruent convex polygons[1] have equal areas.
3. If a convex polygon is *partitioned* into a finite number of other convex polygons (i.e., if the polygon is the union of a finite number of other convex polygons, any two of which without common interior points), then the area of the larger polygon is equal to the sum of the areas of the smaller ones.
4. If a (larger) convex polygon contains another (smaller) convex polygon in its interior, then the area of the larger polygon is greater that that of the smaller one.[2]

---

[1] Although we have not formally defined a notion of congruence for general convex polygons, the idea is pretty much the same as that for triangles: that one polygon can be moved in space, without being deformed, until it coincides with the other. In particular, note that two squares of equal sides are congruent.

[2] A little geometric intuition shows that this item is not independent from the others. In fact, one can prove that, under such a situation, the larger polygon can be partitioned into a finite number

5. The area of a square of side length 1cm is equal to 1cm^2 (one reads 1 *square centimeter*).

From now on, we assume that postulates 1. to 5. do hold. This being so, partition a square of side length $n \in \mathbb{N}$ into n^2 squares of side length 1. Denoting the area of the larger square by A_n, we must have A_n equal to the sum of the areas of the n^2 squares of side length 1, so that

$$A_n = n^2.$$

We now consider a square of side length $\frac{m}{n}$, with $m, n \in \mathbb{N}$, and area $A_{\frac{m}{n}}$. Arrange n^2 copies of it by piling n squares per row, in n rows, thus forming a square of side length $\frac{m}{n} \cdot n = m$. As we already know, such a larger square has area m^2; on the other hand, since it is partitioned into n^2 (congruent) squares of area $A_{\frac{m}{n}}$, its area is equal to the sum of the areas of these n^2 squares, so that

$$m^2 = n^2 \cdot A_{\frac{m}{n}}.$$

Hence,

$$A_{\frac{m}{n}} = \frac{m^2}{n^2} = \left(\frac{m}{n}\right)^2.$$

The above discussion suggests that the area of a square of side length l must be equal to l^2. In order to confirm such a supposition, we argue in a way similar to that of the proof of Thales' theorem: for $k \in \mathbb{N}$, we take rational numbers x_k and y_k such that

$$x_k < l < y_k \text{ and } y_k - x_k < \frac{1}{k}.$$

Then, we construct squares of side lengths x_k and y_k, the first contained in the square of side length l and the second containing it. Since we already know how to compute the areas of square os rational side lengths, Postulate 4. above guarantees that the area A_l of the square of side length l must satisfy the inequalities

$$x_k^2 < A_l < y_k^2.$$

However, since we also have $x_k^2 < l^2 < y_k^2$, we conclude that both numbers A_l and l^2 must belong to the interval (x_k^2, y_k^2), so that

---

of other convex polygons, one of which is the smaller one. Once this has been done, a simple application of items 1. and 3. let us derive 4. as a theorem.

Fig. 5.1 A square of side length l has area l^2

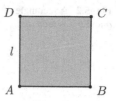

$$|A_l - l^2| < y_k^2 - x_k^2 = (y_k - x_k)(y_k + x_k)$$

$$< \frac{1}{k}(y_k - x_k + 2x_k)$$

$$< \frac{1}{k}\left(\frac{1}{k} + 2l\right).$$

Satisfying the inequality above for every $k \in \mathbb{N}$, we conclude that $|A_l - l^2| = 0$, so that

$$A_l = l^2. \tag{5.1}$$

The coming proposition summarizes the above discussion (Fig. 5.1).

Proposition 5.1 *A square of side length l has area equal to l^2.*

An argument analogous to the one that led to (5.1) allows us to prove that a rectangle of sides with lengths a and b has area equal to ab (cf. Fig. 5.2). Indeed, we start with a rectangle of sides with lengths $m, n \in \mathbb{N}$, and decompose it into mn squares of side lengths equal to 1 to conclude that the area of the rectangle is equal to mn. Then, we take a rectangle whose sides have lengths $\frac{m_1}{n_1}$ and $\frac{m_2}{n_2}$, with $m_1, m_2, n_1, n_2 \in \mathbb{N}$, and assemble $n_1 n_2$ copies of it to form a rectangle of sides m_1 and m_2. Adding the equal areas of the small rectangles, we conclude that the area of the initial rectangle is equal to

$$\frac{m_1 m_2}{n_1 n_2} = \frac{m_1}{n_1} \cdot \frac{m_2}{n_2}.$$

Finally, we consider a rectangle of sides equal to two arbitrary positive real numbers a and b; given $k \in \mathbb{N}$, we take rational numbers x_k, y_k, u_k, v_k such that $x_k < a < y_k$, $u_k < b < v_k$ and $y_k - x_k, v_k - u_k < \frac{1}{k}$. Letting A denote the area of the rectangle of sides a and b, an argument analogous to the one given in the case of a square assures that both A and ab lie in the open interval $(u_k x_k, y_k v_k)$. Therefore, for every $k \in \mathbb{N}$, we have

$$|A - ab| < v_k y_k - u_k x_k = (v_k - u_k)y_k + u_k(y_k - x_k)$$

$$< \frac{1}{k}(y_k + u_k) < \frac{1}{k}((y_k - x_k) + 2x_k + (v_k - u_k) + 2u_k)$$

$$< \frac{1}{k}\left(\frac{2}{k} + 2a + 2b\right).$$

Fig. 5.2 A rectangle of sides
a and b has area ab

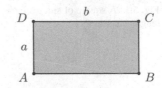

Fig. 5.3 Area of a
parallelogram

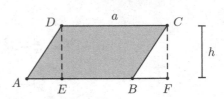

Also as before, the fact that the inequality above must hold for every $k \in \mathbb{N}$ guarantees that $A = ab$. We have thus proved the coming

Proposition 5.2 *A rectangle of sides a and b has area equal to ab.*

We compute the area of a parallelogram as a corollary of the former proposition. To this end, we fix an edge of the parallelogram, which we shall call its **basis**, and shall refer to the distance from it to the opposite side as the **height** or **altitude** of the parallelogram (relative to the chose basis). The desired result is as follows.

Proposition 5.3 *The area of a parallelogram of basis a and height h is equal to ah.*

Proof Let $ABCD$ be a parallelogram of diagonals AC and BD (cf. Fig. 5.3), and E and F be the feet of the perpendiculars dropped from D and C, respectively, to $\overleftrightarrow{AB}$. Moreover, suppose, without loss of generality, that $E \in AB$. It is immediate to verify that triangles ADE and BCF are congruent by HL, so that $\overline{AE} = \overline{BF}$ and (by Postulate 2.) $A(ADE) = A(BCF)$. Therefore, we have

$$A(ABCD) = A(ADE) + A(BEDC)$$
$$= A(BCF) + A(BEDC)$$
$$= A(CDEF).$$

On the other hand, $CDEF$ is a rectangle of height h and basis

$$\overline{EF} = \overline{EB} + \overline{BF} = \overline{EB} + \overline{AE} = \overline{AB} = a.$$

Hence, Proposition 5.3 gives $A(ABCD) = A(EFCD) = ah$.   □

From the formula for the computation of the area of a parallelogram we can easily deduce the usual formula for the area of triangles. We do this next.

Fig. 5.4 Area of a triangle

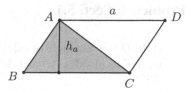

Proposition 5.4 *Let ABC be a triangle with sides $\overline{AB} = c$, $\overline{AC} = b$ and $\overline{BC} = a$, and heights h_a, h_b and h_c, relative to the sides BC, AC and AB, respectively. Then,*

$$A(ABC) = \frac{ah_a}{2} = \frac{bh_b}{2} = \frac{ch_c}{2}. \tag{5.2}$$

In particular, $ah_a = bh_b = ch_c$.

Proof Let $S = A(ABC)$ and D be the intersection of the parallel to $\overleftrightarrow{BC}$ passing through A with the parallel to $\overleftrightarrow{AB}$ passing through C (cf. Fig. 5.4). Then, ASA implies $ABC \equiv CDA$ (for $B\widehat{A}C = D\widehat{C}A$, AC is a common side of both triangles and $B\widehat{C}A = D\widehat{A}C$), so that Postulate 1. gives $A(ABC) = A(CDA)$. However, since $ABCD$ is a parallelogram with basis a and altitude h_a, the previous proposition furnishes

$$2S = A(ABC) + A(CDA) = A(ABCD) = ah_a.$$

Hence, $A(ABC) = S = \frac{1}{2}ah_a$, and the other two equalities can be likewise deduced.   □

In view of the material discussed so far, to compute the area of a convex polygon is, in principle, an easy task: since the diagonals of the polygon drawn from one of its vertices partition it into triangles, it suffices to compute the area of each one of these triangles with the aid of the last result, and then to add the results thus obtained.

We finish this section by establishing, for future use, the following convention: if two convex polygons have equal areas, we shall say that they are (**area-**)**equivalent**. For instance, according to Proposition 5.3, a parallelogram of basis a and height h is equivalent to a rectangle of sides a and h.

Remark 5.5 With respect to the concept above, a theorem of Bolyai and Gerwien[3] (cf. Chapter 1 of [4], for instance) shows that if the convex polygons $\mathcal{P}_1$ and $\mathcal{P}_2$ are area-equivalent, then one can split $\mathcal{P}_1$ into a finite number of polygonal pieces, rearrange them in a different way and assemble $\mathcal{P}_2$.

---

[3]Farkas Bolyai, Hungarian mathematician, and Paul Gerwien, amateur German mathematician, both of the nineteenth century.

Problems: Sect. 5.1

1. $ABCD$ is a rectangle of sides $\overline{AB} = 32$ and $\overline{BC} = 20$. Points E and F are the midpoints of sides AB and AD, respectively. Compute the area of quadrilateral $AECF$.

2. In parallelogram $ABCD$ of diagonals AC and BD, we mark point E on side AD such that $\overleftrightarrow{BE} \perp \overleftrightarrow{AD}$. If $\overline{BE} = 5$, $\overline{BC} = 12$ and $\overline{AE} = 4$, compute the area of triangle CDE.

3. Let $ABCD$ be a square of side length 1, let E be the midpoint of BC and F that of CD. If G is the intersection point of DE and AF, compute the area of triangle DFG.

4. * If ABC is an equilateral triangle of side length a, prove that:

    (a) The altitudes of ABC are equal to $\frac{a\sqrt{3}}{2}$.

    (b) $A(ABC) = \frac{a^2\sqrt{3}}{4}$.

5. Let $ABCD$ be a square of side 1 and E an interior point of $ABCD$, such that triangle ABE is equilateral. Compute the area of triangle BCE.

6. $ABCD$ is a square of side length 1cm and AEF is an equilateral triangle, with $E \in BC$ and $F \in CD$. Compute the area of triangle AEF.

7. Let ABC be an equilateral triangle. Prove that the sum of the distances of a point lying in the interior of ABC to its sides does not depend on the position of the point and is equal to the length of the altitudes of ABC.

8. Triangle ABC has sides a, b and c. The altitudes corresponding to the sides are respectively equal to h_a, h_b and h_c. If $a + h_a = b + h_b = c + h_c$, prove that ABC is equilateral.

9. (Brazil) Let ABC be a right triangle of area 1m^2. Compute the area of triangle $A'B'C'$, with A' being the symmetric of A with respect to $\overleftrightarrow{BC}$, B' the symmetric of B with respect to $\overleftrightarrow{AC}$ and C' the symmetric of C with respect to $\overleftrightarrow{AB}$.

10. Given a triangle ABC, prove that the triangle formed by the midpoints of its sides has area equal to $\frac{1}{4}$ of that of ABC.

11. Let $ABCD$ be a convex quadrilateral and M, N, P and Q be the midpoints of AB, BC, CD and DA, respectively. Prove that

$$A(MNPQ) = \frac{1}{2}A(ABCD).$$

12. We are given in the plane two squares of side 1cm each, such that the center of one of them coincides with a vertex of the other. Compute all possible values for the area of the portion of the plane common to both squares.

13. Prove the following theorem of Clairaut[4]: let ABC be a given triangle, and $ABDE$ and $ACFG$ be parallelograms constructed outside ABC and having disjoint interiors. Let also H be the point of intersection of lines $\overleftrightarrow{DE}$ and $\overleftrightarrow{FG}$, and $BCIJ$ be a parallelogram for which $\overline{CI} = \overline{AH}$ and $\overleftrightarrow{CI} \parallel \overleftrightarrow{AH}$. Then

$$A(ABDE) + A(ACFG) = A(BCIJ).$$

14. Each diagonal of a convex quadrilateral divides it into two triangles of equal areas. Prove that the quadrilateral is a parallelogram.

15. (OIM—adapted.)

   (a) If two triangles have equal heights, show that the ratio of their areas is equal to the ratio of the lengths of the corresponding bases.
   (b) Let ABC be a given triangle and D, E and F be points on the sides BC, CA and AB, respectively, such that cevians AD, BE and CF concur at P. If $A(BDP) = 40$, $A(CDP) = 30$, $A(CEP) = 35$ and $A(AFP) = 84$, compute the area of ABC.

16. In this problem we give a partial proof of Ceva's theorem 4.25 by means of an argument involving areas of triangles. To this end, let be given a triangle ABC and points A', B' and C', respectively on sides BC, AC and AB and distinct from the vertices of ABC. Also, suppose that $\overleftrightarrow{AA'}$, $\overleftrightarrow{BB'}$ and $\overleftrightarrow{CC'}$ concur at a point P. Show that:

   (a) $\frac{A'B}{A'C} = \frac{A(ABP)}{A(ACP)}$.
   (b) $\frac{BA'}{A'C} \cdot \frac{CB'}{B'A} \cdot \frac{AC'}{C'B} = 1$.

   For the next problem we recall (cf. problem 9, page 99) that, as with convex quadrilaterals, a convex polygon is **tangential** provided there exists a circle lying in its interior and tangent to its sides. In this case, this circle is said to be **inscribed** in the polygon.

17. (OIM) Let $\mathcal{P}$ be a tangential polygon. Line r divides $\mathcal{P}$ in two convex polygons of equal areas and perimeters. Show that r passes through the center of the circle inscribed at $\mathcal{P}$.

18. (IMO) In a convex quadrilateral of area 32cm^2, the sum of the lengths of two opposite sides and one diagonal is equal to 16cm. Compute all possible lengths of the other diagonal.

19. Let P be an interior point of triangle ABC and $x = d(P, \overleftrightarrow{BC})$, $y = d(P, \overleftrightarrow{AC})$, $z = d(P, \overleftrightarrow{AB})$. The purpose of this problem is to prove the famous **Erdös-Mordell inequality**[5]:

$$\overline{AP} + \overline{BP} + \overline{CP} \geq 2(x + y + z),$$

---

[4]Alexis Claude Clairaut, French astronomer and mathematician of the eighteenth century.

[5]Paul Erdös, Hungarian mathematician, and Louis Mordell, British mathematician, both of the twentieth century.

with equality if and only if ABC is equilateral and P is its center. To this end, letting $\overline{AB} = c$, $\overline{AC} = b$, $\overline{BC} = a$, show that:

(a) $\overline{AP} \cdot a \geq by + cz$.

(b) $\overline{AP} \cdot a \geq bz + cy$, $\overline{BP} \cdot b \geq cx + az$, $\overline{CP} \cdot c \geq ay + bx$.

(c) $\overline{AP} + \overline{BP} + \overline{CP} \geq \left(\frac{b}{c} + \frac{c}{b}\right)x + \left(\frac{a}{c} + \frac{c}{a}\right)y + \left(\frac{a}{b} + \frac{b}{a}\right)z \geq 2(x + y + z)$.

(d) Equality holds in (c) if and only if ABC is equilateral and P is its center.

5.2   Some Applications

An immediate consequence of Proposition 5.4 is the following criterion for the equivalence of area-equivalence of triangles.

Corollary 5.6 *If ABC and $A'BC$ are two triangles for which $\overleftrightarrow{AA'} \| \overleftrightarrow{BC}$. Then, $A(ABC) = A(A'BC)$.*

Proof Letting d be the distance between lines $\overleftrightarrow{BC}$ and $\overleftrightarrow{AA'}$ (cf. Fig. 5.5), it is clear that d is the length of the altitudes of triangles ABC and $A'BC$ relative to BC. Hence,

$$A(ABC) = \frac{1}{2}\overline{BC} \cdot d = A(A'BC).$$

□

The previous corollary can be used to transform a convex polygon into an area-equivalent one, with a smaller number of sides. The coming example illustrates this point.

Example 5.7 With respect to the figure below, use compass and straightedge to construct point $E \in \overleftrightarrow{BC}$ such that $A(ABE) = A(ABCD)$.

Fig. 5.5 A criterion for the area-equivalence of two triangles

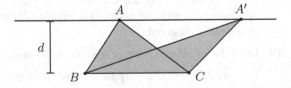

Solution

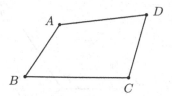

<small>C</small>ONSTRUCTION STEPS

1. Draw, through point D, line r, parallel to line $\overleftrightarrow{AC}$.
2. Mark point E as the intersection of r with $\overleftrightarrow{BC}$.
3. By the former corollary, we have $A(ACD) = A(ACE)$; therefore,

$$
\begin{aligned}
A(ABE) &= A(ABC) + A(ACE) \\
&= A(ABC) + A(ACD) \\
&= A(ABCD)
\end{aligned}
$$

Another interesting consequence of Corollary 5.6 is a proof of Pythagoras' theorem, as well as other metric relations in right triangles, through the computation of certain areas. This is the content of the next

Example 5.8 Let ABC be a triangle right at A, with legs $\overline{AB} = c$, $\overline{AC} = b$ and hypothenuse $\overline{BC} = a$. If H is the foot of the altitude relative to the hypothenuse, $\overline{CH} = m$, $\overline{BH} = n$ and $\overline{AH} = h$, use area computations to establish the metric relations below:

(a) $ah = bc$.
(b) $c^2 = an$ and $b^2 = am$.
(c) $a^2 = b^2 + c^2$.

Proof

(a) It suffices to see that both ah and bc are equal to twice the area of ABC. Indeed,

$$
A(ABC) = \frac{1}{2}\overline{BC} \cdot \overline{AH} = \frac{ah}{2} \quad \text{and} \quad A(ABC) = \frac{1}{2}\overline{AC} \cdot \overline{AB} = \frac{bc}{2}.
$$

(b) Construct, externally to ABC, squares $ABDE$, $BCFG$ and $ACJK$ (cf. Fig. 5.6) and let I be the point of intersection of the half-line $\overrightarrow{AH}$ with FG. Since $\overleftrightarrow{AI} \parallel \overleftrightarrow{BG}$, it follows from Corollary 5.6 that

$$
A(BGA) = A(BGH) = \frac{1}{2}\overline{BG} \cdot \overline{BH} = \frac{an}{2}.
$$

Fig. 5.6 Pythagoras' theorem through the computation of areas

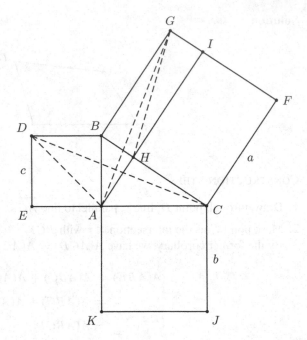

On the other hand, since $\overline{BD} = \overline{AB}$, $\overline{BC} = \overline{BG}$ and $D\widehat{B}C = 90° + \widehat{B} = A\widehat{B}G$, triangles BCD and BGA are congruent by SAS. Hence, $A(BCD) = A(BGA) = \frac{an}{2}$ (I). However, since $\overleftrightarrow{AC} \parallel \overleftrightarrow{BD}$, by applying again Corollary 5.6 we obtain $A(BCD) = A(ABD) = \frac{c^2}{2}$ (II). It thus follows from (I) and (II) that $c^2 = an$. The proof of $b^2 = am$ is analogous and will be left to the reader.

(c) We can certainly do as we did in the proof of item (c) of Theorem 4.10, termwise adding both relations derived in (b). Alternatively, the argument in the proof of (b) guarantees that

$$c^2 = A(ABDE) = 2A(ABD) = 2A(BGH) = A(BGIH);$$

on the other hand, by reasoning likewise, we obtain

$$b^2 = A(ACJK) = 2A(ACJ) = 2A(BCJ)$$
$$= 2A(FCA) = 2A(FCH) = A(FCHI),$$

so that

$$b^2 + c^2 = A(BGIH) + A(FCHI) = A(BCFG) = a^2.$$

$\square$

From the formula for the area of a triangle we can derive a useful expression for the area of a trapezoid. To this end, we let the **altitude** or **height** of a trapezoid be the distance between (the lines containing) its bases.

Proposition 5.9 *If ABCD is a trapezoid of bases* $\overline{AB} = a$, $\overline{CD} = b$ *and altitude h, then*

$$A(ABCD) = \frac{(a+b)h}{2}.$$

Proof Assume, without loss of generality, that $a > b$ (cf. Fig. 5.7). If $E \in AB$ is such that $\overline{AE} = b$, then the quadrilateral $AECD$ has two parallel and equal sides, so that it is a parallelogram. Since $\overline{BE} = a - b$, we get

$$A(ABCD) = A(AECD) + A(EBC)$$
$$= bh + \frac{(a-b)h}{2} = \frac{(a+b)h}{2}.$$

$\square$

The next result is another important consequence of Proposition 5.4.

Proposition 5.10 *If ABCD is a rhombus of diagonals AC and BD, then*

$$A(ABCD) = \frac{1}{2}\overline{AC} \cdot \overline{BD}.$$

Proof In the notations of Fig. 5.8, we have

Fig. 5.7 Area of a trapezoid

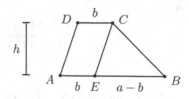

Fig. 5.8 Area of a rhombus

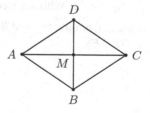

Fig. 5.9 Areas of similar
triangles

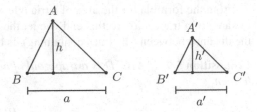

$$A(ABCD) = A(ABC) + A(ACD)$$

$$= \frac{1}{2}\overline{AC} \cdot \overline{BM} + \frac{1}{2}\overline{AC} \cdot \overline{DM}$$

$$= \frac{1}{2}\overline{AC} \cdot \overline{BD}.$$

□

We now show that the ratio between the areas of two similar triangles is equal to
the square of the corresponding similitude ratio.

Proposition 5.11 *Let ABC and $A'B'C'$ be two similar triangles. If k stands for the
similitude ratio from ABC to $A'B'C'$, then*

$$\frac{A(ABC)}{A(A'B'C')} = k^2.$$

Proof Let $\overline{BC} = a$, $\overline{B'C'} = a'$ and h and h' be the altitudes of ABC and
$A'B'C'$ relative to BC and $B'C'$, respectively (cf. Fig. 5.9). Since $a = ka'$ and
(by Problem 3, page 115) $h = kh'$, we conclude that

$$\frac{A(ABC)}{A(A'B'C')} = \frac{ah}{a'h'} = \frac{ka' \cdot kh'}{a'h'} = k^2.$$

□

The coming example brings a classical application of the previous result.

Example 5.12 With respect to the figure below, use compass and straightedge to
construct points $D \in AB$ and $E \in AC$ such that $\overleftrightarrow{DE} \| \overleftrightarrow{BC}$ and $A(ADE) =
A(DBCE)$.

Solution

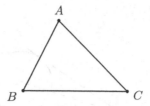

Assuming that the problem has been solved, we have $A(ADE) = \frac{1}{2}A(ABC)$. On the other hand, since $ADE \sim ABC$, the previous proposition guarantees that

$$\frac{\overline{AE}}{\overline{AC}} = \sqrt{\frac{A(ADE)}{A(ABC)}} = \frac{1}{\sqrt{2}}.$$

It now suffices to perform the construction below:

CONSTRUCTION STEPS

1. Draw, externally to ABC, the semicircle Γ of diameter AC.
2. Letting M be the midpoint of AC, mark point $P \in \Gamma$ such that $PM \perp AC$. Pythagoras' theorem, applied to triangle APC, assures that $\overline{AP} = \frac{1}{\sqrt{2}}\overline{AC}$.
3. Finally, obtain point E as the intersection of AC with the circle of center A and radius $\overline{AP}$.

We finish this section by presenting some interesting applications of the formula for the area of triangles, used in conjunction with other previously studied results.

Proposition 5.13 *Let ABC be a triangle of sides $\overline{BC} = a$, $\overline{AC} = b$, $\overline{AB} = c$ and semiperimeter p. If r and r_a respectively denote the inradius and the exradius of ABC relative to BC, then*

$$A(ABC) = pr = (p - a)r_a. \tag{5.3}$$

Proof Let I be the incenter and I_a the excenter of ABC relative to BC (cf. Fig. 5.10). Since the altitudes of triangles AIB, AIC and BIC, respectively relative to AB, AC and BC, are all equal to r, we have

$$A(ABC) = A(AIB) + A(AIC) + A(BIC)$$

$$= \frac{cr}{2} + \frac{br}{2} + \frac{ar}{2} = pr.$$

On the other hand, since the altitudes of $A(AI_aB)$, $A(AI_aC)$ and $A(BI_aC)$, respectively relative to AB, AC and BC, are all equal to r_a, we obtain

Fig. 5.10 Formulas for the
area of a triangle

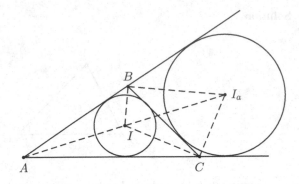

Fig. 5.11 Distances from the
circumcenter to the sides

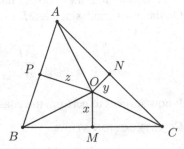

$$A(ABC) = A(AI_aB) + A(AI_aC) - A(BI_aC)$$

$$= \frac{cr_a}{2} + \frac{br_a}{2} - \frac{ar_a}{2} = (p-a)r_a.$$

□

We are now in position to state and prove yet another important consequence
of Ptolemy's theorem, which is known in mathematical literature as **Carnot's
theorem.**[6]

Theorem 5.14 (Carnot)   *Let ABC is an acute triangle of circumcenter O. If x, y
and z denote the distances from O to the sides BC, AC and AB, respectively, then*

$$x + y + z = R + r,$$

where r and R respectively denote the inradius and the circumradius of ABC.

Proof Let M, N and P denote the midpoints of sides BC, AC and AB, respectively,
so that $OM \perp BC$, $ON \perp CA$ and $OP \perp AB$ (cf. Fig. 5.11). Since quadrilaterals
$BMOP$, $CNOM$ and $APON$ have a pair of opposite right angles, they are cyclic.

---

[6]Lazare Carnot, French mathematician of the eighteenth and nineteenth centuries, the first to
systematically use oriented line segments in Geometry.

Letting $\overline{BC} = a$, $\overline{AC} = b$ and $\overline{AB} = c$, and noticing that $\overline{OM} = x$, $\overline{ON} = y$ and $\overline{OP} = z$, we obtain from Ptolemy's theorem the equalities

$$x \cdot \frac{c}{2} + z \cdot \frac{a}{2} = R \cdot \frac{b}{2},$$

$$x \cdot \frac{b}{2} + y \cdot \frac{a}{2} = R \cdot \frac{c}{2}, \tag{5.4}$$

$$y \cdot \frac{c}{2} + z \cdot \frac{b}{2} = R \cdot \frac{a}{2}.$$

On the other hand, since triangles OBC, OCA and OAB partition ABC, we also have

$$A(ABC) = \frac{xa}{2} + \frac{yb}{2} + \frac{zc}{2}.$$

Now, letting p stand for the semiperimeter of ABC, we know from the previous proposition that $A(ABC) = pr$; in turn, substituting such a relation in the last equality above, we obtain

$$\frac{xa}{2} + \frac{yb}{2} + \frac{zc}{2} = pr.$$

Finally, by termwise adding the last equality above with those in (5.4), we get

$$(x + y + z)p = (R + r)p,$$

whence Carnot's theorem follows.   □

A direct inspection of the above proof shows that Carnot's theorem continues to hold for right triangles. For a generalization of it to obtuse triangles, see Problem 16, page 169.

For our last application we shall need a preliminary definition. At this point, the reader may find it useful to recall the content of Theorem 4.28.

Definition 5.15 The **symmedians** of a triangle are the cevians isogonal to the medians of the triangle. Their point of concurrence is the **symmedian point** or the **Lemoine point**[7] of the triangle.

Out of several interesting properties of the Lemoine point, the most striking one is perhaps that collected in Theorem 5.17. However, before we can prove it, we need to establish an auxiliary result which is also important to the analysis of other properties of the Lemoine point.

---

[7]Émile Lemoine, French mathematician of the nineteenth century.

Fig. 5.12 A fundamental
property of symmedians

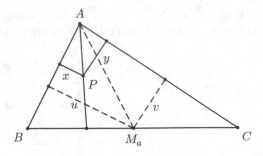

Proposition 5.16 *Let P be a point lying in the interior of a triangle ABC. If x and
y denote the distances from P to the sides AB and AC, respectively, then*

$$\overrightarrow{AP} \text{ is a symmedian } \Leftrightarrow \frac{x}{y} = \frac{\overline{AB}}{\overline{AC}}.$$

Proof Let M_a be the midpoint of BC. In the notations of Fig. 5.12, Lemma 4.27
guarantees that

$$\overrightarrow{AP} \text{ is a symmedian } \Leftrightarrow \frac{x}{y} = \frac{v}{u}. \tag{5.5}$$

On the other hand, letting h denote the length of the altitude of ABC relative to BC,
we have

$$A(ABM_a) = \frac{1}{2}\overline{BM_a} \cdot h = \frac{1}{2}\overline{CM_a} \cdot h = A(ACM_a).$$

Hence,

$$\frac{1}{2}\overline{AB} \cdot u = A(ABM_a) = A(ACM_a) = \frac{1}{2}\overline{AC} \cdot v,$$

so that

$$\frac{v}{u} = \frac{\overline{AB}}{\overline{AC}}. \tag{5.6}$$

It now suffices to combine (5.5) and (5.6) to get the desired result.   □

The final result of this section characterizes the symmedian point as the only
solution of a minimization problem.

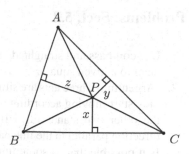

Fig. 5.13 The symmedian point as a point of minimum

Theorem 5.17 *Let ABC be any triangle and P be a point lying inside it. The sum of the squares of the distances from P to the lines $\overleftrightarrow{AB}$, $\overleftrightarrow{AC}$ and $\overleftrightarrow{BC}$ is minimal if and only if P is the Lemoine point of ABC.*

Proof Let x, y and z be the distances from P to the lines $\overleftrightarrow{BC}$, $\overleftrightarrow{AC}$ and $\overleftrightarrow{AB}$, respectively (cf. Fig. 5.13). We want to show that the sum $x^2 + y^2 + z^2$ is minimal if and only if P is the symmedian point of ABC.

To this end, let $\overline{BC} = a$, $\overline{AC} = b$ and $\overline{AB} = c$. Also, writing $S = A(ABC)$ for the area of ABC, we have

$$S = A(ABP) + A(BCP) + A(CAP) = \frac{ax + by + cz}{2},$$

so that $2S = ax + by + cz$. Now, by applying Cauchy's inequality (cf. Section 5.2 of [5], for instance), we obtain

$$(x^2 + y^2 + z^2)(a^2 + b^2 + c^2) \geq (ax + by + cz)^2 = 4S^2,$$

and hence

$$x^2 + y^2 + z^2 \geq \frac{4S^2}{a^2 + b^2 + c^2}.$$

The condition for equality in Cauchy's inequality assures that it holds if and only if we have

$$\frac{x}{a} = \frac{y}{b} = \frac{z}{c}.$$

In turn, in accordance with the previous proposition, the above relations take place if and only if P is the Lemoine point of ABC. $\qquad\square$

Problems: Sect. 5.2

1. Use compass and straightedge to construct an equilateral triangle (area) equivalent to a given square.
2. (Argentina) Three ants are situated on three of the vertices of a rectangle. Then,
   they start moving according with the following rules: (i) when one ant moves,
   the other two stand still; (ii) at any time, the moving ant moves along the
   direction parallel to the line passing through the positions of the other two ants.
   Is it possible that, at some future instant, the ants are situated at the midpoints
   of three of the sides of the original rectangle?
3. (Hungary) Let $ABCD$ be a parallelogram and EFG a triangle whose vertices
   lie on the sides of $ABCD$. Prove that

$$A(ABCD) \geq 2A(EFG).$$

4. Let ABC be any triangle.

   (a) Prove that one can use the medians of ABC to form a triangle DEF.
   (b) Compute the ratio between the areas of triangles ABC and DEF.

5. (TT) In a convex hexagon $ABCDEF$, we have $AB \parallel CF$, $CD \parallel BE$ and
   $EF \parallel AD$. Prove that triangles ACE and BDF have equal areas.
6. Trapezoid $ABCD$, with bases AB and CD and legs AD and BC, is rectangle
   at A. If $\overline{BC} = \overline{CD} = 13$cm and $\overline{AB} = 18$cm, compute the height and the area
   of $ABCD$, as well as the distance from A to $\overleftrightarrow{BC}$.
7. (TT) Find all positive integers n for which it is possible to partition an
   equilateral triangle of side length n into trapezoids of side lengths 1, 1, 1 and 2.
8. $ABCD$ is a trapezoid of bases BC and AD and legs AB and CD. Let E be the
   midpoint of the side CD, and assume that the area of triangle AEB is equal to
   360cm^2. Compute the area of the trapezoid.
9. $ABCD$ is a trapezoid with bases AB and CD and legs AD and BC. If the
   diagonals of $ABCD$ intersect at E, prove that

$$\sqrt{A(ABCD)} = \sqrt{A(ABE)} + \sqrt{A(CDE)}.$$

10. Through a point P in the interior of a triangle ABC we draw the parallels
    to the sides of ABC. Such lines divide ABC into three triangles and three
    parallelograms. If the areas of the triangles are equal to 1cm^2, 4cm^2 and 9cm^2,
    compute the area of ABC.
11. ABC is a triangle of semiperimeter p, inradius r and exradii r_a, r_b and r_c.
    Prove that

$$\frac{1}{r} = \frac{1}{r_a} + \frac{1}{r_b} + \frac{1}{r_c}.$$

12. Prove that the symmedians of an acute triangle pass through the midpoints of the sides of its orthic triangle.

13. Prove that the symmedian point of a right triangle is the midpoint of the altitude relative to the hypothenuse.

14. If P is a point on the side BC of triangle ABC, show that AP is the symmedian relative to BC if and only if

$$\frac{\overline{BP}}{\overline{CP}} = \frac{\overline{AB}^2}{\overline{AC}^2}.$$

15. (OIM) Let ABC be a triangle of incenter I and barycenter G, such that $2\overline{BC} = \overline{AC} + \overline{AB}$. Prove that $\overleftrightarrow{IG} \parallel \overleftrightarrow{BC}$.

16. * Generalize Carnot's theorem for obtuse triangles. More precisely, show that if a triangle ABC is obtuse at A, and x, y and z are the distances of the circumcenter O of ABC to $\overleftrightarrow{BC}$, $\overleftrightarrow{AC}$ and $\overleftrightarrow{AB}$, respectively, then

$$-x + y + z = R + r,$$

where r and R respectively denote the inradius and the circumradius of ABC.

17. $ABCDE$ is a convex pentagon inscribed in a circle. We partition $ABCDE$ into three triangles, by drawing two diagonals of it, which intersect only at vertices of $ABCDE$. Prove that, no matter which diagonals are chosen, the sum of the inradii of the three triangles in which $ABCDE$ gets divided is always the same.

18. (Bulgaria) Trapezoid $ABCD$ has bases $\overline{AB} > \overline{CD}$ and legs BC and AD. The diagonals AC and BD intersect at O, and K, L, M and N are the points of intersection of the angle bisectors of $\angle AOB$, $\angle BOC$, $\angle COD$ and $\angle DOA$ with sides AB, BC, CD and DA, respectively. Point P is the intersection of lines $\overleftrightarrow{KL}$ and $\overleftrightarrow{MN}$, and Q is the intersection of lines $\overleftrightarrow{KN}$ and $\overleftrightarrow{ML}$. Find all values of $k = \frac{\overline{AB}}{\overline{CD}}$ for which quadrilateral $ABCD$ and triangle POQ have equal areas.

5.3   Area and Circumference of a Circle

We finish this chapter by presenting ways of defining the area of a disk and the length of a circle that are more elementary than those considered in [5]. To this end, we first need to discuss a few more facts about convex polygons.

We say that a convex polygon is **regular** if all of its sides and all of its internal angles have equal measures. In particular, it follows from Problem 11, page 38, that, in a regular polygon of n sides, each internal angle measures $\frac{180(n-2)}{n}$ degrees. Note further that regular polygons of three and four sides are precisely equilateral triangles and squares.

Fig. 5.14 Establishing the
existence of regular polygons

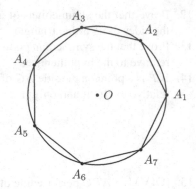

In order to assure the existence of regular polygons of n sides, divide a circle
of center O into n equal arcs, thus obtaining points $A_1, A_2, \ldots, A_n$ (cf. Fig. 5.14,
for the case $n = 7$). Since equal arcs correspond to equal chords, we conclude
that polygon $A_1 A_2 \ldots A_n$ has equal sides. On the other hand, the inscribed angle
theorem gives

$$A_1 \widehat{A_2} A_3 = \frac{1}{2} A_1 \widehat{A_n} A_3 = \frac{1}{2}(360° - A_1 \widehat{A_2} A_3)$$

$$= \frac{1}{2}(360° - A_1 \widehat{O} A_2 - A_2 \widehat{O} A_3)$$

$$= \frac{1}{2}\left(360° - 2 \cdot \frac{360°}{n}\right)$$

$$= \frac{180°(n-2)}{n}$$

and, analogously,

$$A_i \widehat{A_{i+1}} A_{i+2} = \frac{180°(n-2)}{n},$$

for $2 \leq i \leq n$ (here, we make the convention that $A_{n+1} = A_1$ and $A_{n+2} =
A_2$). Therefore, all of the internal angles of $A_1 A_2 \ldots A_n$ are also equal, so that
$A_1 A_2 \ldots A_n$ is a regular polygon of n sides.

We now consider any regular polygon $A_1 A_2 \ldots A_n$, with $n \geq 4$, and draw the
bisectors of the internal angles $\angle A_2 A_1 A_n$, $\angle A_1 A_2 A_3$ and $\angle A_2 A_3 A_4$. Let O_1 denote
the intersection point of the first two bisectors and O_2 that of the last two. Since
$\overline{A_1 A_2} = \overline{A_2 A_3}$ and

$$A_2 \widehat{A_1} O_1 = A_1 \widehat{A_2} O_1 = A_3 \widehat{A_2} O_2 = A_2 \widehat{A_3} O_2 = \frac{90°(n-2)}{n},$$

we have $O_1A_1A_2 \equiv O_2A_2A_3$ by ASA; in particular, $\overline{A_2O_1} = \overline{A_2O_2}$, and hence $O_1 = O_2$. By arguing analogously for each triple of consecutive internal angles of the polygon, we conclude that all of the bisectors of the internal angles of the polygon concur at a single point O, and the result of Problem 9, page 99, guarantees that $A_1A_2 \ldots A_n$ is tangential (in this respect, look at the comments made right after the statement of Problem 16, page 157. Also, notice that the above argument is also valid for $n = 3$; it suffices to change angles $\angle A_2A_1A_n$ and $\angle A_2A_3A_4$ by $\angle A_2A_1A_3$ and $\angle A_2A_3A_1$, respectively).

If O is the center of the circle inscribed in $A_1A_2 \ldots A_n$, the discussion of the previous paragraph has shown that, for $1 \leq i \leq n$, we have $\overline{A_1O} = \overline{A_2O} = \ldots = \overline{A_nO}$. Therefore, the circle of center O and radius equal to such a common distance passes through all of the vertices of the polygon.

We summarize these remarks in the coming

Proposition 5.18 *Every regular polygon is cyclic and tangential, and the corresponding circumscribed and inscribed circles are concentric.*

Let $\Gamma(O; 1)$ be a given circle, of radius 1. For natural numbers $m, n \geq 3$, we consider regular polygons $\mathcal{P}_n$ and $\mathcal{Q}_m$, with n and m sides, respectively, with $\mathcal{P}_n$ being inscribed and $\mathcal{Q}_m$ being circumscribed to Γ. Then, $\mathcal{P}_n$ lies in the interior of $\mathcal{Q}_m$, and writing $A(\mathcal{P}_n)$ and $A(\mathcal{Q}_m)$ to denote their areas, postulate 4 at the beginning of this chapter assures that

$$A(\mathcal{P}_n) < A(\mathcal{Q}_m).$$

Therefore, elementary facts on the supremum (resp. infimum) of sets of real numbers bounded above (resp. below) give (cf. Proposition 7.7 of [5], for instance)

$$\sup\{A(\mathcal{P}_n); \mathcal{P}_n \text{ is inscribed in } \Gamma\} \leq \inf\{A(\mathcal{Q}_m); \mathcal{Q}_m \text{ circumscribes } \Gamma\}.$$

We would like to show that both numbers above are equal and, then, to define the area of Γ by

$$A(\Gamma) = \sup\{A(\mathcal{P}_n); \mathcal{P}_n \text{ is inscribed in } \Gamma\} \tag{5.7}$$

(or, which would then be the same, $A(\Gamma) = \inf\{A(\mathcal{Q}_m); \mathcal{Q}_m \text{ circumscribes } \Gamma\}$). To this end, suppose we have shown that

$$A(\mathcal{Q}_{2^k}) - A(\mathcal{P}_{2^k}) < \left(\frac{3}{4}\right)^{k-3} c \tag{5.8}$$

for $k \geq 3$, where $c = A(\mathcal{Q}_8) - A(\mathcal{P}_8)$ is the difference between the areas of regular octagons respectively circumscribed and inscribed in Γ. Then, a standard fact on the convergence of sequences (cf. Example 7.12 of [5], for instance) assures that

$$A(\mathcal{Q}_{2^k}) - A(\mathcal{P}_{2^k}) \xrightarrow{k} 0.$$

The desired equality (5.7) is now straightforward.

Fig. 5.15 Computing Δ_n

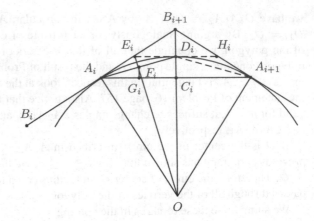

We thus need to estimate the difference between the areas of regular 2^k-gons circumscribed and inscribed in Γ. We start by establishing a slightly more general estimate.

Consider regular polygons $\mathcal{P}_n = A_1 A_2 \ldots A_n$, inscribed in Γ, and $\mathcal{Q}_n = B_1 B_2 \ldots B_n$, circumscribed to Γ, such that, for $1 \leq i \leq n$, vertex A_i is the point of tangency of side $B_i B_{i+1}$ with Γ (cf. Fig. 5.15; here again, we convention that $B_{n+1} = B_1$).

Let $l_n = \overline{A_i A_{i+1}}$ be the side length of $\mathcal{P}_n$, let C_i be the intersection point of the line segments $A_i A_{i+1}$ and $O B_{i+1}$, and $b_n = \overline{C_i B_{i+1}}$ (such a length does not depend on the chosen natural number $1 \leq i \leq n$, thanks to the congruence of the isosceles triangles $O A_i A_{i+1}$). The difference $\Delta_n = A(\mathcal{Q}_n) - A(\mathcal{P}_n)$ can be computed as follows:

$$\Delta_n = \sum_{i=1}^n A(A_i A_{i+1} B_{i+1}) = \sum_{i=1}^n \frac{1}{2} \overline{A_i A_{i+1}} \cdot \overline{C_i B_{i+1}} = \frac{n}{2} l_n b_n. \qquad (5.9)$$

Now, let D_i be the intersection point of the line segment $O B_{i+1}$ with Γ, and E_i be the intersection of $A_i B_{i+1}$ with the perpendicular bisector of $A_i D_i$. It is immediate to check that $A_i D_i$ is an edge of $\mathcal{P}_{2n}$ and E_i is a vertex of $\mathcal{Q}_{2n}$ (the line segment $E_i H_i$ of Fig. 5.15 is an edge of $\mathcal{Q}_{2n}$). Therefore,

$$\Delta_{2n} = \sum_{i=1}^n \left(A(A_i D_i E_i) + A(D_i A_{i+1} H_i) \right) = 2 \sum_{i=1}^n A(A_i D_i E_i)$$

$$= \sum_{i=1}^n \overline{A_i D_i} \cdot \overline{E_i F_i} = n l_{2n} b_{2n}, \qquad (5.10)$$

where F_i is the midpoint of $A_i D_i$.

In order to continue, we need the coming auxiliary result.

Lemma 5.19 *Let ABC be a right triangle at B. If P is the foot of the internal bisector relative to BC, then $\overline{BP} < \frac{1}{2}\overline{BC}$.*

Proof Letting Q be the foot of the perpendicular dropped from P to the hypotenuse AC, the characterization of angle bisectors as locus gives

$$\overline{PB} = \overline{PQ} < \overline{PC}.$$

Hence, $\overline{PB} < \frac{1}{2}\overline{BC}$. $\square$

Now, the inscribed angle theorem gives (cf. Fig. 5.15)

$$B_{i+1}\widehat{A}_i D_i = D_i\widehat{A}_i A_{i+1} = D_i\widehat{A}_i C_i = \frac{180°}{n}$$

(check the last equality above). Thus, $A_i D_i$ is the internal bisector of the right triangle $A_i C_i B_{i+1}$ with respect to the leg $C_i B_{i+1}$. Hence, the characterization of angle bisectors as a locus and the previous lemma give us

$$b_{2n} = \overline{E_i F_i} = \overline{F_i G_i} < \overline{C_i D_i} < \frac{1}{2}\overline{C_i B_{i+1}} = \frac{b_n}{2}. \tag{5.11}$$

On the other hand, for $n \geq 8$ we have

$$B_{i+1}\widehat{A}_{i+1}C_i = \frac{360°}{n} \leq 45°,$$

so that

$$\overline{C_i B_{i+1}} \leq \overline{A_{i+1}C_i} = \overline{A_i C_i} = \frac{l_n}{2}.$$

Hence, by successively applying the triangle inequality, Lemma 5.19 and the last inequality above to triangle $A_i C_i D_i$, we obtain (again for $n \geq 8$)

$$l_{2n} = \overline{A_i D_i} < \overline{A_i C_i} + \overline{C_i D_i} < \frac{l_n}{2} + \frac{1}{2}\overline{C_i B_{i+1}} \leq \frac{l_n}{2} + \frac{1}{2} \cdot \frac{l_n}{2} = \frac{3l_n}{4}.$$

It thus follows from (5.9), (5.10) and (5.11) that

$$\Delta_{2n} = nl_{2n}b_{2n} < n \cdot \frac{3l_n}{4} \cdot \frac{b_n}{2} = \frac{3}{4}\Delta_n$$

for $n \geq 8$.

Fig. 5.16 Approximating π

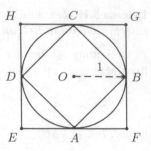

Finally, we prove (5.8): the last inequality above, together with the formula for telescoping products (cf. Section 3.3 of [5], for instance) gives

$$\frac{\Delta_{2^k}}{\Delta_8} = \prod_{j=3}^{k-1} \frac{\Delta_{2^{j+1}}}{\Delta_{2^j}} < \prod_{j=3}^{k-1} \frac{3}{4} = \left(\frac{3}{4}\right)^{k-3},$$

as wished.

The previous discussion allows us to give the following important

Definition 5.20 The real number π is the area of a circle of radius 1.

Departing from the above definition, a simple argument (see Problem 4) shows that the area of a circle of radius R is given by πR^2.

In order to obtain numerical approximations for the value of π, we rely on the reasoning presented so far, which established the well definiteness of the area of a circle of radius 1: we take such a circle, say Γ, and consider squares $ABCD$ and $EFGH$, the first one inscribed and the second one circumscribed to Γ (cf. Fig. 5.16). It is immediate that $\overline{EF} = 2$ and, by Pythagoras' theorem, $\overline{AB} = \sqrt{2}$, so that

$$2 = A(ABCD) < \pi < A(EFGH) = 4.$$

We can easily refine the above reasoning with the aid of Problems 5 and 6, as well as with their analogues for regular polygons circumscribed to Γ, thus getting better and better numerical approximations for π. For future references, the value of π, with five correct decimal places, is

$$\pi \cong 3.14159.$$

We also observe that π is an *irrational* number,[8] and we refer the reader to Section 10.5 of [5] for a proof of this fact.

---

[8]Actually, π is *transcendental*, i.e., cannot be obtained as a root of a nonzero polynomial of rational coefficients. A proof of this fact, which is far beyond the scope of these notes, can be found in [8] or [17]. We shall have more to say on transcendental numbers on Chapter 20 of [6].

Given a circle Γ of center O and radius R, and an arc $\overset{\frown}{AB}$ of Γ, we let the **circular sector** $A\widehat{O}B$ be defined as the portion of Γ formed by the union of all of the arcs OC, as C varies along $\overset{\frown}{AB}$. If $A\widehat{O}B = \alpha$, we say that $A\widehat{O}B$ is a sector of **angle** (or **opening**) α. In what follows we shall define the area of a circular sector $A\widehat{O}B$, showing that

$$A(A\widehat{O}B) = \frac{\alpha}{360°} \cdot \pi R^2,$$

provided α is the angle (in degrees) of it.

If $\alpha = 360° \cdot \frac{m}{n}$, with $m, n \in \mathbb{N}$ and $m < n$, our previous discussion on the area of a circle naturally leads us to define the area of the circular sector $A\widehat{O}B$ as being equal to the supremum of the areas of the polygons $A_1 A_2 \ldots A_{km} A_{km+1} O$, where $k \in \mathbb{N}$ and $A_1 A_2 \ldots A_{kn}$ is a regular polygon of kn sides, inscribed in Γ and such that $A_1 = A$, $A_{km+1} = B$. Then,

$$\frac{A(A_1 A_2 \ldots A_{km} A_{km+1} O)}{A(A_1 A_2 \ldots A_{kn-1} A_{kn})} = \frac{\sum_{i=1}^{km} A(A_i O A_{i+1})}{\sum_{i=1}^{kn-1} A(A_i O A_{i+1})}$$

$$= \frac{km \cdot A(A_1 O A_2)}{kn \cdot A(A_1 O A_2)} = \frac{m}{n}.$$

Since the above computations are valid for every $k \in \mathbb{N}$, we conclude that

$$A(A\widehat{O}B) = \frac{m}{n} \cdot \pi R^2.$$

Now, assume that $\alpha = 360° \cdot x$, with $x \in (0, 1)$ being irrational. It is a standard fact (cf. problem 3 of Section 7.1 of [5], for instance) that we can take rational numbers $0 < r_1 < r_2 < \cdots < x$ with $\sup\{r_1, r_2, \ldots\} = x$. Let $B_n \in \overset{\frown}{AB}$ be such that $A\widehat{O}B_n = \alpha_n$, with $\alpha_n = 360° \cdot r_n$. Then, the circular sector $A\widehat{O}B_n$ is contained in the circular sector $A\widehat{O}B$, albeit the discrepancy between the angles of such sectors gets smaller and smaller, as n increases without bound. Hence, by naturally extending Postulate 4 at the beginning of this chapter, we define the area of $A\widehat{O}B$ as being equal to the supremum of the areas of the sectors $A\widehat{O}B_n$. However, what we have already done above gives

$$A(A\widehat{O}B_n) = r_n \pi R^2,$$

so that

$$\sup\{A(A\widehat{O}B_n)\} = \sup\{r_n\} \cdot \pi R^2 = x\pi R^2 = \frac{\alpha}{360°} \cdot \pi R^2.$$

We finish this chapter by taking again a circle Γ of radius R and defining and computing its **length**, or **circumference**. This will be defined as the only positive real number $\ell(\Gamma)$ satisfying the following condition: for all regular polygons $\mathcal{P}$ and $\mathcal{Q}$, with $\mathcal{P}$ inscribed in Γ and $\mathcal{Q}$ circumscribed to Γ, we have

$$\ell(\mathcal{P}) \leq \ell(\Gamma) \leq \ell(\mathcal{Q}), \tag{5.12}$$

where $\ell(\cdot)$ stands for the perimeter of the polygon within parentheses.

In what follows, we shall show that $\ell(\Gamma)$ is well defined, with

$$\ell(\Gamma) = 2\pi R.$$

In order to develop some intuition on this value, let us take another circle, concentric with Γ but with radius $R + \frac{1}{n}$, where $n \in \mathbb{N}$. For a sufficiently large n, it is geometrically plausible that a good approximation for the area of the region of the plane in between the two circles (the gray region of Fig. 5.17) is the area of a rectangle of basis $\ell(\Gamma)$ and height $\frac{1}{n}$. This being so, we get

$$\ell(\Gamma) \cdot \frac{1}{n} \cong \pi \left(R + \frac{1}{n} \right)^2 - \pi R^2$$

or, which is the same,

$$\ell(\Gamma) \cong 2\pi R + \frac{1}{n}.$$

Assuming (also plausibly) that the approximation above gets better and better as n increases without bound, we conclude that $\ell(\Gamma) = 2\pi R$.

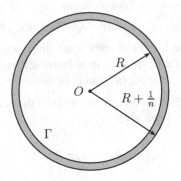

Fig. 5.17 The length of a circle of radius R

From a more formal viewpoint, in order to show that definition (5.12) makes sense and to rigorously compute $\ell(\Gamma)$, let us consider again a regular polygon $\mathcal{P}_n = A_1 A_2 \ldots A_n$ inscribed in $\Gamma(O; R)$ (cf. Fig. 5.15). Let l_n be the length of the edges of $\mathcal{P}_n$ and (by invoking once more the congruence of the isosceles triangles $O A_i A_{i+1}$) a_n be the distance from O the sides of $\mathcal{P}_n$.

Letting $\Gamma'(O; a_n)$, the inclusions $\Gamma' \subset \mathcal{P}_n \subset \Gamma$ furnish the inequalities

$$\pi a_n^2 < A(\mathcal{P}_n) < \pi R^2.$$

On the other hand, in the notations of Fig. 5.15, we have

$$A(\mathcal{P}_n) = \sum_{i=1}^{n} A(O A_i A_{i+1}) = \frac{n}{2} l_n a_n,$$

so that

$$\pi a_n^2 < \frac{n}{2} l_n a_n < \pi R^2$$

or, which is the same,

$$2\pi a_n < n l_n = \ell(\mathcal{P}_n) < \frac{2\pi R^2}{a_n}.$$

Applying Pythagoras' theorem to the right triangle $O A_i C_i$, we obtain

$$a_n = \sqrt{R^2 - \left(\frac{l_n}{2}\right)^2},$$

and hence

$$2\pi \sqrt{R^2 - \frac{l_n^2}{4}} < \ell(\mathcal{P}_n) < \frac{2\pi R^2}{\sqrt{R^2 - \frac{l_n^2}{4}}} < 2\pi R.$$

Therefore,

$$2\pi R \sqrt{1 - \frac{l_n^2}{4R^2}} < \ell(\mathcal{P}_n) < 2\pi R. \tag{5.13}$$

Now, it follows from Problem 5 that

$$l_{2n}^2 = \frac{l_n^2}{4} + \left(R - \sqrt{R^2 - \frac{l_n^2}{4}} \right)^2 = \frac{l_n^2}{4} + \left(\frac{l_n^2}{4\left(R + \sqrt{R^2 - \frac{l_n^2}{4}} \right)} \right)^2$$

$$= \frac{l_n^2}{4} + \frac{l_n^2}{16} \left(\frac{l_n}{R + \sqrt{R^2 - \frac{l_n^2}{4}}} \right)^2 .$$

If $n = 6$, then all of the triangles $O A_i A_{i+1}$ are equilateral, so that $l_6 = R$. If $n \geq 6$, Corollary 2.25 assures that $l_n \leq l_6 = R$, and hence

$$l_{2n}^2 = \frac{l_n^2}{4} + \frac{l_n^2}{16} \left(\frac{l_n}{R + \sqrt{R^2 - \frac{l_n^2}{4}}} \right)^2 < \frac{l_n^2}{4} + \frac{l_n^2}{16} = \frac{5 l_n^2}{16} < \frac{9 l_n^2}{16}.$$

The above computations furnish, for every integer $k \geq 4$ and with the aid of the formula for telescoping products,

$$l_{2^k} = l_8 \cdot \frac{l_{2^k}}{l_8} = l_8 \cdot \prod_{j=4}^{k} \frac{l_{2^j}}{l_{2^{j-1}}} < l_8 \cdot \prod_{j=4}^{k} \frac{3}{4} = \left(\frac{3}{4} \right)^{k-3} l_8 < \frac{3 l_8}{k}.$$

In view of such an inequality and letting $n = 2^k$ into (5.13), we conclude that

$$2\pi R \sqrt{1 - \frac{9 l_8^2}{4 R^2 k^2}} < 2\pi R \sqrt{1 - \frac{l_{2^k}^2}{4 R^2}} < \ell(\mathcal{P}_{2^k}) < 2\pi R.$$

Therefore,

$$\sup \ell(\mathcal{P}_n) = 2\pi R.$$

Finally, by arguing in a similar way (cf. Problem 7), we obtain

$$\inf \ell(\mathcal{Q}_n) = 2\pi R, \tag{5.14}$$

as wished.

Problems: Sect. 5.3

1. * Generalize the result of Problem 1, page 179, showing that if a point P lies in the interior of a regular polygon $A_1A_2 \ldots A_n$, then the sum of the distances of P to the lines containing the sides of $A_1A_2 \ldots A_n$ does not depend on the position of P.

2. * Let $\mathcal{P}$ and $\mathcal{Q}$ be two regular n-gons, with side lengths respectively equal to l_1 and l_2. Prove that

$$\frac{A(\mathcal{P})}{A(\mathcal{Q})} = \left(\frac{l_1}{l_2}\right)^2.$$

3. Two regular 20-gons have side lengths equal to 5cm and 12cm. Compute the side length of a third regular 20-gon, knowing that its area is equal to the sum of the areas of the two given ones.

4. * Prove that the area of a circle of radius R is equal to πR^2.

5. * Let l_n be the side length of a regular n-gon inscribed in a circle of radius R. Prove that

$$2R^2 - l_{2n}^2 = R\sqrt{4R^2 - l_n^2}.$$

   Then, apply the above relation to compute l_8 and l_{16}.

6. * Let l_n be the side length of a regular n-gon $\mathcal{P}_n$, inscribed in a circle of radius R. Prove that

$$A(\mathcal{P}_n) = \frac{n l_n}{4}\sqrt{4R^2 - l_n^2}.$$

7. * Complete the argument for the well definiteness of the definition of the circumference of a circle, by proving (5.14).

8. Points A, B and C are collinear, with $B \in AC$, and Γ, Γ_1 and Γ_2 are semicircles of diameters respectively equal to AC, AB and BC, all lying in a single half-plane, of those determined by $\overleftrightarrow{AC}$. Line r, perpendicular to AC and passing through B, intersects Γ at D. Letting S denote the area of the portion of Γ which is exterior to Γ_1 and Γ_2, compute the ratio $\frac{\overline{BD}^2}{S}$.

9. Triangle ABC is right at A. Semicircles Γ_1 and Γ_2 have AB and AC as diameters, respectively, and lie outside ABC. If Γ denotes the circumcircle of ABC, prove that the sum of the areas of the portions of Γ_1 and Γ_2 which are exterior to Γ is equal to the area of ABC itself.[9]

---

[9]Such portions of Γ_1 and Γ_2 are usually referred to as the **lunes of Hippocrates**, in honor of the Greek astronomer and mathematician of the fourth century BC Hippocrates of Chios.

10. If Γ is the circumcircle of a triangle ABC, prove that the area of ABC is less that half of the area of Γ.

11. * Let be given a circle Γ, of center O and radius R, and an arc $\overparen{AB}$ of Γ, such that $A\widehat{O}B = \alpha$. Adapt the steps of the discussion that led to the definition and computation of the area of circular sectors to define and compute the length of the arc $\overparen{AB}$. Arguing this way, show that

$$\ell(\overparen{AB}) = \frac{\alpha}{360°} \cdot 2\pi R.$$

12. Is is a curious fact that $\pi \cong \sqrt{2} + \sqrt{3}$; indeed, with five correct decimal places, we have $\sqrt{2} + \sqrt{3} \cong 3.14626$. Use this information to *rectify* a circle of radius 1, with an absolute error less than 0.01. In other words, use compass and straightedge to construct a line segment of length approximately equal to 2π, with absolute error less than 0.01.

13. We are given a circle Γ and convex polygons $\mathcal{P}$ and $\mathcal{Q}$, with $\mathcal{P}$ lying in the open disk bounded by Γ and Γ lying in the interior of $\mathcal{Q}$. Prove that $\ell(\mathcal{P}) < \ell(\Gamma) < \ell(\mathcal{Q})$, where $\ell(\cdot)$ stands for the perimeter of the curve within parentheses.

14. (IMO shortlist) We are given $n > 1$ line segments in the plane, such that the sum of its lengths is 1. Prove that one can choose a line r such that the sum of the lengths of the orthogonal projections of the n segments into r is less than $\frac{1}{\pi}$.

Chapter 6
The Cartesian Method

This chapter is devoted to the study of Plane Euclidean Geometry through the use of the *analytic* or *cartesian method*,[1] in contraposition to the synthetic method—around which the exposition has been built in the previous chapters—, and to the trigonometric method, which will be presented in the next chapter. As a result of such an approach, we will be able to develop certain aspects of the theory which have been postponed until here, the most notable of them being a first acquaintance with *conics*.

6.1 The Cartesian Plane

Draw in the plane two perpendicular lines x and y, intersecting at point O. Then, look at x and y as copies of $\mathbb{R}$, by choosing the same unit of measure and letting O correspond to 0 in both of them. This way, each of x and y is divided into two half-lines, one being *positive* (i.e., the one containing the positive reals) and the other being *negative*, with the convention that, in each line, the positive half-line is indicated by a small arrow (in Fig. 6.1, for the sake of simplicity we assumed x to be horizontal and y to be vertical with respect to the usual reading position).

Lines x and y divide the plane into four angular regions, each of which being determined by the half-lines of x and y which form their boundaries. We shall refer to such four regions as the **quadrants**, and shall number them from 1 to 4, under the convention shown in Fig. 6.1.

Given any point in the plane, say A, draw through it the perpendiculars to x and to y, which intersect such lines at the points A_x and A_y, respectively. Conversely,

---

[1]It is quite common that authors refer to the analytic method by using the expression *Analytic Geometry*. We shall systematically avoid this expression along these notes, to emphasize that it is just a set of methods, and not a new kind of geometry.

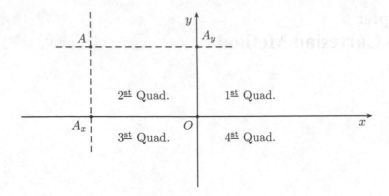

Fig. 6.1 Constructing the cartesian plane

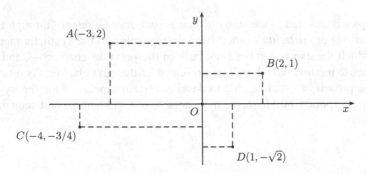

Fig. 6.2 Plotting some points in the cartesian plane

once points A_x and A_y are arbitrarily chosen on lines x and y, respectively, the perpendicular to x through A_x and to y through A_y intersect at a point A. Therefore, to give a point A in the plane is the same as to give its orthogonal projections A_x and A_y on lines x and y, respectively.

On the other hand, since lines x and y are being looked at as copies of $\mathbb{R}$, the orthogonal projections A_x and A_y of point A on x and y correspond to real numbers x_A and y_A, respectively, which completely determine the point A (for x_A and y_A completely determine points A_x and A_y). This being so, we write $A = (x_A, y_A)$, or $A(x_A, y_A)$. We illustrate this correspondence by plotting points $A(-3, 2)$, $B(2, 1)$, $C(-4, -3/4)$ and $D(1, -\sqrt{2})$ in Fig. 6.2.

In general, whenever we fix in the plane two lines x and y perpendicular at a point O and choose in each of them a *positive* half-line of origin O, we shall say that the plane is furnished with a **cartesian coordinate system**[2] xOy, or also that

---

[2]René Descartes, French mathematician, philosopher and scientist of the seventeenth century. Descartes' legacy to Mathematics and science is a huge one, and came mainly from his landmarking book *Discours de la Méthode* (Discourse on the Method) and its three corresponding

it has turned into a **cartesian plane**. This way, given a point $A(x_A, y_A)$ in it, we shall say that x_A and y_A are the **cartesian coordinates** of A. In this context, the real number x_A is the *x-coordinate* or the **abscissa** of A, whereas the real number y_A is the *y-coordinate* or the **ordinate**[3] of A. Lines x are y respectively called the *x-axis* or **axis of abscissas** and *y-axis* or **axis of ordinates** of the cartesian system upon consideration. In particular, the points situated on the x and y axes have cartesian coordinates respectively of the form $(x_0, 0)$ and $(0, y_0)$; point O, which represents 0 in both axes, has both of its coordinates equal to zero.

Hereafter, whenever we refer to the coordinates of one or more points in the plane, we shall tacitly assume, unless stated otherwise, that a cartesian coordinate system xOy has been chosen in the plane.

Now that we have defined cartesian coordinate systems, our first task is to translate, into the language of coordinates, some of the geometric concepts and results we have developed so far. In this respect, our first result gives formulas for the computation of the coordinates of a point P of a line segment AB, such that P divides AB in two other segments whose lengths form a certain ratio.

Proposition 6.1 *We are given a real number* $t \in (0, 1)$ *and points* $A(x_A, y_A)$ *and* $B(x_B, y_B)$. *If* $P(x_P, y_P)$ *is the point along segment AB for which* $\overline{AP} = t \cdot \overline{AB}$, *then*

$$x_P = (1 - t)x_A + tx_B \text{ and } y_P = (1 - t)y_A + ty_B. \tag{6.1}$$

Proof We shall prove that $x_P = (1-t)x_A + tx_B$, the proof of the other given relation being totally analogous. If $x_A = x_B$, then segment AB is vertical and, hence,

$$x_P = x_A = (1 - t)x_A + tx_A = (1 - t)x_A + tx_B.$$

If $x_A \neq x_B$, assume, without loss of generality, that $x_A < x_B$ (the case $x_A > x_B$ is analogous). Letting A', P' and B' be the orthogonal projections of points A, P and B on the x-axis (cf. Fig. 6.3), it follows from Thales' theorem 4.2 that

$$\frac{\overline{A'P'}}{\overline{A'B'}} = \frac{\overline{AP}}{\overline{AB}} = t.$$

However, since $A'(x_A, 0)$, $P'(x_P, 0)$ and $B'(x_B, 0)$, with $P' \in A'B'$ (for $P \in AB$) we have $\overline{A'P'} = x_P - x_A$ and $\overline{P'B'} = x_B - x_P$. Substituting such equalities in the above relation, we obtain

$$\frac{x_P - x_A}{x_B - x_A} = t$$

---

appendices. This book marks a turning point on the way of doing science, for, along it, Descartes strongly rejected the scholastic tradition of using speculation, instead of deduction, as the central strategy for the investigation of natural phenomena. On the other hand, its appendix *La Géométrie* layed down the foundations of the analytic method.

[3] We call the reader's attention so as not to confuse *ordinate* with *coordinate*; the ordinate of a point of the cartesian plane is one of its coordinates.

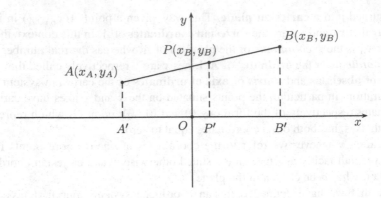

Fig. 6.3 Dividing a line segment in a given ratio

or, which is the same, $x_P - x_A = t(x_B - x_A)$. Therefore,

$$x_P = x_A + t(x_B - x_A) = (1 - t)x_A + tx_B,$$

as we wished to show.   □

For an important generalization of the previous result, see Problem 8, page 201. For the time being, let us isolate a relevant particular case.

Corollary 6.2 *Given points A and B in the cartesian plane, the coordinates of the midpoint M of segment AB are the arithmetic means of the corresponding coordinates of A and B. More precisely, if $A(x_A, y_A)$ and $B(x_B, y_B)$, then $M\left(\frac{x_A + x_B}{2}, \frac{y_A + y_B}{2}\right)$.*

Proof Since $\overline{AM} = \frac{1}{2}\overline{AB}$, it suffices to let $t = \frac{1}{2}$ in (6.1) to obtain the coordinates of M.   □

Remark 6.3 In the notations of Proposition 6.1, it is suggestive to write

$$P = (1 - t)A + tB \qquad\qquad (6.2)$$

as a shorthand for both relations in (6.1). In particular, we abbreviate the coordinates of the midpoint M of line segment AB by writing $M = \frac{A+B}{2}$. Hereafter, we shall stick to this usage whenever convenient.

As the coming example illustrates, one of the highlights of the cartesian method is its use to obtain euclidean geometry theorems that, yet accessible by other means, become trivial when analytically approached.

Example 6.4 Let us apply the analytic method to show that, in every triangle, the medians concur at a single point, the barycenter of the triangle, and that such a point divides each median, from the corresponding vertex, in the ratio 2 : 1.

To this end, let ABC be any triangle, M be the midpoint of side BC and G be the point along median AM such that $\overline{AG} = 2\,\overline{GM}$. Then, $\overline{AG} = \frac{2}{3}\overline{AM}$, and (6.2) (with $t = \frac{2}{3}$), together with Corollary 6.2, give

$$G = \left(1 - \frac{2}{3}\right)A + \frac{2}{3}M = \frac{1}{3}A + \frac{2}{3}\left(\frac{A + B}{2}\right) = \frac{A + B + C}{3}. \qquad (6.3)$$

Analogous computations performed with the other two medians show that they pass through this point too.

We continue the exposition of the theory by deriving an extremely useful formula for the computation of the distance between two points of the cartesian plane in terms of their coordinates.

Proposition 6.5 *For points $A(x_A, y_A)$ and $B(x_B, y_B)$ in the cartesian plane, we have*

$$\overline{AB} = \sqrt{(x_A - x_B)^2 + (y_A - y_B)^2}. \qquad (6.4)$$

Proof We need to consider four distinct cases: $x_A \le x_B$ and $y_A \le y_B$; $x_A \le x_B$ and $y_A > y_B$; $x_A > x_B$ and $y_A \le y_B$; $x_A > x_B$ and $y_A > y_B$. Nevertheless, since the analysis of each of these four cases is essentially equivalent to the remaining ones, we shall concentrate ourselves in the case $x_A \le x_B$ and $y_A \le y_B$ (cf. Fig. 6.4).

For the sake of simplicity of notation, let $A(a, b)$ and $B(c, d)$, so that $a \le c$ and $b \le d$. If $a = c$ (the possibility at left, in Fig. 6.4), we clearly have

$$\overline{AB} = d - b = |b - d| = \sqrt{0^2 + (b - d)^2} = \sqrt{(a - c)^2 + (b - d)^2}.$$

Since the case $b = d$ can be dealt with in pretty much the same way, from now on we shall assume that $a < b$ and $c < d$ (possibility at right in Fig. 6.4). Draw through

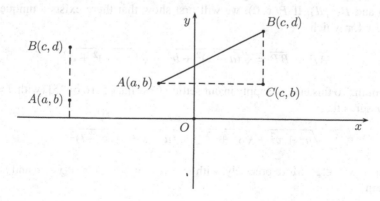

Fig. 6.4 Distance between two points of the cartesian plane

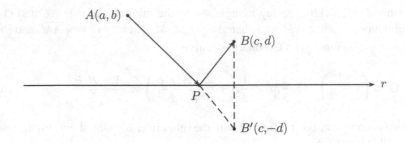

Fig. 6.5 Example 2.28 via the analytic method

A the parallel to the x-axis and through B the parallel to the y-axis, and mark their intersection point C. Since C has the same ordinate as A and the same abscissa as B, we conclude that $C(c, b)$. Moreover, since the cartesian axes are perpendicular, triangle ABC is right at C. Therefore, it follows from Pythagoras' theorem, together with the two particular cases above, that

$$\overline{AB}^2 = \overline{AC}^2 + \overline{BC}^2 = (c - a)^2 + (d - b)^2 = (a - c)^2 + (b - d)^2,$$

as we wished to show. $\square$

We shall now use formula (6.4) and a suitable corollary of Cauchy-Schwarz inequality to analyse Example 2.28 with the aid of the analytic method.

Example 6.6 In the notations of Fig. 6.5, show that there exists a single point $P \in r$ for which the sum $\overline{AP} + \overline{BP}$ is minimum. Moreover, show also that such a point is obtained as the intersection of the straightlines $\overleftrightarrow{AB'}$ and r, where B' stands for the symmetric of B with respect to r.

Proof We consider a cartesian system in which r corresponds to the x-axis, and let $A(a, b)$ and $B(c, d)$. If $P(x, 0)$, we will first show that there exists a unique real value of x for which

$$\overline{AP} + \overline{BP} = \sqrt{(a - x)^2 + b^2} + \sqrt{(x - c)^2 + d^2}$$

is minimum. To this end, we apply in our setting Corollary 5.16 of [5] (with $n = 2$), which assures that

$$\sqrt{u^2 + v^2} + \sqrt{s^2 + t^2} \geq \sqrt{(u + s)^2 + (v + t)^2}$$

for all $u, v, s, t \in \mathbb{R}$. More precisely, with $u = a - x$, $v = b$, $s = x - c$ and $t = d$, we obtain

$$\overline{AP} + \overline{BP} = \sqrt{(a-x)^2 + b^2} + \sqrt{(x-c)^2 + d^2}$$
$$\geq \sqrt{((a-x) + (x-c))^2 + (b+d)^2}$$
$$= \sqrt{(a-c)^2 + (b+d)^2}$$
$$= \overline{AB'},$$

where $B'(c, -d)$ is the symmetric of B with respect to r. However, since $\overline{BP} = \overline{B'P}$, equality holds if and only if $\overline{AP} + \overline{B'P} = \overline{AB'}$, i.e., if and only if points A, P and B' are collinear. ☐

The next example uses the formula for the distance between two points to establish the nonexistence, in the cartesian plane, of equilateral triangles whose vertices have integer coordinates. Yet with respect to this problem, a beautiful theorem of Minkowski[4] (cf. Problem 11, page 287) states that squares are the only regular polygons, in the cartesian plane, all of whose vertices can have integer coordinates.

Example 6.7 Does there exist, in the cartesian plane, points A, B, C of integer coordinates and such that ABC is an equilateral triangle? Justify your answer!

Solution No! For the sake of contradiction, assume that $A(x_A, y_A)$, $B(x_B, y_B)$, $C(x_C, y_C)$, with x_A, y_A, x_B, y_B, x_C, $y_C \in \mathbb{Z}$, was such a triple of points. Then, on the one hand, according to Problem 11, we would have $A(ABC) = \frac{m}{2}$, for some $m \in \mathbb{N}$.

On the other hand, by applying the result of Problem 4, page 156, together with formula (6.4), we conclude that

$$A(ABC) = \overline{AB}^2 \cdot \frac{\sqrt{3}}{4} = [(x_B - x_A)^2 + (y_B - y_A)^2]\frac{\sqrt{3}}{4}.$$

The two expressions obtained above for the area of ABC would then give

$$\sqrt{3} = \frac{2m}{(x_B - x_A)^2 + (y_B - y_A)^2} \in \mathbb{Q},$$

which is a contradiction. ☐

We finish this section by using formula (6.4) to analyse a certain locus which generalizes the notion of perpendicular bisector of a line segment (in the notations of the coming proposition, the perpendicular bisector amounts to the case $k = 0$). By doing so, we complete the proof of Theorem 4.35.

---

[4]Hermann Minkowski, German mathematician of the nineteenth century.

Proposition 6.8 *Given a real number k and distinct points A and B in the plane, the locus of the points P in the plane for which $\overline{AP}^2 - \overline{BP}^2 = k^2$ is a line perpendicular to $\overleftrightarrow{AB}$. More precisely, if M is the midpoint of AB, then such a line intersects $\overrightarrow{MB}$ at a distance $\frac{k^2}{2AB}$ of M.*

Proof Choose a cartesian system in which $A(0, 0)$ and $B(a, 0)$. If $P(x, y)$, then a straightforward application of the formula for the distance between two points of the cartesian plane gives

$$\overline{AP}^2 - \overline{BP}^2 = k^2 \Leftrightarrow (x^2 + y^2) - ((x - a)^2 + y^2) = k^2$$

$$\Leftrightarrow x = \frac{a^2 + k^2}{2a}.$$

In turn, this last equality is the same as saying that P belongs to the vertical line (thus, perpendicular to $\overleftrightarrow{AB}$) formed by those points whose abscissas are equal to $\frac{a^2+k^2}{2a}$. $\qquad\square$

Problems: Sect. 6.1

1. * Given real numbers a and b, prove that points (a, b) and $(-a, -b)$ of a cartesian plane are symmetric with respect to the origin.

2. * We are given cartesian systems xOy and $x'O'y'$, such that the x'-axis is the horizontal line $y = y_0$ of the first cartesian system, whereas the y'-axis is the vertical line $x = x_0$ of the first system. If a point A has coordinates (x, y) in the first system and (x', y') in the second, show that $x' = x - x_0$ and $y' = y - y_0$.

3. In the cartesian plane, let $P = \left(\frac{1}{3}, \sqrt{2}\right)$. If A and B are distinct points of integer coordinates, prove that $\overline{AP} \neq \overline{BP}$. Then, given $n \in \mathbb{N}$, use this fact to show that there exists a disk in the plane containing exactly n points of integer coordinates.

4. Show that, in the cartesian plane, there are infinitely many regular polygons, of pairwise distinct numbers of edges, and such that each of then contains exactly 2018 points of integer coordinates in its interior.

5. Let ABC be a triangle with sides $\overline{AB} = c$, $\overline{AC} = b$, $\overline{BC} = a$. If m_a is the length of the median relative to BC, prove that

$$a^2 + 4m_a^2 = 2(b^2 + c^2).$$

6. * If $ABCD$ is a convex quadrilateral of diagonals AC and BD, and M and N respectively denote the midpoints of them, we say that line segment MN is **Euler's median** of $ABCD$. Prove **Euler's median theorem**: in the above notations, we have

$$\overline{AB}^2 + \overline{BC}^2 + \overline{CD}^2 + \overline{DA}^2 = \overline{AC}^2 + \overline{BD}^2 + 4\overline{MN}^2.$$

7. If R stands for the circumradius of a triangle ABC of circumcenter O and barycenter G, prove that

$$\overline{OG}^2 = R^2 - \frac{1}{9}(\overline{AB}^2 + \overline{BC}^2 + \overline{CA}^2).$$

8. (OIM) ABC is an equilateral triangle of side length 2 and incircle Γ. Given a point P in Γ, show that:

   (a) $\overline{AP}^2 + \overline{BP}^2 + \overline{CP}^2 = 5$.
   (b) AP, BP, CP are the sides of a triangle of area equal to $\frac{\sqrt{3}}{4}$.

9. (Romania) Let $m, n > 1$ be given integers and $A_1, A_2, \ldots, A_n, B_1, B_2, \ldots, B_m$ be given points in the plane, with $\overline{A_i B_j} = \sqrt{i+j}$ for $1 \le i \le n$, $1 \le j \le m$. Prove that all of the A_i's are collinear, all of the B_j's are collinear, and that the two straightlines containing these sets of points are perpendicular.

10. * Given circles Γ and Σ in the plane, we say that Γ intersects Σ *along a diameter* if $\Gamma \cap \Sigma = \{A, B\}$, with AB being a diameter of Σ. Let two nonconcentric circles Γ_1 and Γ_2 be given. Prove that the locus of the points of the plane which are centers of circles that intersect both Γ_1 and Γ_2 along diameters is the line symmetric to the radical axis of Γ_1 and Γ_2 with respect to the midpoint of the line segment joining the centers of the circles.

11. * In the cartesian plane, let $A_1 A_2 A_3$ be a triangle with vertices $A_i(x_i, y_i)$, for $1 \le i \le 3$. Show that

$$A(A_1 A_2 A_3) = \frac{1}{2} \left| \sum_{i=1}^{3} (x_i y_{i+1} - x_{i+1} y_i) \right|.$$

Then, conclude that, in the cartesian plane, every triangle whose vertices have integer coordinates has area at least $\frac{1}{2}$.

12. (IMO) Prove that, for every integer $n > 2$, there exists in the plane a set of n points satisfying the two following conditions:

   (a) The distance between any two of them is an irrational number.
   (b) Each three of them are noncollinear and form a triangle of rational area.

   Problems 13 to 18 are devoted to the proof of the famous *Theorem of Pick*[5] on *simple polygons* in the cartesian plane having **lattice points** (i.e., points of integer coordinates) as vertices. We follow Essay 5 of the beautiful [15], to which we refer for more details.

13. * Let A, B and C be noncollinear lattice points in the cartesian plane. If $A(ABC) = \frac{1}{2}$, show that ABC does not contain other lattice points, either along its sides or in its interior.

---

[5]Georg A. Pick, Austrian mathematician of the nineteenth and twentieth centuries.

We now need to generalize the concept of convex polygon, in the following way: we say that $\mathcal{P} = A_1 A_2 \ldots A_n$, with $n \geq 3$, is a **simple polygon** of n edges, or simply an **n-gon**, if the following conditions are satisfied (under the convention that $A_0 = A_n$ and $A_{n+1} = A_1$):

(a) points $A_1, A_2, \ldots, A_n$ are pairwise distinct.
(b) line segments $A_i A_{i+1}$ and $A_j A_{j+1}$ intersect if and only if $|i - j| \leq 1$.
(c) for $1 \leq i \leq n$, points A_{i-1}, A_i and A_{i+1} are not collinear.

We shall assume without proof that $\mathcal{P}$ divides the plane in two regions, exactly one of which is bounded[6]; in turn, such a bounded region is called the **interior** of $\mathcal{P}$. For $1 \leq i \leq n$, the **interior angle** of $\mathcal{P}$ at vertex A_i is the angle $\angle A_{i-1} A_i A_{i+1}$, of vertex A_i, sides $\overrightarrow{A_i A_{i-1}}$ and $\overrightarrow{A_i A_{i+1}}$ and which intersects the interior of $\mathcal{P}$.

14. * The purpose of this problem is to show that any simple n-gon $\mathcal{P}$ can be partitioned into $n - 2$ triangles by drawing diagonals which intersect only at vertices of $\mathcal{P}$. To this end, do the following items:

(a) Choose a cartesian system in which no edge of $\mathcal{P}$ is parallel to the ordinate axis. Show that exactly one point in the boundary of $\mathcal{P}$ has maximum abscissa, and that it is one of the vertices of the n-gon.

(b) From now on, let A denote the vertex of $\mathcal{P}$ whose existence was established in (a), and B and C be the vertices of $\mathcal{P}$ adjacent to A. If $BC \subset \mathcal{P}$, write $\mathcal{P} = ABC \cup \mathcal{P}'$; then, apply an adequate inductive hypothesis to $\mathcal{P}'$ to obtain the desired result.

(c) If $BC \not\subset \mathcal{P}$, show that there exists a vertex D of $\mathcal{P}$ lying in the interior of ABC with maximum abscissa. Conclude that $AD \subset ABC$.

(d) *By cutting* $\mathcal{P}$ *along* AD, show that we get a partition $\mathcal{P} = \mathcal{P}' \cup \mathcal{P}''$, with both $\mathcal{P}'$ and $\mathcal{P}''$ having fewer edges that $\mathcal{P}$. Then, apply an adequate inductive hypothesis to $\mathcal{P}'$ and $\mathcal{P}''$ to obtain the desired result also in this case.

15. If $\mathcal{P}$ is a simple n-gon, show that:

(a) The sum of the measures of its internal angles is equal to $180°(n - 2)$.

(b) If $\mathcal{P}$ is partitioned into $n - 2$ triangles as in the previous problem, then the sum of the areas of these triangles do not depend on the way the partition was carried out.

Note that item (b) of the previous problem allows us to *define* the **area** of a simple n-gon $\mathcal{P}$ as the sum of the areas of the $n - 2$ triangles obtained from a partition of $\mathcal{P}$ as in Problem 14.

16. If A, B and C are noncollinear lattice points, such that ABC does not contain other lattice points neither in its interior nor along its edges, we shall say that

---

[6]A proof of this result is beyond the scope of these notes. It is known as the **Jordan curve theorem**, in honor of its discoverer, the French mathematician of the nineteenth century Camille Jordan.

ABC is a *fundamental triangle* in the cartesian plane. This being said, do the following items:

(a) Show that every triangle whose vertices are lattice points can be partitioned into fundamental triangles.
(b) For the remaining items, let $\mathcal{P}$ be a simple n-gon in the cartesian plane, with lattice points as vertices.

     i. Show that $\mathcal{P}$ can be partitioned into fundamental triangles.
     ii. Assume that $\mathcal{P}$ has been partitioned into k fundamental triangles, and let I and B stand for the numbers of points of integer coordinates lying respectively in the interior and along the edges of $\mathcal{P}$. Show that the sum of the interior angles of these k fundamental triangles is equal to

$$180°(n - 2) + 180°(B - n) + 360°I,$$

     iii. Conclude that $k = B + 2I - 2$, so that k does not depend on the way the partition was carried out.

17. Let ABC be a fundamental triangle in the cartesian plane, and $\mathcal{R}$ be the smallest rectangle with sides parallel to the cartesian axes and containing ABC.

(a) Show that A, B, C lie along the edges of $\mathcal{R}$, so that its vertices are lattice points.
(b) If $\mathcal{R}$ has area S, show that it can be partitioned into $2S$ fundamental triangles, one of which is ABC.
(c) Show that $A(ABC) = \frac{1}{2}$.

18. Prove **Pick's theorem**: if $\mathcal{P}$ is a simple polygon in the cartesian plane, with lattice points as vertices, then

$$A(\mathcal{P}) = I + \frac{B}{2} - 1,$$

where $A(\mathcal{P})$ stands for the area of $\mathcal{P}$ and I and B denote the numbers of lattice points respectively in the interior and along the edges of $\mathcal{P}$.
19. Given a positive integer n, a square of side length n covers $(n + 1)^2$ lattice points if its vertices are also lattice points and its sides are parallel to the axes. Prove that, no matter how the square is situated in the cartesian plane, it covers at most $(n + 1)^2$ points of integer coordinates.

6.2 Lines and Circles

As we have previously stressed, in all that follows we assume a cartesian system of coordinates given in the plane. In this section, we study the problem of how to algebraically represent straightlines and circles in such a system. In what concerns lines, we have the following fundamental result.

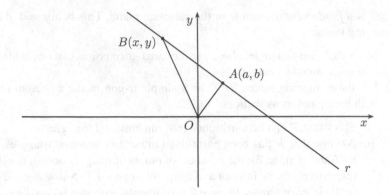

Fig. 6.6 A line in the cartesian plane not passing through the origin

Theorem 6.9 *Every straightline in the cartesian plane can be sees as the set of points (x, y) satisfying an equation of the form*

$$ax + by + c = 0, \tag{6.5}$$

with a, b and c being real numbers such that $a \neq 0$ or $b \neq 0$.

Proof Let O be the origin and r be a line in the plane. We first assume that $O \notin r$. If $A(a, b)$ is the foot of the perpendicular dropped from O to r (cf. Fig. 6.6), it is clear that a point $B(x, y)$ lies in r if and only if $O\widehat{A}B = 90°$. In turn, Pythagoras' theorem and its converse assure that

$$(x, y) \in r \Leftrightarrow \overline{OA}^2 + \overline{AB}^2 = \overline{OB}^2$$

$$\Leftrightarrow (a^2 + b^2) + [(x - a)^2 + (y - b)^2] = x^2 + y^2$$

$$\Leftrightarrow ax + by - (a^2 + b^2) = 0,$$

with (thanks to $O \notin r$) $a \neq 0$ or $b \neq 0$. It thus follows that r is the set of solutions (x, y) of equation $ax + by + c = 0$, with $c = -(a^2 + b^2)$.

Let us now suppose that $O \in r$ (cf. Fig. 6.7) and, on the perpendicular to r drawn through O, mark any point $A(a, b)$, distinct from O itself. As in the first case, a point $B(x, y)$ lies in r if and only if $A\widehat{O}B = 90°$. Hence, again from Pythagoras' theorem and its converse, we have

$$(x, y) \in r \Leftrightarrow \overline{OA}^2 + \overline{OB}^2 = \overline{AB}^2$$

$$\Leftrightarrow (a^2 + b^2) + (x^2 + y^2) = [(x - a)^2 + (y - b)^2]$$

$$\Leftrightarrow ax + by = 0.$$

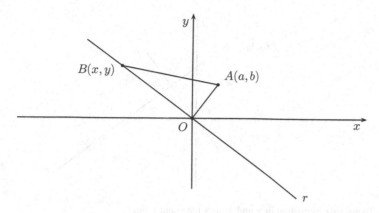

Fig. 6.7 A line in the cartesian plane passing through the origin

Finally, since $A \notin r$, we have $a \neq 0$ or $b \neq 0$, so that r is the set of solutions (x, y) of equation $ax + by + c = 0$, with $c = 0$. □

In the notations of the previous result, we say that (6.5) is the **linear equation** of the straightline r, which we indicate by writing

$$r : \{ax + by + c = 0\}.$$

The term *linear* here refers to the fact that $ax+by+c$ is a *first degree real polynomial in x and y*.

Remark 6.10 For every real $k \neq 0$, it is of course also licit to say that

$$(ka)x + (kb)y + (kc) = 0$$

is the linear equation of the straightline r. Actually, it is not difficult to algebraically verify that every linear equation of r is of this form. More precisely, let a', b' and c' be real numbers, with at least one of a' and b' being nonzero; if the solution set of $a'x + b'y + c' = 0$ is r, then there exists a nonzero real number k such that $a' = ka$, $b' = kb$ and $c' = kc$. Indeed, this follows from the fact that the linear system of equations

$$\begin{cases} ax + by = -c \\ a'x + b'y = -c' \end{cases}.$$

has infinitely many solutions if and only if $a' = ka$, $b' = kb$ and $c' = kc$, for some nonzero k (see Section 2.4 of [5], for instance).

Henceforth, we shall use the above remark without further comments.

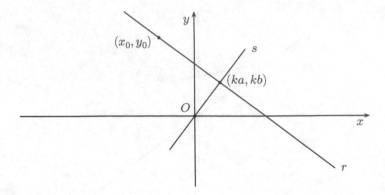

Fig. 6.8 Every linear equation in x and y does represent a line

We now collect a few consequences of Theorem 6.9, which will shed light on the relations between the linear equation of a straightline and its position in the cartesian plane.

Corollary 6.11 *Let r be a line in the cartesian plane, with linear equation $ax + by + c = 0$. If s is the perpendicular to r passing through the origin, then s has linear equation $-bx + ay = 0$.*

Proof According to the previous theorem, line s has a linear equation of the form $cx + dy = 0$, for some $c, d \in \mathbb{R}$, not both zero. However, since r has linear equation $ax + by + c = 0$, the proof of Theorem 6.9 assures that $(a, b) \in s$. Hence, $ca + db = 0$ and, by the remark above, we can take $c = -b$ and $d = a$. $\qquad\square$

Let us establish the converse of Theorem 6.9.

Corollary 6.12 *If a, b and c are given real numbers, with $a \neq 0$ or $b \neq 0$, then the solution set of equation $ax + by + c = 0$ is a line in the cartesian plane.*

Proof Suppose that $c \neq 0$, the analysis of the case $c = 0$ being totally analogous. Choose an arbitrary solution $x = x_0$, $y = y_0$ of equation $ax + by + c = 0$, so that $ax_0 + by_0 + c = 0$. If s is the line passing through $(0, 0)$ and (a, b), then it has linear equation $-bx + ay = 0$. Let r be the line passing through (x_0, y_0) and perpendicular to s, say at point (ka, kb) (cf. Fig. 6.8. It is pretty straightforward to show that all points on s are of this form). By the previous corollary, r has a linear equation of the form $(ka)x + (kb)y + c' = 0$, for some $c' \in \mathbb{R}$, or also $ax + by + c'' = 0$, with $c'' = \frac{c'}{k}$ (since $c \neq 0$, we have $(0, 0) \notin r$, so that $k \neq 0$). However, since $(x_0, y_0) \in r$, we have $ax_0 + by_0 + c'' = 0$, and it follows from what we did above that

$$c'' = -ax_0 - by_0 = c.$$

Thus, $ax + by + c = 0$ is a linear equation of r. $\qquad\square$

Thanks to the previous corollary, whenever convenient, hereafter we shall refer to a linear equation $ax + by + c = 0$, with $a \neq 0$ or $b \neq 0$, simply as *a line*. We then notice that a line $ax + by + c = 0$ is vertical in the chosen cartesian plane if and only if $b = 0$. Hence, if a line r of equation $ax + by + c = 0$ is not vertical, then we can write it as $y = -\frac{a}{b}x - \frac{c}{b}$ or, more succinctly (writing a in place of $-\frac{a}{b}$ and b in place of $-\frac{c}{b}$), as

$$y = ax + b. \tag{6.6}$$

Setting $x = 0$ in the equation above, we see that r intersects the ordinate axis at the point $(0, b)$; also, we shall prove in Proposition 7.12 that the coefficient a is intimately related with the angle formed by r and the horizontal axis. For this reason, one usually refers to (6.6) as the **slope-intercept** equation of r.

We now apply the comments above to present the usual necessary and sufficient condition for the parallelism of two straightlines in terms of their slope-intercept equations.

Proposition 6.13 *In the cartesian plane, two non vertical lines r and s are parallel if and only if have slope-intercept equations of the form $y = ax + b$ and $y = ax + b'$, for some real numbers a, b and b', with $b \neq b'$.*

Proof First assume that $r \parallel s$, with r of equation $y = ax + b$ or, which is the same, $ax - y + b = 0$. Corollary 6.11 assures that line t, of equation $x + ay = 0$ is perpendicular to r. Now, since $r \parallel s$, we also have $s \perp t$, so that a further application of Corollary 6.11 guarantees that s has linear equation of the form $ax - y + b' = 0$, for some $b' \in \mathbb{R}$.

Conversely, if r and s have slope-intercept equations of the forms $y = ax + b$ and $y = ax + b'$, with $b \neq b'$, it is immediate to see that the linear system of equations

$$\begin{cases} y = ax + b \\ y = ax + b' \end{cases}$$

does not have solutions, so that $r \parallel s$ (in this respect, see also Problem 5). $\square$

We finish the study of lines in the cartesian plane by establishing the usual necessary and sufficient condition for the perpendicularity of two straightlines. We shall only look at the nontrivial case in which none of the two lines is either vertical or horizontal.

Proposition 6.14 *Two non vertical lines r and s, of slope-intercept equations $y = ax + b$ and $y = a'x + b'$, are perpendicular if and only if $aa' = -1$.*

Proof We know from Corollary 6.11 that line t of equation $x + ay = 0$ (or, equivalently, $y = -\frac{1}{a}x$) is perpendicular to r. On the other hand, since $r \perp s \Leftrightarrow t \parallel s$, the previous proposition assures that $r \perp s$ if and only if t has slope-intercept equation of the form $y = a'x + c$, for some $c \in \mathbb{R}$. Since $c = 0$ (for t passes through the origin) and t has slope-intercept equation $y = -\frac{1}{a}x$, we conclude that t has equation $y = a'x$ if and only if $a' = -\frac{1}{a}$, i.e., if and only if $aa' = -1$. $\square$

The coming example exercises the results of Propositions 6.13 and 6.14.

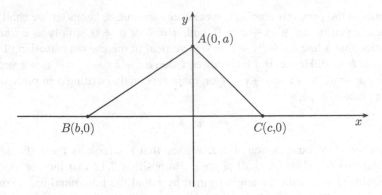

Fig. 6.9 The orthocenter of a triangle through the analytic method

Example 6.15 We are given in the cartesian plane line r, of equation $2x - 3y + \sqrt{2} = 0$, and point $A(2, -1)$. Find the equations of lines s and t, which pass through A and are such that $s \parallel r$ and $t \perp r$.

Solution Since r has slope-intercept equation $y = \frac{2}{3}x + \frac{\sqrt{2}}{3}$, Propositions 6.13 and 6.14 assure that the corresponding equations for s and t are $y = \frac{2}{3}x + c$ and $y = -\frac{3}{2}x + d$, respectively, with $c, d \in \mathbb{R}$ being chosen in such a way that $A \in s, t$. Substituting the coordinates of A in the equations for s and t, we obtain $c = -\frac{7}{3}$ and $d = 2$.   □

We now present yet another instance on how one can profitably use the analytic method in Euclidean Geometry, in the spirit of Example 6.4.

Example 6.16 Prove that, in every triangle, the three altitudes pass through a single point.

Proof Let ABC be a given (without loss of generality) non right triangle. Choose a cartesian system in which $A(0, a)$, $B(b, 0)$ and $C(c, 0)$, with $b, c \neq 0$ and $b \neq c$ (cf. Fig. 6.9). One of the altitudes of ABC is the ordinate axis. In order to obtain the equations of the other two altitudes, we shall first get the equations of lines $\overleftrightarrow{AB}$ and $\overleftrightarrow{AC}$; for the sake of simplicity, we shall resort to the result of Problem 6, according to which:

$$\overleftrightarrow{AB} : \left\{ \frac{x}{b} + \frac{y}{a} = 1 \right\} \quad \text{and} \quad \overleftrightarrow{AC} : \left\{ \frac{x}{c} + \frac{y}{a} = 1 \right\}$$

or, which is the same, $\overleftrightarrow{AB} : \{y = -\frac{a}{b}x + a\}$ and $\overleftrightarrow{AC} : \{y = -\frac{a}{c}x + a\}$.

Letting r and s denote the lines containing the altitudes of ABC through B and C, respectively, Proposition 6.14 gives

$$r : \left\{ y = \frac{c}{a}x + \beta \right\} \quad \text{and} \quad s : \left\{ y = \frac{b}{a}x + \gamma \right\},$$

with $\beta, \gamma \in \mathbb{R}$ such that $B \in r$ and $C \in s$. Substituting the coordinates of B and C in the equations above, we obtain $\beta = \gamma = -\frac{bc}{a}$, so that

$$r : \{ay = cx - bc\} \quad \text{and} \quad s : \{ay = bx - bc\}.$$

Finally, the solution of the linear system

$$\begin{cases} ay = cx - bc \\ ay = bx - bc \end{cases}$$

gives the coordinates of the point H of intersection of r and s. Since $b \neq c$, it is clear from the system that $x = 0$, i.e., point H also belongs to the altitude of ABC passing through A. □

A direct application of formula (6.4) allows us to look at circles in the cartesian plane. To this end, recall that the circle $\Gamma(C; R)$ is defined as the set of points A whose distance from the center C is equal to R.

Given in the plane a cartesian system of coordinates as in Fig. 6.10, let $C(x_0, y_0)$ and $A(x, y)$ be any point. It follows from (6.4) that

$$A \in \Gamma \Leftrightarrow \overline{AC} = R \Leftrightarrow \overline{AC}^2 = R^2$$
$$\Leftrightarrow (x - x_0)^2 + (y - y_0)^2 = R^2$$
$$\Leftrightarrow x^2 + y^2 - 2x_0 x - 2y_0 y + (x_0^2 + y_0^2 - R^2) = 0.$$

Therefore, we say that

$$x^2 + y^2 - 2x_0 x - 2y_0 y + (x_0^2 + y_0^2 - R^2) = 0 \tag{6.7}$$

is the **equation of the circle** of center C and radius R.

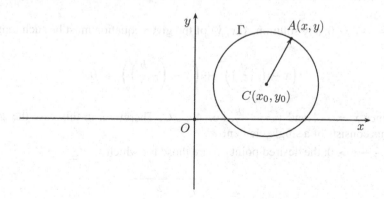

Fig. 6.10 The equation of a circle in the cartesian plane

Conversely, it is natural to ask whether every equation of the form

$$x^2 + y^2 + ax + by + c = 0, \tag{6.8}$$

has, as solution set in the cartesian plane, the set of points of a circle. We shall now see that, apart from the case in which there is at most one solution (x, y) for such an equation, the answer is yes! To this end, we start by *completing squares* in $x^2 + ax$ and $y^2 + by$ to obtain

$$x^2 + y^2 + ax + by + c =$$

$$= \left(x^2 + 2 \cdot \frac{a}{2} x + \frac{a^2}{4} \right) + \left(y^2 + 2 \cdot \frac{b}{2} x + \frac{b^2}{4} \right) + \left(c - \frac{a^2}{4} - \frac{b^2}{4} \right)$$

$$= \left(x + \frac{a}{2} \right)^2 + \left(y + \frac{b}{2} \right)^2 + \left(c - \frac{a^2}{4} - \frac{b^2}{4} \right).$$

Hence, (6.8) is equivalent to the equation

$$\left(x - \left(-\frac{a}{2} \right) \right)^2 + \left(y - \left(-\frac{b}{2} \right) \right)^2 = \frac{a^2 + b^2}{4} - C.$$

Letting $C\left(-\frac{a}{2}, -\frac{b}{2} \right)$, it follows from the relation above, together with formula (6.4) for the distance between two points in the cartesian plane, that the set of points $A(x, y)$ satisfying (6.8) coincides with the set

$$\left\{ A; \ \overline{AC}^2 = \frac{a^2 + b^2}{4} - c \right\}.$$

Then, there are three possibilities:

- $\frac{a^2+b^2}{4} - c < 0$: since the square of a real number is always nonnegative, in this case there is no point A such that $\overline{AC}^2 = \frac{a^2+b^2}{4} - c$, and the set of points is empty.
- $\frac{a^2+b^2}{4} - c = 0$: any solution $A(x, y)$ of the given equation must be such that

$$\left(x - \left(-\frac{a}{2} \right) \right)^2 + \left(y - \left(-\frac{b}{2} \right) \right)^2 = 0,$$

  whence $x = -\frac{a}{2}$ and $y = -\frac{b}{2}$, i.e. $A = C$. Therefore, in this case the set of points consists of a single element.
- $\frac{a^2+b^2}{4} - c > 0$: the desired points A are those for which

$$\overline{AC} = \sqrt{\frac{a^2 + b^2}{4} - c},$$

i.e., the solution set of (6.8) is the circle of center $C\left(-\frac{a}{2}, -\frac{b}{2}\right)$ and radius $R = \sqrt{\frac{a^2+b^2}{4} - c}$.

The above discussion has proved the following

Theorem 6.17 *In the cartesian plane, equation*

$$x^2 + y^2 + ax + by + c = 0$$

represents the empty set, a single point or a circle, according to whether $\frac{a^2+b^2}{4} - c$ is negative, zero or positive, respectively. Moreover, in the last case the center of the circle is the point $\left(-\frac{a}{2}, -\frac{b}{2}\right)$ and its radius is equal to $\sqrt{\frac{a^2+b^2}{4} - c}$.

If we are given an equation of the form (6.8), the easiest way to find out whether it represents an empty set, a single point or a circle is to reconstruct the steps of the proof of the above result. Note also that, upon doing this, in the case of a circle we will eventually find its center and radius. In any event, for a specific equation this procedure is quite simple: it suffices to complete squares! Let us take a look at an example.

Example 6.18 Identify the set of points (x, y) of the cartesian plane that satisfy equation $x^2 + y^2 - 2x + 6y + 5 = 0$.

Proof It suffices to successively write

$$x^2 + y^2 - 2x + 6y + 5 = (x^2 - 2x + 1) + (y^2 + 6y + 3^2) - 5$$
$$= (x-1)^2 + (y+3)^2 - 5$$

to conclude that the desired set of points is

$$\{(x, y); (x-1)^2 + (y-(-3))^2 = \sqrt{5}^2\},$$

i.e., the circle of center $(1, -3)$ and radius $\sqrt{5}$. ☐

The next example illustrates once more how one can use the analytical method in establishing results in Plane Euclidean Geometry.

Example 6.19 ABC is an isosceles triangle of equal sides AC and AB. If M is the midpoint of basis BC, prove analytically that $AM \perp BC$.

Proof Let $\overline{AB} = \overline{AC} = a$ and $\overline{BC} = b$. Choose a cartesian system in which $B(0, 0)$, $C(b, 0)$ and A lies in the first quadrant (cf. Fig. 6.11). Corollary 6.2 gives $M\left(\frac{b}{2}, 0\right)$. On the other hand, since $\overline{AB} = a$, vertex A lies in the circle of center B and radius a; analogously, A is in the circle of center C and radius a, and is the only point that satisfies both these conditions.

Hence, the coordinates of A form the single solution (x, y), with $y > 0$ (for A is located in the first quadrant), of the system formed by the equations of the above circles, namely,

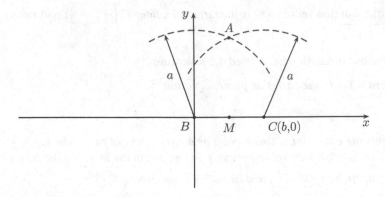

Fig. 6.11 Median relative to the basis of an isosceles triangle

$$\begin{cases} x^2 + y^2 = a^2 \\ (x-b)^2 + y^2 = a^2 \end{cases}.$$

In order to solve such a system, we expand the left hand side of the second equation and, then, subtract the first equation from the result. By proceeding this way, we arrive at the equality $-2bx + b^2 = 0$, so that $x = \frac{b}{2}$. This already suffices to establish the desired result: since A and M both have abscissas equal to $\frac{b}{2}$, we conclude that $\overleftrightarrow{AM}$ is a vertical line. Therefore, $\overleftrightarrow{AM}$ is perpendicular to the horizontal axis and, hence, to the side BC. $\square$

Problems: Sect. 6.2

1. * Given real numbers a and b, prove that points $A(a, b)$ and $B(b, a)$ of the cartesian plane are symmetric with respect to the bisector of odd quadrants.
2. * We are given points $A(a, b)$, $B(c, d)$ and the vertical line r, formed by the points of the cartesian plane whose abscissas are equal to x_0. Prove that A and B are symmetric with respect to r if and only if $b = d$ and $x_0 = \frac{a+c}{2}$.
3. * We are given points $A(a, b)$, $B(c, d)$ and the horizontal line r, formed by the points of the cartesian plane whose ordinates are equal to $y = y_0$. Prove that A and B are symmetric with respect to r if and only if $a = c$ and $y_0 = \frac{b+d}{2}$.
4. * In a cartesian system with origin O, we have points $A(a, b)$ and B, with A lying in the first quadrant and B in the second. If $A\widehat{O}B = 90°$ and $\overline{OA} = \overline{OB}$, prove that $B(-b, a)$.
5. * Let a, b, c, d be given real numbers, with $a^2 + b^2, c^2 + d^2 > 0$. Give a geometric interpretation of the application of gaussian elimination (cf. Section 2.4 of [5], for instance) to the linear system

$$\begin{cases} ax + by = e \\ cx + dy = f \end{cases}.$$

More precisely, show that the system of equations above is possible determined, possible undetermined or impossible, according to whether the straightlines of equations $ax + by = e$ and $cx + dy = f$ are respectively concurrent, parallel or *coincident* (i.e., equal).

6. * Let x_0 and y_0 be nonzero real numbers. Show that the equation of the line passing through $(x_0, 0)$ and $(0, y_0)$ is $\frac{x}{x_0} + \frac{y}{y_0} = 1$.

7. We are give real numbers a, b and c, with $c > 0$. Compute the area of the bounded region of the cartesian plane whose boundary is formed by the ordered pairs (x, y) for which $|x - a| + |y - b| = c$.

8. * Give distinct points A and B in the cartesian plane, let $P_t = (1 - t)A + tB$, with $t \in \mathbb{R}$.

   (a) Prove that $P_t \in \overleftrightarrow{AB}$ and, for every $P \in \overleftrightarrow{AB}$, there exists a unique $t \in \mathbb{R}$ such that $P = P_t$.

   (b) Discuss the position of P_t along $\overleftrightarrow{AB}$ in terms of the value of t.

   The above description of the points of line $\overleftrightarrow{AB}$ is usually referred to as the **parametric equation** of $\overleftrightarrow{AB}$, with **parameter** t.

9. * In the cartesian plane, we are given a triangle ABC and points P, Q and R, lying in the lines $\overleftrightarrow{BC}$, $\overleftrightarrow{CA}$ and $\overleftrightarrow{AB}$, respectively, and all of them distinct from A, B and C. According to the result of the previous problem, we can write

   $$P = (1 - s)B + sC, \quad Q = (1 - t)C + tA, \quad R = (1 - u)A + uB, \qquad (6.9)$$

   for some $s, t, u \in \mathbb{R}$. Prove the following *analytic versions* of the theorems of Ceva and Menelaus (cf. Theorems 4.22 and 4.25):

   (a) Lines $\overleftrightarrow{AP}$, $\overleftrightarrow{BQ}$ and $\overleftrightarrow{CR}$ are concurrent if and only if

   $$stu = (1 - s)(1 - t)(1 - u).$$

   (b) Points P, Q and R are collinear if and only if

   $$stu = -(1 - s)(1 - t)(1 - u).$$

10. Given distinct points B and C, let P and Q be points on line $\overleftrightarrow{BC}$, distinct from B and C and such that $P = (1 - s)B + sC$ and $Q = (1 - t)B + tC$. Prove that P and Q are harmonic conjugates with respect to B and C (cf. Problem 5, page 136) if and only if $\frac{1}{s} + \frac{1}{t} = 2$.

11. In the cartesian plane, let ABC be the triangle with vertices $A(0, 0)$, $B(2, 1)$ and $C(1, 5)$. Find the coordinates of the point P such that the sum of the squares of the distances from P to the vertices of ABC is as small as possible, and compute the corresponding minimum sum.

12. * In the cartesian plane, let Γ be the circle of equation $x^2 + y^2 = 1$ and r be the straightline of equation $ax + by = c$, with a and b being given nonzero real numbers and c being a real variable. Among all lines r that intersect Γ, find the one with the largest possible value of c. Then, use the obtained result to give another proof of Cauchy's inequality (cf. Theorem 5.13 of [5], for instance) for $n = 2$.

13. We are given in the plane a circle Γ and a point A, exterior to the disk bounded by Γ. Show how to draw through A a line r intersecting Γ at points B and P, such that P is the midpoint of AB. Under what circumstances is there a solution?

14. Given a positive real number k and distinct points A and B in the plane, find the locus of points P for which $\overline{AP}^2 + \overline{BP}^2 = k^2$.

15. Give another proof of Theorem 4.16, this time using the cartesian method.

6.3 A First Look on Conics

In this section we will study the most elementary properties of *conics*, namely, *ellipses*, *hyperbolas* and *parabolas*. As we shall see in Sect. 10.4, the name *conic* comes from the fact that those curves can be obtained as the intersections of a circular right cone with planes in specific positions.

Four our purposes, the following definition of conic will suffice.

Definition 6.20 Let a point F and a straightline d be given in the plane, with $F \notin d$; let a real number $\epsilon > 0$ be also given.[7] The **conic** of **focus** F, **directrix** d and **eccentricity** ϵ is the plane curve formed by all points P such that

$$\overline{PF} = \epsilon \cdot \text{dist}(P; d), \qquad (6.10)$$

where $\text{dist}(P; d)$ stands for the distance from P to d. The **parameter** of the conic is the distance p from F to d.

We can identify the points of a conic by invoking the cartesian method. In the notations above, we choose a cartesian system in which $F(c, 0)$ and d has equation $x = x_0$, with $x_0 > c$, so that $p = x_0 - c$ (see Fig. 6.12, where, for the sake of convenience, we have taken $c > 0$ and have sketched a portion of a conic, in order to ease the reader's understanding).

---

[7]This is the lowercase Greek letter *epsilon*.

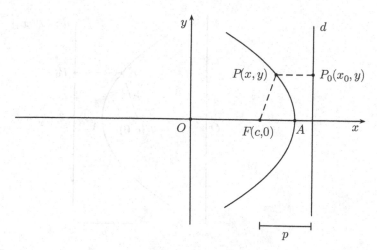

Fig. 6.12 The conic of focus F, directrix d and eccentricity ϵ

Let C denote the conic of focus F, directrix d and eccentricity ϵ. For $A(a, 0)$, with $c < a < x_0$, to lie in C, we must have $a - c = \epsilon(x_0 - a)$, whence $a = \frac{\epsilon x_0 + c}{1+\epsilon}$. Therefore, one of the points of C is

$$A\left(\frac{\epsilon x_0 + c}{1 + \epsilon}, 0\right). \tag{6.11}$$

More generally, by taking any point $P(x, y)$ of the cartesian plane (plotted at the left of d in Fig. 6.12, for the sake of convenience), we have

$$P \in C \Leftrightarrow \overline{PF} = \epsilon \cdot \text{dist}(P; d)$$

$$\Leftrightarrow \sqrt{(x - c)^2 + y^2} = \epsilon |x - x_0|$$

$$\Leftrightarrow (x - c)^2 + y^2 = \epsilon^2 (x - x_0)^2 \tag{6.12}$$

$$\Leftrightarrow x^2 - 2cx + c^2 + y^2 = \epsilon^2(x^2 - 2x_0 x + x_0^2)$$

$$\Leftrightarrow (1 - \epsilon^2)x^2 + 2(\epsilon^2 x_0 - c)x + y^2 = (\epsilon x_0)^2 - c^2.$$

We first deal with the case of a conic of eccentricity $\epsilon = 1$, which is called a **parabola**. In such a case, equation (6.12) reduces to $2(x_0 - c)x + y^2 = x_0^2 - c^2$ or, which is the same, to

$$x = -\frac{1}{2(x_0 - c)}y^2 + \frac{x_0 + c}{2}.$$

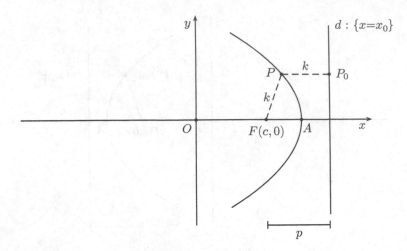

Fig. 6.13 Parabola of focus $F(c, 0)$, directrix d and parameter p

Recalling that $p = x_0 - c$ is the parameter of the parabola and writing $x_0 + c = p + 2c$, the equation above turns into

$$x = -\frac{1}{2p}y^2 + x_0 + \frac{p}{2} + c. \tag{6.13}$$

Figure 6.13 sketches the parabola of focus $F(c, 0)$, with $c > 0$, directrix d and parameter p by looking at it through its definition, namely, as the set of points P in the cartesian plane that are at the same distance from F and d.

Changing the roles of x and y in (6.13) (which amounts to a reflection of the coordinate axes along line $x = y$) we obtain, for our parabola, equation

$$y = -\frac{1}{2p}x^2 + \frac{p}{2} + c,$$

with focus $F(0, c)$ and directrix $d : \{y = x_0\}$. Inverting the positive orientation of the y-axis (which is equivalent to changing y by $-y$ in the equation above, we arrive at the equation

$$y = \frac{1}{2p}x^2 - \left(\frac{p}{2} + c\right); \tag{6.14}$$

now, the focus is $F(0, -c)$ and the directrix is the line $d : \{y = -x_0\}$.

If we translate the x-axis to the position of the line $y = \alpha$, we get a cartesian system in which the new ordinate y' relates to the old one by means of $y' = y - \alpha$. The focus now is $F(0, -c - \alpha)$, while the directrix is the straightline $d : \{y = -x_0 - \alpha\}$. In this new system, the equation of the parabola is obtained from (6.14) changing y by $y' + \alpha$. Since the name we give to the variable is irrelevant, we shall simply write $y + \alpha$ in place of y in (6.14), thus arriving at the equation

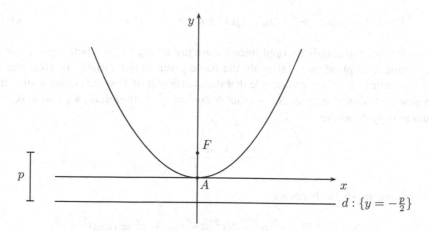

Fig. 6.14 The canonical form of a parabola

$$y + \alpha = \frac{1}{2p}x^2 - \left(\frac{p}{2} + c\right).$$

Finally, choosing $\alpha = -\left(\frac{p}{2} + c\right)$, we obtain the **canonical form**

$$y = \frac{1}{2p}x^2 \tag{6.15}$$

of the equation of the parabola, with respect to which the focus is $F\left(0, \frac{p}{2}\right)$ and the directrix is the horizontal line of ordinate

$$-x_0 - \alpha = -(p + c) + \left(\frac{p}{2} + c\right) = -\frac{p}{2},$$

as should be the case. At this point, the situation is that of Fig. 6.14.

It is obvious from (6.15) that the parabola is symmetric with respect to the line $x = 0$, which is called its **axis**. For future reference, the point $A(0, 0)$ is the **vertex**.

Before we move on to eccentricities $0 < \epsilon \neq 1$, we point out that (6.15) is the graph of the second degree function $x \mapsto \frac{1}{2p}x^2$. Actually, as Problem 7 shows, every such graph is a parabola.

We now turn to the case $0 < \epsilon \neq 1$. Our strategy will be to try to change the cartesian system in order to simplify (6.12), and hence to better understand the properties of C.

By translating the ordinate axis to the position of line $x = \alpha$, we obtain a new cartesian system $x'O'y'$, in which the new coordinates (x', y') of a point relate to the old ones, say (x, y), by $x' = x - \alpha$ and $y' = y$ (see Problem 2, page 188). Hence, the equation of C in this new system is obtained from (6.12), by writing $x' + \alpha$ in place of x:

$$(1 - \epsilon^2)(x' + \alpha)^2 + 2(\epsilon^2 x_0 - c)(x' + \alpha) + y^2 = (\epsilon x_0)^2 - c^2. \tag{6.16}$$

For the sake of simplicity (and since we are just trying to get a better glimpse on C), writing x in place of x' (after all, the name given to the variable is irrelevant!) and expanding (6.16)), we conclude that the coefficient of x in the equation of C in the new cartesian system is $2(1 - \epsilon^2)\alpha + 2(\epsilon^2 x_0 - c)$; therefore, we can make it equal to 0 by choosing

$$\alpha = -\frac{\epsilon^2 x_0 - c}{1 - \epsilon^2}.$$

Upon doing so, (6.16) becomes

$$(1 - \epsilon^2)x^2 + (1 - \epsilon^2)\alpha^2 + 2(\epsilon^2 x_0 - c)\alpha + y^2 = (\epsilon x_0)^2 - c^2,$$

or, substituting the value of α,

$$(1 - \epsilon^2)x^2 + \frac{(\epsilon^2 x_0 - c)^2}{1 - \epsilon^2} - 2\frac{(\epsilon^2 x_0 - c)^2}{1 - \epsilon^2} + y^2 = (\epsilon x_0)^2 - c^2.$$

After simplifying the last expression above and replacing $x_0 - c$ by p, we finally arrive at the equation

$$(1 - \epsilon^2)x^2 + y^2 = \frac{\epsilon^2 p^2}{1 - \epsilon^2}. \tag{6.17}$$

Now, note that, in the new cartesian system, we have $F(c - \alpha, 0)$ and $d : \{x = x_0 - \alpha\}$, with

$$c - \alpha = c + \frac{\epsilon^2 x_0 - c}{1 - \epsilon^2} = \frac{\epsilon^2 p}{1 - \epsilon^2}$$

and

$$x_0 - \alpha = x_0 + \frac{\epsilon^2 x_0 - c}{1 - \epsilon^2} = \frac{p}{1 - \epsilon^2}.$$

As we did before, for the sake of simplicity we shall henceforth write simply $F(c, 0)$ and $d : \{x = x_0\}$, with

$$c = \frac{\epsilon^2 p}{|1 - \epsilon^2|} \quad \text{and} \quad x_0 = \frac{p}{|1 - \epsilon^2|}. \tag{6.18}$$

Notice that (6.17) remains unchanged if we replace c and x_0 respectively by $-c$ and $-x_0$. Hence, had we begun our computations in (6.12) taking $F'(-c, 0)$ and $d' : \{x = -x_0\}$, we would have arrived at the same equation (6.17), and it would be equally fair to call F' and d' respectively the *focus* and the *directrix* of the conic C. Therefore, in the new cartesian system C has foci $F_1(-c, 0)$ and $F_2(c, 0)$, and directrices $d_1 : \{x = -x_0\}$ and $d_2 : \{x = x_0\}$, with c and x_0 given as in (6.18).

Dividing both sides of (6.17) by $\frac{\epsilon^2 p^2}{1-\epsilon^2}$, that equation turns into

$$\frac{x^2}{\left(\frac{\epsilon p}{|1-\epsilon^2|}\right)^2} \pm \frac{y^2}{\left(\frac{\epsilon p}{\sqrt{|1-\epsilon^2|}}\right)^2} = 1,$$

with the chosen sign being $+$ or $-$, according to whether $0 < \epsilon < 1$ or $\epsilon > 1$, respectively. (Check this claim!) Letting

$$a = \frac{\epsilon p}{|1 - \epsilon^2|} \quad \text{and} \quad b = \frac{\epsilon p}{\sqrt{|1 - \epsilon^2|}} \tag{6.19}$$

and looking separately at the cases $0 < \epsilon < 1$ and $\epsilon > 1$, we obtain

$$a^2 \mp b^2 = \frac{\epsilon^2 p^2}{(1 - \epsilon^2)^2} \mp \frac{\epsilon^2 p^2}{|1 - \epsilon^2|}$$

$$= \frac{\epsilon^2 p^2}{(1 - \epsilon^2)^2}(1 \mp |1 - \epsilon^2|) \tag{6.20}$$

$$= \frac{\epsilon^4 p^2}{(1 - \epsilon^2)^2} = c^2.$$

Hence, the equation of C takes the form

$$\frac{x^2}{a^2} \pm \frac{y^2}{b^2} = 1, \tag{6.21}$$

with the chosen sign being $+$ or $-$, according to whether $0 < \epsilon < 1$ or $\epsilon > 1$, respectively. Equation (6.21) is said to be the **canonical form** of a conic of eccentricity $\epsilon \neq 1$.

We summarize part of the above discussion in the following

Theorem 6.21 *Every conic of eccentricity $\epsilon \neq 1$ has two foci F_1 and F_2, two directrices d_1 and d_2 and satisfies the following properties:*

(a) *d_1 and d_2 are parallel lines, perpendicular to $\overleftrightarrow{F_1 F_2}$ and symmetric with respect to the midpoint of $F_1 F_2$.*

(b) *The conic is symmetric with respect to $\overleftrightarrow{F_1 F_2}$, to the perpendicular bisector of $F_1 F_2$ and to the midpoint of $F_1 F_2$.*

Proof The first part follows from what we did before. For what is left, choose the cartesian system in such a way that the foci are $F_1(-c, 0)$ and $F_2(c, 0)$, the directrices are $d_1 : \{x = -x_0\}$ and $d_1 : \{x = -x_0\}$ and the equation of the conic has the form (6.21), with a, b, c and x_0 given in terms of p and ϵ as before.

(a) Immediate from the above.
(b) Since (6.21) remains unchanged if we replace x by $-x$ or y by $-y$, Problem 2, page 200, assures that the conic is symmetric with respect to the line $y = 0$, which is the perpendicular bisector of $F_1 F_2$, and problem 3, page 200, assures that the conic is symmetric with respect to the line $x = 0$, which coincides with $\overleftrightarrow{F_1 F_2}$. Finally, being symmetric relatively to two perpendicular lines, it is also symmetric with respect to their intersection point, which is exactly the midpoint of $F_1 F_2$.

$\square$

In the notations of the statement of the previous theorem, we say that the midpoint O of $F_1 F_2$ is the **center** of the conic, and that a conic of eccentricity $\epsilon \neq 1$ is **central**.

From now on, we specialize our discussion, looking separately at the cases $0 < \epsilon < 1$ and $\epsilon > 1$.

A conic of eccentricity $0 < \epsilon < 1$ is called an **ellipse**. In the notations of the previous discussion (and in the chosen cartesian system), our ellipse has equation

$$\frac{x^2}{a^2} + \frac{y^2}{b^2} = 1, \tag{6.22}$$

with

$$a = \frac{\epsilon p}{1 - \epsilon^2}, \quad b = \frac{\epsilon p}{\sqrt{1 - \epsilon^2}} \tag{6.23}$$

and p being the parameter of the ellipse. In particular, it immediately follows that $a > b$ and that the ellipse passes through the points $A_1(-a, 0)$, $A_2(a, 0)$, $B_1(0, -b)$ and $B_2(0, b)$. The line segments $A_1 A_2$ and $B_1 B_2$ are called the **major axis** and the **minor axis** of the ellipse (6.22).

Since $\left|\frac{x}{a}\right| \leq 1$ and $\left|\frac{y}{b}\right| \leq 1$ for any point (x, y) in the ellipse, we conclude that it is entirely contained in the rectangle with edges parallel to the coordinate axes and passing through A_1, A_2, B_1 and B_2. Figure 6.15 presents the approximate geometric shape of the ellipse of equation (6.22). This will be justified by Theorem 6.22 below.

From (6.18), the ellipse of equation (6.22) has foci $F_1(-c, 0)$ and $F_2(c, 0)$ and directrices $d_1 : \{x = x_0\}$ and $d_2 : \{x = -x_0\}$, with

$$c = \frac{\epsilon^2 p}{1 - \epsilon^2} \quad \text{and} \quad x_0 = \frac{p}{1 - \epsilon^2};$$

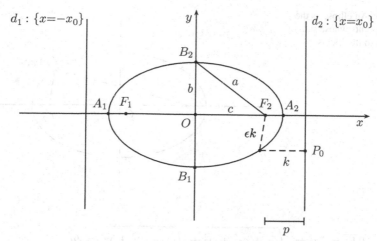

Fig. 6.15 Axes, foci and directrices of the ellipse $\frac{x^2}{a^2} + \frac{y^2}{b^2} = 1$

hence,

$$\frac{c}{a} = \frac{a}{x_0} = \epsilon.$$

Thus, $c < a < x_0$, so that $F_1, F_2 \in A_1 A_2$ but d_1 and d_2 do not intersect $A_1 A_2$. The length $\overline{F_1 F_2} = 2c$ is called the **focal distance** of the ellipse. Finally, it follows from (6.20) that

$$a^2 = b^2 + c^2. \tag{6.24}$$

As promised above, we now prove a result which will provide us with an accurate idea of the geometric shape of an ellipse. It also presents an ellipse as the solution of another locus problem.

Theorem 6.22 *Given real numbers $a > c > 0$, let $b = \sqrt{a^2 - c^2}$. The ellipse of major axis $\overline{A_1 A_2} = 2a$, minor axis $\overline{B_1 B_2} = 2b$ and focal distance $\overline{F_1 F_2} = 2c$ coincides with the locus of points P in the plane for which*

$$\overline{PF_1} + \overline{PF_2} = 2a. \tag{6.25}$$

Proof Choose a cartesian system in which $F_1(-c, 0)$ and $F_2(c, 0)$, and let $P(x, y)$ (cf. Fig. 6.16). Then, formula (6.4) for the distance between two points guarantees that

Fig. 6.16 An ellipse as the
locus of points whose sum of
distances to the foci is
constant

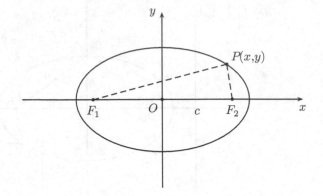

$$\overline{PF_1} + \overline{PF_2} = 2a \Leftrightarrow \sqrt{(x+c)^2 + y^2} + \sqrt{(x-c)^2 + y^2} = 2a$$

$$\Leftrightarrow \sqrt{(x+c)^2 + y^2} = 2a - \sqrt{(x-c)^2 + y^2}$$

$$\Rightarrow (x+c)^2 + y^2 = [2a - \sqrt{(x-c)^2 + y^2}]^2$$

$$\Leftrightarrow (x+c)^2 + y^2 = 4a^2 - 4a\sqrt{(x-c)^2 + y^2} + (x-c)^2 + y^2$$

$$\Leftrightarrow a\sqrt{(x-c)^2 + y^2} = a^2 - cx$$

$$\Rightarrow a^2[(x-c)^2 + y^2] = (a^2 - cx)^2$$

$$\Leftrightarrow (a^2 - c^2)x^2 + a^2y^2 = a^2(a^2 - c^2)$$

$$\Leftrightarrow b^2x^2 + a^2y^2 = a^2b^2$$

$$\Leftrightarrow \frac{x^2}{a^2} + \frac{y^2}{b^2} = 1.$$

We thus conclude that every point P that satisfies relation $\overline{PF_1} + \overline{PF_2} = 2a$ belongs
to the ellipse $\frac{x^2}{a^2} + \frac{y^2}{b^2} = 1$.

Conversely, if we take a point $P(x, y)$ in such an ellipse, we shall conclude that
$\overline{PF_1} + \overline{PF_2} = 2a$ if we are able to show that, in the sequence of implications and
equivalencies above, the implications are actually equivalencies. To this end, note
that $\frac{x^2}{a^2} + \frac{y^2}{b^2} = 1$ implies (as we already know!) $|x| \le a$, so that $a^2 - cx \ge a^2 - ca = a(a - c) > 0$. Hence,

$$a^2[(x-c)^2 + y^2] = (a^2 - cx)^2 \Leftrightarrow a\sqrt{(x-c)^2 + y^2} = |a^2 - cx|$$

$$\Rightarrow a\sqrt{(x-c)^2 + y^2} = a^2 - cx,$$

and we have succeeded in transforming the second implication above into an equivalence. In order to assure that the first implication is also an equivalence, note firstly that the equality $(x + c)^2 + y^2 = [2a - \sqrt{(x - c)^2 + y^2}]^2$ is equivalent to

$$\sqrt{(x + c)^2 + y^2} = |2a - \sqrt{(x - c)^2 + y^2}|.$$

Hence, if we show that $2a - \sqrt{(x - c)^2 + y^2} \geq 0$ whenever $P(x, y)$ lies in the ellipse $\frac{x^2}{a^2} + \frac{y^2}{b^2} = 1$, the last equality above will give

$$\sqrt{(x + c)^2 + y^2} = 2a - \sqrt{(x - c)^2 + y^2},$$

as wished. For what is left to do, just compute

$$\sqrt{(x - c)^2 + y^2} = \sqrt{(x - c)^2 + b^2\left(1 - \frac{x^2}{a^2}\right)}$$

$$= \sqrt{x^2\left(1 - \frac{b^2}{a^2}\right) - 2cx + (b^2 + c^2)}$$

$$= \frac{1}{a}\sqrt{(cx)^2 - 2cxa^2 + a^4} = \frac{1}{a}|cx - a^2|$$

$$\leq \frac{1}{a}(c|x| + a^2) \leq c + a < 2a.$$

$\square$

We now turn to the case of an eccentricity $\epsilon > 1$, in which the conic is called a **hyperbola**. In the notations of the discussion that led to (6.21) (and in the corresponding cartesian system), our hyperbola has equation

$$\frac{x^2}{a^2} - \frac{y^2}{b^2} = 1, \tag{6.26}$$

with

$$a = \frac{\epsilon p}{\epsilon^2 - 1}, \quad b = \frac{\epsilon p}{\sqrt{\epsilon^2 - 1}} \tag{6.27}$$

and p being the parameter of the hyperbola. Thus, the hyperbola passes through the points $A_1(-a, 0)$, $A_2(a, 0)$, $B_1(0, -b)$ and $B_2(0, b)$. As with ellipses, the line segments A_1A_2 and B_1B_2 are called the **major axis** and the **minor axis** of the hyperbola (6.26).

Since $\left|\frac{x}{a}\right| \geq 1$ for any point (x, y) in the hyperbola, we conclude that it is contained in the union of the half-planes $\{x \geq a\}$ and $\{x \leq -a\}$. The portions of the hyperbola contained in each such half-plane (which, by Theorem 6.21, are symmetric with respect to the vertical axis) are its **branches**.

Fig. 6.17 Axes, foci, directrices and asymptotes of the hyperbola $\frac{x^2}{a^2} - \frac{y^2}{b^2} = 1$

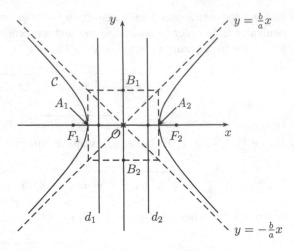

The discussion that led to (6.18) assures that the hyperbola of equation (6.26) has foci $F_1(-c, 0)$ and $F_2(c, 0)$ and directrices $d_1 : \{x = x_0\}$ and $d_2 : \{x = -x_0\}$, with

$$c = \frac{\epsilon^2 p}{\epsilon^2 - 1} \quad \text{and} \quad x_0 = \frac{p}{\epsilon^2 - 1};$$

hence, as with ellipses we have

$$\frac{c}{a} = \frac{a}{x_0} = \epsilon.$$

In this case, however, the equalities above give $c > a > x_0$, so that $A_1, A_2 \in F_1 F_2$ and d_1 and d_2 intersect $A_1 A_2$. The length $\overline{F_1 F_2} = 2c$ is called the **focal distance** of the hyperbola. Finally, it follows from (6.20) that

$$a^2 + b^2 = c^2. \tag{6.28}$$

Figure 6.17 presents the approximate geometric shape of the hyperbola of equation (6.26). In this respect, see also the coming result.

An important geometric interpretation of the length of the minor axis of a hyperbola comes from its **asymptotes**.[8] For the hyperbola of equation (6.26), these are the lines $y = \pm \frac{b}{a} x$. A simple computation shows that none of these lines intersect the hyperbola. On the other hand, the lines and the hyperbola come closer and closer as $|x| \to \infty$. In order to prove this, write the portion of the hyperbola situated in the first and third quadrants as

$$y = y(x) = \frac{b}{a}\sqrt{x^2 - a^2},$$

---

[8]This concept was already dealt with, in a more general context, in Section 9.2 of [5].

with $|x| \geq a$; then, note that

$$\lim_{|x| \to +\infty} \left| \frac{b}{a}x - \frac{b}{a}\sqrt{x^2 - a^2} \right| = \lim_{|x| \to +\infty} \frac{b}{|x + \sqrt{x^2 - a^2}|} = 0.$$

For a simple geometric construction of the asymptotes of a hyperbola, see Problem 4.

The theorem below presents a hyperbola as the solution of another locus problem. Since its proof is pretty much similar to that of Theorem 6.22, we leave it as an exercise for the reader (see Problem 2).

Theorem 6.23 *Given real numbers $c > a > 0$, let $b = \sqrt{c^2 - a^2}$. The hyperbola of major axis $\overline{A_1 A_2} = 2a$, minor axis $\overline{B_1 B_2} = 2b$ and focal distance $\overline{F_1 F_2} = 2c$ coincides with the locus of the points P in the plane for which*

$$\left| \overline{PF_1} - \overline{PF_2} \right| = 2a. \tag{6.29}$$

Problems: Sect. 6.3

1. Prove that the eccentricity of the ellipse $\frac{x^2}{a^2} + \frac{y^2}{b^2} = 1$ measures its *flattening* in the vertical direction or, which is the same, its *stretching* in the horizontal direction. More precisely, prove that, for a fixed major axis $A_1 A_2$, as $\epsilon \to 1-$ the ellipse will come closer and closer to the line segment $A_1 A_2$, whereas as $\epsilon \to 0+$ it will come closer and closer to the circle of diameter $A_1 A_2$.
2. * Prove Theorem 6.23.
3. In the notations of Fig. 6.15, mark point C such that $O F_2 C B_2$ is a rectangle. Prove that the straightline perpendicular to OC through C passes through the intersection point of $\overleftrightarrow{A_1 A_2}$ and d_2.
4. Show that the asymptotes of the hyperbola $\frac{x^2}{a^2} - \frac{y^2}{b^2} = 1$ are the diagonals of the rectangle defined by the lines $x = \pm a$ and $y = \pm b$. Then, use this fact to:

   (a) Show that the asymptotes are perpendicular if and only if $a = b$.
   (b) Construct the asymptotes and the minor axis of the hyperbola, given its major axis and foci.

5. Given a nonzero real number c, let $\mathcal{H}$ be the set of points (x, y) of the cartesian plane for which $xy = c$. Prove that $\mathcal{H}$ is a hyperbola.
6. In the hyperbola $\mathcal{H}$ of equation $xy = 1$ (see the previous problem) we choose three distinct points A, B and C. Prove that the orthocenter of ABC is also in $\mathcal{H}$.

7. We are given real numbers a, b and c, with $a \neq 0$. Show that the graph of the second degree function $x \mapsto ax^2 + bx + c$ is a parabola.

8. (Spain) Let p and q be real numbers such that $p^2 > q \neq 0$. Suppose that the parabola $y = x^2 + 2px + q$ intersects the horizontal axis at the points $A(x_1, 0)$ and $B(x_2, 0)$, and let $C(0, q)$. Show that, independently of the values of p and q, the circumcircle of ABC always passes through a fixed point.

9. Through a point P of the hyperbola $\frac{x^2}{a^2} - \frac{y^2}{b^2} = 1$ we draw parallels to the major and minor axes, which meet the asymptotes respectively at the points Q_1 and Q_2, R_1 and R_2. Show that $\overline{PQ_1} \cdot \overline{PQ_2} = a^2$ and $\overline{PR_1} \cdot \overline{PR_2} = b^2$.

10. We are given an ellipse $\mathcal{E}$ and a line r. For each line s, parallel to r and meeting $\mathcal{E}$ at distinct points P and Q, we mark the midpoint M of the line segment PQ. Show that, as s varies, the locus of the point M is a line segment.

11. (Germany) A sequence $(\Gamma_n)_{n \geq 1}$ of distinct circles in the cartesian plane is defined in the following way:

   (a) Γ_1 is the circle of equation $x^2 + y^2 = 1$.
   (b) For every $k \geq 1$, Γ_{k+1} is externally tangent to Γ_n and also tangent to both branches of the hyperbola $x^2 - y^2 = 1$.

   Compute the radius of Γ_n as a function of n.

6.4   A Second Look on Conics

In this last section we use the apparatus developed so far to study a number of interesting properties of conics, related in a way or another to the problem of *tangency* of lines and conics. We therefore begin with the following

Definition 6.24 A parabola $\mathcal{P}$ and a straightline r are tangent if $\mathcal{P} \cap r$ consists of a single point and r is not parallel to the axis of $\mathcal{P}$. An ellipse $\mathcal{E}$ (resp. a hyperbola $\mathcal{H}$) and a straightline r are **tangent** if $\mathcal{E} \cap r$ (resp. $\mathcal{H} \cap r$) consists of a single point. In any such case, the single common point of the conic and r is their **tangency point**.

In face of this definition, a natural problem that poses itself is that of finding (if any) the tangent(s) to a conic $\mathcal{C}$ through a point P of it. The following result gives a first answer.

Theorem 6.25 *If $\mathcal{C}$ is a conic and P is one of its points, then there exists a unique line r, tangent to $\mathcal{C}$ and passing through P.*

Solution We shall consider the case of a parabola, leaving those of an ellipse or a hyperbola to the reader (cf. Problem 1).

We can assume, without loss of generality, that the parabola is in its canonical form $y = ax^2$, and let $P(x_0, y_0)$. A non vertical line r passing through P has equation $y = mx + n$, with $y_0 = mx_0 + n$. Hence, we can write such an equation as

$$y = mx + y_0 - mx_0 = m(x - x_0) + y_0.$$

Our task is to show that there exists a unique value of m for which the system of equations

$$\begin{cases} y = ax^2 \\ y = m(x - x_0) + y_0 \end{cases}.$$

has (x_0, y_0) as its single solution.

Since we are assuming that P lies in the parabola, (x_0, y_0) is indeed a solution of the system. On the other hand, if (x, y) is any solution, then

$$ax^2 = m(x - x_0) + y_0 = m(x - x_0) + ax_0^2,$$

so that $a(x^2 - x_0^2) = m(x - x_0)$. Hence, either $x = x_0$ or $a(x + x_0) = m$, and this last equality gives $x = \frac{m}{a} - x_0$. We therefore must have $\frac{m}{a} - x_0 = x_0$, whence $m = 2ax_0$. (Note that this value of m coincides with the one obtained from the derivative of $y = y(x) = ax^2$ at x_0, as should be the case.)   □

Now that we know that tangents do exist, we shall present an important *synthetic characterization* of tangents in the coming theorem. As the subsequent discussion and the proposed problems will show, such a result is the source of a number of important properties of conics.

Theorem 6.26

(a) *Let* $\mathcal{P}$ *be a parabola of focus* F *and directrix* d. *If* $P \in \mathcal{P}$ *and* T *is the foot of the perpendicular dropped from* P *to* d, *then the tangent to* $\mathcal{P}$ *passing through* P *is the bisector of the angle* $\angle FPT$.

(b) *Let* $\mathcal{C}$ *be an ellipse (resp. a hyperbola) of foci* F_1 *and* F_2. *If* $P \in \mathcal{C}$, *then the tangent to* $\mathcal{C}$ *passing through* P *is the external (resp. internal) bisector of angle* $\angle F_1 P F_2$.

Proof For the case of a parabola, we refer the reader to Problem 2.

For the remaining cases, we first point out that the proofs of both of them are quite similar, so that we shall stick to the case of a hyperbola. Since we already know that the tangent to $\mathcal{C}$ through P is unique, it suffices to show that the internal bisector r of $\angle F_1 P F_2$ is tangent to $\mathcal{C}$. We shall do this by showing that $r \cap \mathcal{C} = \{P\}$ (see the figure below).

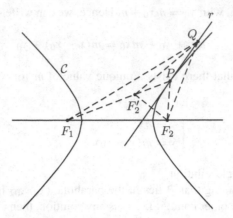

Since r is the internal bisector of $\angle F_1 P F_2$, the symmetric F_2' of F_2 with respect to r lies in $F_2 P$. Therefore, if $Q \in r \setminus \{P\}$, then the congruence of triangles QPF_2 and QPF_2' gives $\overline{QF_2} = \overline{QF_2'}$. Thus, letting $2a$ denote the major axis of the hyperbola and applying the triangle inequality to $QF_1 F_2'$, we get

$$\left| \overline{QF_1} - \overline{QF_2} \right| = \left| \overline{QF_1} - \overline{QF_2'} \right| > \overline{F_1 F_2'} = \left| \overline{PF_1} - \overline{PF_2'} \right|$$
$$= \left| \overline{PF_1} - \overline{PF_2} \right| = 2a.$$

This shows that $Q \notin C$, as wished.   □

The previous result can be used, together with compass and straightedge, to provide a systematic way of constructing an arbitrary finite number of points of a given conic.[9] To this end, it is worth starting with the following

Definition 6.27 If C is an ellipse (resp. a hyperbola) of foci F_1, F_2 and major axis $2a$, its **director circles** are those centered at F_1, F_2 and with radii $2a$.

In the notations of the previous definition, since $\overline{F_1 F_2} < 2a$ for an ellipse, we conclude that each of its director circles contains both foci. Accordingly, since $\overline{F_1 F_2} > 2a$ for a hyperbola, each of its director circles contains exactly one of its foci.

We can now state and prove the following important consequence of Theorem 6.26.

Proposition 6.28

(a) *Let P be a parabola of focus F and directrix d. A line r is tangent to P if and only if the symmetric of F with respect to r lies in d.*

---

[9]For another such procedure for ellipses, see Problem 5, page 239.

(b) Let C be an ellipse (resp. a hyperbola) of foci F_1 and F_2. A line r is tangent to C if and only if the symmetric of F_1 (resp. F_2) with respect to r lies in the director circle of C centered at F_2 (resp. F_1).

Proof As in the proof of the previous result, the case of a parabola is dealt with in Problem 2. Also as in the previous result, the proof of the other two cases are quite similar, so we only consider the case of an ellipse.

Firstly, let $P \in C$ and r be tangent to C at P (in the figure below, $r = \overleftrightarrow{PX}$). We shall show that the symmetric Q of F_2 with respect to r lies in the director circle with respect to F_1 (to show that the symmetric of F_1 with respect to r lies in the director circle with respect to F_2 is entirely analogous).

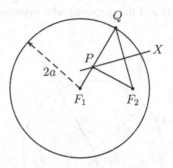

Theorem 6.26 assures that r is the external bisector of angle $\angle F_1 P F_2$. If Q is the symmetric of F_2 with respect to r, then

$$F_2\widehat{P}X = Q\widehat{P}X = \frac{1}{2}(180° - F_1\widehat{P}F_2),$$

so that

$$F_1\widehat{P}F_2 + F_2\widehat{P}Q = F_1\widehat{P}F_2 + 2 \cdot \frac{1}{2}(180° - F_1\widehat{P}F_2) = 180°.$$

Thus, F_1, P and Q are collinear.

Now, if $2a$ stands for the major axis of C, Theorem 6.22 assures that $\overline{F_1P} + \overline{F_2P} = 2a$. Thus, we compute

$$\overline{F_1Q} = \overline{F_1P} + \overline{PQ} = \overline{F_1P} + \overline{F_2P} = 2a,$$

which shows that Q lies in the director circle of C centered at F_1.

Conversely, suppose that the symmetric Q of F_2 with respect to r lies in the director circle centered at F_1. If P is the intersection of F_1Q and r, then $\overline{F_2P} = \overline{PQ}$, so that

$$\overline{F_1P} + \overline{F_2P} = \overline{F_1P} + \overline{PQ} = \overline{F_1Q} = 2a.$$

Hence, $P \in C \cap r$ and r is the external bisector of $\angle F_1 P F_2$, and the previous theorem assures that r is tangent to C at P. $\square$

As promised before, we now use the above result to devise a systematic procedure for constructing points on a conic. Again, we do this for an ellipse, leaving the cases of a hyperbola and of a parabola to Problems 4 and 5.

Example 6.29 From an ellipse $\mathcal{E}$ we know the positions of its foci F_1, F_2 and the length $2a$ of its major axis. If a half-line $\overrightarrow{F_1 X}$ intersects $\mathcal{E}$ at P, show how to construct P with compass and straightedge.

Solution Draw the director circle Γ of $\mathcal{E}$ with center F_1, and mark the intersection point Q of $\overrightarrow{F_1 X}$ and Γ (see the figure below). According to the previous proposition, P is the intersection of $F_1 Q$ and the perpendicular bisector of $F_2 Q$.

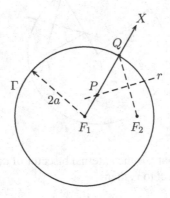

$\square$

Proposition 6.28 also allows us to solve the problem of drawing the tangent to a given conic passing through a given point, and we do this next. To this end, it is convenient to introduce the following

Definition 6.30 If $\mathcal{P}$ is a parabola of vertex A and directrix d, its **auxiliary line** is the line passing through A and parallel to d. If C is an ellipse (resp. hyperbola) of center O and major axis $2a$, its **auxiliary circle** is the circle of center O and radius a.

Note that the major axis $A_1 A_2$ of an ellipse (resp. hyperbola) is a diameter of its auxiliary circle. Also, in the case of an ellipse, the auxiliary circle contains both foci, whereas in the case of a hyperbola it contains none of them.

Corollary 6.31

(a) *Let $\mathcal{P}$ be a parabola of focus F. If a line r is tangent to $\mathcal{P}$, then the foot of the perpendicular dropped from F to r lies in the auxiliary line.*

(b) *Let C be an ellipse (resp. a hyperbola) of foci F_1, F_2 and center O. If a line r is tangent to C, then the feet of the perpendiculars dropped from F_1 and F_2 to r lie in the auxiliary circle.*

Proof For item (a), if S denotes the foot of the perpendicular dropped from F to r, then item (a) of Proposition 6.28 assures that $\overrightarrow{FS}$ intersects d at the point T, which is the symmetric of F with respect to r (see the figure below).

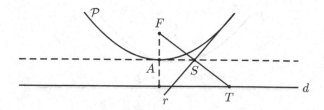

Since $\overleftrightarrow{AF} \perp d$ and A is at equal distances from F and d, the result of Problem 10, page 66, assures that $\overleftrightarrow{AS} \parallel d$, hence is the auxiliary line of $\mathcal{P}$.

In item (b) we look at the case of a hyperbola, that of an ellipse being entirely analogous. To this end, in the figure below line r is tangent to the hyperbola C, and Q is the symmetric of F_2 with respect to r.

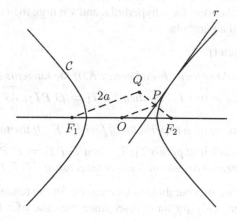

According to item (b) of Proposition 6.28, Q lies in the director circle of C centered at F_1, whence $\overline{F_1 Q} = 2a$. Now, if F_2Q intersects r at P, then P is the midpoint of F_2Q. Since O is the midpoint of F_1F_2, the midsegment theorem applied to triangle F_1F_2Q assures that $\overline{OP} = \frac{1}{2}\overline{F_1Q} = a$, and P lies in the auxiliary circle of C.   □

We are finally in position to solve the tangency problem alluded above. We do this in the coming example for a parabola, referring to Problem 7 for the case of a hyperbola.

Example 6.32 In the figure below, $\mathcal{P}$ is the parabola of focus F and directrix d. Use compass and straightedge to draw all tangents to $\mathcal{P}$ passing through the given point Q.

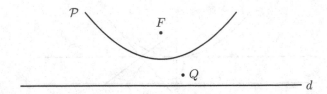

Solution If r is one of the tangents we are looking for and S be the foot of the perpendicular dropped from F to r, then item (a) of the previous corollary assures that S lies in the auxiliary line of $\mathcal{P}$. Since $F\widehat{S}Q = 90°$, the point S can be constructed as the intersection of the auxiliary line with the circle of diameter FQ. In general, there are two distinct solutions.   □

We finish this section by establishing two beautiful results on conics, the first of which is due to Poncelet. For a better understanding of the statement below, Fig. 6.18 illustrates Poncelet's theorem for a hyperbola, and we urge the reader to sketch the cases of an ellipse and a parabola.

Theorem 6.33 (Poncelet)

(a) *Let $\mathcal{P}$ be a parabola of focus F and vertex A. If the tangents to $\mathcal{P}$ drawn from a point P touch it at points T_1, T_2, then $F\widehat{P}T_1 = \angle(\overrightarrow{PT_2},\ \overrightarrow{AF})$ and $\overleftrightarrow{FP}$ bisects angle $\angle T_1FT_2$.*

(b) *Let $\mathcal{C}$ be an ellipse (resp. a hyperbola) of foci F_1, F_2. If the tangents to $\mathcal{C}$ drawn from a point P touch it at points T_1, T_2, then $F_1\widehat{P}T_1 = F_2\widehat{P}T_2$ and $\overrightarrow{F_iP}$ is an internal (resp. internal or external) angle bisector of $\angle T_1F_iT_2$, for $i = 1, 2$.*

Proof We leave the case of a parabola as an exercise for the reader (cf. Problem 13), and prove the theorem only for an ellipse, since the case of a hyperbola is quite similar.

In the figure below, $t_1 = \overleftrightarrow{PT_1}$ and $t_2 = \overleftrightarrow{PT_2}$ are the tangents to the ellipse and R_1, R_2 are the symmetrics of F_2 with respect to t_1, t_2, respectively. Hence, they lie in the director circle Γ centered at F_1, and $\overline{PF_2} = \overline{PR_1} = \overline{PR_2}$, so that the circle Σ of center P and radius PF_2 passes through R_1 and R_2.

Fig. 6.18 Poncelet's theorem
for a hyperbola

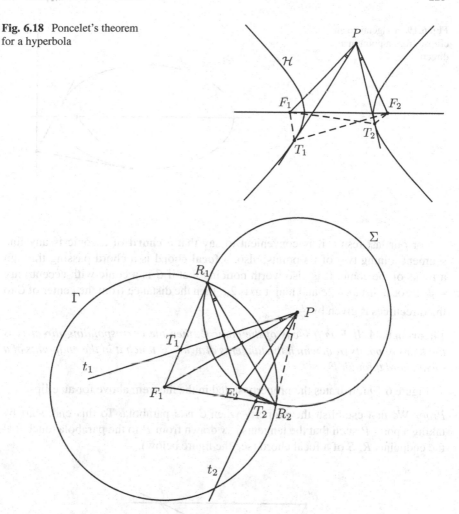

For the first part, note that $\overleftrightarrow{PF_1} \perp \overleftrightarrow{R_1R_2}$ and $\overleftrightarrow{PT_1} \perp \overleftrightarrow{F_2R_1}$ imply $F_1\widehat{P}T_1 = F_2\widehat{R}_1R_2$, as displayed in the figure above. Also, since t_2 is the perpendicular bisector of F_2R_2, we have $F_2\widehat{P}T_2 = R_2\widehat{P}T_2$. Hence, it suffices to show that $F_2\widehat{R}_1R_2 = \frac{1}{2}F_2\widehat{P}R_2$, which follows from the inscribed angle theorem.

For the second part, we first establish that $T_1\widehat{F}_1P = T_2\widehat{F}_1P$. To this end, it suffices to notice that $T_1\widehat{F}_1P = R_1\widehat{F}_1P = R_2\widehat{F}_1P = T_2\widehat{F}_1P$, where the second equality comes from the fact that $\overleftrightarrow{F_1P}$ is the perpendicular bisector of R_1R_2.

We are finally left to showing that $T_1\widehat{F}_2P = T_2\widehat{F}_2P$. This can be accomplished through an argument similar to the one above, starting by taking the symmetrics of F_1 with respect to t_1, t_2, observing that they lie in the director circle centered at F_2, etc.   □

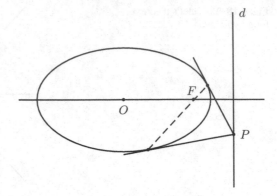

For our last result, it is convenient to say that a **chord** of a conic is any line
segment joining two of its points; also, a **focal chord** is a chord passing through
a focus of the conic. It is also worth noticing that if C is a conic with eccentricity
$\epsilon \neq 1$, focal distance $2c$ and major axis $2a$, then the distance from the center of C to
the directrices is given by $\frac{a^2}{c}$.

Theorem 6.34 *If F is a focus of the conic C, then the corresponding directrix is
the locus of points from which the tangents drawn to C touch it in the endpoints of a
focal chord through F.*

Figure 6.19 illustrates the property stated in the theorem above for an ellipse.

Proof We first establish the property when C is a parabola. To this end, start by
taking a point P such that the tangents r, s drawn from P to the parabola touch it at
the endpoints R, S of a focal chord (see the figure below).

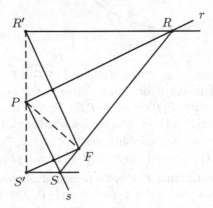

Let R' and S' are the symmetrics of F with respect to r and s, respectively. Since
r is tangent to the parabola and $\overleftrightarrow{RR'}$ is symmetric of $\overleftrightarrow{FR}$ with respect to r, we know

from Proposition 6.28 that $R' \in d$ and $\overleftrightarrow{RR'} \perp d$. Analogously, $S' \in d$ and $\overleftrightarrow{SS'} \perp d$; in particular, $R'S' = d$ and $\overleftrightarrow{RR'} \parallel \overleftrightarrow{SS'}$.

Now, let $P\widehat{R}R' = P\widehat{R}F = \alpha$ and $P\widehat{S}S' = P\widehat{S}F = \beta$. On the one hand, the alluded parallelism gives $\alpha + \beta = 90°$. On the other, since triangles $RR'F$ and $SS'F$ are isosceles of bases $R'F$ and $S'F$, we obtain $R'\widehat{R}F = 90° - \alpha$ and $S'\widehat{S}F = 90° - \beta$, so that

$$R'\widehat{F}S' = 180° - R'\widehat{R}F - S'\widehat{S}F = \alpha + \beta = 90°.$$

Hence,

$$R'\widehat{P}F + S'\widehat{P}F = (180° - 2P\widehat{F}R') + (180° - 2P\widehat{F}S')$$
$$= 360° - 2R'\widehat{F}S' = 180°,$$

and P, R', S' are collinear. Since this is the same as saying that $P \in \overleftrightarrow{R'S'}$ we have $P \in d$.

Conversely, if $P \in d$, one can easily rework the argument above backwards to show that the tangents drawn from P to the parabola touch it at the endpoints of a focal chord.

We now turn to the case of an ellipse, leaving the (quite similar) case of a hyperbola to the reader. Let F_1 and F_2 be its foci, and P be a point from which the tangents r and s drawn to C touch it at R and S, respectively, with $F_2 \in RS$ (see the figure below).

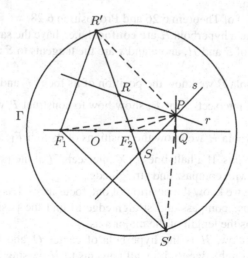

If R' and S' denote the symmetrics of F_1 with respect to r and s, respectively, Proposition 6.28 assures that R' and S' lie in the director circle Γ centered at F_2.

In turn, since $\overline{PR'} = \overline{PF_1} = \overline{PS'}$, triangle $PR'S'$ is isosceles of basis $R'S'$; since PF_2 is the median relative to the basis, it is also the altitude. Therefore, we compute

$$\overline{PF_1}^2 - \overline{PF_2}^2 = \overline{PR'}^2 - \overline{PF_2}^2 = \overline{F_2R'}^2 = 4a^2.$$

Thus, if Q stands for the foot of the perpendicular dropped from P to $\overleftrightarrow{F_1F_2}$, Proposition 6.8 assures that $\overline{OQ} = \frac{4a^2}{2F_1F_2} = \frac{a^2}{c}$. Therefore, $\overrightarrow{PQ}$ is nothing but the directrix of the ellipse relative to F_2.

Conversely, if P lies in the directrix relative to F_2, then the argument above, worked backwards, assures that the tangents drawn from P to the ellipse touch it at the endpoints of a focal chord.   □

In the case of a parabola, the previous result has the following

Corollary 6.35 *If d is the directrix of the parabola $\mathcal{P}$, then the tangents to $\mathcal{P}$ drawn from a point of d are perpendicular.*

Proof In the notations of the proof of the theorem, note that $\overleftrightarrow{PR} \perp \overleftrightarrow{FR'}$ and $\overleftrightarrow{PS} \perp \overleftrightarrow{FS'}$, together with $R'\widehat{F}S' = 90°$, give $R\widehat{P}S = 90°$.   □

Problems: Sect. 6.4

1. * Complete the proof of Theorem 6.25 by considering the cases of an ellipse and a hyperbola.
2. * Prove items (a) of Theorem 6.26 and Proposition 6.28.
3. An ellipse $\mathcal{E}$ and a hyperbola $\mathcal{H}$ are **confocal**, i.e., have the same foci. If P is a common point of $\mathcal{E}$ and $\mathcal{H}$, and r and s are the tangents to $\mathcal{E}$ and $\mathcal{H}$ at P show that $r \perp s$.
4. * From a parabola $\mathcal{P}$ we know the position of its focus F and its directrix d. If a half-line $\overrightarrow{FX}$ intersects $\mathcal{P}$ at P, show how to construct P with compass and straightedge.
5. * From a hyperbola $\mathcal{H}$ we know the positions of its foci F_1, F_2 and the length $2a$ of its major axis. If a half-line $\overrightarrow{F_1X}$ intersects $\mathcal{H}$ at the point P, show how to construct P with compass and straightedge.
6. From an ellipse we know the position of one focus as well as of three tangents. Show how to use compass and straightedge to find the position of the other focus, as well as the length of the major axis.
7. In the figure below, $\mathcal{H}$ is the hyperbola of center O and foci F_1, F_2. Use compass and straightedge to draw all tangents to $\mathcal{H}$ passing through the given point Q.

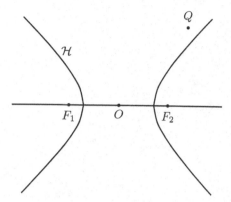

8. Let C be an ellipse (resp. hyperbola) of foci F_1, F_2 and minor axis $2b$. Let r be a straightline tangent to C, and P_1, P_2 be the feet of the perpendiculars dropped from F_1, F_2 to r, respectively. Show that $\overline{F_1 P_1} \cdot \overline{F_2 P_2} = b^2$.

9. (Brazil) Three tangents to a parabola pairwise intersect at points A, B and C. Prove that the circumcircle of ABC always passes through the focus of the parabola.

10. We are given in the plane lines r, s, t, u, all of which are tangent to a parabola $\mathcal{P}$. Show how to use compass and straightedge to find the focus of $\mathcal{P}$.

11. Let $\mathcal{E}$ be an ellipse of center O and major and minor axes $2a$ and $2b$, respectively. The **circle of Monge** of $\mathcal{E}$ is the circle Γ of center O and radius $\sqrt{a^2 + b^2}$. Show that Γ is the locus of the points P in the plane from which one can draw perpendicular tangents to $\mathcal{E}$.

12. Let $\mathcal{H}$ be a hyperbola of center O and major and minor axes $2a$ and $2b$, respectively, with $a > b$. The **circle of Monge** of $\mathcal{H}$ is the circle Γ of center O and radius $\sqrt{a^2 - b^2}$. Show that Γ is the locus of the points P in the plane from which one can draw perpendicular tangents to $\mathcal{H}$.

13. Finish the proof of Theorem 6.33 by establishing Poncelet's theorem for a parabola.

Chapter 7
Trigonometry and Geometry

In this chapter we present the rudiments of Trigonometry, along with several applications to Plane Euclidean Geometry.[1] As in the previous chapter, we aim at developing a set of computational tools that allow us to successfully approach metric problems for which the methods developed so far are useless. We shall refer to the systematic use of such tools in geometric problems as the *trigonometric method*.

7.1 Trigonometric Arcs

In the cartesian plane, the **unit circle** is the circle Γ of Fig. 7.1, centered at the origin $O(0, 0)$ and with radius 1 (hence, length 2π).

Given a real number c, we mark on Γ, starting from A, an arc of length $|c|$ (possibly with $|c| > 2\pi$), in the counterclockwise sense if $c > 0$ and in the clockwise sense if $c < 0$. Letting P be the final endpoint of such an arc, we say that $\overset{\frown}{AP}$ is an arc of c **radians**.

Remark 7.1 It is worth stressing that *radian* is not a unit of measure; it is simply a name we use when we wish to refer to lengths of arcs drawn along the unit circle. Nevertheless, by the sake of tradition we shall frequently refer, without further notice, to *"an arc whose measure is c radians"*

Along Γ, we use to refer to the *counterclockwise direction* as the **trigonometric sense**. This way, the arc of 2π radians in Γ gives a complete turn around Γ in the trigonometric sense, starting and ending at A. On the other hand, the arc of -2π radians in Γ gives a complete turn around Γ in the clockwise sense, also

---

[1]For a purely analytical presentation of Trigonometry, thus avoiding any geometric considerations, we refer the reader to Section 11.5 of [5], for instance.

© Springer International Publishing AG, part of Springer Nature 2018

A. Caminha Muniz Neto, *An Excursion through Elementary Mathematics, Volume II*, Problem Books in Mathematics, https://doi.org/10.1007/978-3-319-77974-4_7

Fig. 7.1 The unit circle

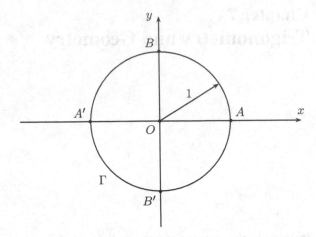

starting and ending at A. Yet in another way, we have the following fundamental correspondences:

- 2π radians correspond to an arc of $360°$, measured in the trigonometric sense and starting from A.
- -2π radians correspond to an arc of $360°$, measured in the clockwise sense and starting from A.

More generally, letting θ denote the measure in degrees and c denote the measure in radians of some arc, with $c > 0$, Problem 11, page 180, gives

$$\frac{\theta}{360} = \frac{c}{2\pi}. \tag{7.1}$$

Let us take a look at a couple of simple applications of the formula above.

Example 7.2 Mark on the unit circle the endpoints of the arcs of $2\pi/3$, $-2\pi/3$, $\pi/4$ and π radians.

Solution The arc of $2\pi/3$ radians is arc $\overset{\frown}{AP}$ such that $\ell(\overset{\frown}{AP}) = 2\pi/3$, measured from A in the trigonometric sense. Letting θ denote the measure in degrees corresponding to such an arc, we have

$$\frac{\theta}{360} = \frac{2\pi/3}{2\pi},$$

whence $\theta = 120°$. We have plot point P in Fig. 7.2.

The arc of $-2\pi/3$ radians is arc $\overset{\frown}{AP'}$ for which $\ell(\overset{\frown}{AP'}) = 2\pi/3$, measured from A in the clockwise sense. Thus, it is immediate to see that P' is the symmetric of P with respect to the x-axis.

Fig. 7.2 The arcs of $\pm 2\pi/3$ radians

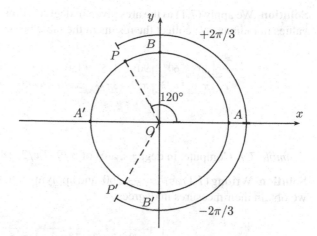

Fig. 7.3 The arc of $\pi/4$ radians

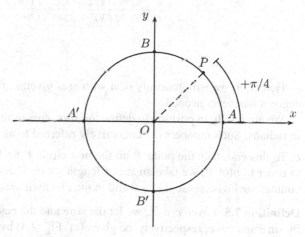

Since π is half of 2π, the arc of π radians is, in Fig. 7.2, arc $\overset{\frown}{AA'}$, measured in the counterclockwise sense.

Finally, β stands for the measure (in degrees) of the arc of $\pi/4$ radians, then

$$\frac{\beta}{360} = \frac{\pi/4}{2\pi},$$

so that $\beta = 45°$. Since the arc is positive, its endpoin is point P of Fig. 7.3.

□

Example 7.3 Compute in radians the following arcs, given in degrees: 30°, 60°, 90°, 135°, 150°, 240°, 270°, 300°.

Solution We apply (7.1) to the arcs given in degrees, thus getting the corresponding values in radians. We collect the results in the table below:

θ	30°	60°	90°	135°	150°	240°	270°	300°
c	$\pi/6$	$\pi/3$	$\pi/2$	$3\pi/4$	$5\pi/6$	$4\pi/3$	$3\pi/2$	$5\pi/3$

□

Example 7.4 Compute, in degrees, arcs of $\pi/9$, $7\pi/2$, 18π and $11\pi/5$ radians.

Solution Writing (7.1) as $\theta = \frac{c}{2\pi} \cdot 360$, and applying it to the given arcs (in radians), we obtain their measures in degrees:

c	$\pi/9$	$7\pi/2$	18π	$11\pi/5$
θ	20°	630°	3240°	396°

□

Hereafter, we will generally deal with arcs given in radians, converting them to degrees whenever needed.

We are finally in position to define the *sine*, *cosine* and *tangent* of an arc c given in radians. Such numbers are collectively referred to as the **trigonometric arcs** of c. To this end, take the point P on the unit circle Γ such that $\overset{\frown}{AP} = c$ (recall that, in order to plot P, we take an arc of length $|c|$ on Γ and departing from A, in the counterclockwise sense if $c > 0$ and in the clockwise sense if $c < 0$).

Definition 7.5 Given $c \in \mathbb{R}$, we let the **sine** and the **cosine** of c (radians), denoted by $\sin c$ and $\cos c$, respectively, be given (cf. Fig. 7.4) by:

$$\cos c = \text{abscissa of } P; \quad \sin c = \text{ordinate of } P.$$

The greatest ordinate of a point of Γ is that of $B(0, 1)$, which is equal to 1, whereas the smallest such ordinate is that of $B'(0, -1)$, which is equal to -1. Analogously, the greatest abscissa of a point on Γ is 1, attained solely at $A(1, 0)$; the smallest abscissa is -1, attained only at $A'(-1, 0)$. Therefore,

$$\begin{cases} -1 \leq \sin c \leq 1 \\ -1 \leq \cos c \leq 1 \end{cases}. \tag{7.2}$$

Conversely, given a real number $\alpha \in [-1, 1]$, the horizontal line passing through point $(0, \alpha)$ intersect Γ in at least one point P; letting $P(\sin c, \cos c)$, it is immediate that $\sin c = \alpha$. In other words, every real number of the interval $[-1, 1]$ is the sine of some arc. Arguing with the vertical line passing through $(\alpha, 0)$, we conclude that an analogous statement is also true for cosines.

Fig. 7.4 Sine and cosine of an arc

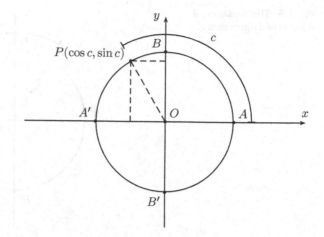

Given $k \in \mathbb{Z}$ arbitrarily, it is immediate that an arc of $2k\pi$ radians departing from $A(1, 0)$ ends also at A. More generally, given $c \in \mathbb{R}$, the arcs of c and $c + 2k\pi$ radians departing from A have equal endpoints; therefore,

$$\begin{cases} \sin(c + 2k\pi) = \sin c \\ \cos(c + 2k\pi) = \cos c \end{cases} \tag{7.3}$$

for every $k \in \mathbb{Z}$.

Relation (7.4) below is known in mathematical literature as the **fundamental relation of Trigonometry**.

Proposition 7.6 *For every $c \in \mathbb{R}$, we have*

$$\sin^2 c + \cos^2 c = 1. \tag{7.4}$$

Proof Let $\overset{\frown}{AP} = c$ (cf. Fig. 7.5). Since $P(\cos c, \sin c)$ and $O(0, 0)$, formula (6.4), together with $\overline{AP} = 1$, give

$$\sqrt{(\cos c - 0)^2 + (\sin c - 0)^2} = 1.$$

This is exactly the desired relation. □

Example 7.7 The sine and cosine of an arc allow us to describe a circle **parametrically**. More precisely, the equation of the circle Γ of center (x_0, y_0) and radius R is $(x - x_0)^2 + (y - y_0)^2 = R^2$ (cf. Fig. 7.6). A generic point $P(x, y)$ on Γ can be thought of as a particle that traverses the circle in the counterclockwise sense, say, starting from $A(x_0 + R, y_0)$.

Fig. 7.5 The fundamental
relation of Trigonometry

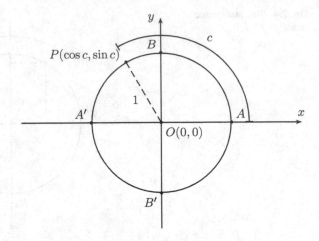

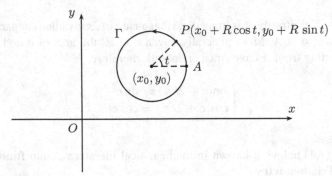

Fig. 7.6 Parametric equations of a circle

Since $P(x, y) \in \Gamma$ if and only if

$$\left(\frac{x - x_0}{R}\right)^2 + \left(\frac{y - y_0}{R}\right)^2 = 1,$$

we can write, for such a P on Γ,

$$\frac{x - x_0}{R} = \cos t \quad \text{and} \quad \frac{y - y_0}{R} = \sin t,$$

with $0 \le t \le 2\pi$ (here, the variable t—the *parameter*—is thought of as time),
whence

$$x = x_0 + R \cos t \quad \text{and} \quad y = y_0 + R \sin t.$$

Fig. 7.7 Sine, cosine and tangent of $\frac{\pi}{3}$

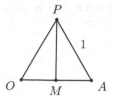

Whenever convenient, we can ask the movement to start from some point $Q(x_0 + R\cos t_0, y_0 + R\sin t_0)$ on Γ, thus writing $x = x_0 + R\cos(t - t_0)$ and $y = y_0 + R\sin(t - t_0)$; we can further ask for a whole turn around the circle to be completed when t varies in the closed interval $[0, 1]$; it suffices to write $x = x_0 + R\cos 2\pi(t - t_0)$ and $y = y_0 + R\sin 2\pi(t - t_0)$.

Definition 7.8 If $c \in \mathbb{R}$ is such that $\cos c \neq 0$, then the **tangent** of c, abbreviated $\tan c$, is defined by

$$\tan c = \frac{\sin c}{\cos c}.$$

In the notations of Fig. 7.4, note that $\cos c = 0$ exactly when the endpoint P of arc $\overset{\frown}{AP} = c$ coincides with B or B'. On the other hand, it is pretty clear that this happens precisely when $c = \frac{\pi}{2} + 2k\pi$ or $c = \frac{3\pi}{2} + 2k\pi$, for some $k \in \mathbb{Z}$. This is the same as having $c = \frac{\pi}{2} + k\pi$, for some $k \in \mathbb{Z}$. Thus:

$$\tan c \text{ is defined if and only if } c \neq \frac{\pi}{2} + k\pi \text{ for every } k \in \mathbb{Z}.$$

In the two coming examples we compute the trigonometric arcs of $\frac{\pi}{3}$ and $\frac{\pi}{4}$ radians (Fig. 7.7).

Example 7.9 Let us calculate the sine, cosine and tangent of $\frac{\pi}{3}$ radians (an arc of 60°). To this end, we let Γ be the unit circle and mark on it arc $\overset{\frown}{AP} = \frac{\pi}{3}$. Triangle OAP is isosceles of basis AP, with $A\widehat{O}P = 60°$, hence equilateral. Letting M be the midpoint of BC, we have $\cos\frac{\pi}{3} = \overline{OM} = \frac{1}{2}$ and, by Corollary 4.12, $\sin\frac{\pi}{3} = \overline{PM} = \frac{\sqrt{3}}{2}$. Hence, $\tan\frac{\pi}{3} = \frac{\sqrt{3}/2}{1/2} = \sqrt{3}$.

Example 7.10 Likewise the previous chapter, in order to compute the sine, cosine and tangent of $\frac{\pi}{4}$ (which is equivalent in degrees to 45°), mark point P on the first quadrant of the unit circle Γ, such that $\overset{\frown}{AP} = 45°$ (cf. Fig. 7.8), and let Q be the foot of the perpendicular dropped from P to OA.

Fig. 7.8 Sine, cosine and tangent of $\frac{\pi}{4}$

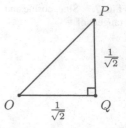

Since OPQ is a right and isosceles triangle of hyphotenuse $\overline{OP} = 1$, Pythagoras' theorem furnishes

$$\sin \frac{\pi}{4} = \overline{PQ} = \frac{1}{\sqrt{2}} \quad \text{and} \quad \cos \frac{\pi}{4} = \overline{OQ} = \frac{1}{\sqrt{2}}.$$

Therefore, $\tan \frac{\pi}{4} = \frac{1/\sqrt{2}}{1/\sqrt{2}} = 1$.

Back to the general discussion, if $\overset{\frown}{AP} = c$, with $\frac{\pi}{2} < c < \pi$ (cf. Fig. 7.4), then P lies in the portion of the unit circle Γ situated in the second quadrant of the cartesian plane, so that the abscissa of P is negative and its ordinate is positive. Hence,

$$\cos c < 0, \quad \sin c > 0 \quad \text{and} \quad \tan c = \frac{\sin c}{\cos c} < 0.$$

In this respect, see also Problem 2, page 239.

The tangent of an arc of c radians has a quite useful geometric interpretation, which we present in the coming proposition. Henceforth, whenever the endpoint P of an arc $\overset{\frown}{AP}$ of the unit circle belongs to the k-th quadrant of the cartesian plane, we shall simply say that $\overset{\frown}{AP}$ is an arc of the k-quadrant.

Proposition 7.11 *In the notations of Figure 7.9, if $\overset{\frown}{AP} = c$ is an arc of the first or third quadrants of the cartesian plane, then $\tan c = \overline{AP'}$; if $\overset{\frown}{AP} = c$ is an arc of the second or fourth quadrants, then $\tan c = -\overline{AP'}$.*

Proof We shall look at the case in which c is an arc of the second quadrant (the proof in the remaining cases is completely analogous). Let P'' be the foot of the perpendicular dropped from P to the horizontal axis. The (obvious) similarity of triangles $PP''O$ and $P'AO$ allows us to write

$$\frac{\overline{PP''}}{\overline{P''O}} = \frac{\overline{P'A}}{\overline{AO}}.$$

Fig. 7.9 Geometric
interpretation of the tangent

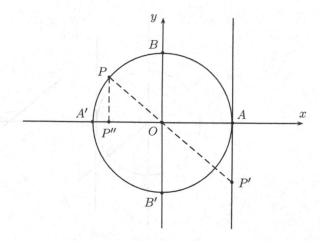

Substituting $\overline{PP''} = \sin c$, $\overline{P''O} = -\cos c$ and $\overline{AO} = 1$ in the equality above, we
obtain

$$\overline{P'A} = \frac{\overline{PP''} \cdot \overline{AO}}{\overline{P'A}} = -\tan c,$$

which is precisely what we wished to prove.

□

An immediate consequence of the proposition above is the fact that if two arcs
differ from and integer multiple of π, then their tangents are equal. In symbols,
$\tan(\pi + c)$ is defined if and only if so is $\tan c$; moreover, if this is so, then we have

$$\tan(\pi + c) = \tan c. \tag{7.5}$$

For what comes next, we adopt the following convention: given an angle θ, with
$0° \leq \theta \leq 360°$, we define the sine, cosine and tangent of θ as being respectively
equal to the sine, cosine and tangent of the arc corresponding to θ in radians, i.e.,
of the arc $c = 2\pi \cdot \frac{\theta}{360}$. For instance, for $\theta = 20°$, the corresponding arc is $c = \frac{\pi}{9}$
radians; thus, by definition we have

$$\sin 20° = \sin \frac{\pi}{9}, \quad \cos 20° = \cos \frac{\pi}{9} \quad \text{and} \quad \tan 20° = \tan \frac{\pi}{9}.$$

The coming proposition brings and important consequence of the previous result.

Proposition 7.12 *Given in the cartesian plane a non vertical line of equation* $y =
ax + b$, *the real number* $a \neq 0$ *is equal to the tangent of the trigonometric angle
the horizontal axis forms with the line in question. More precisely, in the notations
of Fig. 7.10, we have* $a = \tan \alpha$.

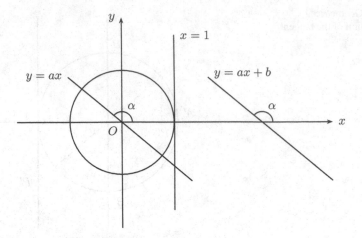

Fig. 7.10 The slope of a straightline

Proof In Fig. 7.10, the depicted circle represents the unit circle. According to Proposition 6.13, the parallel to line $y = ax + b$ passing through the origin has equation $y = ax$. Since the trigonometric angle the horizontal axis forms with such a line is also equal to α, the previous proposition (together with the above conventions) assure that $\tan \alpha$ is equal to the ordinate of the common point of lines $y = ax$ and $x = 1$.

For such a point, we have $x = 1$ and $y = a \cdot 1 = a$, so that $\tan \alpha = a$, as wished.   □

Thanks to the previous result, if a non vertical line r has equation $y = ax + b$, we say that a is its **angular coefficient**. Thus, a measures the slope of r, which explains the terminology *slope-intercept equation* attached to (6.6).

The coming example collects an important application of the notion of angular coefficient.

Example 7.13 In a cartesian system, we are given point (x_0, y_0). If an angle α is such that $0° < \alpha < 180°$, find the equation of the straightline passing through the given point and forming, with the horizontal axis, a trigonometric angle of measure α.

Solution If $\alpha = 90°$, then the desired line is vertical, hence has equation $x = x_0$. If $\alpha \neq 90°$ (cf. Fig. 7.11), we can assume that its equation is $y = ax + b$, for some $a, b \in \mathbb{R}$. Proposition 7.12 furnishes $a = \tan \alpha$. On the other hand, since (x_0, y_0) lies in the straightline, we must have

$$y_0 = ax_0 + b = (\tan \alpha)x_0 + b,$$

Fig. 7.11 A line with prescribed slope and passing through a given point

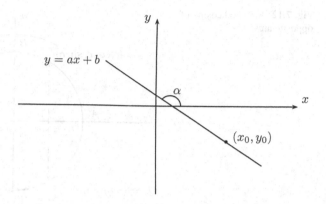

so that $b = y_0 - (\tan \alpha)x_0$. Thus, the desired equation is

$$y = (\tan \alpha)x + y_0 - (\tan \alpha)x_0.$$

$\square$

In the notations of the above example, it is worth pointing out that one uses to write the equation obtained above as

$$\frac{y - y_0}{x - x_0} = \tan \alpha. \qquad (7.6)$$

(Even though, strictly speaking, we cannot substitute x by x_0 in this last expression.)

We shall finish this initial discussion on trigonometric arcs by relating the sines, cosines and tangents of arcs that hold some simple relations with each other.

Proposition 7.14 *For every $c \in \mathbb{R}$ we have*

$$\sin(-c) = -\sin c \quad and \quad \cos(-c) = \cos c. \qquad (7.7)$$

Proof In the unit circle, let P and Q be the endpoints of arcs $\overset{\frown}{AP} = c$ and $\overset{\frown}{AQ} = -c$ (Fig. 7.12 illustrates the case $\frac{\pi}{2} < c < \pi$; the other cases are entirely analogous). Since the arcs of c and $-c$ radians have equal lengths but run in opposite directions (one in the counterclockwise sense and the other in the clockwise sense), it is immediate to see that P and Q are symmetric with respect to the horizontal axis. Hence, P and Q have equal abscissas but opposite ordinates. However, since $P(\cos c, \sin c)$ and $Q(\cos(-c), \sin(-c))$, it thus follows that $\cos c = \cos(-c)$ (equal abscissas) and $\sin c = -\sin(-c)$ (opposite ordinates). $\square$

Proposition 7.15 *For every $c \in \mathbb{R}$ we have*

$$\sin\left(\frac{\pi}{2} - c\right) = \cos c \quad and \quad \cos\left(\frac{\pi}{2} - c\right) = \sin c. \qquad (7.8)$$

Fig. 7.12 Sine and cosine of
opposite arcs

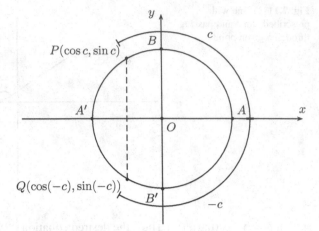

Fig. 7.13 Sine and cosine of
complementary arcs

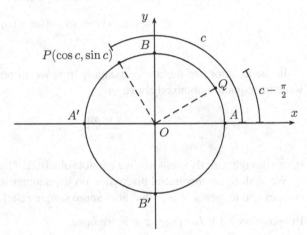

Proof By the previous result, it suffices to show that $\sin\left(c - \frac{\pi}{2}\right) = -\cos c$ and $\cos\left(c - \frac{\pi}{2}\right) = \sin c$. To this end, we argue also as in the previous proof: we consider, in the unit circle, the endpoints P and Q of the arcs $\overset{\frown}{AP} = c$ and $\overset{\frown}{AQ} = c - \frac{\pi}{2}$ (in this case too, we illustrate the case $\frac{\pi}{2} < c < \pi$ in Fig. 7.13, leaving the—totally analogous—verification of the remaining cases to the reader).

Since $\overline{OP} = \overline{OQ}$ and $Q\widehat{O}P = 90°$, Problem 4, page 200, guarantees that, if $Q(x_0, y_0)$, then $P(-y_0, x_0)$. However, since

$$P(\cos c, \sin c) \quad \text{and} \quad Q\left(\cos(c - \frac{\pi}{2}), \sin(c - \frac{\pi}{2})\right),$$

we must have

$$-\cos c = \sin\left(c - \frac{\pi}{2}\right) \quad \text{and} \quad \sin c = \cos\left(c - \frac{\pi}{2}\right),$$

as wished.   □

Corollary 7.16 *For every $c \in \mathbb{R}$, we have*

$$\sin(\pi - c) = \sin c \ \ and \ \ \cos(\pi - c) = -\cos c. \tag{7.9}$$

Proof Let us prove the sine formula, the deduction of the cosine formula being quite similar. Writing $\pi - c = \frac{\pi}{2} - \left(c - \frac{\pi}{2}\right)$ and successively applying (7.8), (7.7) and (7.8), we obtain

$$\sin(\pi - c) = \cos\left(c - \frac{\pi}{2}\right) = \cos\left(\frac{\pi}{2} - c\right) = \sin c.$$

$\square$

Example 7.17 In view of the proposition above, we can compute the trigonometric arcs of $\frac{\pi}{6}$ quite easily. Indeed, since $\frac{\pi}{2} - \frac{\pi}{6} = \frac{\pi}{3}$, we get

$$\sin\frac{\pi}{6} = \cos\frac{\pi}{3} = \frac{1}{2}, \ \ \cos\frac{\pi}{6} = \sin\frac{\pi}{3} = \frac{\sqrt{3}}{2} \ \ \text{and} \ \ \tan\frac{\pi}{3} = \frac{1/2}{\sqrt{3}/2} = \frac{1}{\sqrt{3}}.$$

Problems: Sect. 7.1

1. Mark, in the unit circle, the endpoints of the arcs of $\pi/3$, $3\pi/4$, $3\pi/2$, $-\pi$, $-2\pi/3$, $-3\pi/2$, $-4\pi/3$ and $-5\pi/2$ radians.

2. * Mark, in the unit circle, the signs of the trigonometric arcs of $\overset{\frown}{AP} = c$, according to the quadrant of the cartesian plane to which the point P belongs.

3. Compute the sine, cosine and (if it exists) tangent of the arcs of π, $-\pi/2$, 4π, $7\pi/2$ and $-3\pi/2$ radians.

4. Compute the sine, cosine and tangent of $2\pi/3$, $3\pi/4$, $5\pi/6$, $7\pi/6$, $4\pi/3$, $5\pi/4$, $5\pi/3$, $3\pi/2$ and $7\pi/4$ radians.

5. We are given, in the cartesian plane, the ellipse $\mathcal{E}$ of equation $\frac{x^2}{a^2} + \frac{y^2}{b^2} = 1$, do the following items:

   (a) If (x, y) is a point of $\mathcal{E}$, prove that there exists a unique $\theta \in [0, 2\pi)$ for which $x = a\cos\theta$ and $y = a\sin\theta$, and conversely. Equations $x = a\cos\theta$ and $y = a\sin\theta$, with $\theta \in [0, 2\pi)$, are the **parametric equations** of the ellipse, and we can look at them as a description of the *trajectory* of a particle that, departing from the point $A(a, 0)$ at instant $\theta = 0$, makes a complete turn arount $\mathcal{E}$ in the counterclockwise sense, returning to the point A at instant $\theta = 2\pi$.

   (b) Define the **auxiliary circles** of $\mathcal{E}$ as the circles of equations $x^2 + y^2 = a^2$ and $x^2 + y^2 = b^2$. Let a half-line of origin O form a trigonometric angle β with the horizontal axis, and suppose that it intersects the auxiliary circles above at points S and Q, respectively. If $P(x, y)$ is the intersection point of the parallel through Q to the horizontal axis with the parallel through S to the vertical axis, prove that P lies in the ellipse (note that the construction just described allows us to plot as many points of the ellipse as we wish).

6. (OCM) We are given in the cartesian plane the ellipse $\mathcal{E}$ of equation $\frac{x^2}{a^2} + \frac{y^2}{b^2} = 1$.

   (a) Let ABC be a triangle inscribed in $\mathcal{E}$, and Γ be the auxiliary circle of $\mathcal{E}$, of radius a. If $A', B', C' \in \Gamma$ are such that $\overleftrightarrow{AA'}$, $\overleftrightarrow{BB'}$, $\overleftrightarrow{CC'}$ are vertical, compute the area of $A'B'C'$ in terms of the area of ABC.
   (b) Given a line r in the plane, find all triangles ABC inscribed in $\mathcal{E}$, of largest possible area and having one side parallel to r.

7. We are given in the plane an ellipse $\mathcal{E}$ of major axis $2a$ and minor axis $2b$. Show that the supremum of the areas of the convex polygons inscribed in $\mathcal{E}$ is equal to πab. This way, it is pretty reasonable to *define*[2] the area of $\mathcal{E}$ by setting $A(\mathcal{E}) = \pi ab$.

7.2 Some Useful Identities

In this section, we continue our introduction to Trigonometry by deducing some quite useful formulas, starting with those for the trigonometric arcs of the sum and difference of two given arcs.

Proposition 7.18 *For $a, b \in \mathbb{R}$, we have:*

(a) $\cos(a \pm b) = \cos a \cos b \mp \sin a \sin b$.
(b) $\sin(a \pm b) = \sin a \cos b \pm \cos a \sin b$.
(c) $\tan(a \pm b) = \frac{\tan a \pm \tan b}{1 \mp \tan a \tan b}$, *whenever* $\tan a$, $\tan b$ *and* $\tan(a \pm b)$ *are defined.*

Proof We first derive the formula for $\cos(a+b)$. Assume, without loss of generality (why?), that $a, b > 0$ and mark, in the unit circle Γ, the endpoints P, Q and R of the arcs $\overset{\frown}{AP} = a$, $\overset{\frown}{AQ} = -b$ and $\overset{\frown}{AR} = a + b$ (cf. Fig. 7.14). Then, $P(\cos a, \sin a)$, $Q(\cos(-b), \sin(-b)) = (\cos b, -\sin b)$ and $R(\cos(a + b), \sin(a + b))$.

Since the arcs $\overset{\frown}{AR}$ and $\overset{\frown}{QP}$ (drawn in the counterclockwise sense) are both equal to $a + b$ radians, the corresponding chords AR and PQ have equal lengths. Hence, the equalities

$$\overline{AR}^2 = (1 - \cos(a + b))^2 + \sin^2(a + b)$$

and

$$\overline{PQ}^2 = (\cos a - \cos b)^2 + (\sin a + \sin b)^2$$

---

[2]Note that this definition agrees with the one usually seen in Calculus courses—cf. Chapter 10 of [5], for instance.

Fig. 7.14 Trigonometric arcs
of the sum and difference of
two arcs

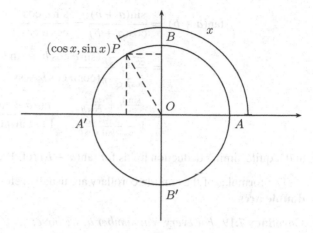

give us

$$(1 - \cos(a + b))^2 + \sin^2(a + b) = (\cos a - \cos b)^2 + (\sin a + \sin b)^2.$$

By expanding both sides of the above relation and substituting the equalities $\cos^2 a + \sin^2 a = 1$, $\cos^2 b + \sin^2 b = 1$ and $\cos^2(a + b) + \sin^2(a + b) = 1$, we obtain

$$-2\cos(a + b) = -2\cos a \cos b + 2 \sin a \sin b,$$

as wished.

In what concerns the remaining formulas, note that

$$\cos(a - b) = \cos(a + (-b)) = \cos a \cos(-b) - \sin a \sin(-b)$$
$$= \cos a \cos b + \sin a \sin b$$

and

$$\sin(a + b) = \cos\left(\frac{\pi}{2} - a - b\right)$$
$$= \cos\left(\frac{\pi}{2} - a\right)\cos b + \sin\left(\frac{\pi}{2} - a\right)\sin b$$
$$= \sin a \cos b + \cos a \sin b;$$

morcover (see Problem 1), an analogous reasoning gives us the formula for $\sin(a - b)$. Finally,

$$\tan(a+b) = \frac{\sin(a+b)}{\cos(a+b)} = \frac{\sin a \cos b + \sin b \cos a}{\cos a \cos b - \sin a \sin b}$$

$$= \frac{\frac{1}{\cos a \cos b}(\sin a \cos b + \sin b \cos a)}{\frac{1}{\cos a \cos b}(\cos a \cos b - \sin a \sin b)}$$

$$= \frac{\frac{\sin a}{\cos a} + \frac{\sin b}{\cos b}}{1 - \frac{\sin a}{\cos a}\frac{\sin b}{\cos b}} = \frac{\tan a + \tan b}{1 - \tan a \tan b},$$

and a quite similar deduction holds for $\tan(a - b)$ (cf. Problem 1).   □

The formulas of the coming corollary are usually referred to as the **formulas for double arcs**.

Corollary 7.19 *For every real number a, we have:*

(a) $\cos 2a = \cos^2 a - \sin^2 a$.
(b) $\sin 2a = 2 \sin a \cos a$.
(c) $\tan 2a = \frac{2 \tan a}{1 - \tan^2 a}$, *whenever* $\tan a$ *and* $\tan 2a$ *are defined.*

Proof It suffices to led $b = a$ in the formulas with the $+$ sign in Proposition 7.18.
                                                                                        □

It is worth observing that we can apply the formulas of the proposition and corollary above, as well as those of Proposition 7.22, if the involved arcs are measured in *degrees*, instead of radians. This is due to the (easily established) fact that if arcs of α and β degrees correspond to a and b radians, then an arc of $\alpha \pm \beta$ degrees corresponds to $a \pm b$ radians. From now on, we shall use this remark without further comments.

Example 7.20 A classical application of the formulas above is the computation of the trigonometric arcs of $75°$. To this end, note that $75° = 30° + 45°$, whence item (a) of Proposition 7.18 gives

$$\cos 75° = \cos(30° + 45°) = \cos 30° \cos 45° - \sin 30° \sin 45°$$

$$= \frac{\sqrt{3}}{2} \cdot \frac{\sqrt{2}}{2} - \frac{1}{2} \cdot \frac{\sqrt{2}}{2} = \frac{\sqrt{6} - \sqrt{2}}{4}.$$

A quite useful application of the formulas for addition of arcs is collected in the proposition below.

Proposition 7.21 *If a and b are given positive reals and θ is a real variable, then*

$$|a \cos \theta + b \sin \theta| \leq \sqrt{a^2 + b^2}.$$

Proof We write

$$a \cos \theta + b \sin \theta = \sqrt{a^2 + b^2} \left(\frac{a}{\sqrt{a^2 + b^2}} \cos \theta + \frac{b}{\sqrt{a^2 + b^2}} \sin \theta \right).$$

Now, letting $x_0 = \frac{a}{\sqrt{a^2+b^2}}$ and $y_0 = \frac{b}{\sqrt{a^2+b^2}}$, we have $x_0^2 + y_0^2 = 1$, so that (x_0, y_0) belongs to the first quadrant of the unit circle. Hence, there exists $\alpha \in (0, \pi/2)$ for which

$$\cos \alpha = x_0 = \frac{a}{\sqrt{a^2 + b^2}} \quad \text{and} \quad \sin \alpha = y_0 = \frac{b}{\sqrt{a^2 + b^2}}.$$

It thus follows from item (a) of Proposition 7.18 that

$$a \cos \theta + b \sin \theta = \sqrt{a^2 + b^2}(\cos \alpha \cos \theta + \sin \alpha \sin \theta)$$

$$= \sqrt{a^2 + b^2} \cos(\theta - \alpha). \tag{7.10}$$

However, since $|\cos(\theta - \alpha)| \le 1$, we obtain

$$|a \cos \theta + b \sin \theta| = \sqrt{a^2 + b^2}|\cos(\theta - \alpha)| \le \sqrt{a^2 + b^2}.$$

$\square$

We finish this quick look at Trigonometry by deducing the usual formulas for **transformation in product**, according to the next

Proposition 7.22 *For all* $a, b \in \mathbb{R}$, *we have:*

(a) $\sin a \pm \sin b = 2 \sin \left(\frac{a \pm b}{2}\right) \cos \left(\frac{a \mp b}{2}\right)$.
(b) $\cos a + \cos b = 2 \cos \left(\frac{a+b}{2}\right) \cos \left(\frac{a-b}{2}\right)$.
(c) $\cos a - \cos b = -2 \sin \left(\frac{a+b}{2}\right) \sin \left(\frac{a-b}{2}\right)$.
(d) $\tan a \pm \tan b = \frac{\sin(a \pm b)}{\cos a \cos b}$.

Proof Let us transform $\sin a + \sin b$ and $\tan a + \tan b$ into products, the remaining cases being entirely analogous (cf. Problem 10).

Setting $x = \frac{a+b}{2}$ and $y = \frac{a-b}{2}$, we have $a = x + y$ and $b = x - y$, so that

$$\sin a + \sin b = \sin(x + y) + \sin(x - y)$$

$$= (\sin x \cos y + \sin y \cos x) + (\sin x \cos y - \sin y \cos x)$$

$$= 2 \sin x \cos y,$$

as wished.

For what is left to do, it suffices to see that

$$\tan a + \tan b = \frac{\sin a}{\cos a} + \frac{\sin b}{\cos b} = \frac{\sin a \cos b + \sin b \cos a}{\cos a \cos b}$$

$$= \frac{\sin(a + b)}{\cos a \cos b}.$$

$\square$

Fig. 7.15 Sine, cosine and tangent in right triangles

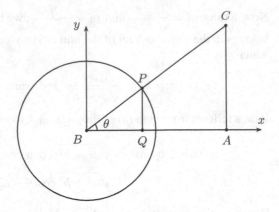

We finish the section by presenting a first link between Trigonometry and the geometry of triangles. In order to do this, let $0° < \theta < 90°$ be given. The corresponding arc c (in radians) satisfies $0 < c < \frac{\pi}{2}$, so that $\sin\theta > 0$, $\cos\theta > 0$ and $\tan\theta > 0$. We consider a triangle ABC, right at A and such that $A\widehat{B}C = \theta$ (cf. Fig. 7.15). Then, we take a cartesian system of origin B, such that the half-line $\overrightarrow{AB}$ coincides with the positive part of the x-axis and the leg AC is situated in the first quadrant. By drawing the circle of center B and radius 1, we can mark its intersection P with the hypotenuse BC of ABC (or with its extension, if $\overline{BC} < 1$). If Q denotes the foot of the perpendicular dropped from P to the horizontal axis, it is pretty clear that $BPQ \sim BCA$. Therefore,

$$\frac{\overline{AC}}{\overline{BC}} = \frac{\overline{QP}}{\overline{BQ}}, \quad \frac{\overline{AB}}{\overline{BC}} = \frac{\overline{QB}}{\overline{BP}}, \quad \frac{\overline{AC}}{\overline{AB}} = \frac{\overline{QP}}{\overline{QB}}.$$

Now, since

$$\overline{PQ} = \sin\theta, \quad \text{and} \quad \overline{QB} = \cos\theta,$$

the relations above give us

$$\frac{\overline{AC}}{\overline{BC}} = \sin\theta, \quad \frac{\overline{AB}}{\overline{BC}} = \cos\theta, \quad \frac{\overline{AC}}{\overline{AB}} = \tan\theta.$$

In short, for a triangle ABC, right at A and such that $A\widehat{B}C = \theta$, we have:

$$\sin\theta = \frac{\text{leg opposite to } \theta}{\text{hypotenuse}}, \quad \cos\theta = \frac{\text{leg adjacent to } \theta}{\text{hypotenuse}},$$
$$\tan\theta = \frac{\text{leg opposite to } \theta}{\text{leg adjacent to } \theta}. \tag{7.11}$$

The equalities just deduced, together with the formulas derived in the first part of this section, allow us to examine some interesting problems.

Example 7.23 (IMO—Shortlist) Let triangle ABC be right at A, and M be the midpoint of AC. Prove that

$$\tan M\widehat{B}C \le \frac{1}{2\sqrt{2}}.$$

Proof Let $\overline{AB} = c$, $\overline{AC} = 2b$, $M\widehat{B}A = \alpha$ and $M\widehat{B}C = \beta$.

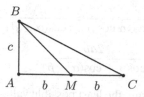

From

$$\tan(\alpha + \beta) = \frac{\tan\alpha + \tan\beta}{1 - \tan\alpha \cdot \tan\beta}$$

we get

$$\frac{2b}{c} = \frac{\frac{b}{c} + \tan\beta}{1 - \frac{b}{c} \cdot \tan\beta},$$

and hence $\tan\beta = \frac{bc}{2b^2+c^2}$. However, since $2b^2 + c^2 \ge 2\sqrt{2}bc$, it comes that

$$\tan\beta = \frac{bc}{2b^2 + c^2} \le \frac{bc}{2\sqrt{2}bc} = \frac{1}{2\sqrt{2}}.$$

$\square$

Example 7.24 We are given an angle $\angle XOY$ and a point P inside it. Points $A \in \overrightarrow{OX}$ and $B \in \overrightarrow{OY}$ are chosen in such a way that $P \in AB$ and $O\widehat{A}B, O\widehat{B}A < 90°$. Find the least possible value of $\overline{AP} \cdot \overline{BP}$.

Solution Let $X\widehat{O}Y = \theta$, and R and S be the feet of the perpendiculars dropped from P to $\overrightarrow{OX}$ and $\overrightarrow{OY}$, respectively. Let also $\overline{PR} = a$ and $\overline{PS} = b$, $R\widehat{P}A = \alpha$, $S\widehat{P}B = \beta$.

Since $R\widehat{P}S + R\widehat{O}S = \pi$ and $\alpha + \beta + R\widehat{P}S = \pi$, we get $\alpha + \beta = \theta$. On the other hand,

$$\cos\alpha = \frac{\overline{PR}}{\overline{AP}} = \frac{a}{\overline{AP}} \text{ and } \cos\beta = \frac{\overline{PS}}{\overline{BP}} = \frac{b}{\overline{BP}},$$

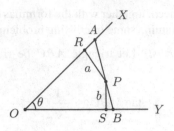

so that

$$\overline{AP} \cdot \overline{BP} = \frac{ab}{\cos\alpha\cos\beta} = \frac{2ab}{\cos(\alpha+\beta)+\cos(\alpha-\beta)}$$

$$= \frac{2ab}{\cos\theta+\cos(\alpha-\beta)}.$$

Since a, b and θ are constant, the least possible value of $\overline{AP} \cdot \overline{BP}$ is $\frac{2ab}{\cos\theta+1}$, and is attained exactly when $\cos(\alpha-\beta) = 1$, i.e., when $\alpha = \beta$. ☐

Problems: Sect. 7.2

1. * Prove the formulas of Proposition 7.18 for the expansions of $\sin(a-b)$ and $\tan(a-b)$.
2. * Compute the sine, cosine and tangent of $15°$.
3. * For $a \in \mathbb{R}$, prove that $\cos 2a = \begin{cases} 2\cos^2 a - 1 \\ 1 - 2\sin^2 a \end{cases}$.
4. Use the formulas of Corollary 7.19 and those of the previous problem to compute the sine, cosine and tangent of $22°30'$.
5. * In each case below, solve the given equations for $x \in \mathbb{R}$:

   (a) $\sin x = 0$.
   (b) $\cos x = 0$.
   (c) $\tan x = 0$.

6. * Given $\alpha \in \mathbb{R}$, solve the equations below for $x \in \mathbb{R}$:

   (a) $\sin x = \sin\alpha$.
   (b) $\cos x = \cos\alpha$.
   (c) $\tan x = \tan\alpha$.

7. * In each of the following items, solve the given equation for $x \in \mathbb{R}$, when $A = -\frac{1}{2}$, $A = \frac{\sqrt{3}}{2}$ and $A = -1$:

   (a) $\sin x = A$.
   (b) $\cos x = A$.

8. Solve the equation $\tan x = A$ for $A = \sqrt{3}$ and $A = 1$.

9. Compute $\sin(x - y)$ in terms of $a = \sin x + \sin y$ and $b = \cos x + \cos y$, knowing that $ab \neq 0$.

10. Deduce the other formulas of Proposition 7.22.

11. * Given $a, b \in \mathbb{R}$, with $b \neq 2k\pi$ for every $k \in \mathbb{Z}$, prove that

$$\sum_{k=0}^{n} \sin(a + kb) = \frac{\sin\left(a + \frac{nb}{2}\right) \sin \frac{(n+1)b}{2}}{\sin \frac{b}{2}}$$

and

$$\sum_{k=0}^{n} \cos(a + kb) = \frac{\cos\left(a + \frac{nb}{2}\right) \sin \frac{(n+1)b}{2}}{\sin \frac{b}{2}}.$$

12. * For $k \in \mathbb{N}$, show that

$$\sum_{k=1}^{n} \sin \frac{2k\pi}{n} = \sum_{k=1}^{n} \cos \frac{2k\pi}{n} = 0.$$

13. $ABCD$ is a square of side length 1, and E is a point on the edge CD, such that $\overline{AE} = \overline{AB} + \overline{CE}$. Letting F be the midpoint of lado CD, show that $E\widehat{A}B = 2 \cdot F\widehat{A}D$.

14. Lines r and s are parallel and are at a distance of 5cm from each other. A point P is located in the strip bounded by them, at a distance of 1cm from r. If points $A \in r$ and $B \in s$ are such that $\angle APB = 90°$, compute the length of AP for the area of APB to be as small as possible.

15. (OCM) Triangle ABC is right at A and such that $\overline{AB} = 1$ and $\overline{AC} = 3$. If D and E are points in AC for which $\overline{AD} = \overline{DE} = \overline{EC}$, prove that $A\widehat{D}B + A\widehat{E}B = 45°$.

16. * For a regular n-gon inscribed in a circle of radius R, let l_n and a_n denote the lengths of the sides and the distances from the center of the circle to them (one usually says that a_n is the **apothem** of the polygon). Compute l_n and a_n for $n = 3, 4, 6$ and 8.

17. We have n circles of radii r, all placed in the interior of a regular n-gon whose sides have length l. Each circle is tangent to two consecutive sides of the polygon, as well as to two other circles. Compute r in terms of l and n.

18. (Austria) We are give a cyclic pentagon $ABCDE$. If a, b and c denote the distances from A to the lines $\overleftrightarrow{BC}$, $\overleftrightarrow{CD}$ and $\overleftrightarrow{DE}$, respectively, find the distance from A to $\overleftrightarrow{BE}$.

19. (Estonia) ABC is an acute triangle of orthocenter H and barycenter G. Show that $\overleftrightarrow{HG} \parallel \overleftrightarrow{BC} \Leftrightarrow \tan \widehat{B} \cdot \tan \widehat{C} = 3$.

20. (Bulgaria) Triangle ABC is right at C and isosceles, and M is the midpoint of the hypotenuse AB. We choose a point P on BC and let N be the foot of the perpendicular dropped from C to AP; then, we mark $L \in AP$ such that $\overline{AL} = \overline{CN}$. If $A(ABC) = 4A(LMN)$, find all possible values of $C\widehat{A}P$.

21. (Belarus) Let ABC be a right triangle of hypothenuse BC, and P a point in the interior of ABC, with $\overline{BP} = \overline{AP}$. Let H stand for the foot of the altitude and M for the midpoint of BC. Prove that PM bisects $\angle HPC$ if and only if $\widehat{B} = 60°$.

    For our last problem the reader may find it convenient to recal Girard-Viète's formula for the sum of the roots of a polynomial equation (cf. Proposition 16.6 of [6], for instance). Specifically, you shall need the following result: if a_0, a_1, a_2, a_3, a_4 are given real numbers, with $a_4 \neq 0$, and x_1, x_2, x_3, x_4 are distinct real roots of the polynomial equation $a_4x^4 + a_3x_3 + a_2x^2 + a_1x + a_0 = 0$, then $x_1 + x_2 + x_3 + x_4 = -\frac{a_3}{a_4}$.

22. (Putnam) Two parabolas of perpendicular axes intersect at four distinct points A, B, C and D. Prove that $ABCD$ is a cyclic quadrilateral.

7.3 The Cosine Law

In this section we deepen the relation between Trigonometry and the geometry of triangles, by deducing the **cosine law** and presenting a number of interesting and useful applications of it. Such a formula can be seen as a generalization of Pythagoras' theorem for right triangles, and will be an almost indispensable tool from now on.

Theorem 7.25 (Cosine Law) *If triangle ABC has side lengths $\overline{AB} = c$, $\overline{AC} = b$ and $\overline{BC} = a$, then*

$$a^2 = b^2 + c^2 - 2bc \cos \widehat{A}. \tag{7.12}$$

Proof Let H denote the foot of the altitude relative to side AC, and let h stand for its length. We look separately at the cases $\widehat{A} < 90°$, $\widehat{A} = 90°$ and $\widehat{A} > 90°$:

(i) $\widehat{A} < 90°$: in this case (cf. Fig. 7.16), points H and C are in a single half line, of those determined on $\overleftrightarrow{AC}$ by point A, and (7.11) gives us

$$\overline{AH} = c \cos \widehat{A} \text{ and } h = c \sin \widehat{A}.$$

On the other hand, by applying Pythagoras' theorem to triangle BCH, we obtain

Fig. 7.16 The cosine law for $\widehat{A} < 90°$

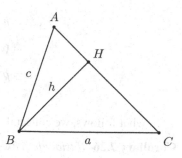

Fig. 7.17 The cosine law for $\widehat{A} = 90°$

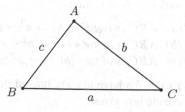

Fig. 7.18 The cosine law for $\widehat{A} > 90°$

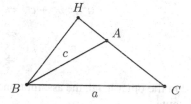

$$a^2 = h^2 + \overline{CH}^2 = h^2 + (b - \overline{AH})^2$$
$$= (c \sin \widehat{A})^2 + (b - c \cos \widehat{A})^2$$
$$= b^2 + c^2(\sin^2 \widehat{A} + \cos^2 \widehat{A}) - 2bc \cos \widehat{A}$$
$$= b^2 + c^2 - 2bc \cos \widehat{A},$$

where, in the last equality, we use the fundamental relation of Trigonometry (7.4).

(ii) $\widehat{A} = 90°$: in this case $\cos \widehat{A} = 0$, and Pythagoras' theorem (cf. Fig. 7.17) gives

$$a^2 = b^2 + c^2 = b^2 + c^2 - 2bc \cos \widehat{A}.$$

(iii) $\widehat{A} > 90°$: here (cf. Fig. 7.18), vertex A lies in the line segment CH. Since $B\widehat{A}H = 180° - \widehat{A}$, applying (7.11) to triangle BHA and invoking (7.9), we get

$$\overline{AH} = c \cos(180° - \widehat{A}) = -c \cos \widehat{A} \text{ and } h = c \sin(180° - \widehat{A}) = c \sin \widehat{A}.$$

Now, by Pythagoras' theorem applied to triangle BCH, and reasoning as in (i), we get

$$a^2 = h^2 + \overline{CH}^2 = h^2 + (b + \overline{AH})^2$$
$$= (c \sin \widehat{A})^2 + (b - c \cos \widehat{A})^2$$
$$= b^2 + c^2 - 2bc \cos \widehat{A}.$$

□

In what follows, we collect three important consequences of the cosine law.

Corollary 7.26 *If triangle ABC has side lengths* $\overline{AB} = c$, $\overline{AC} = b$ *and* $\overline{BC} = a$, *with* $a > b > c$, *then:*

(a) ABC is right (at A) $\Leftrightarrow a^2 = b^2 + c^2$.
(b) ABC is acute $\Leftrightarrow a^2 < b^2 + c^2$.
(c) ABC is obtuse (at A) $\Leftrightarrow a^2 > b^2 + c^2$.

Proof We have already proved item (a) in Sect. 4.2. In order to establish (b), the cosine law gives

$$a^2 < b^2 + c^2 \Leftrightarrow b^2 + c^2 - 2bc \cos A < b^2 + c^2$$
$$\Leftrightarrow -2bc \cos A < 0 \Leftrightarrow \cos A > 0$$
$$\Leftrightarrow \widehat{A} < 90°.$$

On the other hand, the assumption $a > b > c$ implies $\widehat{A} > \widehat{B} > \widehat{C}$, so that ABC is acute.

Finally, for item (c), we reason as above:

$$a^2 > b^2 + c^2 \Leftrightarrow b^2 + c^2 - 2bc \cos A > b^2 + c^2$$
$$\Leftrightarrow -2bc \cos A > 0 \Leftrightarrow \cos A < 0$$
$$\Leftrightarrow \widehat{A} > 90°.$$

□

Formula (7.13) below is known as **Stewart's relation**.[3]

Theorem 7.27 (Stewart) *Let ABC be a triangle of side lengths* $\overline{AB} = c$, $\overline{AC} = b$ *and* $\overline{BC} = a$. *If P is a point on the side BC, for which* $\overline{BP} = x$, $\overline{CP} = y$ *and* $\overline{AP} = z$, *then*

$$b^2 x + c^2 y = a(xy + z^2). \tag{7.13}$$

Proof If $A\widehat{P}C = \theta$, then $A\widehat{P}B = 180° - \theta$ (cf. Fig. 7.19). By applying the cosine law to triangle APC to compute $\overline{AC} = b$, we obtain

$$b^2 = z^2 + y^2 - 2yz \cos \theta.$$

---

[3]The formula is named after Matthew Stewart, Scottish mathematician of the XVIII century.

Fig. 7.19 Stewart's relation

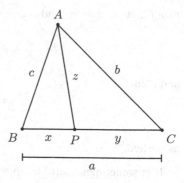

On the other hand, since $\cos(180° - \theta) = -\cos\theta$, cosine law, applied to triangle APB to compute $\overline{AB} = c$, give us

$$c^2 = z^2 + x^2 - 2xz\cos(180° - \theta)$$

$$= z^2 + x^2 + 2xz\cos\theta.$$

Solving both relations above for $\cos\theta$ and equating the results, we arrive at

$$\frac{z^2 + y^2 - b^2}{2yz} = \frac{c^2 - z^2 - x^2}{2xz}$$

or, which is the same, at

$$x(z^2 + y^2 - b^2) = y(c^2 - z^2 - x^2).$$

We can rewrite the last equality above as

$$xz^2 + yz^2 + xy^2 + x^2y = b^2x + c^2y,$$

after which grouping summands at the left hand side gives

$$(x + y)z^2 + xy(x + y) = b^2x + c^2y.$$

Finally, substituting $x + y = a$ furnishes us the desired result.   □

Formula (7.14) below has already appeared in Problem 5, page 188.

Corollary 7.28 *Let ABC be a triangle of side lengths $\overline{AB} = c$, $\overline{AC} = b$ and $\overline{BC} = a$. If m_a stands for the length of the median of ABC relative to BC, then*

$$a^2 + 4m_a^2 = 2(b^2 + c^2). \tag{7.14}$$

Proof Setting $z = m_a$ and $x = y = \frac{a}{2}$ in Stewart's relation (cf. Fig. 7.19), we obtain

$$b^2 \cdot \frac{a}{2} + c^2 \cdot \frac{a}{2} = a \left(\left(\frac{a}{2} \right)^2 + m_a^2 \right),$$

and then

$$\frac{b^2 + c^2}{2} = \frac{a^2}{4} + m_a^2,$$

as wished.   □

It is sometimes useful to apply the cosine law in the following disguised form: given a triangle ABC with $\widehat{A} < 90°$, let H denote the foot of the perpendicular dropped from B to $\overleftrightarrow{AC}$ and $\overline{AH} = m$ (cf. Fig. 7.16). Then, regardless of whether $\widehat{C}$ is acute, right or obtuse, we have

$$a^2 = h^2 + \overline{CH}^2 = h^2 + (b - \overline{AH})^2$$
$$= (c^2 - m^2) + (b - m)^2 \qquad (7.15)$$
$$= b^2 + c^2 - 2bm.$$

Accordingly, if $\widehat{C} > 90°$ and H and m are as above, a computation quite similar to the one above gives

$$a^2 = b^2 + c^2 + 2bm.$$

The following example uses the first of the two formulas above, together with the following simple fact on rational numbers (cf. Example 6.24 of [6]): if $r \in \mathbb{Q}$ is positive and $r^2 \in \mathbb{N}$, then $r \in \mathbb{N}$.

Example 7.29 (Brazil) Given a natural number h, show that there exists at most a finite number of acute triangles of integer sides, such that h is one of its heights.

Proof We stick to the notations of the figure below.

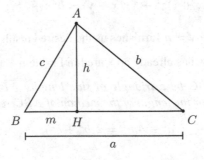

The triangle inequality gives $a + b - c > 0$, so that $a + b - c \in \mathbb{N}$. However, since $c = m + n = \sqrt{a^2 - h^2} + \sqrt{b^2 - h^2}$, we have

$$a + b - c = a + b - \sqrt{a^2 - h^2} + \sqrt{b^2 - h^2}$$

$$= \frac{h^2}{a + \sqrt{a^2 - h^2}} + \frac{h^2}{b + \sqrt{b^2 - h^2}}.$$

Now, a simple computation shows that

$$\frac{h^2}{a + \sqrt{a^2 - h^2}} < \frac{1}{2} \Leftrightarrow a > h^2 + \frac{1}{4},$$

and analogously for $\frac{h^2}{b + \sqrt{b^2 - h^2}}$. Therefore, we cannot have $a, b > h^2 + \frac{1}{4}$, so that either $a \le h^2 + \frac{1}{4}$ or $b \le h^2 + \frac{1}{4}$.

Suppose $a \le h^2 + \frac{1}{4}$ (the other case being entirely analogous). Then, we have at most a finite number of possibilities for a. The triangle inequality applied to ABH and ACH gives $a + b < c + h$, so that $0 < a + b - c < h$. We then conclude that $a + b - c \in \mathbb{N}$ and assumes at most a finite number of values, and since we already have at most a finite number of possibilities for a, the same holds for $b - c$.

On the other hand, (7.15) gives $b^2 = a^2 + c^2 - 2cm$, so that

$$m = \frac{a^2 + c^2 - b^2}{2c} \in \mathbb{Q}.$$

However, since $m^2 = a^2 - h^2 \in \mathbb{N}$, we conclude that $m \in \mathbb{N}$. Since $m < a$, this shows that m assumes at most a finite number of values, too.

Finally, letting $b - c = n$, it follows that

$$m = \frac{a^2 + c^2 - b^2}{2c} = \frac{a^2 - n(b + c)}{2c},$$

whence $nb + (n + 2m)c = a^2$. It is now immediate to note from the system of equations

$$\begin{cases} b - c = n \\ nb + (n + 2m)c = a^2 \end{cases},$$

together with the fact that we have at most a finite number of values for m and n, that we also have at most a finite number of values for b and c. □

Problems: Sect. 7.3

1. (OCM) From triangle ABC we know the lengths b and c of AC and AB, respectively. If $\overline{BC} = \sqrt{b^2 + c^2 + bc}$, compute, in degrees, the measure of $\angle ACB$.

2. * Do the following items:

   (a) In the notations of Problem 16, prove that $l_{10} = R\left(\frac{\sqrt{5}-1}{2}\right)$.

   (b) Show that $l_5 = R\sqrt{\frac{5-\sqrt{5}}{2}}$.

   (c) Prove that there exists a triangle with sides l_5, l_6 and l_{10}, and that such a triangle is necessarily a right one.

3. * Use the result of the previous problem to show that $\cos 36° = \frac{1+\sqrt{5}}{4}$.

4. (Slovenia) The three musketeers, Athos, Porthos and Aramis, had a terrible argument in a pub. After the fight, each one of them followed his own path, along directions that formed 120° with each other. If their speeds speeds were 10km/h, 20km/h and 40km/h, respectively. prove that at each instant after the fight their positions formed the vertices of a right triangle.

5. (Baltic Way) The internal bisectors of the angles $\angle A$ and $\angle B$ of a triangle ABC meet sides BC and AC respectively at points D and E. If $\overline{BD} + \overline{AE} = \overline{AB}$, compute all possible values of $A\widehat{C}B$.

6. (OIM) ABC is an equilateral triangle of side length ℓ. Point P inside ABC is such that $\overline{AP}, \overline{BP}$ and $\overline{CP}$ are the lengths of the sides of a right triangle of acute angles 30° and 60°. If $\overline{AP}$ is the length of the hypothenuse, compute it in terms of ℓ.

7. A convex hexagon has three consecutive sides of length a and the other three of length b. If it is inscribed in a circle of radius r, do the following items:

   (a) Compute r as a function of a and b.

   (b) Compute its area as a function of a and b.

8. Use the material of this section to prove the formula for the Euler median of a convex quadrilateral (cf. Problem 6, page 188).

9. Prove that, in every parallelogram, the sum of the squares of the lengths of the sides equals the sum of the squares of the diagonals. Then, use this fact to deduce formula (7.14) for the length of a median of a triangle.

10. (EKMC) If the angles at the vertices of the larger basis of a trapezoid are unequal, show that the diagonal departing from the vertex of the largest angle is smaller than the other one.

11. ABC is a right triangle of legs $\overline{AB} = 3$ and $\overline{AC} = 4$, and circumcircle Γ. Find the radius of the circle Σ, tangent to the legs AB and AC and internally tangent to Γ.

12. In triangle ABC, we have $\overline{AB} = c$, $\overline{AC} = b$ and $\overline{BC} = a$. If P stands for the foot of the internal bisector of $\angle BAC$, show that

$$\overline{AP} = \frac{2\sqrt{bcp(p-a)}}{b+c},$$

where p is the semiperimeter of ABC.

13. In triangle ABC, we have $\overline{AB} = c$, $\overline{AC} = b$ and $\overline{BC} = a$. If I denotes the incenter, show that $\overline{AI} < \sqrt{bc}$.

14. We are given a regular polygon $A_1 A_2 \ldots A_{2n}$ and a point P in the plane. Show that

$$\sum_{k=1}^{n} \overline{A_{2k}P}^2 = \sum_{k=1}^{n} \overline{A_{2k-1}P}^2.$$

15. (Russia) ABC is a triangle with sides $\overline{AB} = c$, $\overline{AC} = b$ and $\overline{BC} = a$, and medians relative to AB, AC and BC of lengths respectively equal to m_c, m_b and m_a. If R stands for the circumradius of ABC, prove that

$$\frac{a^2 + b^2}{2m_c} + \frac{a^2 + c^2}{2m_b} + \frac{b^2 + c^2}{2m_a} \leq 6R,$$

with equality if and only if ABC is equilateral.

16. (EKMC) We are given two exterior circles and draw an inner and an outer common tangent. The resulting points of tangency define a chord in each circle. Prove that the intersection point of these two chords is collinear with the centers of the circles.

17. (EKMC) Consider six points in the plane, no three of which are collinear. Prove that the ratio between the lenghts of the greatest and smallest segments determined by two of these points is at least $\sqrt{3}$.

18. Let P be a point in the interior of the inscribed circle Γ of triangle ABC. The lines $\overleftrightarrow{AP}$, $\overleftrightarrow{BP}$, $\overleftrightarrow{CP}$ meet Γ respectively at points X, Y, Z. If Γ touches the sides BC, AC, AB respectively at D, E and F, show that lines $\overleftrightarrow{DX}$, $\overleftrightarrow{EY}$ and $\overleftrightarrow{FZ}$ are concurrent.

7.4   The Sine Law

We begin this section by deriving another fundamental relation involving Trigonometry and the geometry of triangles, namely, the **sine law**. This is formula (7.16) below.

Fig. 7.20 The sine law

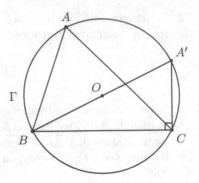

Theorem 7.30 (Sine Law) *If R denotes the circumradius of a triangle of side lengths a, b and c, then*

$$\frac{a}{\sin A} = \frac{b}{\sin B} = \frac{c}{\sin C} = 2R. \tag{7.16}$$

Proof Let ABC be such a triangle, with $\overline{AB} = c$, $\overline{AC} = b$ and $\overline{BC} = a$ (cf. Fig. 7.20). By assuming that ABC is acute (the analysis of the remaining cases is entirely analogous), we shall prove that $\sin \widehat{A} = \frac{a}{2R}$ (the equalities $\sin \widehat{B} = \frac{b}{2R}$ and $\sin \widehat{C} = \frac{c}{2R}$ can also be dealt with in exactly the same way).

Let O and Γ be the circumcenter and the circumcircle of ABC, respectively, and A' be the symmetric of B with respect to O. Then $A' \in \Gamma$, so that the inscribed angle theorem gives $B\widehat{A'}C = B\widehat{A}C = \widehat{A}$. On the other hand, since $A'B$ is a diameter of Γ, we have $A'\widehat{C}B = 90°$. Therefore, in the right triangle $A'BC$, we have

$$\sin \widehat{A} = \sin B\widehat{A'}C = \frac{\overline{BC}}{\overline{BA'}} = \frac{a}{2R}.$$

$\square$

The coming corollary is, almost surely, the most important consequence of sine law. The formula given by the first equality in (7.17) is usually referred to as the **sine formula for the area** of a triangle.

Corollary 7.31 *If triangle ABC has side lengths $\overline{AB} = c$, $\overline{AC} = b$, $\overline{BC} = a$, internal angles $\widehat{A}$, $\widehat{B}$, $\widehat{C}$ and circumradius R, then*

$$A(ABC) = \frac{1}{2}bc \sin A = \frac{abc}{4R}. \tag{7.17}$$

Proof Suppose that ABC is acute (the proof in the cases of a right or obtuse ABC are pretty much the same), and let H_b and h_b stand for the foot and the length of the altitude relative to AC, respectively (cf. Fig. 7.21). Then, in the right triangle ABH_b

Fig. 7.21 Useful formulas
for the area of a triangle

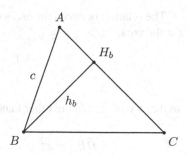

we have

$$\sin \widehat{A} = \frac{\overline{BH_b}}{\overline{AB}} = \frac{h_b}{c},$$

so that $h_b = c \sin \widehat{A}$. It thus follows that

$$A(ABC) = \frac{1}{2} b h_b = \frac{1}{2} bc \sin \widehat{A}.$$

In order to obtain the other formula in (7.17), it suffices to apply the sine law to the first formula:

$$A(ABC) = \frac{1}{2} bc \sin \widehat{A} = \frac{1}{2} bc \cdot \frac{a}{2R} = \frac{abc}{4R}.$$

$\square$

The sine formulas for the area, together with the cosine law, allows us to solve the following

Example 7.32 (Brazil) Given a triangle ABC, find points $D \in \overrightarrow{AB}$ and $E \overrightarrow{AC}$ such that $A(ADE) = \frac{1}{2} A(ABC)$ and $\overline{DE}$ is as small as possible.

Solution In the notations of the figure below, let $\overline{AB} = c$, $\overline{AC} = b$, $B\widehat{A}C = \alpha$, $\overline{AD} = x$, $\overline{AE} = y$ (note that the fact that D and E lie along the sides AB and AC—instead of their extensions—does not interfere in the solution).

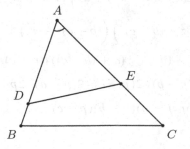

The relation between the areas of ADE and ABC give, through the sine formula for the area,

$$\frac{1}{2}xy \sin \alpha = \frac{1}{4}bc \sin \alpha,$$

so that $xy = \frac{bc}{2}$. On the other hand, it follows from the cosine law that

$$\overline{DE}^2 = x^2 + y^2 - 2xy \cos \alpha = x^2 + y^2 - bc \cos \alpha.$$

However, since b, c and α are constant, $\overline{DE}$ will be minimal if and only if the $x^2 + y^2$ is minimal. But the arithmetic-geometric means inequality gives $x^2 + y^2 \geq 2xy = bc$, with equality if and only if $x = y = \sqrt{\frac{bc}{2}}$.

$\square$

Yet another useful expression for the area of a triangle is **Heron's formula**, due to the Greek mathematician of the 1st century a.D. Heron of Alexandria.

Proposition 7.33 *If triangle ABC has side lengths a, b, c and semiperimeter p, then*

$$A(ABC) = \sqrt{p(p-a)(p-b)(p-c)}. \tag{7.18}$$

Proof By successively applying the sine formula for the area of ABC, the fundamental relation of Trigonometry and the cosine law, we obtain

$$16A(ABC)^2 = 4b^2c^2 \sin^2 \widehat{A} = 4b^2c^2(1 - \cos^2 \widehat{A})$$

$$= 4b^2c^2\left(1 - \left(\frac{b^2+c^2-a^2}{2bc}\right)^2\right).$$

Therefore, several applications of the formula for factorizing the difference of two squares give

$$16A(ABC)^2 = 4b^2c^2\left(1 - \frac{b^2+c^2-a^2}{2bc}\right)\left(1 + \frac{b^2+c^2-a^2}{2bc}\right)$$

$$= \left(2bc - (b^2+c^2) + a^2\right)\left(2bc + (b^2+c^2) - a^2\right)$$

$$= \left(a^2 - (b-c)^2\right)\left((b+c)^2 - a^2\right)$$

$$= (a - (b-c))(a + (b-c))((b+c) - a)((b+c) + a)$$

$$= 2(p-b) \cdot 2(p-c) \cdot 2(p-a) \cdot 2p$$

$$= 16p(p-a)(p-b)(p-c).$$

$\square$

We finish this section by discussing two interesting examples, the first of which is a neat application of Heron's formula which solves the **isoperimetric problem** for triangles.

Example 7.34 Show that, among all triangles with the same perimeter, the equilaterals are those of largest area.

Proof Let p stand for the (known) semiperimeter, A for the area and a, b, c for the lengths of the sides of the triangle. Heron's formula, applied in conjunction with the inequality between the arithmetic and geometric means of three positive reals (cf. Section 5.1 of [5], for instance), give us

$$\frac{A^2}{p} = (p-a)(p-b)(p-c) \le \left(\frac{(p-a)+(p-b)+(p-c)}{3}\right)^3 = \frac{p^3}{27},$$

with equality holding if and only if $p - a = p - b = p - c$, i.e., if and only if the triangle is equilateral.   □

Example 7.35 (IMO) Let $\widehat{A}$ be the smallest angle of triangle ABC, and U be a point in the smallest arc $\overset{\frown}{BC}$ of the circumcircle of ABC. The perpendicular bisectors of AB and AC intersect AU at T and W, respectively. If $\overleftrightarrow{BT}$ and $\overleftrightarrow{CW}$ meet at V, prove that $\overline{AU} = \overline{BV} + \overline{CV}$

Proof In the notations of the figure below, ACW and ABT are isosceles, so that $A\widehat{C}W = \beta$ and $A\widehat{B}T = \gamma$. Moreover, the inscribed angle theorem gives $C\widehat{B}U = C\widehat{A}U = \beta$ and $B\widehat{C}U = B\widehat{A}U = \gamma$. Thus, the exterior angle theorem gives

$$C\widehat{W}U = C\widehat{A}W + A\widehat{C}W = 2\beta, \quad B\widehat{T}U = B\widehat{A}T + A\widehat{B}T = 2\gamma$$

and

$$B\widehat{V}C = T\widehat{W}V + V\widehat{T}W = 2\beta + 2\gamma = 2(\beta + \gamma) = 2\widehat{A}.$$

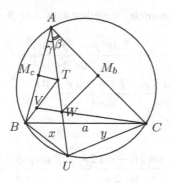

Now, let $\overline{AC} = b$ and $\overline{AB} = c$. Ptolemy's theorem applied to the quadrilateral $ABUC$, together with the sine law, give

$$\overline{AU} = b \cdot \frac{x}{a} + c \cdot \frac{y}{a} = b \cdot \frac{\sin \gamma}{\sin \widehat{A}} + c \cdot \frac{\sin \beta}{\sin \widehat{A}}. \tag{7.19}$$

Also,

$$\overline{BV} + \overline{CV} = \overline{BT} - \overline{TV} + \overline{CW} + \overline{VW}$$

$$= \frac{c}{2 \cos \gamma} - \overline{TV} + \frac{b}{2 \cos \beta} + \overline{VW},$$

where in the second equality we have computed $\cos \beta$ and $\cos \gamma$ in the right triangles CWM_b and BTM_c.

In order to compute $\overline{TV}$, apply the sine law to triangle TVW to obtain

$$\overline{TV} = \overline{TW} \cdot \frac{\sin 2\beta}{\sin 2\widehat{A}} = \left(\overline{AW} - \overline{AT} \right) \cdot \frac{\sin 2\beta}{\sin 2\widehat{A}}$$

$$= \left(\overline{CW} - \overline{BT} \right) \cdot \frac{\sin 2\beta}{\sin 2\widehat{A}} = \left(\frac{b}{2 \cos \beta} - \frac{c}{2 \cos \gamma} \right) \cdot \frac{\sin 2\beta}{\sin 2\widehat{A}}.$$

Likewise, $\overline{VW} = \left(\frac{b}{2 \cos \beta} - \frac{c}{2 \cos \gamma} \right) \cdot \frac{\sin 2\gamma}{\sin 2\widehat{A}}$, and substituting these expressions for $\overline{TV}$ and $\overline{VW}$ into that for $\overline{BV} + \overline{CV}$, we get

$$\overline{BV} + \overline{CV} = \frac{c}{2 \cos \gamma} - \left(\frac{b}{2 \cos \beta} - \frac{c}{2 \cos \gamma} \right) \cdot \frac{\sin 2\beta}{\sin 2\widehat{A}}$$

$$+ \frac{b}{2 \cos \beta} + \left(\frac{b}{2 \cos \beta} - \frac{c}{2 \cos \gamma} \right) \cdot \frac{\sin 2\gamma}{\sin 2\widehat{A}}$$

$$= \frac{b}{2 \cos \beta} \left(1 + \frac{\sin 2\gamma - \sin 2\beta}{\sin 2\widehat{A}} \right) + \frac{c}{2 \cos \gamma} \left(1 + \frac{\sin 2\beta - \sin 2\gamma}{\sin 2\widehat{A}} \right).$$

We now use some Trigonometry, together with $\widehat{A} = \beta + \gamma$, to compute

$$1 + \frac{\sin 2\gamma - \sin 2\beta}{\sin 2\widehat{A}} = 1 + \frac{2 \sin(\gamma - \beta) \cos(\gamma + \beta)}{2 \sin \widehat{A} \cos \widehat{A}} = 1 + \frac{\sin(\gamma - \beta)}{\sin \widehat{A}}$$

$$\frac{\sin(\gamma + \beta) + \sin(\gamma - \beta)}{\sin \widehat{A}} = \frac{2 \sin \gamma \cos \beta}{\sin \widehat{A}}.$$

Analogously,

$$1 + \frac{\sin 2\beta - \sin 2\gamma}{\sin 2\widehat{A}} = \frac{2 \sin \beta \cos \gamma}{\sin \widehat{A}}.$$

Finally, it follows from the formulas above and (7.19) that

$$\overline{BV} + \overline{CV} = \frac{b}{2\cos\beta} \cdot \frac{2\sin\gamma\cos\beta}{\sin\widehat{A}} + \frac{c}{2\cos\gamma} \cdot \frac{2\sin\beta\cos\gamma}{\sin\widehat{A}}$$

$$= b \cdot \frac{\sin\gamma}{\sin\widehat{A}} + c \cdot \frac{\sin\beta}{\sin\widehat{A}} = \overline{AU}.$$

$\square$

Problems: Sect. 7.4

1. The lengths of the sides of triangle ABC are a, b and c. Letting p denote the semiperimeter and R the circumradius of ABC, prove that

$$\sin\widehat{A} + \sin\widehat{B} + \sin\widehat{C} = \frac{p}{R}.$$

2. (OCM) ABC is a triangle in which $A\widehat{B}C = 2A\widehat{C}B$. Show that $b^2 = c(a+c)$.
3. The lengths of the sides of a triangle form an arithmetic progression with midterm ℓ. If the measure of the larger angle exceeds that of the smaller one in $90°$, compute the ratio between the lengths of the larger and smaller sides.
4. Use compass and straightedge to construct a triangle ABC, given the lengths b of side AC, c of side AB and knowing that the median relative to BC divides angle $\angle BAC$ in two angles such that one is twice the other.
5. Triangle ABC is such that $\overline{AB} = 13$, $\overline{AC} = 14$ and $\overline{BC} = 15$. A semicircle of radius R has its center along side BC and is tangent to sides AB and AC. Compute the value of R.
6. Let ABC be a triangle of sides 4, 5 and 6. We choose a point D along one of the sides of ABC, and then drop the perpendiculars DP and DQ to the other two sides. Compute the least possible length of line segment PQ.
7. * (IMO—adapted) Given a triangle ABC of sides a, b and c, prove **Weitzenböck's inequality**[4]:

$$A(ABC) \le \frac{\sqrt{3}}{12}(a^2 + b^2 + c^2),$$

with equality if and only if ABC is equilateral.
8. (Poland) Let ABC be a triangle with circumcircle Γ, and D be the midpoint of the arc $\overset{\frown}{BC}$ of Γ not containing vertex A. If K and L denote the feet of the perpendiculars dropped from B and C to $\overleftrightarrow{AD}$, respectively, prove that $\overline{BK} + \overline{CL} \le \overline{AD}$.

---

[4]Roland Weitzenböck, austrian mathematician of the XX century.

9. (OIM) Starting from a triangle ABC, we construct a convex hexagon $A_1A_2B_1B_2C_1C_2$ in the following way:

   (a) $A \in BB_2, CC_1, B \in AA_1, CC_2$ and $C \in AA_2, BB_1$.
   (b) $\overline{AB_2} = \overline{AC_1} = \overline{BC}, \overline{BA_1} = \overline{BC_2} = \overline{AC}$ and $\overline{CA_2} = \overline{CB_1} = \overline{AB}$.

   Prove that

   $$A(A_1A_2B_1B_2C_1C_2) \geq 13 \cdot A(ABC).$$

10. Prove *Archimedes' theorem* (cf. Problem 28, page 84) with the aid of the theorey developed in this chapter: let ABC be a triangle in which $AB > AC$. In the circumcircle of ABC we mark the midpoint M of the arc $\overset{\frown}{BC}$ that contains A. If N is the foot of the perpendicular dropped from M to the line segment AB, then $\overline{BN} = \overline{AN} + \overline{AC}$.

11. Let ABC be a triangle in which $\overline{AB} = c$, $\overline{AC} = b$ and $B\widehat{A}C = 2\alpha$. If P is the foot of the internal bisector relative to A, prove that

    $$\overline{AP} = \frac{2bc \cos\alpha}{b+c}.$$

12. (Canada) ABC is a triangle of sides $\overline{BC} = a$ and $\overline{AC} = b$. Knowing that its area equals $\frac{1}{4}(a^2 + b^2)$, find all possible values for the measure of $\angle ACB$.

13. * Prove the **sine formula** for the area of a convex quadrilateral: if $ABCD$ is a convex quadrilateral with diagonals AC and BD, forming an angle α with each other, then

    $$A(ABCD) = \frac{1}{2}\overline{AC} \cdot \overline{BD}\sin\alpha.$$

14. (Belarus) A trapezoid has area 18cm^2 and sum of diagonals 12cm. Show that the diagonals are perpendicular and have equal lengths.

15. (Austria) In a convex quadrilateral $ABCD$, let E be the intersection point of the diagonals AC and BD. If $A(ABE) = S_1$, $A(CDE) = S_2$ and $A(ABCD) = S$, prove that

    $$\sqrt{S} \geq \sqrt{S_1} + \sqrt{S_2}.$$

    When does the equality occur?

16. $ABCD$ is a cyclic quadrilateral for which there exists a point M along side CD such that the perimeter and the area of triangle ADM respectively coincide with those of quadrilateral $ABCM$. Prove that $ABCD$ has two sides of equal lengths.

17. (Brazil) $ABCD$ is a cyclic quadrilateral of sides a, b, c and d. Prove **Brahmagupta's formula**[5] for the area of $ABCD$:

$$A(ABCD) = \sqrt{(p-a)(p-b)(p-c)(p-d)}.$$

18. Prove **Hypparchus' theorem**[6]: if $ABCD$ is a cyclic quadrilateral of sides $\overline{AB} = a$, $\overline{BC} = b$, $\overline{CD} = c$, $\overline{DA} = d$ and diagonals $\overline{AC} = x$, $\overline{BD} = y$, then

$$\frac{x}{y} = \frac{ad + bc}{ab + cd}.$$

19. (IMO) Use compass and straightedge to construct a cyclic quadrilateral $ABCD$, knowing the lengths a, b, c and d of sides AB, BC, CD and DA, respectively.

20. (Soviet Union) Given a convex polygon in the plane, we can draw a line cutting it into two other convex polygons, choose one of them, turn it upside down and glue it back to the other one, along the original cut. Does there exist a finite sequence of such operations that transform a square into a triangle?

21. * We are given a triangle ABC and cevians AP, BQ and CR, with $P \in BC$, $Q \in AC$ and $R \in AB$. If $B\widehat{A}P = \alpha$, $C\widehat{A}P = \alpha'$, $C\widehat{B}Q = \beta$, $A\widehat{B}Q = \beta'$, $A\widehat{C}R = \gamma$ and $B\widehat{C}R = \gamma'$, prove that

$$\frac{\overline{BP}}{\overline{PC}} = \frac{\sin\alpha \cdot \sin\widehat{C}}{\sin\alpha' \cdot \sin\widehat{B}},$$

with likewise equalities for the ratios in which Q divides AC and R divides AB. Then, use such relations to establish the following *trigonometric version* of Ceva's theorem:

$$AP, \; BQ \text{ and } CR \text{ concur } \Leftrightarrow \frac{\sin\alpha}{\sin\alpha'} \cdot \frac{\sin\beta}{\sin\beta'} \cdot \frac{\sin\gamma}{\sin\gamma'} = 1.$$

Finally, use such a formulation of Ceva's theorem to give an alternative proof of Theorem 4.28.

22. Let ABC be an acute triangle of incenter I. Mark point $A_1 \in \overrightarrow{AI}$ such that $A_1 \neq A$ and the midpoint M of AA_1 lies in the circumcircle of ABC; then, define $N \in \overrightarrow{BI}$ and $P \in \overrightarrow{CI}$ in analogous ways.

   (a) Prove that $A(A_1B_1C_1) = (4R+r)p$, where r and R respectively denote the inradius and the circumradius of ABC, and p stands for its semiperimeter.
   (b) Conclude that $A(A_1B_1C_1) \geq 9 \cdot A(ABC)$, with equality if and only if ABC is equilateral.

---

[5] Brahmagupta, Indian astronomer and mathematician of the VII century.
[6] Hypparchus of Nicaea, Greek astronomer and mathematician of the II century.

7.5 Ptolemy's Inequality

For what comes next, it is worth recalling the discussion in the paragraph preceding the Simson-Wallace theorem (cf. Theorem 3.40): given a triangle ABC and a point P, with A, B, C and P in *general position* (i.e., such that P is noncollinear with any two of A, B, C), the **pedal triangle** of P with respect to ABC is the (possibly degenerated) triangle whose vertices are the feet of the perpendiculars dropped from P to the lines $\overleftrightarrow{AB}$, $\overleftrightarrow{AC}$, $\overleftrightarrow{BC}$.

The above mentioned result assures that triangle DEF is degenerated if and only if P lies in the circumcircle of ABC. In any case, the following proposition teaches us to compute the lengths of the sides of DEF (Fig. 7.22).

Proposition 7.36 *We are given in the plane a triangle ABC and a point P, with A, B, C and P in general position. Let DEF be the (possibly degenerated) pedal triangle of P with respect to ABC, with $D \in \overleftrightarrow{BC}$, $E \in \overleftrightarrow{AC}$ and $F \in \overleftrightarrow{AB}$. Then,*

$$\overline{DE} = \overline{PC} \cdot \sin \widehat{C}, \quad \overline{EF} = \overline{PA} \cdot \sin \widehat{A} \ and \ \overline{FD} = \overline{PB} \cdot \sin \widehat{B}.$$

Proof There are three essentially distinct cases to consider: P lies in the interior of ABC; P lies in the angular region $\angle ABC$ but outside the triangle ABC; P lies in the angular region OPP to $\angle ABC$. Let us analyse the first case (cf. Fig. 3.38), the analysis of the two remaining ones being quite similar.

It suffices to show that $\overline{DE} = \overline{PC} \cdot \sin \widehat{C}$ (the remaining equalities can be deduces through analogous reasonings). Since $P\widehat{E}C = P\widehat{D}C = 90°$, quadrilateral

Fig. 7.22 Computing the
sides of a pedal triangle

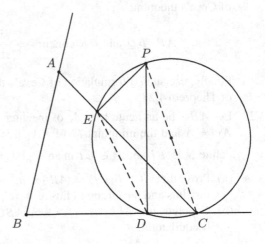

Fig. 7.23 Generalizing
Ptolemy's theorem

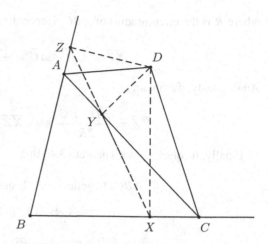

$PEDC$ is cyclic and its circumscribed circle has diameter PC. Hence, by applying the sine law to triangle DEC, we obtain

$$\frac{\overline{DE}}{\sin E\widehat{C}D} = \overline{PC}.$$

Finally, taking into account that $\angle ECD = \angle ACB = \angle C$, we get the desired result.
□

The proposition above, together with the Simson-Wallace theorem allows us to generalize Ptolemy's theorem 4.18 as described below.

Theorem 7.37 *If* $ABCD$ *is a convex quadrilateral of diagonals* AC *and* BD, *then*

$$\overline{AB} \cdot \overline{CD} + \overline{AD} \cdot \overline{BC} \geq \overline{AC} \cdot \overline{BD}, \qquad (7.20)$$

with equality holding if and only if $ABCD$ *is cyclic.*

Proof Let X, Y and Z denote the feet of the perpendiculars dropped from D respectively to $\overleftrightarrow{BC}$, $\overleftrightarrow{AC}$ and $\overleftrightarrow{AB}$ (cf. Fig. 7.23). As in the proof of the Simson-Wallace theorem, we can assume with no loss of generality that X and Y lie in the line segments BC and AC, respectively, while Z lies in the extension of AB.

Proposition 7.36 gives

$$\frac{\overline{XY}}{\sin \widehat{C}} = \overline{CD};$$

on the other hand, the sine law applied to triangle ABC furnishes

$$\frac{\overline{AB}}{\sin \widehat{C}} = 2R,$$

where R is the circumradius of ABC. Hence, from the two relations above we get

$$\overline{XY} = \overline{CD} \cdot \sin \widehat{C} = \frac{\overline{CD} \cdot \overline{AB}}{2R}.$$

Analogously, we obtain

$$\overline{YZ} = \frac{\overline{AD} \cdot \overline{BC}}{2R} \text{ and } \overline{XZ} = \frac{\overline{BD} \cdot \overline{AC}}{2R}.$$

Finally, it follows from Theorem 3.40 that

$$ABCD \text{ cyclic} \Leftrightarrow X, Y \text{ and } Z \text{ collinear}$$

$$\Leftrightarrow \overline{XY} + \overline{YZ} = \overline{XZ}.$$

Since in any event triangle inequality gives $\overline{XY} + \overline{YZ} \geq \overline{XZ}$, the expressions above for $\overline{XY}$, $\overline{XZ}$ and $\overline{YZ}$ furnish

$$\frac{\overline{AB} \cdot \overline{CD}}{2R} + \frac{\overline{AD} \cdot \overline{BC}}{2R} \geq \frac{\overline{AC} \cdot \overline{BD}}{2R},$$

with equality if and only if $ABCD$ is cyclic. $\square$

Given a triangle ABC in the plane, **Steiner's problem**[7] for ABC asks one to find the point(s) P inside it for which the sum

$$\overline{PA} + \overline{PB} + \overline{PC}$$

is as small as possible. In what comes next, we shall use the general version of Ptolemy's theorem to solve this problem for a certain class of triangles.

In order to properly state the result below recall (cf. Problem 6, page 126) that if ABC is a triangle whose internal angles are all less than $120°$, then its *point of Fermat* is the only point P inside ABC such that

$$A\widehat{P}B = A\widehat{P}C = B\widehat{P}C = 120°.$$

Theorem 7.38 (Steiner) *If ABC is a triangle with internal angles all less than $120°$, then its point of Fermat is the only solution for Steiner's problem for ABC.*

Proof We first note that the construction delineated in Problem 6, page 126 (cf. Fig. 7.24), guarantees that if P is the intersection point of the circumcircles of the equilateral triangles ACE, BCD and ABF, then

$$\overline{AD} = \overline{BE} = \overline{CF} = \overline{AP} + \overline{BP} + \overline{CP}. \tag{7.21}$$

---

[7] Jakob Steiner, Swiss mathematician of the XIX century.

Fig. 7.24 Steiner's problem

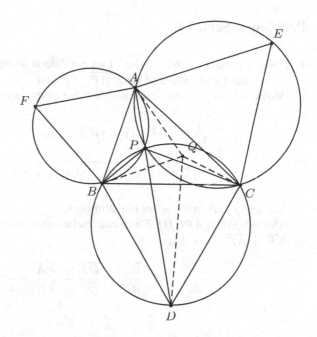

On the other hand, if $Q \neq P$ is another point in the interior of ABC, then Q cannot lie in all of these three circles (for otherwise Q would coincide with P). Assume, without loss of generality, that Q is not in the circumcircle of BCD. Then, Ptolemy's inequality (7.20) applied to quadrilateral $QBDC$ (which is not cyclic) gives us

$$\overline{BQ} \cdot \overline{CD} + \overline{CQ} \cdot \overline{BD} > \overline{DQ} \cdot \overline{BC}.$$

However, since $\overline{BC} = \overline{BD} = \overline{CD}$, it follows that

$$\overline{BQ} + \overline{CQ} > \overline{DQ}.$$

Finally, it suffices to use triangle inequality and (7.21):

$$\overline{AQ} + \overline{BQ} + \overline{CQ} > \overline{AQ} + \overline{DQ} \geq \overline{AD} = \overline{AP} + \overline{BP} + \overline{CP},$$

so that the only solution of Steiner's problem is, indeed, the point of Fermat P of triangle ABC. $\qquad\square$

Problems: Sect. 7.5

1. Given a point P in the interior of an equilateral triangle ABC, show that there always exist a triangle with sides $\overline{AP}$, $\overline{BP}$ and $\overline{CP}$.
2. (Vietnam) Find the smallest possible value of the expression

$$\sqrt{(x+1)^2 + (y-1)^2} + \sqrt{(x-1)^2 + (y+1)^2} +$$

$$+ \sqrt{(x+2)^2 + (y+2)^2},$$

as x and y vary in the set of real numbers.
3. (IMO shortlist) $ABCDEF$ is a convex hexagon with $\overline{AB} = \overline{BC}$, $\overline{CD} = \overline{DE}$, $\overline{EF} = \overline{AF}$. Show that

$$\frac{\overline{BC}}{\overline{BE}} + \frac{\overline{DE}}{\overline{DA}} + \frac{\overline{FA}}{\overline{FC}} \geq \frac{3}{2}.$$

Chapter 8
Vectors in the Plane

This chapter is devoted to a systematic study of vectors in the plane, as well as to the presentation of some applications of them to Euclidean Geometry. In this sense, we shall try to emphasize the use of vectors as being, at the same time, alternative and complementary to the synthetic and cartesian methods. It is within this spirit that we shall use vectors to revisit several previously obtained results; in particular, we call the reader's attention to the variety of uses of the concept of *scalar product of two vectors*, in Sect. 8.3.

8.1 Geometric Vectors

A **geometric vector**, or simply a **vector**, in the plane is an *oriented segment*, i.e., a line segment in which one of its endpoints is the **initial point** and the other is the **terminal point** (cf. Fig. 8.1).

Generally, we shall represent vectors by using boldface lower case Latin letters. If a vector **v** (cf. Fig. 8.1) has initial point A and terminal point B, we shall also write $\mathbf{v} = \overrightarrow{AB}$. Note that vectors $\overrightarrow{AB}$ and $\overrightarrow{BA}$ are distintcs; indeed, although they have the same endpoints, the direction of $\overrightarrow{BA}$ is opposite to that of $\overrightarrow{AB}$.

The notion of equality for vectors is, in a certain sense, *weaker* than that of equality of line segments: we say that vectors $\mathbf{v} = \overrightarrow{AB}$ and $\mathbf{w} = \overrightarrow{CD}$ are **equal** provided the quadrilateral $ACDB$ (i.e., the quadrilateral of sides AC, CD, DB and BA) is a parallelogram (cf. Fig. 8.1). In this case, we also say that $\overrightarrow{AB}$ and $\overrightarrow{CD}$ **represent** the same vector and write $\mathbf{v} = \mathbf{w}$ or $\overrightarrow{AB} = \overrightarrow{CD}$. A single vector has infinitely many representatives: given a vector $\mathbf{v} = \overrightarrow{AB}$ and an arbitrary point X, if Y is such that $AXYB$ is a parallelogram, then $\overrightarrow{XY}$ is another representative for **v**.

Fig. 8.1 Vectors in the plane

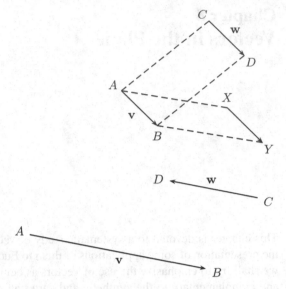

Fig. 8.2 Vectors with
opposite directions

The **modulus** of vector **v**, denoted $||\mathbf{v}||$, is the length of any of its representative.
For instance, if $\mathbf{v} = \overrightarrow{AB}$, then $||\mathbf{v}|| = \overline{AB}$ (cf. Fig. 8.2).

For what comes, it is useful to extend the definitions above to consider any point in the plane as a vector in its own, called the **zero, or null, vector** and denoted **0** (note that such a convention is equivalent to considering a point as a line segment of zero length). Thus, $||\mathbf{0}|| = 0$. A **nonzero** or **non null** vector is a vector **v** such that $\mathbf{v} \neq \mathbf{0}$.

We say that two nonzero vectors **v** and **w** have the same **direction** provided the following condition is satisfied: if $\mathbf{v} = \overrightarrow{AB}$ and $\mathbf{w} = \overrightarrow{CD}$, then $\overleftrightarrow{AB} \parallel \overleftrightarrow{CD}$ and $AD \cap CB \neq \emptyset$. It is not difficult to check that this condition does not depend on the representatives of **v** and **w**. Also in the above notations, if $\overleftrightarrow{AB} \parallel \overleftrightarrow{CD}$ but $AD \cap CB = \emptyset$, then we say that **v** and **w** have **opposite directions**. This is the case of vectors **v** and **w** of Fig. 8.2. Moreover, if one of these cases happens, we shall sometimes say that **v** and **w** are **parallel**.

One of the great advantages of working with vectors is the possibility of performing with them *operations* quite similar to the addition and subtraction of real numbers. Specifically, given nonzero vectors **v** and **w** (cf. Fig. 8.3), choose representatives $\mathbf{v} = \overrightarrow{AB}$ and $\mathbf{w} = \overrightarrow{BC}$ (i.e., such that the terminal point of **v** coincides with the initial point of **w**) and define the **sum** of **v** and **w** as being the vector $\mathbf{v} + \mathbf{w} = \overrightarrow{AC}$. Define also $\mathbf{v} + \mathbf{0} = \mathbf{v}$, for every vector **v**.

If vectors **v** and **w** do not lie in parallel lines (as is the case in Fig. 8.3), we can also compute its sum by means of the **parallelogram rule**: we choose a representative $\overrightarrow{AD}$ for **w** (with the same initial point as that of **v**) and observe that $\mathbf{v} + \mathbf{w}$ is the diagonal vector $\overrightarrow{AC}$ of parallelogram $ABCD$.

Fig. 8.3 Addition of vectors

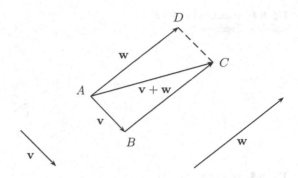

It is immediate to verify (cf. Problem 1) that the addition of vectors is *well defined*, i.e., that the sum vector $\mathbf{v} + \mathbf{w}$ of $\mathbf{v}$ and $\mathbf{w}$ does not depend on the chosen representatives for $\mathbf{v}$ and $\mathbf{w}$. Moreover, it clearly follows from the parallelogram rule that the addition of vectors is **commutative**, i.e., that

$$\mathbf{v} + \mathbf{w} = \mathbf{w} + \mathbf{v},$$

for all vectors $\mathbf{v}$ and $\mathbf{w}$.

Another important property of the addition of vectors is its **associativity**, which is the object of the coming result.

Proposition 8.1 *Given vectors* $\mathbf{a}$, $\mathbf{b}$ *and* $\mathbf{c}$, *we have*

$$\mathbf{a} + (\mathbf{b} + \mathbf{c}) = (\mathbf{a} + \mathbf{b}) + \mathbf{c}.$$

Proof Suppose, for the sake of convenience, that we have already chosen representatives for $\mathbf{a}$, $\mathbf{b}$ and $\mathbf{c}$ in such a way that $\mathbf{a} = \overrightarrow{OA}$, $\mathbf{b} = \overrightarrow{OB}$ and $\mathbf{c} = \overrightarrow{OC}$ (cf. Fig. 8.4). Successively mark points D and E such that $OADB$ and $ODEC$ are parallelograms. Then, $\overrightarrow{OD} = \mathbf{a} + \mathbf{b}$, so that

$$(\mathbf{a} + \mathbf{b}) + \mathbf{c} = \overrightarrow{OD} + \overrightarrow{OC} = \overrightarrow{OE}.$$

Now, mark point F such that $OBFC$ is also a parallelogram. Then, on the one hand, $\overrightarrow{OF} = \mathbf{b} + \mathbf{c}$; on the other, the fact that $ODEC$ and $OBFC$ are parallelograms assures that the line segments BF and DE are both parallel and equal to the line segmen OC. Hence, quadrilateral $DEFB$ is a parallelogram too (since is opposite sides DE and BF are equal and parallels). In particular, as vectors we have

$$\overrightarrow{FE} = \overrightarrow{BD} = \overrightarrow{OA},$$

Fig. 8.4 Associativity of the
addition of vectors

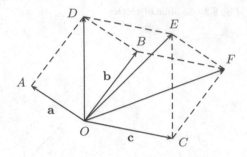

Fig. 8.5 Vectors **v** and $-\frac{1}{2}\mathbf{v}$

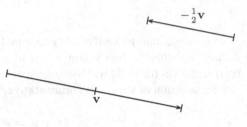

where in the last equality we used the fact that $OADB$ is also a parallelogram. Thus,
$\overrightarrow{OA} = \overrightarrow{FE}$, so that $OAEF$ is a parallelogram too. But this gives us

$$\mathbf{a} + (\mathbf{b} + \mathbf{c}) = \overrightarrow{OA} + \overrightarrow{OF} = \overrightarrow{OE} .$$

□

A second useful operation with vectors is the **multiplication of vectors by
scalars** (in the context of operations with vectors, we will generally refer to a real
number as a **scalar**). Thus, given a vector **v** and a scalar k, we define the *product
of **v** by the scalar* k, or simply the **product by scalar** $k\mathbf{v}$, in the following way: if
$k = 0$ or $v = \mathbf{0}$, then $k\mathbf{v} = \mathbf{0}$; if $k \neq 0$ and $\mathbf{v} \neq \mathbf{0}$, then (look at Fig. 8.5 for an
example) $k\mathbf{v}$ is the only vector satisfying the following conditions:

  (*i*) $k\mathbf{v}$ is parallel to **v**.
  (*ii*) $\|k\mathbf{v}\| = |k| \cdot \|\mathbf{v}\|$.
  (*iii*) $k\mathbf{v}$ and **v** have the same (resp. opposite) directions if $k > 0$ (resp. $k < 0$).

We shall also say that $k\mathbf{v}$ is a **scalar multiple** of **v**.
In the particular case of $k = -1$, we shall usually write $-\mathbf{v}$ to denote $(-1)\mathbf{v}$, and
say that $-\mathbf{v}$ is **opposite** to **v**.
The operation of multiplication by scalars has the properties listed in the proposition below, whose proof we leave as an exercise for the reader (cf. Problem 3).

Proposition 8.2 *Given scalars k, k_1 and k_2, and vectors **v**, $\mathbf{v_1}$ and $\mathbf{v_2}$, we have:*

  (*a*) $(k_1 k_2)\mathbf{v} = k_1 (k_2 \mathbf{v})$.
  (*b*) $(k_1 + k_2)\mathbf{v} = (k_1 \mathbf{v}) + (k_2 \mathbf{v})$.
  (*c*) $k(\mathbf{v} + \mathbf{w}) = (k\mathbf{v_1}) + (k\mathbf{v_2})$.

With the operation of multiplication of vectors by scalars at our disposal, we are in position to define the **difference** $\mathbf{v} - \mathbf{w}$ of the vectors $\mathbf{v}$ and $\mathbf{w}$ (in this order) as the sum of $\mathbf{v}$ with the opposite of $\mathbf{w}$:

$$\mathbf{v} - \mathbf{w} = \mathbf{v} + (-\mathbf{w}). \tag{8.1}$$

In particular, it is immediate that $\mathbf{v} - \mathbf{v} = \mathbf{0}$.

Since the difference between two vectors is defined as a particular case of addition, it should be no surprise that we can obtain it geometrically with the aid of the parallelogram rule. Indeed, Fig. 8.6 shows how this can be done: if the parallelogram $ABCD$ is such that $\overrightarrow{AB} = \mathbf{v}$ and $\overrightarrow{BC} = \mathbf{w}$, then the diagonal vector $\overrightarrow{AC}$ represents $\mathbf{v} + \mathbf{w}$, whereas the diagonal vector $\overrightarrow{DB}$ represents $\mathbf{v} - \mathbf{w}$.

Problems: Sect. 8.1

1. * Given non null vectors $\mathbf{v}$ and $\mathbf{w}$, choose representants $\overrightarrow{AB}$, $\overrightarrow{A'B'}$ for $\mathbf{v}$ and $\overrightarrow{AC}$, $\overrightarrow{A'C'}$ for $\mathbf{w}$. If D and D' are the points chosen so that $ACDB$ and $A'C'D'B'$ are parallelograms, prove that $\overrightarrow{AD} = \overrightarrow{A'D'}$.
2. * Prove that two nonzero parallel vectors can be seen as scalar multiples from one another. More precisely, given parallel vectors $\mathbf{v}, \mathbf{w} \neq \mathbf{0}$, prove that

$$\begin{cases} \mathbf{v} = \frac{\|v\|}{\|w\|}\mathbf{w}, & \text{if } \mathbf{v} \text{ and } \mathbf{w} \text{ have the same direction} \\ \mathbf{v} = -\frac{\|v\|}{\|w\|}\mathbf{w}, & \text{if } \mathbf{v} \text{ and } \mathbf{w} \text{ have opposite directions} \end{cases}.$$

3. * Prove Proposition 8.2.
4. * We are given in the plane distinct points O, A, B and C, such that $\overrightarrow{AC} = k \cdot \overrightarrow{AB}$ for some $k \in \mathbb{R}$. Prove that

$$\overrightarrow{OC} = (1 - k)\,\overrightarrow{OA} + k\,\overrightarrow{OB}.$$

Fig. 8.6 The difference of two vectors

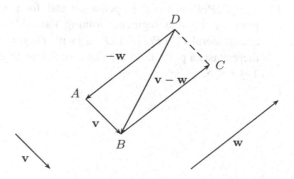

5. Let M, N, P, Q be the midpoints of sides AB, BC, CD, AD of quadrilateral $ABCD$. Prove that $MNPQ$ is a parallelogram.

6. * Let ABC be a triangle of incenter I and $\overline{AB} = c$, $\overline{AC} = b$, $\overline{BC} = a$. If O is any point in the plane, prove that

$$\overrightarrow{OI} = \frac{a\ \overrightarrow{OA} + b\ \overrightarrow{OB} + c\ \overrightarrow{OC}}{a+b+c}.$$

7. (TT) Given a triangle ABC, let H_a, H_b and H_c stand for the feet of the altitudes corresponding to A, B and C, respectively. Prove that the triangle formed by the orthocenters of triangles AH_bH_c, BH_aH_c, CH_aH_b is congruent to triangle $H_aH_bH_c$.

8. * If ABC is a triangle of circumcenter O and orthocenter H, prove that

$$\overrightarrow{OH} = \overrightarrow{OA} + \overrightarrow{OB} + \overrightarrow{OC}.$$

9. (France) Let $A_1A_2 \ldots A_n$ be a regular n-gon.

   (a) If n is even, show that for every point P in the plane there exists a choice of signs $+$ and $-$ that turns true the equality

   $$\pm\ \overrightarrow{PA_1} \pm\ \overrightarrow{PA_2} \pm \cdots \pm\ \overrightarrow{PA_n} = \mathbf{0}.$$

   (b) If n is odd, show that there is at most a finite set of points P in the plane for which the equality above is true for some choice of signs $+$ and $-$.

10. $A_1A_2 \ldots A_{2n}$ is a simple polygon in the plane (cf. discussion preceding Problem 14, page 190) and, for $1 \leq i \leq 2n$, M_i is the midpoint of A_iA_{i+1} (with $A_{2n+1} = A_1$). For a fixed point B_1 in the plane and for $2 \leq i \leq 2n + 1$, let B_i be the symmetric of B_{i-1} with respect to M_{i-1}. Prove that $B_{2n+1} = B_1$.

11. (Soviet Union) When each side of a convex quadrilateral $ABCD$ is extended by its own length, we obtain a new quadrilateral $A'B'C'D'$, in which B is the midpoint of AA', C is the midpoint of BB', D is the midpoint of CC' and A that of DD'. Show how to construct $ABCD$ from $A'B'C'D'$.

12. $P_1P_2P_3P_4P_5$ is a convex pentagon and, for $1 \leq i \leq 5$, Q_i is the intersection point of the line segments joining the midpoints of the opposite sides of quadrilateral $P_{i+1}P_{i+2}P_{i+3}P_{i+4}$ (with $P_{i+5} = P_i$ for $1 \leq i \leq 5$). Prove that there exists a point O in the plane such that $O \in P_iQ_i$ and $\overline{OP_i} = 4\overline{OQ_i}$, for $1 \leq i \leq 5$.

Fig. 8.7 Equality of vectors
in coordinates

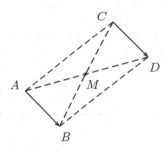

8.2 Vectors in the Cartesian Plane

If we furnish the plane with a cartesian system, it is useful to examine the relationship between the corresponding cartesian coordinates and the operations on vectors defined in the previous section. The following lemma is the key result.

Lemma 8.3 *Given points $A(x_A, y_A)$, $B(x_B, y_B)$, $C(x_C, y_C)$, $D(x_D, y_D)$ in the cartesian plane, we have*

$$\overrightarrow{AB} = \overrightarrow{CD} \Leftrightarrow \begin{cases} x_B - x_A = x_D - x_C \\ y_B - y_A = y_D - y_C \end{cases}. \tag{8.2}$$

Proof We already know that $\overrightarrow{AB} = \overrightarrow{CD}$ if and only if quadrilateral $ABDC$ of sides AB, BC, CD, DA is a parallelogram. Moreover, we also know that this is the case if and only if its diagonals AD and BC have the same midpoint M (cf. Fig. 8.7). On the other hand, it follows from Corollary 6.2 that this happens if and only if

$$\frac{A + D}{2} = \frac{B + C}{2},$$

i.e., if and only if

$$x_A + x_D = x_B + x_C \text{ and } y_A + y_D = y_B + y_C.$$

□

The previous result assures, in particular, that in a fixed cartesian system xOy each vector $\mathbf{v}$ has a single representative of the form $\overrightarrow{OV}$ (cf. Fig. 8.8). Indeed, if $\mathbf{v} = \overrightarrow{AB}$, with $A(x_A, y_A)$ and $B(x_B, y_B)$, then the lemma guarantees that $\overrightarrow{AB} = \overrightarrow{OV}$, with $V(x_B - x_A, y_B - y_A)$; yet in another way, $\mathbf{v} = \overrightarrow{OV}$. On the other hand, if $\overrightarrow{OV_1} = \overrightarrow{OV_2}$ for certain $V_1(x_1, y_1)$ and $V_2(x_2, y_2)$, then again from the lemma we get

$$(x_1 - 0, y_1 - 0) = (x_2 - 0, y_2 - 0),$$

so that $V_1 = V_2$.

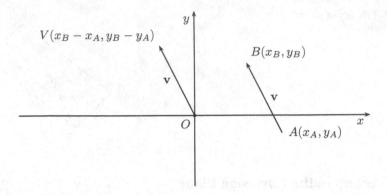

Fig. 8.8 Representing vectors in a cartesian plane

Hence, whenever we deal with vectors, there is a huge advantage in taking a cartesian system in the plane: once such a system is chosen, every vector automatically admits a *canonical representative*, whose initial point coincides with the origin of the cartesian system. In this sense, if $\mathbf{v} = \overrightarrow{OV}$, with $V(x, y)$, we hereafter adopt the convention of writing

$$\mathbf{v} = (x, y).$$

The usefulness of this notation will become clear as we proceed with our study of vectors.

Proposition 8.4 *If, in a cartesian system,* $\mathbf{v} = (x_1, y_1)$ *and* $\mathbf{w} = (x_2, y_2)$*, then:*

(a) $\mathbf{v} + \mathbf{w} = (x_1 + x_2, y_1 + y_2)$*.*
(b) $\mathbf{v} - \mathbf{w} = (x_1 - x_2, y_1 - y_2)$*.*
(c) $k\mathbf{v} = (kx_1, ky_1)$*.*

Proof Let O denote the origin of the cartesian system, and $V(x_1, y_1)$, $W(x_2, y_2)$.

(a) Setting $\mathbf{v} + \mathbf{w} = \overrightarrow{OA}$, with $A(a, b)$, we have

$$\overrightarrow{VA} = \overrightarrow{OA} - \overrightarrow{OV} = (\mathbf{v} + \mathbf{w}) - \mathbf{v} = \mathbf{w} = \overrightarrow{OW},$$

and Lemma 8.3 gives $a - x_1 = x_2 - 0$ and $b - y_1 = y_2 - 0$. Hence, $a = x_1 + x_2$, $b = y_1 + y_2$ and $A(x_1 + x_2, y_1 + y_2)$, so that $\mathbf{v} + \mathbf{w} = \overrightarrow{OA} = (x_1 + x_2, y_1 + y_2)$.
(b) The proof of this item is quite similar to that of the previous one and will be left as an exercise for the reader.
(c) Let us look at the case $k > 0$ (the case $k \leq 0$ can be dealt with analogously). Also, assume that $x_1, y_1 \neq 0$ (the cases $x_1 = 0$ or $y_1 = 0$ can be treated quite similarly, too), and let $\mathbf{v} = \overrightarrow{OV}$, $k\mathbf{v} = \overrightarrow{OA}$, with $A(a, b)$. Since $k > 0$, it is

clear that a and x_1 (and likewise b and y_1) have equal signs. Letting $V_x(x_1, 0)$, $V_y(0, y_1)$, $A_x(a, 0)$ and $A_y(0, b)$, it is pretty clear that $AA_xO \sim VV_xO$ and $AA_yO \sim VV_yO$. The first similarity gives

$$\frac{|a|}{|x_1|} = \frac{\overline{OA_x}}{\overline{OV_x}} = \frac{\overline{OA}}{\overline{OV}} = k,$$

whence $|a| = k|x_1|$; however, since a and x_1 have equal signs, we obtain $a = kx_1$. Analogously, $b = ky_1$, and thus

$$k\mathbf{v} = \overrightarrow{OA} = (a, b) = (kx_1, ky_1).$$

$\square$

The previous proposition has an interesting and useful consequence, which we now explore. Fixed a cartesian system in the plane, we denote

$$\mathbf{i} = (1, 0) \text{ and } \mathbf{j} = (0, 1), \tag{8.3}$$

and say that the vectors $\mathbf{i}$ and $\mathbf{j}$ form, in this order, the **canonical basis** of the plane with respect to the chosen cartesian system. This terminology comes from the fact that, for an arbitrary vector $\mathbf{v} = (x, y)$, the proposition allows us to write

$$\mathbf{v} = (x, 0) + (0, y) = x(1, 0) + y(0, 1) = x\mathbf{i} + y\mathbf{j}.$$

In words, the equality above says that every vector can be uniquely written as a sum of multiples of the *canonical vectors* (i.e., the vectors of the canonical basis), with coefficients equal to the coordinates of the endpoint of the canonical representative of the vector. For $\mathbf{v} = (x, y)$, equality

$$\mathbf{v} = x\mathbf{i} + y\mathbf{j} \tag{8.4}$$

is usually referred to as the **orthonormal expansion formula** of $\mathbf{v}$ with respect to the canonical basis.

From a *vectorial* point of view, there are two ways of representing lines, the first one being the object of the coming example (for the second, see Example 8.12).

Example 8.5 The previous proposition allows us to describe a straightline **parametrically**, i.e., as the trajectory of a material particle moving uniformly along it. In order to be precise, fix in the plane a cartesian system of origin O, a point $P(x_0, y_0)$ in the line r a vector $\mathbf{v} = (a, b)$ parallel to it and let $\mathbf{p} = \overrightarrow{OP}$. As the *parameter t* varies in the set of reals, the endpoint of the canonical representative of the vector $\mathbf{p} + t\mathbf{v}$ runs through all points of. Hence, r is composed by the points (x, y) of the form

$$x = x_0 + ta \text{ and } y = y_0 + tb,$$

for some $t \in \mathbb{R}$. The equalities above are called the **parametric equations** of the straightline r.

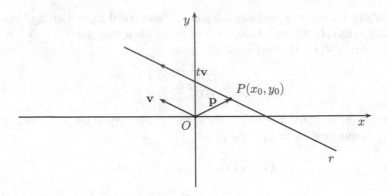

Fig. 8.9 Parametric representation of a straightline

We finish this section by computing a sum of vectors that will be useful several times hereafter (Fig. 8.9).

Proposition 8.6 *If* $A_1 A_2 \ldots A_n$ *is a regular polygon of center* O, *then*

$$\sum_{k=1}^{n} \overrightarrow{OA_k} = \mathbf{0}.$$

Proof Assume, without loss of generality, that the radius of the circle circumscribed to the polygon is equal to 1, and choose a cartesian system of coordinates in which $A_1(1, 0)$ and A_2 lies in the first quadrant. For $2 \leq k \leq n$, the trigonometric arc $\overset{\frown}{A_1 A_k}$ is such that $A_1 \widehat{O} A_k = \frac{2(k-1)\pi}{n}$; therefore,

$$\overrightarrow{OA_k} = \left(\cos \frac{2(k-1)\pi}{n}, \sin \frac{2(k-1)\pi}{n} \right).$$

Now, by applying item (a) of Proposition 8.4 several times, we obtain

$$\sum_{k=1}^{n} \overrightarrow{OA_k} = \left(\sum_{k=1}^{n} \cos \frac{2(k-1)\pi}{n}, \sum_{k=1}^{n} \sin \frac{2(k-1)\pi}{n} \right),$$

so that it suffices to establish the equalities

$$\sum_{k=1}^{n} \cos \frac{2(k-1)\pi}{n} = \sum_{k=1}^{n} \sin \frac{2(k-1)\pi}{n} = 0.$$

These, in turn, follow immediately from Problem 12, page 247. $\qquad\qquad\square$

Fig. 8.10 The unit vector $\mathbf{e}_\theta$

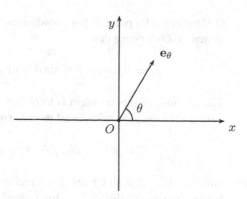

Problems: Sect. 8.2

1. * In a cartesian system of coordinates of origin O, let A and B be points such that $\overline{OA} = \overline{OB}$ and $\overrightarrow{OB}$ can be obtained from $\overrightarrow{OA}$ by means of a trigonometric rotation of $+90°$, centered at O. If $\overrightarrow{OA} = a\mathbf{i} + b\mathbf{j}$, show that $\overrightarrow{OB} = -b\mathbf{i} + a\mathbf{j}$.

2. Triangle ADE is right at D and such that $\overline{AE} = 2\overline{AD}$. Externally to it, we construct squares $ABCD$, $DEFG$ and $AEIJ$. Then, we construct square $BJKL$, such that it has no interior points in common with the other three squares. Prove that points C, G and L are collinear and compute the ratio between the lengths of segments CG and GL.

   In what comes we shall need the following notation: given $\theta \in \mathbb{R}$, we let

   $$\mathbf{e}_\theta = (\cos\theta)\mathbf{i} + (\sin\theta)\mathbf{j}$$

   be the unit vector departing from the origin of a cartesian system and forming a trigonometric angle of θ radians with the positive part of the horizontal axis (cf. Fig. 8.10).

3. * Generalize Problem 1, doing the items below:

   (a) If $\mathbf{u} = a\mathbf{i} + b\mathbf{j}$, with $a, b \in \mathbb{R}$, and $\alpha \in [0, 2\pi)$ is such that $\cos\alpha = \dfrac{a}{\sqrt{a^2+b^2}}$ and $\sin\alpha = \dfrac{b}{\sqrt{a^2+b^2}}$, prove that α is the unique real number in the interval $[0, 2\pi)$ for which $\mathbf{u} = ||\mathbf{u}||\mathbf{e}_\alpha$.

   (b) If $\mathbf{v} = c\mathbf{i} + d\mathbf{j}$ is the vector obtained from $\mathbf{u}$ by means of a trigonometric rotation of angle θ, then $||\mathbf{u}|| = ||\mathbf{v}||$ and $\mathbf{v} = ||\mathbf{v}||\mathbf{e}_{\alpha+\theta}$. From this, conclude that

   $$c = a\cos\theta - b\sin\theta \ \text{ and } \ d = a\sin\theta + b\cos\theta.$$

4. * Let xOy and $x'Oy'$ be cartesian systems of coordinates with the same origin O, such that $x'Oy'$ is obtained from xOy by means of a trigonometric rotation

Fig. 8.10 The unit vector $\mathbf{e}_\theta$

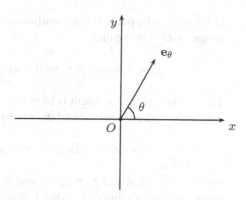

Problems: Sect. 8.2

1. * In a cartesian system of coordinates of origin O, let A and B be points such that $\overline{OA} = \overline{OB}$ and $\overrightarrow{OB}$ can be obtained from $\overrightarrow{OA}$ by means of a trigonometric rotation of $+90°$, centered at O. If $\overrightarrow{OA} = a\mathbf{i} + b\mathbf{j}$, show that $\overrightarrow{OB} = -b\mathbf{i} + a\mathbf{j}$.

2. Triangle ADE is right at D and such that $\overline{AE} = 2\overline{AD}$. Externally to it, we construct squares $ABCD$, $DEFG$ and $AEIJ$. Then, we construct square $BJKL$, such that it has no interior points in common with the other three squares. Prove that points C, G and L are collinear and compute the ratio between the lengths of segments CG and GL.

   In what comes we shall need the following notation: given $\theta \in \mathbb{R}$, we let

   $$\mathbf{e}_\theta = (\cos\theta)\mathbf{i} + (\sin\theta)\mathbf{j}$$

   be the unit vector departing from the origin of a cartesian system and forming a trigonometric angle of θ radians with the positive part of the horizontal axis (cf. Fig. 8.10).

3. * Generalize Problem 1, doing the items below:

   (a) If $\mathbf{u} = a\mathbf{i} + b\mathbf{j}$, with $a, b \in \mathbb{R}$, and $\alpha \in [0, 2\pi)$ is such that $\cos\alpha = \dfrac{a}{\sqrt{a^2+b^2}}$ and $\sin\alpha = \dfrac{b}{\sqrt{a^2+b^2}}$, prove that α is the unique real number in the interval $[0, 2\pi)$ for which $\mathbf{u} = ||\mathbf{u}||\mathbf{e}_\alpha$.

   (b) If $\mathbf{v} = c\mathbf{i} + d\mathbf{j}$ is the vector obtained from $\mathbf{u}$ by means of a trigonometric rotation of angle θ, then $||\mathbf{u}|| = ||\mathbf{v}||$ and $\mathbf{v} = ||\mathbf{v}||\mathbf{e}_{\alpha+\theta}$. From this, conclude that

   $$c = a\cos\theta - b\sin\theta \text{ and } d = a\sin\theta + b\cos\theta.$$

4. * Let xOy and $x'Oy'$ be cartesian systems of coordinates with the same origin O, such that $x'Oy'$ is obtained from xOy by means of a trigonometric rotation

of θ radians. If a point P has coordinates (x_0, y_0) in system xOy and (x_0', y_0') in system $x'Oy'$, prove that

$$x_0' = x_0 \cos \theta + y_0 \sin \theta \quad \text{and} \quad y_0' = -x_0 \sin \theta + y_0 \cos \theta.$$

5. The purpose of this problem is to obtain a partial classification for the *solution curves* $\mathcal{C}$ of the **general second degree equation**

$$ax^2 + 2bxy + cy^2 + dx + ey + f = 0, \tag{8.5}$$

where a, b, c, d, e and f are given real constants, with at least one of a, b and c being nonzero. To this end, do the following items:

(a) Let $x'O'y'$ be the cartesian system obtained from xOy by axes translation, such that the new origin O' has coordinates (x_0, y_0) in the system xOy. If $ac - b^2 \neq 0$, use the result of Problem 2, page 188, to show that there exist unique such x_0 and y_0 for which the equation of $\mathcal{C}$ in system $x'O'y'$ has the form

$$a'(x')^2 + 2b'x'y' + c'(y')^2 + f' = 0.$$

(b) From now one, suppose that $\mathcal{C}$ is the solution curve of a second degree equation of the form

$$ax^2 + 2bxy + cy^2 + f = 0.$$

Show that $\mathcal{C}$ is symmetric with respect to O, so that we call it a **central curve**.

(c) Now, let $x'Oy'$ be the cartesian system obtained from xOy by means of a trigonometric rotation of θ radians. If $a \neq c$, use the result of the previous problem to show that θ can be chosen in such a way that, in the cartesian system $x'Oy'$, the equation defining $\mathcal{C}$ has the form

$$a'(x')^2 + c'(y')^2 + f' = 0.$$

(d) Use the previous items to show that, if the solution curve of (8.5) is central, then it is the empty set, the union of two (distinct) lines, a single line, an ellipse or a hyperbola.

6. (Romania) We partition the plane into regular hexagons of pairwise disjoint interiors and side lengths equal to 1. Prove that there does not exist a square whose vertices coincide with vertices of these hexagons.

7. In this problem, we present an alternative proof for Problem 6, page 126, this time using vectors. To this end, and with notations as in the statement of that problem, do the following items:

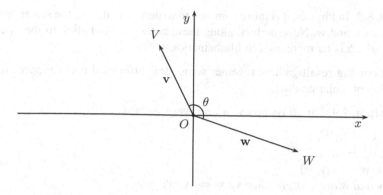

Fig. 8.11 The angle between two vectors

(a) Let P be the point of Fermat of triangle ABC, fix a cartesian system of origin P and let $\overrightarrow{PA} = x\mathbf{e}_\alpha$, $\overrightarrow{PB} = y\mathbf{e}_\beta$ and $\overrightarrow{PC} = z\mathbf{e}_\gamma$, with $0 \le \alpha < \beta < \gamma < 2\pi$ and $\mathbf{e}_\alpha$, $\mathbf{e}_\beta$, $\mathbf{e}_\gamma$ as in Problem 3. Conclude that

$$\overrightarrow{PD} = z\mathbf{e}_\gamma + y\mathbf{e}_{\beta + \frac{\pi}{3}} - z\mathbf{e}_{\gamma + \frac{\pi}{3}},$$

and derive analogous expressions for $\overrightarrow{PE}$ and $\overrightarrow{PF}$.

(b) Show that A, P and D are collinear and $\overrightarrow{AD} = x + y + z$.

8.3   The Scalar Product of Two Vectors

Given nonzero vectors $\mathbf{v}$ and $\mathbf{w}$ in the plane, with canonical representatives $\mathbf{v} = \overrightarrow{OV}$ and $\mathbf{w} = \overrightarrow{OW}$, we define the **angle between $\mathbf{v}$ and $\mathbf{w}$** as the angle (in radians) $0 \le \theta \le \pi$ such that $\theta = V\widehat{O}W$ (cf. Fig. 8.11). A simple geometric argument assures that the angle between $\mathbf{v}$ and $\mathbf{w}$ does not depend on the chosen cartesian system (in this respect, see Problem 1). This being the case, we say that the nonzero vectors $\mathbf{v}$ and $\mathbf{w}$ are **orthogonal** if $\theta(\mathbf{v}, \mathbf{w}) = \frac{\pi}{2}$. In this case, we denote $\mathbf{v} \perp \mathbf{w}$.

With the definition of angle between nonzero vectors at our disposal, we are in position to study the most important operation with two vectors, their *scalar product*.

Definition 8.7 The **scalar product** of the nonzero vectors $\mathbf{v}$ and $\mathbf{w}$ is the scalar $\langle \mathbf{v}, \mathbf{w} \rangle$ given by

$$\langle \mathbf{v}, \mathbf{w} \rangle = ||v|| \cdot ||w|| \cos\theta(\mathbf{v}, \mathbf{w}). \tag{8.6}$$

We extend the above definition by letting $\langle \mathbf{v}, \mathbf{w} \rangle = 0$ whenever $\mathbf{v} = \mathbf{0}$ or $\mathbf{w} = \mathbf{0}$.

Remark 8.8 In Physics, it is more common to write $\mathbf{v} \cdot \mathbf{w}$ to denote the scalar product of vectors $\mathbf{v}$ and $\mathbf{w}$. Nevertheless, along these notes we shall stick to the notation $\langle \mathbf{v}, \mathbf{w} \rangle$, which is far more used in Mathematics.

The coming result gathers together some straightforward consequences of the definition of scalar product.

Proposition 8.9 *For all vectors* $\mathbf{v}$, $\mathbf{w}$ *and every scalar* k, *we have:*

(a) $\langle \mathbf{v}, \mathbf{v} \rangle = ||\mathbf{v}||^2.$
(b) $\langle \mathbf{v}, \mathbf{w} \rangle = \langle \mathbf{w}, \mathbf{v} \rangle.$
(c) $\langle k\mathbf{v}, \mathbf{w} \rangle = k\langle \mathbf{v}, \mathbf{w} \rangle.$
(d) If $\mathbf{v}$ *and* $\mathbf{w}$ *are nonzero, then* $\mathbf{v} \perp \mathbf{w} \Leftrightarrow \langle \mathbf{v}, \mathbf{w} \rangle = 0.$

Proof Firstly, note that (a), (b) and (c) follow immediately if $\mathbf{v} = \mathbf{0}$, $\mathbf{w} = \mathbf{0}$ or $k = 0$. Therefore, assume that $\mathbf{v}, \mathbf{w} \neq \mathbf{0}$ and $k = 0$.

(a) Let $\mathbf{w} = \mathbf{v}$ in the definition of scalar product, noticing that $\theta(\mathbf{v}, \mathbf{w}) = 0$.
(b) It suffices to observe that $\theta(\mathbf{v}, \mathbf{w}) = \theta(\mathbf{w}, \mathbf{v})$ and invoke the definition again.
(c) We look separately at the cases $k > 0$ and $k < 0$:

- If $k > 0$, then $k\mathbf{v}$ has the same direction as $\mathbf{v}$, so that $\theta(k\mathbf{v}, \mathbf{w}) = \theta(\mathbf{v}, \mathbf{w})$. Hence, by the definition of scalar product, we have

$$\langle k\mathbf{v}, \mathbf{w} \rangle = ||k\mathbf{v}|| \cdot ||\mathbf{w}|| \cos(k\mathbf{v}, \mathbf{w})$$
$$= |k| \cdot ||\mathbf{v}|| \cdot ||\mathbf{w}|| \cos(\mathbf{v}, \mathbf{w})$$
$$= k\langle \mathbf{v}, \mathbf{w} \rangle.$$

- If $k < 0$, then $k\mathbf{v}$ and $\mathbf{v}$ have opposite directions, whence $\theta(k\mathbf{v}, \mathbf{w}) = \pi - \theta(\mathbf{v}, \mathbf{w})$ (cf. Problem 2). Thus, $\cos\theta(k\mathbf{v}, \mathbf{w}) = -\cos\theta(\mathbf{v}, \mathbf{w})$, and the definition of scalar product gives

$$\langle k\mathbf{v}, \mathbf{w} \rangle = ||k\mathbf{v}|| \cdot ||\mathbf{w}|| \cos(k\mathbf{v}, \mathbf{w})$$
$$= |k| \cdot ||\mathbf{v}|| \cdot ||\mathbf{w}||(-\cos(\mathbf{v}, \mathbf{w}))$$
$$= k\langle \mathbf{v}, \mathbf{w} \rangle.$$

(d) Since $\mathbf{v}, \mathbf{w} \neq \mathbf{0}$, we have $\langle \mathbf{v}, \mathbf{w} \rangle = 0$ if and only if $\cos\theta(\mathbf{v}, \mathbf{w}) = 0$. However, since $0 \leq \theta(\mathbf{v}, \mathbf{w}) \leq \pi$, we conclude that $\cos\theta(\mathbf{v}, \mathbf{w}) = 0$ if and only if $\theta(\mathbf{v}, \mathbf{w}) = \frac{\pi}{2}$, which is the same as $\mathbf{v} \perp \mathbf{w}$. $\square$

Another immediate consequence of the definition of scalar product is the coming inequality, which is known as **Cauchy's inequality**.[1] After Corollary 8.13, we shall show that it is a particular case of the classical algebraic inequality of Cauchy (cf. Section 5.2 of [5]).

---

[1] Augustin Louis Cauchy, French mathematician of the XIX century.

Proposition 8.10 *Given nonzero vectors* **v** *and* **w**, *we have*

$$|\langle \mathbf{v}, \mathbf{w}\rangle| \leq ||\mathbf{v}|| \cdot |\mathbf{w}||, \tag{8.7}$$

with equality if and only if $\mathbf{v} = k\mathbf{w}$ *for some scalar* $k > 0$.

Proof Since $|\cos\theta(\mathbf{v}, \mathbf{w})| \leq 1$, with equality if and only if $\theta(\mathbf{v}, \mathbf{w}) = 0$, we have

$$|\langle \mathbf{v}, \mathbf{w}\rangle| = ||v|| \cdot ||w|| \cdot |\cos\theta(\mathbf{v}, \mathbf{w})|$$
$$\leq ||v|| \cdot ||w||,$$

with equality if and only if $\theta(\mathbf{v}, \mathbf{w}) = 0$. We are thus left to showing that $\theta(\mathbf{v}, \mathbf{w}) = 0$ if and only if the stated condition for equality holds.

If there exists a scalar $k > 0$ for which $\mathbf{v} = k\mathbf{w}$, then it is pretty clear that $\theta(\mathbf{v}, \mathbf{w}) = 0$. Conversely, if $\theta(\mathbf{v}, \mathbf{w}) = 0$, the it follows from Problem 2, page 273, that $\mathbf{v} = \frac{||v||}{||w||}\mathbf{w}$, and it suffices to take $k = \frac{||v||}{||w||} > 0$. □

In order to further extend the theory, we now need to write the scalar product in coordinates.

Proposition 8.11 *If vectors* **v** *and* **w** *have canonical representatives* $\mathbf{v} = (x_1, y_1)$ *and* $\mathbf{w} = (x_2, y_2)$ *in a certain cartesian system, then*

$$\langle \mathbf{v}, \mathbf{w}\rangle = x_1 x_2 + y_1 y_2. \tag{8.8}$$

Proof Let O stand for the origin of the cartesian system under consideration, and $\mathbf{v} = \overrightarrow{OV}$, $\mathbf{w} = \overrightarrow{OW}$ (cf. Fig. 8.12). Cosine law (7.12) applied to triangle OVW gives us

$$\overline{VW}^2 = \overline{OV}^2 + \overline{OW}^2 - 2\overline{OV}\,\overline{OW}\cos\theta,$$

whence

$$2\langle \mathbf{v}, \mathbf{w}\rangle = 2||\mathbf{v}|| \cdot ||\mathbf{w}||\cos\theta = 2\overline{OV}\,\overline{OW}\cos\theta$$
$$= \overline{OV}^2 + \overline{OW}^2 - \overline{VW}^2$$
$$= (x_1^2 + y_1^2) + (x_2^2 + y_2^2) - [(x_1 - x_2)^2 + (y_1 - y_2)^2]$$
$$= 2(x_1 x_2 + y_1 y_2).$$

□

The coming example gives a first application of scalar products.

Example 8.12 The notion of scalar product trivializes the deduction of the equation of a straightline, provided we know one of its points and a vector perpendicular to

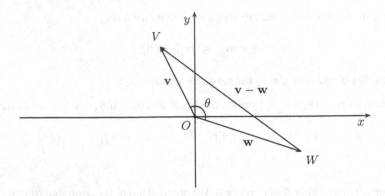

Fig. 8.12 The scalar product in coordinates

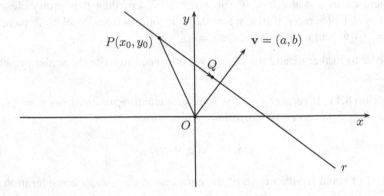

Fig. 8.13 The equation of a line via scalar product

it. Indeed, suppose that $\mathbf{v} = (a, b)$ is perpendicular to line r, and that $P(x_0, y_0)$ is one of its points (cf. Fig. 8.13). If $Q(x, y)$ stands for any point of the plane, then

$$Q \in r \iff \mathbf{v} \perp \overrightarrow{PQ} \iff (a, b) \perp (x - x_0, y - y_0)$$
$$\iff a(x - x_0) + b(y - y_0) = 0$$
$$\iff ax + by - (ax_0 + by_0) = 0.$$

Also with respect to the previous example, we observe that one can use vectors to obtain the other results derived in Sect. 6.2 (see Problem 3).

Back to the development of the theory, the following consequence of the previous proposition is frequently useful.

Corollary 8.13 *For all vectors* $\mathbf{v}_1$, $\mathbf{v}_2$ *and* $\mathbf{w}$, *we have:*

$$\langle \mathbf{v}_1 + \mathbf{v}_2, \mathbf{w} \rangle = \langle \mathbf{v}_1, \mathbf{w} \rangle + \langle \mathbf{v}_2, \mathbf{w} \rangle. \tag{8.9}$$

Proof Choose a cartesian system xOy in the plane and assume that, in such a system, the canonical representatives of v_1, v_2 and w are $v_1 = (x_1, y_1)$, $v_2 = (x_2, y_2)$ and $w = (a, b)$. Propositions 8.4 and 8.11 give

$$\langle v_1 + v_2, w \rangle = \langle (x_1 + x_2, y_1 + y_2), (a, b) \rangle$$
$$= (x_1 + x_2)a + (y_1 + y_2)b$$
$$= (x_1 a + y_1 b) + (x_2 a + y_2 b)$$
$$= \langle v_1, w \rangle + \langle v_2, w \rangle.$$

$\square$

As promised before, we now relate (8.7) with the usual version of Cauchy inequality: letting $v = (x_1, y_1)$ and $w = (x_2, y_2)$ be the canonical representatives of v and w in a fixed cartesian system, Proposition 8.11 assures that Cauchy's inequality (8.7) is precisely

$$|x_1 x_2 + y_1 y_2| \le \sqrt{x_1^2 + y_1^2} \sqrt{x_2^2 + y_2^2}.$$

But this is exactly what Theorem 5.13 of [5] says for $n = 2$.

We finish this section by exhibiting an interesting application of formula (8.8), which establishes a partial converse of Proposition 8.6.

Example 8.14 If the sum of three nonzero vectors of equal moduli is equal to zero, prove that they form angles of 120° with each other.

Proof Let a, b and c denote the vectors under consideration, so that $||a|| = ||b|| = ||c|| = k > 0$ and

$$a + b + c = 0. \tag{8.10}$$

Write α, β, γ to denote the angles between the pairs of vectors a and b, b and c, c and a, respectively, as shown in Fig. 8.14. By computing the scalar product of a with both sides of (8.10), we successively get

$$0 = a \cdot (a + b + c)$$
$$= ||a||^2 + a \cdot b + a \cdot c$$
$$= k^2 + ||a|| \cdot ||b|| \cos \beta + ||a|| \cdot ||c|| \cos \gamma$$
$$= k^2 + k^2 \cos \alpha + k^2 \cos \gamma,$$

so that $\cos \alpha + \cos \gamma = -1$. Analogously, the scalar product of b with both sides of (8.10) results in the equality $k^2 \cos \alpha + k^2 + k^2 \cos \beta = 0$, whence $\cos \alpha + \cos \beta = -1$. Arguing likewise, we also obtain $\cos \beta + \cos \gamma = -1$, and then the system of equations

Fig. 8.14 Vectors with the
same modulus and zero sum

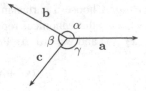

$$\begin{cases} \cos\alpha + \cos\gamma = -1 \\ \cos\alpha + \cos\beta = -1 \\ \cos\beta + \cos\gamma = -1 \end{cases}.$$

Looking at it as a linear system in $\cos\alpha$, $\cos\beta$ and $\cos\gamma$, it is pretty clear that the only possible solution is $\cos\alpha = \cos\beta = \cos\gamma = -\frac{1}{2}$. Therefore, Problem 7, page 246, assures that each of α, β and γ equals 120° or 240°; however, since $\alpha + \beta + \gamma = 360°$, the only actual possibility is $\alpha = \beta = \gamma = 120°$.   □

Problems: Sect. 8.3

1. * Prove that the definition of angle between two nonzero vectors does not depend on the chosen cartesian system.
2. * Given nonzero vectors **v** and **w**, relate the angles $\theta(\mathbf{v}, \mathbf{w})$ and $\theta(-\mathbf{v}, \mathbf{w})$.
3. Use vectors to give another proof of Corollary 6.11, as well as of Propositions 6.13 and 6.14.
4. Given a straightline $r : \{ax + by + c = 0\}$ and a point $P(x_0, y_0)$ not belonging to it, prove that the distance from P to r can be computed by the following formula:

$$d(P; r) = \frac{|ax_0 + by_0 + c|}{\sqrt{a^2 + b^2}}.$$

5. Consider in the plane a regular n-gon $A_1 A_2 \ldots A_n$ of center O, and a circle Γ, also centered at O. For $P \in \Gamma$, show that the value of $\sum_{i=1}^{n} \overline{PA_i}^2$ is independent from the position of P.
6. Use vectors to give another proof for Problem 14, page 255.
7. Let ABC be a triangle of sides $\overline{AB} = c$, $\overline{AC} = b$ and $\overline{BC} = a$. Also, let H be the orthocenter, O the circumcenter and R the circumradius of ABC. Show that

$$\overline{OH}^2 = 9R^2 - (a^2 + b^2 + c^2).$$

8. Let ABC be a triangle of sides $\overline{AB} = c$, $\overline{AC} = b$ and $\overline{BC} = a$. Let also I denote the incenter, O the circumcenter and r and R the inradius and circumradius of ABC, respectively. Use vectors to show that

$$\overline{OI}^2 = R^2 - 2Rr.$$

9. (OCM) ABC is a triangle of incenter I and sides $\overline{AB} = c$, $\overline{AC} = b$ and $\overline{BC} = a$. If P is a point in the plane, show that the sum $a\overline{AP}^2 + b\overline{BP}^2 + c\overline{CP}^2$ is minimum if and only if P coincides with the incenter.

10. (BMO) In a triangle ABC, O is the circumcenter, D is the midpoint of AB and E the barycenter of ACD. Prove that lines $\overleftrightarrow{OE}$ and $\overleftrightarrow{CD}$ are perpendicular if and only if $\overline{AB} = \overline{AC}$.

11. * The purpose of this problem is to prove the following result of H. Minkowski[2]: for an integer $2 < n \neq 4$, there does not exist, in the cartesian plane, a regular n-gon with vertices of integer coordinates. To this end, do the following items:

    (a) Use Example 6.7, to show the result for $n = 6$.
    (b) Adapt the argument presented in the solution of Example 6.7, in conjunction with the result of Problem 3, page 254, to show the result for $n = 5$.
    (c) Let $n > 6$ be and integer and, by the sake of contradiction, let $\mathcal{P}_0 = A_1 A_2 \ldots A_n$ be a regular n-gon with vertices of integer coordinates and side length l. Let O stand for the origin of the cartesian system and $\mathbf{a}_i = \overrightarrow{A_i A_{i+1}}$ for $1 \leq i \leq n$ (with $A_{n+1} = A_1$). If A'_i is such that $\overrightarrow{OA'_i} = \mathbf{a}_i$, show that $\mathcal{P}_1 = A'_1 A'_2 \ldots A'_n$ also has vertices of integer coordinates too inteiras, and has side length $l' = 2l \sin \frac{\pi}{n}$.
    (d) Iterate the construction of the previous item, thus getting a sequence $(\mathcal{P}_k)_{k \geq 0}$ of regular n-gons with vertices of integer coordinates. If l_k stands for the side length of $\mathcal{P}_k$, show that $l_k = (2 \sin \frac{\pi}{n})^k l_0$ for $k \geq 0$.
    (e) Use the previous item to reach a contradiction.

---

[2]Hermann Mikowski, German mathematician of the XIX and XX centuries.

Chapter 9
A First Glimpse on Projective Techniques

This chapter presents the fundamentals of inversion and some techniques emanated from Projective Geometry, along with a number of applications. Instead of trying to convey a comprehensive (and axiomatic) account of Projective Geometry, we center attention on the concepts of cross ratio and polarity, for this leads us more quickly to applications. In doing so, we take the somewhat unusual route of first discussing harmonic conjugation and harmonic pencils, which is all we need for studying polarity, thus postponing the introduction of the more general notions of cross ratio, projective pencils and perspectivity. Although such an approach contains some repetition, we believe it is pedagogically more adequate for the novices. In particular, points at infinity are avoided throughout.

9.1 Inversion

Inversion is a kind of *geometric transformation*[1] of the (punctured) plane which has the remarkable property of *angle invariance*. As we shall see here, this and other notable properties of inversions will allows us to solve problems which would be either inaccessible or extremely difficult by other means. The concepts developed here will also play a role in the next section, in the context of *polarity*. We shall have more to say about inversion (this time in space) in the problems of Sect. 10.3.

Definition 9.1 Given a point O in the plane α and a real number $k > 0$, the **inversion** of **center** (or **pole**) O and **ratio** or **modulus** k is the mapping $I :$ $\alpha \setminus \{O\} \to \alpha \setminus \{O\}$ that sends $A \in \alpha \setminus \{O\}$ to $A' \in \alpha \setminus \{O\}$ such that $A' \in \overrightarrow{OA}$ and $\overline{OA} \cdot \overline{OA'} = k^2$. In this case, A' is said to be the **inverse** of A (with respect to I).

---

[1]For much more on geometric transformations, we refer the reader to the collection [24, 25] and [26] of professor I. M. Yaglom, as well as to [27].

© Springer International Publishing AG, part of Springer Nature 2018
A. Caminha Muniz Neto, *An Excursion through Elementary Mathematics, Volume II*,
Problem Books in Mathematics, https://doi.org/10.1007/978-3-319-77974-4_9

Fig. 9.1 Points A, $A' = I(A)$, and the inversion circle

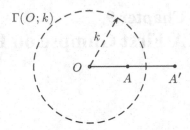

Along all of this section, whenever there is no danger of confusion we shall let I, O and k be as above, and shall systematically write A' for the inverse of a point $A \neq O$.

Figure 9.1 shows a point $A \neq O$, together with its inverse A' and the circle $\Gamma(O; k)$. We will show below that such a circle, which is called the **inversion circle**, plays an important role with respect to I.

In the notations of the definition above, if $I(A) = A'$ and $I(A') = A''$, then

$$\overline{OA'} \cdot \overline{OA''} = k^2 = \overline{OA} \cdot \overline{OA'},$$

so that $\overline{OA} = \overline{OA''}$. Since $A'' \in \overrightarrow{OA'} = \overrightarrow{OA}$, it follows that $A'' = A$. Therefore, A is the inverse of A' and

$$(I \circ I)(A) = I(I(A)) = I(A') = A'' = A,$$

whence $I \circ I : \alpha \setminus \{O\} \to \alpha \setminus \{O\}$ is the identity. From this, it easily follows that I is bijective.[2]

Another straightforward consequence of the definition above is that

$$I(A) = A \Leftrightarrow \overline{OA'} = \overline{OA} \Leftrightarrow \frac{k^2}{\overline{OA}} =, \overline{OA} \Leftrightarrow \overline{OA} = k.$$

Thus, the set of fixed points of I is the inversion circle.

Now, let $D_1 = \{A \in \alpha \setminus \{O\}; \overline{OA} < k\}$ and $D_2 = \{A \in \alpha; \overline{OA} > k\}$. Since

$$\overline{OA} < k \Leftrightarrow \overline{OA'} = \frac{k^2}{\overline{OA}} > k,$$

we conclude that $I(D_1) \subset D_2$ and $I(D_2) \subset D_1$. However, since $\alpha \setminus \{O\} = D_1 \cup D_2 \cup \Gamma(O; k)$ and $I : \alpha \setminus \{O\} \to \alpha \setminus \{O\}$ is a bijection, we conclude that

$$I(D_1) = D_2 \text{ and } I(D_2) = D_1.$$

---

[2]If you need a proof of this fact, see for instance Example 6.37 of [5].

The following simple lemma uses the circle of inversion to give a simple construction (with straightedge and compass) of the inverse of point $A \neq O$.

Lemma 9.2 *Let I be the inversion of center O and ratio k, and Γ be its inversion circle.*

(a) *If $0 < \overline{OA} < k$ and the perpendicular to $\overleftrightarrow{OA}$ at A intersects Γ at T_1 and T_2, then the tangents to Γ at T_1 and T_2 meet $\overleftrightarrow{OA}$ at A'.*

(b) *If $\overline{OB} > k$ and the tangents from B to Γ touch it at T_1 and T_2, then $T_1 T_2$ intersects $\overleftrightarrow{OB}$ at B'.*

Proof For item (a), if t_1 (resp. t_2) stands for the tangent drawn to Γ through T_1 (resp. T_2—see Fig. 9.2), we already know that t_1, t_2 and $\overrightarrow{OA}$ are concurrent. Letting B denote their point of concurrence, we also know that $OB \perp T_1 T_2$.

Then, metric relations in the right triangle $OT_1 B$ gives $\overline{OA} \cdot \overline{OB} = \overline{OT_1}^2 = k^2$, so that $B = A'$.

For item (b), it suffices to observe that, in Fig. 9.2, A is the inverse of B. Thus, we can construct A from B exactly as described in the statement. □

It is pretty clear that if r is a straightline passing through O, then $I(r)$ (i.e., the set of points A', with $A \in r \setminus \{O\}$) is $r \setminus \{O\}$ itself. In order to investigate the inverses of other kinds of figures, item (a) of the coming result will be of fundamental importance (Fig. 9.3).

Proposition 9.3 *Let be given in the plane pairwise distinct and noncollinear points O, A and B. If A' and B' denote the inverses of A and B with respect to the inversion of center O and ratio k, then:*

(a) $AOB \sim B'OA'$.

(b) $ABB'A'$ *is a cyclic quadrilateral.*

(c) $\overline{A'B'} = \frac{k^2 \overline{AB}}{\overline{OA} \cdot \overline{OB}}$.

Fig. 9.2 Constructing the inverse of A, if $0 < \overline{OA} < k$

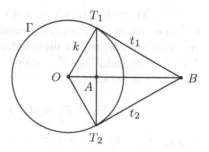

Fig. 9.3 Similarity of AOB and $B'OA'$ and distance from A' to B'

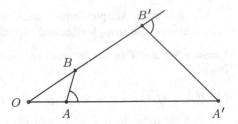

Proof From $\overline{OA} \cdot \overline{OA'} = k^2 = \overline{OB} \cdot \overline{OB'}$, we get $\frac{\overline{OA}}{\overline{OB}} = \frac{\overline{OB'}}{\overline{OA'}}$. Now, since $A' \in \overrightarrow{OA}$ and $B' \in \overrightarrow{OB}$, we obtain $A\widehat{O}B = B'\widehat{O}A'$. Hence, $AOB \sim B'OA'$ by the SAS case of similarity of triangles, which gives item (a).

Now, it follows from (a) that $O\widehat{A}B = O\widehat{B'}A'$. Therefore,

$$B\widehat{A}A' + B\widehat{B'}A' = 180° - O\widehat{A}B + B\widehat{B'}A' = 180°,$$

whence $ABB'A'$ is cyclic.

Finally, as another consequence of (a), $\frac{\overline{A'B'}}{\overline{AB}} = \frac{\overline{OA'}}{\overline{OB}}$, and then

$$\overline{A'B'} = \frac{\overline{OA'} \cdot \overline{AB}}{\overline{OB}} = \frac{k^2 \cdot \overline{AB}}{\overline{OA} \cdot \overline{OB}}.$$

$\square$

For what is to come, it is worth pointing out that the formula of item (c) above remains valid even in the case of collinear points O, A and B. Indeed, in this case $A', B' \in \overleftrightarrow{AB}$; therefore, assuming $\overline{OA} < \overline{OB}$, one computes

$$\overline{A'B'} = \overline{OA'} - \overline{OB'} = \frac{k^2}{\overline{OA}} - \frac{k^2}{\overline{OB}} = \frac{k^2(\overline{OB} - \overline{OA})}{\overline{OA} \cdot \overline{OB}} = \frac{k^2 \cdot \overline{AB}}{\overline{OA} \cdot \overline{OB}}.$$

Our next result, which is actually a corollary of item (b) of the previous one, finds the images of circles and straightlines (not passing through the pole) under an inversion. In order to ease the proof, note that given figures $\mathcal{F}_1$ and $\mathcal{F}_2$, if we know that $\mathcal{F}_1' \subset \mathcal{F}_2$ and $\mathcal{F}_2' \subset \mathcal{F}_1$, then we actually have $\mathcal{F}_1' = \mathcal{F}_2$ and $\mathcal{F}_2' = \mathcal{F}_1$. Indeed, since $I \circ I$ is the identity,

$$\mathcal{F}_1' \subset \mathcal{F}_2 \Rightarrow \mathcal{F}_1 = I\big(I(\mathcal{F}_1)\big) = I\big(\mathcal{F}_1'\big) \subset I(\mathcal{F}_2) = \mathcal{F}_2' \subset \mathcal{F}_1,$$

whence $\mathcal{F}_1 = \mathcal{F}_2'$; likewise, $\mathcal{F}_2 = \mathcal{F}_1'$.

Proposition 9.4 *Let I be an inversion of center O.*

(a) *If r is a straightline not passing through O and A is the foot of the perpendicular dropped from O to r, then $I(r) = \Gamma \setminus \{O\}$, where Γ is the circle of diameter OA'.*

(b) *If Γ is a circle passing through O and OA is diameter of Γ, then $I(\Gamma \setminus \{O\}) = r$, where r is the straightline passing through A' and perpendicular to $\overleftrightarrow{OA}$.*

(c) *If Γ is a circle not passing through O and AB is the diameter of Γ which is collinear with O, then $I(\Gamma) = \Gamma'$, where Γ' is the circle with diameter $A'B'$.*

Proof (a) and (b) Inthe notations of Fig. 9.4, let A be defined as in the statement and A' be the inverse of A (we only consider the case in which $A' \in OA$; the other case is entirely analogous and will be left to the reader). If B is any other point of r, then item(b) of the previous result assures that $ABB'A'$ is cyclic. Since $O\widehat{A}B = 90°$, this gives

$$O\widehat{B'}A' = 180° - A'\widehat{B'}B = B\widehat{A}A' = 90°,$$

so that B' lies in the circle Γ of diameter OA', as desired. Hence, $I(r) \subset \Gamma \setminus \{O\}$.

Now, by exchanging the roles of A, A' and B, B' and arguing in pretty much the same way, we conclude that $I(\Gamma \setminus \{O\}) \subset r$. Thus, the comments we made immediately prior to the statement of the proposition guarantee that $I(r) = \Gamma \setminus \{O\}$ and $I(\Gamma \setminus \{O\}) = r$.

(c) As in the previous items, this item follows at once, provided we show that $I(\Gamma) \subset \Gamma'$ and $I(\Gamma') \subset \Gamma$. We shall establish the first inclusion, the second one being quite similar (Fig. 9.5). We start by recalling, from item (b) of the previous result, that $ACC'A'$ and $BCC'B'$ are cyclic quadrilaterals. Therefore, $A'\widehat{C'}X = A'\widehat{A}C$ and $B'\widehat{C'}C = A\widehat{B}C$, and hence

Fig. 9.4 Inverse of a line not passing through the pole

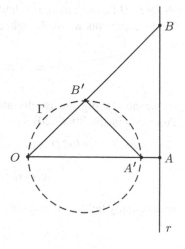

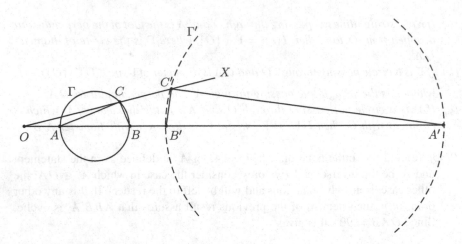

Fig. 9.5 Inverse of a circle not passing through the pole

$$A'\widehat{C'}B' = 180° - A'\widehat{C'}X - B'\widehat{C'}C$$

$$= 180° - A'\widehat{A}C - A\widehat{B}C$$

$$= A\widehat{C}B = 90°.$$

Thus, C' lies in the circle of diameter $A'B'$, namely, Γ'. $\qquad\qquad\qquad\square$

At this point, a few remarks are in order. First of all, in the notations of items (a) and (b) of the proposition, even though the inverse of line r is (strictly speaking) $\Gamma \setminus \{O\}$, whenever there is no danger of confusion we shall simply say that Γ *is the image of r through I*. Likewise, we shall also say that *r is the image of Γ with respect to I*. With respect to item (c), note that if A is closer to O than B, then B' is closer to O than A'; also, if Γ has radius r and Γ' radius r', then we do not have $\frac{k^2}{r}$, for the comments we made right after Proposition 9.3 give

$$2r' = \overline{A'B'} = \frac{k^2\overline{AB}}{OA \cdot OB} = \frac{2k^2 r}{OA \cdot OB}.$$

We now apply the results above to provide a conceptually simpler proof of Ptolemy's inequality (7.20).

Example 9.5 If $ABCD$ is a convex quadrilateral of diagonals AC and BD, then

$$\overline{AB} \cdot \overline{CD} + \overline{AD} \cdot \overline{BC} \geq \overline{AC} \cdot \overline{BD},$$

with equality holding if and only if $ABCD$ is cyclic.

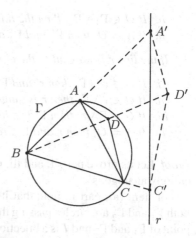

Fig. 9.6 Ptolemy's inequality through inversion

Proof Let Γ be the circumcircle of ABC. Assume (according with the previous proposition) that the inversion of center B and ratio 1 maps Γ into the line r shown in Fig. 9.6, so that A is mapped into A' and C into C'.

Suppose first that $D \notin \Gamma$ (in Fig. 9.6 we consider the case in which D lies in the disk bounded by Γ; the case of D lying outside that disk can be dealt with in pretty much the same way). If D' is its inverse, then $D' \notin \Gamma$, so that triangle inequality gives $\overline{A'D'} + \overline{C'D'} > \overline{A'C'}$. By applying the formula of item (b) of Proposition 9.3 to compute these lengths (with B in place of O), we get

$$\frac{\overline{AD}}{\overline{BA} \cdot \overline{BD}} + \frac{\overline{CD}}{\overline{BC} \cdot \overline{BD}} > \frac{\overline{AC}}{\overline{BA} \cdot \overline{BC}}$$

or, which is the same,

$$\overline{AB} \cdot \overline{CD} + \overline{AD} \cdot \overline{BC} > \overline{AC} \cdot \overline{BD}.$$

Now, if $D \in \Gamma$, then $D' \in r$ as well, whence $\overline{A'D'} + \overline{C'D'} = \overline{A'C'}$. Therefore, arguing exactly as above, we obtain

$$\overline{AB} \cdot \overline{CD} + \overline{AD} \cdot \overline{BC} = \overline{AC} \cdot \overline{BD}.$$

$\square$

The previous result allows us to establish, for circles and lines, a quite useful corollary on *preservation of tangency upon inversion*.

Corollary 9.6 *Let I be an inversion of center O.*

(a) If Γ_1 and Γ_2 are distinct circles tangent at A, then:

  i. If $O \notin \Gamma_1 \cup \Gamma_2$, then Γ_1' and Γ_2' are circles tangent at A'.

ii. If $O \in \Gamma_1 \setminus \Gamma_2$, then the line Γ_1' is tangent to the circle Γ_2' at A'.

iii. If $A = O$, then Γ_1' and Γ_2' are parallel lines.

(b) *If the line r is tangent to the circle Γ at A, then:*

    i. If $O \notin r \cup \Gamma$, then r' and Γ' are circles tangent at A'.

    ii. If $O \in r \setminus \Gamma$, then r is tangent to the circle Γ' at A'.

    iii. If $O \in \Gamma \setminus r$, then Γ' is a line tangent to the circle r' at A'.

    iv. If $A = O$, then r and Γ' are parallel lines.

Proof Let us prove parts i. and iii. of item (a), the proofs of the other claims being entirely analogous.

For item (a), part i., note that item (c) of the previous proposition assures that both Γ_1' and Γ_2' are circles passing through A'. However, since A is the only common point of Γ_1 and Γ_2 and I is a bijection, we conclude that A' is the only common point of Γ_1' and Γ_2'. Finally, it suffices to note that two circles with only one common point must necessarily be tangent.

For part iii. of (a), item (b) of the previous proposition assures that $(\Gamma_1 \setminus \{O\})'$ and $(\Gamma_2 \setminus \{O\})'$ are two straightlines. However, since $\Gamma_1 \setminus \{O\}$ and $\Gamma_2 \setminus \{O\}$ have no common points, the same holds for their inverses (due to the bijectiveness of I once more). Hence, $(\Gamma_1 \setminus \{O\})'$ and $(\Gamma_2 \setminus \{O\})'$ must be parallel. $\square$

This corollary, together with its refinement in Theorem 9.8, are the main reasons for the usefulness of inversion. Examples 9.7 and 9.9 will illustrate their uses.

Example 9.7 A triangle ABC is inscribed in a circle Γ. Point P is the foot of the internal bisector of angle $\angle BAC$, and circles Γ_1 and Γ_2 are tangent to Γ and to the line segments AP and BC. Prove that the points of tangency of Γ_1 and Γ_2 with AP coincide.

Proof If M is the other point of intersection of $\overrightarrow{AP}$ with Γ (i.e., $M \neq A$), then (cf. Fig. 9.7) $\overline{MB} = \overline{MC} := k$. Therefore, the inversion $I(M; k^2)$ leaves B and C fixed, hence transforms $\overleftrightarrow{BC}$ into Γ and vice-versa.

Fig. 9.7 The effect of $I(M; k^2)$

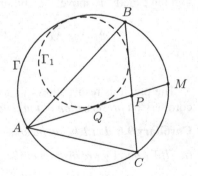

Fig. 9.8 Angle between two circles Γ_1 and Γ_2

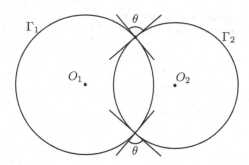

Now, let Q be the point of tangency of Γ_1 and AP. Since $I(\overleftrightarrow{AM}) = \overleftrightarrow{AM}$ and $I(\Gamma_1)$ is a circle tangent to $I(\Gamma)$, $I(\overleftrightarrow{BC})$ and $I(\overleftrightarrow{AM})$, we conclude that $I(\Gamma_1) = \Gamma_1$. In particular,

$$\{I(Q)\} = I(\Gamma_1) \cap I(\overleftrightarrow{AM}) = \Gamma_1 \cap \overleftrightarrow{AM} = \{Q\},$$

so that $I(Q) = Q$. Then, $\overline{MQ}^2 = k^2 = \overline{MC}^2$, whence, $\overline{MQ} = \overline{MC}$; therefore, Proposition 3.37 assures that Q is the incenter of ABC. Accordingly, Γ_1 touches AP at Q. □

The next result establishes, in the particular case of lines and/or circles, the most important property of inversions, called **angle invariance**.[3] For its statement, it is worth defining the **angle** between two concurrent circles (resp. between concurrent circle and line) as the angle formed by their tangents (resp. by the line and the tangent to the circle) through their concurrence point (cf. Fig. 9.8; note that this does not depend on the chosen point of concurrence).

Theorem 9.8 *Let I be an inversion of center O. If two circles (resp. a circle and a line, two lines) pass through $A \neq O$ and form an angle θ at A, then their inverses form an angle θ at A'.*

Proof We shall do the proof[4] in the case of two circles not passing through O. The analysis of the other cases is quite similar and will be left to the reader.

Let Γ_1 and Γ_2 be two circles not passing through O and meeting at A. Figure 9.9 shows portions of Γ_1 and of its inverse Γ_1' near A, as well as the tangents r_1 and s_1 to Γ_1 and Γ_1' at A and A', respectively.

For $B \in \Gamma_1 \setminus \overleftrightarrow{OA}$, the fact that $ABB'A'$ is cyclic gives $B\widehat{A}A' = X\widehat{B}'A'$. Therefore, the External Angle Theorem allows us to compute

---

[3]This is actually true for any two *regular* curves with a common point. Nevertheless, since we shall not need this extra generality, we decided to stick to the cases below.

[4]Although the argument below relies upon *continuity* considerations (and therefore is not entirely self-contained), it will suffice for our purposes.

Fig. 9.9 Angle invariance
upon inversion I

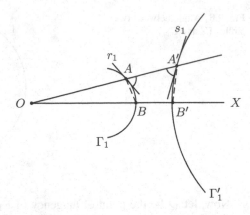

Fig. 9.10 Angle invariance
upon inversion II

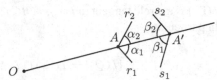

$$B\widehat{A}A' - B'\widehat{A'}A = B\widehat{A}A' - (X\widehat{B'}A' - A'\widehat{O}B') = A'\widehat{O}B'. \qquad (9.1)$$

Let the half-line $\overrightarrow{OB}$ rotate around O towards $\overrightarrow{OA}$. Then, the secants $\overleftrightarrow{AB}$ to Γ_1 and $\overleftrightarrow{A'B'}$ to Γ_1' correspondingly rotate towards r and s, respectively, whereas $A'\widehat{O}B'$ goes to $0°$. Thus, at the limit position, (9.1) turns into $\alpha_1 = \beta_1$, where α_1 (resp. β_1) is the indicated angle between r_1 (resp. s_1) and $\overleftrightarrow{OA}$.

Now, repeat the reasoning above with Γ_2, Γ_2' and $\overleftrightarrow{OA}$, thus proving (cf. Fig. 9.10) that if r_2 and s_2 denote the tangents to Γ_2 and Γ_2' at A, and α_2 (resp. β_2) stands for the angle between r_2 (resp. s_2) and $\overleftrightarrow{OA}$, then $\alpha_2 = \beta_2$. Then, $\alpha_1 + \alpha_2 = \beta_1 + \beta_2$, as wished.

□

Since the case of two tangent circles (resp. of a line tangent to a circle) corresponds to an angle of $0°$, it is now immediate to see that Corollary 9.6 is a particular case of the previous result. Another important particular case is that of two **orthogonal** circles (resp. of a line orthogonal to a circle), i.e., such that they concur and form an angle of $90°$, in the sense of the previous paragraph. In this case (cf. Fig. 9.11 for the case of two circles), since a tangent to a circle is perpendicular to the radius through the tangency point, one readily shows that the tangent to one circle through a common point passes through the center of the other, and vice-versa (resp. the tangent to the circle is perpendicular to the given line).

As we stressed before, our final example makes use of the property of angle invariance (more precisely, of *orthogonality invariance*) of inversions.

Fig. 9.11 Two orthogonal
circles Γ_1 and Γ_2

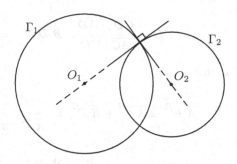

Example 9.9 (IMO—Adapted) Let ABC be an acute triangle and D be an interior point, with $A\widehat{D}B = A\widehat{C}B + 90°$ and $\overline{AC} \cdot \overline{BD} = \overline{AD} \cdot \overline{BC}$. Prove that the tangents to the circumcircles of triangles ACD and BCD at a common point are perpendicular to each other.

Proof We let the figure below represent the described situation, and let Γ be the circumcircle of ACD and Σ be that of BCD.

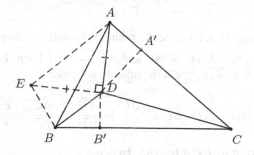

To say that the tangents to Γ and Σ at D are perpendicular is the same as to say that Γ and Σ are orthogonal. In turn, Theorem 9.8 and the subsequent discussion assure that this is so if and only if an inversion of center C transforms these circles into two perpendicular straightlines. As we shall now see, this gives a clue to the solution.

Indeed, upon the inversion of center C and ratio $\overline{CD}$, let A' be the inverse of A and B' be that of B. Since D is its own inverse, Proposition 9.4 guarantees that $\overleftrightarrow{A'D}$ is the inverse of Γ and $\overleftrightarrow{B'D}$ is that of Σ. Thus,

$$\overleftrightarrow{A'D} \perp \overleftrightarrow{B'D} \Leftrightarrow A'B'D \text{ is right at } D \Leftrightarrow \overline{A'B'}^2 = \overline{A'D}^2 + \overline{B'D}^2.$$

Now, Proposition 9.3 gives

$$\overline{A'D} = \frac{\overline{CD}^2 \cdot \overline{AD}}{\overline{AC} \cdot \overline{CD}}, \quad \overline{B'D} = \frac{\overline{CD}^2 \cdot \overline{BD}}{\overline{BC} \cdot \overline{CD}}, \quad \overline{A'B'} = \frac{\overline{CD}^2 \cdot \overline{AB}}{\overline{AC} \cdot \overline{BC}},$$

so that

$$\overline{A'B'}^2 = \overline{A'D}^2 + \overline{B'D}^2 \Leftrightarrow \frac{\overline{CD}^4 \cdot \overline{AB}^2}{\overline{AC}^2 \cdot \overline{BC}^2} = \frac{\overline{CD}^4 \cdot \overline{AD}^2}{\overline{AC}^2 \cdot \overline{CD}^2} + \frac{\overline{CD}^4 \cdot \overline{BD}^2}{\overline{BC}^2 \cdot \overline{CD}^2}$$

$$\Leftrightarrow \frac{\overline{AB}^2}{\overline{AC}^2 \cdot \overline{BC}^2} = \frac{\overline{AD}^2}{\overline{AC}^2 \cdot \overline{CD}^2} + \frac{\overline{BD}^2}{\overline{BC}^2 \cdot \overline{CD}^2}$$

$$\Leftrightarrow \overline{AB}^2 \cdot \overline{CD}^2 = \overline{AD}^2 \cdot \overline{BC}^2 + \overline{BD}^2 \cdot \overline{AC}^2.$$

However, since it is given that $\overline{AC} \cdot \overline{BD} = \overline{AD} \cdot \overline{BC}$, we conclude that

$$\overline{A'B'}^2 = \overline{A'D}^2 + \overline{B'D}^2 \Leftrightarrow \overline{AB}^2 \cdot \overline{CD}^2 = 2\overline{AD}^2 \cdot \overline{BC}^2$$

$$\Leftrightarrow \frac{\overline{AB} \cdot \overline{CD}}{\overline{AD} \cdot \overline{BC}} = \sqrt{2}. \tag{9.2}$$

In order to establish this, let E be the point in the half-plane opposite to C with respect to $\overleftrightarrow{AD}$ such that $A\widehat{D}E = 90°$ and $\overline{AD} = \overline{DE}$. Then, $E\widehat{D}B = A\widehat{D}B - 90° = A\widehat{C}B$ and $\overline{AE} = \overline{AD}\sqrt{2}$. On the other hand, since

$$\overline{AC} \cdot \overline{BD} = \overline{AD} \cdot \overline{BC} \Leftrightarrow \frac{\overline{AC}}{\overline{BC}} = \frac{\overline{AD}}{\overline{BD}} \Leftrightarrow \frac{\overline{AC}}{\overline{BC}} = \frac{\overline{ED}}{\overline{BD}},$$

we conclude that $ACB \sim EDB$ by SAS. Hence,

$$\frac{\overline{AB}}{\overline{BC}} = \frac{\overline{EB}}{\overline{BD}} \quad \text{and} \quad A\widehat{B}C = E\widehat{B}D.$$

The first equality above can be written as $\frac{\overline{AB}}{\overline{EB}} = \frac{\overline{BC}}{\overline{BD}}$, whereas the second one implies

$$E\widehat{B}A = E\widehat{B}D - A\widehat{B}D = A\widehat{B}C - A\widehat{B}D = D\widehat{B}C.$$

Hence,

$$\frac{\overline{AB}}{\overline{EB}} = \frac{\overline{BC}}{\overline{BD}} \quad \text{and} \quad E\widehat{B}A = D\widehat{B}C,$$

so that $ABE \sim CBD$, again by SAS. But this finally gives

$$\frac{\overline{AB}}{\overline{BC}} = \frac{\overline{AE}}{\overline{CD}} = \frac{\overline{AD}\sqrt{2}}{\overline{CD}},$$

which is equivalent to (9.2). □

Problems: Sect. 9.1

1. (England) We are given a circle Γ of radius 1 and a chord AB of Γ. For each point C in AB, draw the circles α_C and β_C, tangent to AB at C, internally tangent to Γ and lying in opposite sides with respect to AB. If r_C and R_C stand for the radii of α_C and β_C, respectively, with $r_C \leq R_C$ for every $C \in AB$, prove that $\frac{r_C}{R_C}$ does not depend on the position of C along B.

2. In a convex quadrilateral $ABCD$ of diagonals AC and BD, the circumcircles of triangles ABC and ACD are orthogonal. Prove that

$$\overline{AB}^2 \cdot \overline{CD}^2 + \overline{AD}^2 \cdot \overline{BC}^2 = \overline{AC}^2 \cdot \overline{BD}^2.$$

3. We are give in the plane a circle Γ and distinct points A and B, both of which lying outside the disk bounded by Γ. Use compass and straightedge compass to construct all circles α, tangent to Γ and passing through A and B.

4. We are given a line r, a point A and a circle Γ, with A and Γ lying in a single half-plane, of those determined by r. Use compass and straightedge to construct all circles Σ, passing through A and tangent to r and Γ.

5. We are given a line r and nonconcentric circles Γ_1 and Γ_2 lying in a single half-plane, of those determined by r. Use compass and straightedge to construct all circles Γ, simultaneously tangent to r, Γ_1 and Γ_2.

6. Let α and β be two exterior circles, and r and s be their common external tangents. Assume that r (resp. s) touch α and β respectively at A and B (resp. At C and D). Let M and N be the midpoints of AB and CD, respectively, and P and Q be the other intersections of AN and BN with α and β, also respectively (s that $P \neq A$ and $Q \neq B$). Prove that $\overleftrightarrow{CP}$, $\overleftrightarrow{DQ}$ and $\overleftrightarrow{MN}$ concur at a point of the circumcircle of triangle NPQ, which is externally tangent to both α and β.

7. The orthogonal circles α and β intersect at A and B. We choose points $C \in \alpha \setminus \{A, B\}$ and $D \in \beta \setminus \{A, B\}$, such that $A, B \notin \overleftrightarrow{CD}$. Prove that the circumcircles of triangles ACD and BCD are also orthogonal.

8. We are given $n \geq 4$ points in the plane, satisfying the following condition: any circle passing through three of the given points contains a fourth one of them. Show that all of the points lie in a single circle.

9. Let I be the inversion of center O and inversion circle Γ. Show that a circle Σ is orthogonal to Γ if and only if $I(\Sigma) = \Sigma$.

10. In an acute triangle ABC, the circle Σ of diameter BC has orthocenter H. If P and Q are the feet of the tangents drawn from A to Σ, show that P, Q and H are collinear.[5]

11. Let A, B, C be given collinear points, with $B \in AC$. Draw semicircles α, β and Γ_0, on the same side of AC and with diameters AC, AC and BC, respectively. For each $k \in \mathbb{N}$, let Γ_k be the circle tangent to α, β and Γ_{k-1}. Show that, for every $n \in \mathbb{N}$, the distance from the center of Γ_n to AC is n times its diameter.

12. (Serbia) ABC is an acute triangle of orthocenter H, and D, E, F are the feet of the altitudes relative to A, B, C, respectively; M and S be the midpoints of BC and AH, also respectively, and G the intersection of EF and AH. If AM intersects the circumcircle of BCH at N, prove that $H\widehat{M}A = G\widehat{N}S$.

For the coming problem, given a point O and $k > 0$, we can extend the notion of inversion by including all maps $I : \alpha \setminus \{O\} \to \alpha \setminus \{O\}$ such that

$$OA \cdot OA = -k^2,$$

where OA and OA stand for *oriented segments* (as in Sect. 4.4). In this sense, we shall say that the inversions studied in this section are *direct*, whereas those as above are *opposite*; in both cases, k is the ratio of the inversion. Whenever we refer to an *inversion* without further qualifying it, we shall always assume that it is a direct one.

13. Let I be an opposite inversion with pole O and ratio k. State and prove analogues of the properties studied in this section for I.

14. We are given in the plane a circle Γ and points A and B not in Γ. We draw a secant r to Γ through B, which cuts Γ at points C and D. Then we draw lines $\overleftrightarrow{AC}$ and $\overleftrightarrow{AD}$, which intersect Γ again at points E and F, respectively. Show that, as the direction of r varies, the circumcenter of triangle AEF varies along a fixed line.

15. Let P be a fixed point in the interior of a given circle Γ. We draw through P two perpendicular chords AA and BB. If M is the foot of the perpendicular dropped from P to AB and N is the midpoint of AB, prove that the product $\overline{PM} \cdot \overline{PN}$ remains constant as the direction of the chords vary.

16. The purpose of this problem is to prove **Feuerbach's theorem**[6]: in every triangle, the nine-point circle is tangent to the incircle and to the excircles. To this end, let the notations be as in Fig. 9.12, with $\overline{AB} = c$, $\overline{AC} = b$, $\overline{BC} = a$, and do the following items:

  (a) If M is the midpoint of BC, show that the inversion of center M and ratio $\frac{1}{2}|b - c|$ fixes D, J, Γ and Σ.

  (b) If P is the foot of the internal bisector relative to A and H_a is the foot of the altitude of ABC dropped from A, compute

---

[5] For another approach to this problem, see Problem 5, page 317.

[6] Karl Wilhelm von Feuerbach, German mathematician of the nineteenth century.

Fig. 9.12 Feuerbach's theorem

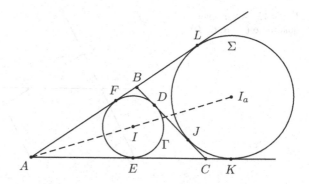

$$\overline{MP} = \frac{a|b-c|}{2(b+c)} \quad \text{and} \quad \overline{MH_a} = \frac{b^2 - c^2}{2a}.$$

Then, show that P and H_a are inverses of each other.

(c) Let R, S be the symmetrics of C, B with respect to $\overleftrightarrow{AI}$, let N be the midpoint of AC and Q be the intersection of $\overleftrightarrow{MN}$ and $\overleftrightarrow{RS}$. Compute $\overline{MQ} = \frac{(b-c)^2}{2c}$, then show that N and Q are inverses of each other.

(d) Conclude the proof of Feuerbach's theorem.

9.2 Harmonic Pencils

In all that follows, whenever convenient we shall make free use of the concept of *oriented segments*, as defined in Sect. 4.4. We shall also make free use of Ceva's and Menelaus' theorems.

We begin by elaborating the discussion initiated at Problem 5, page 136.

Definition 9.10 Given distinct points $C, D \in \overleftrightarrow{AB} \setminus \{A, B\}$, we say that C and D (in this order) are **harmonic conjugates** with respect to A and B (also in this order) if

$$\frac{AC}{CB} = -\frac{AD}{DB}.$$

In this case, we also say that (A, B, C, D) is a **harmonic quadruple**.

It follows immediately from the definition above that, if (A, B, C, D) is a harmonic quadruple, then so is (A, B, D, C). Moreover, it is easy to show that exactly one of C and D lies in segment AB, and none of them coincides with the midpoint of AB. Finally, since

Fig. 9.13 A harmonic
quadruple (A, B, C, D)

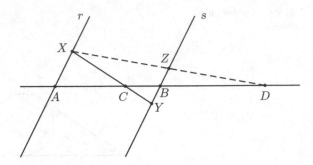

$$\frac{AC}{CB} = -\frac{AD}{DB} \Leftrightarrow \frac{CA}{AD} = -\frac{CB}{BD},$$

we conclude that (C, D, A, B) is also a harmonic quadruple.

Let (A, B, C, D) be a harmonic quadruple as in Fig. 9.13, and r and s be parallel lines drawn from A and B, respectively. Draw a line t through C, intersecting r at X and s at Y. If Z is the symmetric of Y with respect to B, then and application of Thales' theorem gives

$$\frac{\overline{AX}}{\overline{BZ}} = \frac{\overline{AX}}{\overline{BY}} = \frac{\overline{AC}}{\overline{BC}} = \frac{\overline{AD}}{\overline{BD}},$$

so that $XAD \sim ZBD$ by SAS ($X\widehat{A}D = Z\widehat{B}D$). Hence, $X\widehat{D}A = Z\widehat{D}B$, and X, Z and D are collinear points.

Conversely, let M be the midpoint of AB and $C \in AB \setminus \{A, B, M\}$. Choose points $X \in r$ and $Y \in s$ such that X, Y and C are collinear. If Z is the symmetric of Y with respect to B and D is the intersection of $\overleftrightarrow{XY}$ and $\overleftrightarrow{AB}$, then a slight modification of the above reasoning shows that (A, B, C, D) is a harmonic quadruple. This proves the following

Lemma 9.11 *Let M be the midpoint of AB, and $C \in AB \setminus \{A, B, M\}$. Then, there exists a single point $D \in \overleftrightarrow{AB} \setminus \{A, B, M\}$ such that (A, B, C, D) is a harmonic quadruple.*

Thanks to the previous lemma, we shall sometimes say that D (resp. C) is the **harmonic conjugate** of C (resp. of D) with respect to A and B.

Example 9.12 In a triangle ABC with $\overline{AB} \neq \overline{AC}$, let P be the foot of the internal angle bisector and Q that of the external angle bisector relative to A. Then, (B, C, P, Q) is a harmonic quadruple. Indeed, with respect to Fig. 9.14, the internal and external bisector theorems give

$$\frac{BP}{PC} = \frac{\overline{AB}}{\overline{AC}} = -\frac{BQ}{QC}.$$

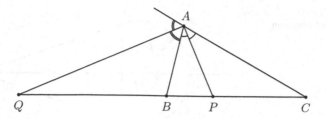

Fig. 9.14 Feet of angle bisectors as harmonic conjugates

Our next result, though very simple, is the key step to link harmonic conjugation to inversion.

Lemma 9.13 *If* (A, B, C, D) *is a harmonic quadruple and* O *is the midpoint of* AB, *then* C *and* D *are inverses of each other with respect to the inversion of center* O *and modulus* $\frac{AB}{2}$.

Proof Assume, with no loss of generality, that $C \in AB$ and $D \in \overrightarrow{AB} \setminus AB$ (cf. Fig. 9.15). Then,

$$\frac{AC}{CB} = -\frac{AD}{DB} \Leftrightarrow \frac{\overline{AC}}{\overline{BC}} = \frac{\overline{AD}}{\overline{BD}}.$$

Now, since $\overline{AD} > \overline{BD}$, we conclude that $\overline{AC} > \overline{BC}$, and hence $C \in BO$. Letting $\overline{AO} = \overline{BO} = k$, we thus have

$$\frac{\overline{AC}}{\overline{BC}} = \frac{\overline{AD}}{\overline{BD}} \Leftrightarrow \frac{\overline{AO} + \overline{OC}}{\overline{BO} - \overline{OC}} = \frac{\overline{AO} + \overline{OD}}{\overline{OD} - \overline{OB}}$$

$$\Leftrightarrow \frac{k + \overline{OC}}{k - \overline{OC}} = \frac{k + \overline{OD}}{\overline{OD} - k}$$

$$\Leftrightarrow (k + \overline{OC})(\overline{OD} - k) = (k + \overline{OD})(k - \overline{OC})$$

$$\Leftrightarrow -k^2 + \overline{OC} \cdot \overline{OD} = k^2 - \overline{OC} \cdot \overline{OD}$$

$$\Leftrightarrow \overline{OC} \cdot \overline{OD} = k^2.$$

$\square$

In the notations of Fig. 9.15, and according to Lemma 9.2, given A, B and $C \in OB \setminus \{O, B\}$, we have the following alternative construction for the harmonic conjugate of C: draw the perpendicular to AB through C and let T_1 and T_2 be its intersections with the circle $\Gamma(O; k)$; then D is the intersection point of the tangents to Γ drawn through T_1 and T_2. By the same token, given $D \in \overrightarrow{AB} \setminus AB$, we can get

Fig. 9.15 Harmonic
conjugation and inversion

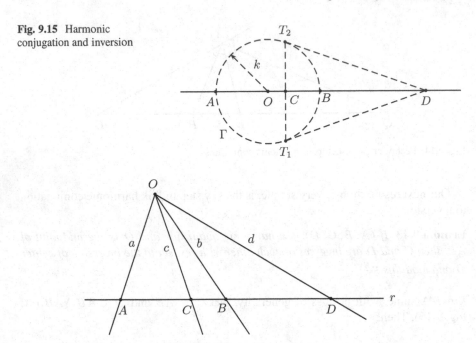

Fig. 9.16 A harmonic pencil

its harmonic conjugate C by first drawing the tangents to Γ through D, then finding
C as the intersection of AB with the line joining the tangency points T_1 and T_2.

In order to continue our presentation, we now need a definition of *harmonicity*
for concurrent lines, instead of for collinear points. This is done as below.

Definition 9.14 Given distinct lines a, b, c, d concurrent at O, we say that
(a, b, c, d) is a **harmonic pencil** if there exists a line r, intersecting a, b, c, d at
the pairwise distinct points A, B, C, D, respectively, such that (A, B, C, D) is a
harmonic quadruple (see Fig. 9.16). We also say that O is the **center** of the harmonic
pencil.

In view of the previous definition, a simple way of constructing a harmonic pencil
(actually, the only one) is to start with a harmonic quadruple (A, B, C, D) and a
point O outside $\overleftrightarrow{AB}$, and to consider $\left(\overleftrightarrow{OA}, \overleftrightarrow{OB}, \overleftrightarrow{OC}, \overleftrightarrow{OD} \right)$.

On the other hand, it is an asthonishing fact that if a straightline intersects the
four lines of a harmonic pencil at distinct points, then the four points of intersection
form a harmonic quadruple. In order to show this we need a preliminary result,
which (thanks to Example 9.12 and the concurrence of the internal bisectors of a
triangle) can be viewed as a generalization of Example 4.23 (Fig. 9.17).

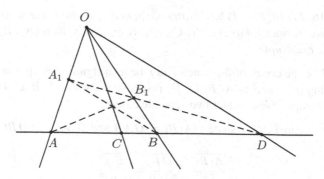

Fig. 9.17 Collinearity and concurrence in harmonic pencils

Fig. 9.18 Complete
quadrangles generate
harmonic pencils

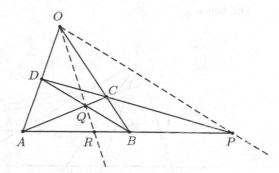

Lemma 9.15 *Let* (A, B, C, D) *be a harmonic quadruple and* $O \notin \overleftrightarrow{AB}$. *If* $A_1 \in$ $\overleftrightarrow{OA} \setminus \{O, A\}$ *and* $B_1 \in \overleftrightarrow{OC} \setminus \{O, C\}$, *then* A_1, B_1, D *are collinear if and only if* $\overleftrightarrow{AB_1}, \overleftrightarrow{A_1B}, \overleftrightarrow{OC}$ *are concurrent.*

Proof In the notations of Fig. 9.18, we apply Menelaus' theorem 4.22 to triangle OAB and points $A_1 \in \overleftrightarrow{OA}$, $B_1 \in \overleftrightarrow{OB}$, $D \in \overleftrightarrow{AB}$ to conclude that

$$A_1, B_1, D \text{ are collinear} \Leftrightarrow \frac{OA_1}{A_1A} \cdot \frac{AD}{DB} \cdot \frac{BB_1}{B_1O} = -1.$$

Now, we apply Ceva's theorem 4.25 to OAB and $A_1 \in \overleftrightarrow{OA}$, $B_1 \in \overleftrightarrow{OB}$, $C \in \overleftrightarrow{AB}$, we obtain that

$$\overleftrightarrow{A_1B}, \overleftrightarrow{AB_1}, \overleftrightarrow{OC} \text{ are concurrent} \Leftrightarrow \frac{OA_1}{A_1A} \cdot \frac{AC}{CB} \cdot \frac{BB_1}{B_1O} = 1.$$

Finally, since $\frac{AC}{CB} = -\frac{AD}{DB}$, it is now immediate that one of the claims above is true if and only if so is the other. $\qquad\square$

Theorem 9.16 *Let (a, b, c, d) be a harmonic pencil. If a line s intersects a, b, c, d at the pairwise distinct points A_1, B_1, C_1, D_1, respectively, then (A_1, B_1, C_1, D_1) is a harmonic quadruple.*

Proof Let O be the center of the pencil and r be a straightline not passing through O and meeting a, b, c, d at A, B, C, D, respectively, with (A, B, C, D) being a harmonic quadruple. We consider two separate cases:

(i) $s \parallel r$ (see figure below): since $OA_1B_1 \sim OAB$ and $OB_1C_1 \sim OBC$, we have

$$\frac{\overline{A_1B_1}}{\overline{AB}} = \frac{\overline{OB_1}}{\overline{OB}} = \frac{\overline{B_1C_1}}{\overline{BC}},$$

and hence $\frac{A_1B_1}{B_1C_1} = \frac{AB}{BC}$.

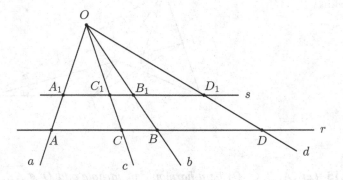

Likewise, $\frac{A_1D_1}{D_1C_1} = \frac{AD}{DC}$. However, since $\frac{AB}{BC} = -\frac{AD}{DC}$, it follows that $\frac{A_1B_1}{B_1C_1} = -\frac{A_1D_1}{D_1C_1}$, and (A_1, B_1, C_1, D_1) is a harmonic quadruple.

(ii) $s \nparallel r$: draw through D the parallel t to s, with $t \cap a = \{A_2\}$, $t \cap b = \{B_2\}$, $t \cap c = \{C_2\}$ and $D_2 = D$. By the previous case, (A_1, B_1, C_1, D_1) is a harmonic quadruple if and only if (A_2, B_2, C_2, D_2) is a harmonic quadruple. It thus suffices to establish this last claim.

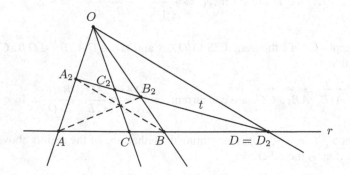

For what is left to do, since (A, B, C, D) is a harmonic quadruple and A_2, B_2, D are collinear, we conclude that $\overleftrightarrow{A_2B}$, $\overleftrightarrow{B_2A}$, $\overleftrightarrow{OC}$ are concurrent. Therefore, Ceva's theorem (applied to triangle OA_2B_2) gives

$$\frac{OA}{AA_2} \cdot \frac{A_2C_2}{C_2B_2} \cdot \frac{B_2B}{BO} = 1. \tag{9.3}$$

Also with respect to triangle OA_2B_2, since A, B, D are collinear, Menelaus' theorem gives

$$\frac{OA}{AA_2} \cdot \frac{A_2D}{DB_2} \cdot \frac{B_2B}{BO} = -1. \tag{9.4}$$

By comparing (9.3) and (9.4) and recalling that $D = D_2$, we therefore obtain

$$\frac{A_2C_2}{C_2B_2} = -\frac{A_2D}{DB_2} = -\frac{A_2D_2}{D_2B_2},$$

as we wished to show.

$\square$

Corollary 9.17 *If (a, b, c, d_1) and (a, b, c, d_2) are harmonic pencils, then $d_1 = d_2$.*

Proof Let O be the common center of the harmonic pencils and r be a line meeting a, b, c, d_1 and d_2 respectively at distinct points A, B, C, D_1 and D_2. The previous result assures that both (A, B, C, D_1) and (A, B, C, D_2). Hence, Lemma 9.11 gives $D_1 = D_2$, whence $d_1 = \overleftrightarrow{OD_1} = \overleftrightarrow{OD_2} = d_2$. $\square$

Example 9.18 Let (a, b, c, d) be a harmonic pencil. A line r intersects lines a, b, c at the distinct points A, B, C, respectively, with $C \in AB$. Show that C is the midpoint of AB if and only if $r \parallel d$.

Proof By contraposition, it suffices to show that C is not the midpoint of AB if and only if r is not parallel to d.

If C is not the midpoint of AB, we can choose $D \in \overleftrightarrow{AB}$ such that (A, B, C, D) is a harmonic quadruple. Letting $d' = \overleftrightarrow{OD}$, we conclude that (a, b, c, d') is a harmonic pencil. Therefore, the previous corollary assures that $d = d'$, whence $D \in \overleftrightarrow{AB} \cap d' = r \cap d$, and r is not parallel to d. Conversely, if r is not parallel to d, let $r \cap d = \{D\}$. Since (a, b, c, d) is a harmonic pencil, (A, B, C, D) is a harmonic quadruple. This being the case, we already know that C cannot be the midpoint of AB. $\square$

Our last result on harmonic pencils can be seen as a simple way of producing them. It can also be seen as a sort of converse of Lemma 9.15, in the sense that (in the notations of Fig. 9.18) if A_1, B_1, D are collinear and $\overleftrightarrow{OA}$, $\overleftrightarrow{A_1B}$, $\overleftrightarrow{AB_1}$ are concurrent, then (a, b, c, d) is a harmonic pencil (Fig. 9.19).

Fig. 9.19 A property of the
parallel to one of the lines of
a harmonic pencil

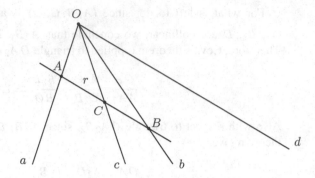

Theorem 9.19 *Let ABCD be a given quadrilateral. If $\overleftrightarrow{AD} \cap \overleftrightarrow{BC} = \{O\}$, $\overleftrightarrow{AB}$
$\cap \overleftrightarrow{CD} = \{P\}$ and $\overleftrightarrow{AC} \cap \overleftrightarrow{BD} = \{Q\}$, then $(\overleftrightarrow{OA}, \overleftrightarrow{OB}, \overleftrightarrow{OP}, \overleftrightarrow{OQ})$ is a harmonic
pencil.*

Proof Let (cf. Fig. 9.18) $\overleftrightarrow{OQ} \cap \overleftrightarrow{AB} = \{R\}$. Applying Ceva's and Menelaus'
theorems to triangle OAB with respect to the transversal $\overleftrightarrow{CD}$ and the cevians $\overleftrightarrow{OR}$,
$\overleftrightarrow{AC}$, $\overleftrightarrow{BD}$, respectively, we obtain

$$\frac{OD}{DA} \cdot \frac{AR}{RB} \cdot \frac{BC}{CO} = 1 \text{ and } \frac{OD}{DA} \cdot \frac{AP}{PB} \cdot \frac{BC}{CO} = -1.$$

Hence, $\frac{AR}{RB} = -\frac{AP}{PB}$, so that (A, B, P, R) is a harmonic quadruple. Therefore,
$(\overleftrightarrow{OA}, \overleftrightarrow{OB}, \overleftrightarrow{OP}, \overleftrightarrow{OR})$ is a harmonic pencil, with $\overleftrightarrow{OR} = \overleftrightarrow{OQ}$.   □

In the context of Projective Geometry one usually says that four straightlines,
pairwise concurrent at (six) distinct points define a **complete quadrangle** (in the
notations of Fig. 9.18, this is the case of $\overleftrightarrow{AA_1}$, $\overleftrightarrow{AB}$, $\overleftrightarrow{A_1B_1}$ and $\overleftrightarrow{B_1B}$). Thus, the
theorem above can be stated more simply by saying that *complete quadrangles
define harmonic pencils*. Lemma 9.15, together with Theorem 9.16, show that the
converse also holds.

Problems: Sect. 9.2

1. Let a, b, c, d be distinct lines concurrent at O, and Γ be a circle passing through
   O and not tangent to any of a, b, c, d. If Γ meets a, b, c, d again at A, B, C, D,
   prove that the following assertions are equivalent:

(a) (a, b, c, d) is a harmonic pencil.

(b) $AB \cap CD \neq \emptyset$ and $\frac{\overline{AC}}{\overline{CB}} = \frac{\overline{AD}}{\overline{DB}}$.

2. Let (a, b, c, d) be a harmonic pencil of center O, and let O_1 be a point outside of a, b, c, d and such that $\overleftrightarrow{O_1 O}$ is not perpendicular to any of these lines. If a_1, b_1, c_1, d_1 are the perpendiculars dropped from O_1 to a, b, c, d, respectively, show that (a_1, b_1, c_1, d_1) is also a harmonic pencil.

3. We are given three collinear points A, B and C, such that C is not the midpoint of AB. Using only a straightedge, construct the harmonic conjugate D of C with respect to AB.

9.3 Polarity

In this section we develop the important concept of *polar* of a point with respect to a circle, along with some of its most important properties. As we shall see, *polarity* is intimately related to inversion and harmonic conjugation, and will give rise to some interesting applications.

Definition 9.20 Given a circle Γ of center O and a point $P \neq O$, the **polar of P with respect to** Γ is the straightline perpendicular to $\overleftrightarrow{OP}$ and passing through the inverse of P with respect to the inversion I of center O and inversion circle Γ.

In the notations of the previous definition, we shall systematically denote the polar of a point by the corresponding lowercase Latin letter. Figure 9.20 shows a point P lying outside the disk bounded by Γ, together with its polar p.

Also with respect to the previous definition, since all points of Γ are fixed by I, if $Q \in \Gamma$ then its polar q is the tangent to Γ passing through Q.

Fig. 9.20 A point P and its polar p

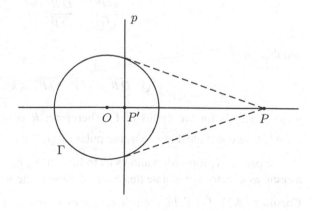

Note also that, since $I(P') = P$, the polar of P' is the line p' passing through P and perpendicular to $\overleftrightarrow{OP'} = \overleftrightarrow{OP}$; in particular, $p \parallel p'$. The following result sharpens this remark.

Proposition 9.21 *Let Γ be a circle of center O, and P, Q be points distinct from O. Then, $Q \in p \Leftrightarrow P \in q$.*

Proof For the sake of simplicity, we take P outside the disk bounded by Γ (the other cases can be dealt with analogously).

Take a point $Q \in p$. If $Q = P'$ or Q is one of the points of contact of the tangents drawn from P to Γ, there is nothing to do (check this!). Then, we can assume that Q is distinct from these three points. In the figure below, let r be the perpendicular dropped from P to $\overleftrightarrow{OQ}$, and R be the intersection point of r and $\overleftrightarrow{OQ}$.

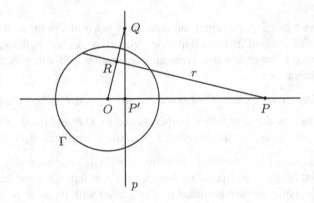

Since $Q\widehat{O}P' = P\widehat{O}R$ and $Q\widehat{P'}O = P\widehat{R}O$, we have $QP'O \sim PRO$. Hence,

$$\frac{\overline{OP'}}{\overline{OQ}} = \frac{\overline{OR}}{\overline{OP}},$$

and this gives

$$\overline{OQ} \cdot \overline{OR} = \overline{OP} \cdot \overline{OP'} = k^2,$$

where k stands for the radius of Γ. Therefore, R is the inverse of Q, and since $r \perp \overleftrightarrow{OQ}$ we conclude that $r = q$, the polar of Q. Thus, $P \in r \Rightarrow P \in q$. $\square$

The previous proposition allows us to look at the polar of a point with respect to a circle as a locus. We isolate this point of view in the following

Corollary 9.22 *Let Γ be a circle of center O and P be a point distinct from O, with polar p with respect to Γ. The locus of the points Q in the plane such that the*

tangents drawn from Q to Γ have tangency points collinear with P is the portion of p lying outside the disk bounded by Γ.

Proof As in the proof of the proposition, we only look at the case in which P lies outside the disk bounded by Γ, the other case being entirely analogous.

If $Q \in p$ lies outside the disk bounded by Γ (see figure below), then the previous proposition assures that $P \in q$, the polar of Q. However, by construction of the inverse of Q with respect to the inversion of center O and inversion circle Γ, if $q \cap \Gamma = \{T_1, T_2\}$ then $q = \overleftrightarrow{T_1 T_2}$. Therefore, P, T_1 and T_2 are collinear.

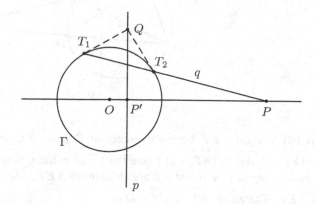

Conversely, let Q be a point outside the disk bounded by Γ, and $T_1, T_2 \in \Gamma$ be such that $\overleftrightarrow{QT_1}$ and $\overleftrightarrow{QT_2}$ are tangent to Γ. If T_1, T_2 and P are collinear, then $P \in \overleftrightarrow{T_1 T_2} = q$, the polar of Q with respect to Γ. Therefore, by the previous proposition, $Q \in p$. $\qquad \square$

We already know that a complete quadrangle gives rise to a harmonic pencil. In turn, the coming result relates complete quadrangles to polarity.

Theorem 9.23 *Let ABCD be a cyclic quadrilateral with circumcircle Γ, whose diagonals AC and BD meet at R. If the support lines of the pairs of opposite sides of ABCD meet at the points P and Q, then $\overleftrightarrow{QR}$ is the polar of P and $\overleftrightarrow{PR}$ is the polar of Q with respect to Γ.*

Proof First of all (and in the notations of Fig. 9.21), even if $\overleftrightarrow{QR}$ is not the polar of P with respect to Γ, Theorem 9.19 assures that $(\overleftrightarrow{QC}, \overleftrightarrow{QD}, \overleftrightarrow{QR}, \overleftrightarrow{QP})$ is a harmonic pencil. Therefore, from Theorem 9.16 we conclude that (A, B, S, P) and (C, D, T, P) are harmonic quadruples, so that S and T are the harmonic conjugates of P with respect to A, B and C, D, respectively. It thus suffices to show that $\overleftrightarrow{ST}$ is the polar of P with respect to Γ.

Fig. 9.21 Polarity and
complete quadrangles

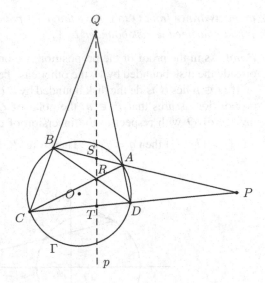

For what is left to do, let EF be the diameter of Γ passing through P, let $\overleftrightarrow{BE} \cap \overleftrightarrow{AF} = \{X\}$ and $\overleftrightarrow{AE} \cap \overleftrightarrow{BF} = \{Y\}$ (see the figure below). Since $E\widehat{A}F = E\widehat{B}F = 90°$, the line segments EA and FB are altitudes of XEF, whence Y is the orthocenter of XEF. Therefore, $\overleftrightarrow{XY} \perp \overleftrightarrow{EF} = \overleftrightarrow{OP}$.

On the other hand, Theorem 9.19 guarantees that $\left(\overleftrightarrow{XE}, \overleftrightarrow{XF}, \overleftrightarrow{XY}, \overleftrightarrow{XP} \right)$ is a harmonic pencil. Thus, Theorem 9.16 guarantees that the point of intersection of $\overleftrightarrow{XY}$ and $\overleftrightarrow{AB}$ is the harmonic conjugate of P with respect to A, B, whence it is S. By the same token, the intersection of $\overleftrightarrow{XY}$ and $\overleftrightarrow{EF}$ is the harmonic conjugate of P with respect to E, F. Lemma 9.13 thus shows that it is exactly the inverse P' of P with respect to Γ.

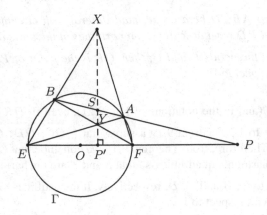

So far, our reasoning has shown that $\overleftrightarrow{XY}$ is the polar p of P with respect to Γ, and that it passes through the harmonic conjugate S of P with respect to A, B. If we now argue exactly as above with $CDFE$, we conclude that p also passes through the harmonic conjugate T of P with respect to C, D. However, $S, T \in p$ implies $p = \overleftrightarrow{ST}$, as wished. □

The theorem also holds, with suitable modifications, if $\overleftrightarrow{AD} \| \overleftrightarrow{BC}$, for example. See Problem 2. We now list two important corollaries of it.

Corollary 9.24 *Let Γ be a given circle and P be a point distinct from the center of Γ and with polar P with respect to Γ. If a line r passing through P intersects Γ and p at points A, B, S, then S is the harmonic conjugate of P with respect to AB.*

Proof In the notations of the proof of the theorem, this follows immediately from the fact that $(\overleftrightarrow{QA}, \overleftrightarrow{QB}, \overleftrightarrow{QS}, \overleftrightarrow{QP})$ is a harmonic pencil. □

The following consequence of the theorem above is due to Pierre Brocard, French mathematician of the nineteenth century, and is usually referred to as **Brocard's theorem**.

Corollary 9.25 (Brocard) *$ABCD$ is a cyclic quadrilateral with circumcenter O, whose diagonals AC and BD meet at the point R. If the support lines of the pairs of opposite sides of $ABCD$ meet at the points P and Q, then R is the orthocenter of OPQ. In particular, $\overleftrightarrow{OR} \perp \overleftrightarrow{PQ}$.*

Proof If Γ is the circumcircle of $ABCD$ (cf. Fig. 9.22), then Theorem 9.23 assures that $\overleftrightarrow{QR}$ is the polar of P with respect to Γ, so that $\overleftrightarrow{QR} \perp \overleftrightarrow{OP}$. Likewise,

Fig. 9.22 R is the orthocenter of OPQ

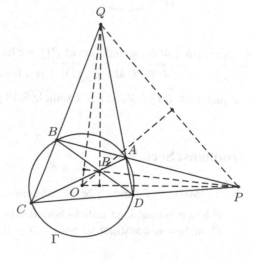

$\overleftrightarrow{PR} \perp \overleftrightarrow{OQ}$, so that R lies in two altitudes of triangle OPQ, hence is its orthocenter. Finally, since the three altitudes of a triangle are always concurrent, we obtain the last claim.   □

We finish this section by presenting a beautiful application of Theorem 9.23, known in mathematical literature as the **butterfly theorem**.

Example 9.26 PQ is a chord of a circle Γ, with midpoint M. We draw two other chords AC and BD of Γ, both passing through M and such that A and B lie in the same side of $\overleftrightarrow{PQ}$. If $AD \cap MP = \{R\}$ and $BC \cap MQ = \{S\}$, show that M is also the midpoint of RS.

Proof We can assume, without loss of generality, that M is not the center of Γ and that $\overleftrightarrow{AB}$ is not parallel to $\overleftrightarrow{PQ}$. (Check these claims!) This way, and letting $\overleftrightarrow{AD} \cap \overleftrightarrow{BC} = \{U\}$ and $\overleftrightarrow{AB} \cap \overleftrightarrow{CD} = \{V\}$ (cf. figure below), Theorem 9.23 guarantees that $\overleftrightarrow{UV}$ is the polar of M with respect to Γ. Therefore, if O is the center of Γ, then $\overleftrightarrow{OM} \perp \overleftrightarrow{UV}$.

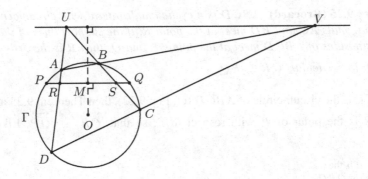

Now, since M is the midpoint of PQ, we have $\overleftrightarrow{OM} \perp \overleftrightarrow{PQ}$, whence $\overleftrightarrow{RS} = \overleftrightarrow{PQ} \parallel \overleftrightarrow{UV}$. Hence, $\left(\overleftrightarrow{UD}, \overleftrightarrow{UM} \, \overleftrightarrow{UC}, \overleftrightarrow{UV} \right)$ is a harmonic pencil and $R \in \overleftrightarrow{UD}$, $S \in \overleftrightarrow{UC}$ are such that $\overleftrightarrow{RS} \parallel \overleftrightarrow{UV}$, so that Example 9.18 gives $\overleftrightarrow{RM} = \overleftrightarrow{SM}$.   □

Problems: Sect. 9.3

1. (OBMU) Let Γ_1 and Γ_2 be circles centered at O_1 and O_2, respectively, and let P be a point outside the disks bounded by Γ_1 and Γ_2, and such that $P \notin \overleftrightarrow{O_1 O_2}$. Show how to construct all points Q in the plane with the following property:

for $i = 1, 2$, if C_i, $D_i \in \Gamma_i$ are such that $\overleftrightarrow{QC_i}$ and $\overleftrightarrow{QD_i}$ are tangent to Γ_i, with $C_i \neq D_i$, then both $\overleftrightarrow{C_1 D_1}$ and $\overleftrightarrow{C_2 D_2}$ pass through P.

2. State and prove a version of Theorem 9.23 in case $\overleftrightarrow{AD} \| \overleftrightarrow{BC}$.

3. We are given a circle Γ and a point P, outside the disk bounded by Γ. Show how to draw the tangents to Γ passing through P, using only a straightedge.

4. ABC is an acute triangle with $\overline{AB} \neq \overline{AC}$. The circle Γ, of diameter BC, meets the support line of the altitude dropped from A at points X and Y. Show that the tangents to Γ drawn from X and Y meet at an excenter of the orthic triangle of ABC.

5. In an acute triangle ABC, the circle Σ of diameter BC has orthocenter H. If P and Q are the feet of the tangents drawn from A to Σ, show that P, Q and H are collinear.

6. The incircle of a scalene triangle ABC touches sides AB, AC, BC respectively at points F, E, D. The straightlines $\overleftrightarrow{AD}$ and $\overleftrightarrow{BE}$ meet at P, whereas the straightlines $\overleftrightarrow{AC}$ and $\overleftrightarrow{DF}$ meet at M and the straightlines $\overleftrightarrow{BC}$ and $\overleftrightarrow{EF}$ meet at N. If I stands for the incenter of ABC, show that $\overleftrightarrow{IP} \perp \overleftrightarrow{MN}$.

7. In a cyclic quadrilateral $ABCD$ which is not a rectangle, $\angle A = 90°$. Show how to use compass and straightedge to construct it, knowing the positions of the following three points: the circumcenter O, the point R of intersection of its diagonals AC, BD and the point Q of intersection of $\overleftrightarrow{AB}$ and $\overleftrightarrow{CD}$.

8. We are given a quadrilateral $MNPQ$, inscribed in the circle Γ and having no right angles. If $\overleftrightarrow{MQ} \cap \overleftrightarrow{NP} = \{U\}$, $\overleftrightarrow{MN} \cap \overleftrightarrow{PQ} = \{V\}$ and the tangents to Γ drawn through M and P (resp. N and Q) meet at W (resp. at X), show that U, V, W, X are collinear.

9. Let $ABCD$ be a tangential quadrilateral and M, N, P, Q be the points where sides AB, BC, CD, AD meet the inscribed circle. Prove that $\overleftrightarrow{AC}$, $\overleftrightarrow{BD}$, $\overleftrightarrow{MP}$, $\overleftrightarrow{NQ}$ are concurrent.

10. (IMO) We are given a triangle ABC and a circle γ of center O, passing through A and C and meeting AB and BC again at points K and N, respectively, with $K \neq A$ and $N \neq C$. The circumcircles of triangles ABC and BKN meet again at M, with $M \neq B$. Show that $O\widehat{M}B = 90°$.

9.4 Cross Ratio and Perspectivities

This section concludes our first walk through projective techniques in Plane Euclidean Geometry by generalizing harmonic quadruples and pencils, as well as introducing a change in our point of view by discussing *perspectivities*. Later, in Sect. 10.4, the results of this section and the previous one will be extended to conics.

We begin by recalling that if (A, B, C, D) is a harmonic quadruple, then $\frac{AC}{CB} = -\frac{AD}{DB}$, a relation which can also be written as $\frac{AC}{CB} \cdot \frac{DB}{AD} = -1$. Also, if (a, b, c, d) is a harmonic pencil and $A \in a$, $B \in b$, $C \in c$, $D \in d$ are pairwise distinct and collinear, then (A, B, C, D) is a harmonic quadruple, so that $\frac{AC}{CB} \cdot \frac{DB}{AD} = -1$.

The discussion above shows that the quantity at the left hand side of this last equality is preserved in harmonic pencils, and we shall now show that it is preserved in a more general setting. To this end, it is convenient to first state the following

Definition 9.27 Given pairwise distinct collinear points A, B, C, D, the **cross ratio** of the oriented segment AB with respect to CD (in this order) is the real number $(A, B; C, D)$ given by

$$(A, B; C, D) = \frac{AC}{CB} \cdot \frac{DB}{AD}.$$

In view of this definition, our previous discussion can be summarized by saying that, for distinct collinear points A, B, C, D,

$$(A, B, C, D) \text{ is a harmonic quadruple} \Leftrightarrow (A, B; C, D) = -1.$$

The coming result can be seen as the analogue of Theorem 9.16 for general cross ratios.

Theorem 9.28 *Let a, b, c, d be pairwise distinct lines, concurrent at a point O. If lines r_1 and r_2 do not pass through O and intersect a, b, c, d respectively at the quadruples of points A_1, B_1, C_1, D_1 and A_2, B_2, C_2, D_2, then*

$$(A_1, B_1; C_1, D_1) = (A_2, B_2; C_2, D_2).$$

Proof Let a line r, not passing through O, meet a, b, c, d at A, B, C, D, respectively. Moreover, assume that the configuration is as displayed in the figure below (the other cases can be dealt with analogously).

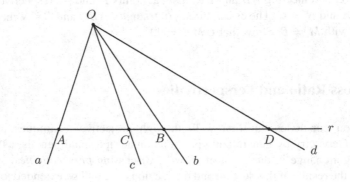

We apply the sine law to the triangles OAC, OCB, OAD, ODB to get

$$\frac{\overline{AC}}{\overline{OC}} = \frac{\sin A\widehat{O}C}{\sin O\widehat{A}C}, \quad \frac{\overline{CB}}{\overline{OC}} = \frac{\sin C\widehat{O}B}{\sin O\widehat{B}C}, \quad \frac{\overline{AD}}{\overline{OD}} = \frac{\sin A\widehat{O}D}{\sin O\widehat{A}D}, \quad \frac{\overline{DB}}{\overline{OD}} = \frac{\sin D\widehat{O}B}{\sin O\widehat{B}D}.$$

Therefore,

$$\frac{\overline{AC}}{\overline{CB}} = \frac{\sin A\widehat{O}C \cdot \sin O\widehat{B}C}{\sin O\widehat{A}C \cdot \sin C\widehat{O}B}, \quad \frac{\overline{DB}}{\overline{AD}} = \frac{\sin D\widehat{O}B \cdot \sin O\widehat{A}D}{\sin O\widehat{B}D \cdot \sin A\widehat{O}D},$$

and taking into account that $\sin O\widehat{B}C = \sin O\widehat{B}D$ and $\sin O\widehat{A}C = \sin O\widehat{A}D$, we obtain

$$\frac{\overline{AC}}{\overline{CB}} \cdot \frac{\overline{DB}}{\overline{AD}} = \frac{\sin A\widehat{O}C}{\sin C\widehat{O}B} \cdot \frac{\sin D\widehat{O}B}{\sin A\widehat{O}D}. \tag{9.5}$$

Now, by the choice of A_i, B_i, C_i, D_i we clearly have $A_1\widehat{O}C_1 = A_2\widehat{O}C_2$, $C_1\widehat{O}B_1 = C_2\widehat{O}B_2$, $D_1\widehat{O}B_1 = D_2\widehat{O}B_2$ and $A_1\widehat{O}D_1 = A_2\widehat{O}D_2$. Therefore, (9.5) applied to A_1, B_1, C_1, D_1 and A_2, B_2, C_2, D_2 gives

$$\frac{\overline{A_1C_1}}{\overline{C_1B_1}} \cdot \frac{\overline{D_1B_1}}{\overline{A_1D_1}} = \frac{\overline{A_2C_2}}{\overline{C_2B_2}} \cdot \frac{\overline{D_2B_2}}{\overline{A_2D_2}}.$$

We are then left to showing that $\frac{A_1C_1}{C_1B_1} \cdot \frac{D_1B_1}{A_1D_1}$ and $\frac{A_2C_2}{C_2B_2} \cdot \frac{D_2B_2}{A_2D_2}$ have equal signs, which follows immediately from the fact that A_i, B_i, C_i, D_i are collinear, with $A_i \in a$, $B_i \in b, C_i \in c, D_i \in d$. □

For what follows, it will be convenient to say that four pairwise distinct lines a, b, c, d, concurrent at a point O, define a **projective pencil** (a, b, c, d) of **center** O. Moreover, if $A \in a, B \in b, C \in c, D \in d$ are pairwise distinct and collinear, we say that the quadruple (A, B, C, D) of points is **incident** with the projective pencil (a, b, c, d), or vice-versa.

If two quadruples of points (A_1, B_1, C_1, D_1) and (A_2, B_2, C_2, D_2) are incident with the projective pencil (a, b, c, d) of center O, we say that (A_1, B_1, C_1, D_1) and (A_2, B_2, C_2, D_2) are **in perspective from** O, and denote

$$A_1B_1C_1D_1 \overset{O}{\barwedge} A_2B_2C_2D_2.$$

Thus, the result of the previous theorem can be stated in symbols by writing (see also Fig. 9.23)

$$A_1B_1C_1D_1 \overset{O}{\barwedge} A_2B_2C_2D_2 \Rightarrow (A_1, B_1; C_1, D_1) = (A_2, B_2; C_2, D_2). \tag{9.6}$$

We now extend the notion of cross ratio to projective pencils.

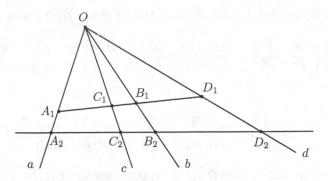

Fig. 9.23 $A_1B_1C_1D_1 \overset{O}{\barwedge} A_2B_2C_2D_2$

Definition 9.29 Given a projective pencil (a, b, c, d), the **cross ratio** of (a, b) with respect to (c, d) is the cross ratio $(A, B; C, D)$, where (A, B, C, D) is any quadruple of points incident with (a, b, c, d).

The consistency of the above definition follows at once from the discussion that led to (9.6). Thus, if $(a, b; c, d)$ stands for the cross ratio of (a, b) with respect to (c, d), then

$$(a, b; c, d) = (A, B; C, D) \text{ if } (A, B, C, D) \text{ is incident with } (a, b, c, d). \quad (9.7)$$

This being said, we can now state and prove two important consequences of Theorem 9.28. For the first of them, let be given two projective pencils (a_1, b_1, c_1, d_1), (a_2, b_2, c_2, d_2) and a line r not passing through any of their centers. We say that the two pencils are **in perspective from the line r** if they are incident to the same quadruple of points of r, i.e., if $a_1 \cap a_2, b_1 \cap b_2, c_1 \cap c_2, d_1 \cap d_2 \in r$ (cf. Fig. 9.24). Also, we denote this in symbols by writing

$$a_1b_1c_1d_1 \overset{r}{\barwedge} a_2b_2c_2d_2.$$

Corollary 9.30 $a_1b_1c_1d_1 \overset{r}{\barwedge} a_2b_2c_2d_2 \Rightarrow (a_1, b_1; c_1, d_1) = (a_2, b_2; c_2, d_2).$

Proof This is immediate from (9.7). Indeed, in the notations of Fig. 9.24, the fact that (A, B, C, D) is incident to both (a_1, b_1, c_1, d_1) and (a_2, b_2, c_2, d_2) implies

$$(a_1, b_1; c_1, d_1) = (A, B; C, D) = (a_2, b_2; c_2, d_2).$$

$\square$

The second consequence of Theorem 9.28 is the analogue of Corollary 9.17 for general cross ratios.

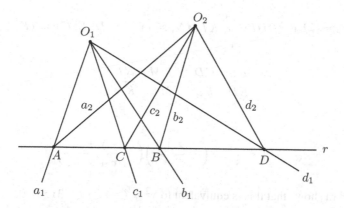

Fig. 9.24 $a_1 b_1 c_1 d_1 \overset{r}{\barwedge} a_2 b_2 c_2 d_2$

Corollary 9.31 *If* (a, b, c, d_1) *and* (a, b, c, d_2) *are projective pencils for which* $(a, b; c, d_1) = (a, b; c, d_2)$, *then* $d_1 = d_2$.

Proof Firstly, note that the two projective pencils have the same center O. Now, draw a line r, not passing through O and intersecting a, b, c, d_1, d_2 respectively at A, B, C, D_1, D_2. Since (A, B, C, D_1) is incident with $(a, b; c, d_1)$, we have $(a, b; c, d_1) = (A, B; C, D_1)$; analogously, $(a, b; c, d_2) = (A, B; C, D_2)$. Therefore, $(a, b; c, d_1) = (a, b; c, d_2)$ implies $(A, B; C, D_1) = (A, B; C, D_2)$, and hence (cf. Problem 1) $D_1 = D_2$. Thus, $d_1 = \overleftrightarrow{OD_1} = \overleftrightarrow{OD_2} = d_2$. □

It is time for us to present an example justifying the developments above.

Example 9.32 Given a triangle ABC, let $D, E \in BC$ be distinct from B, C and such that $\overline{BD} = \overline{DE} = \overline{EC}$. If a line r intersects the line segments AB, AD, AE, AC respectively at points K, L, M, N, show that $\overline{KN} \geq 3\overline{LM}$, with equality if and only if $\overleftrightarrow{KN} \parallel \overleftrightarrow{BC}$.

Proof Letting $x = \overline{KL}$, $y = \overline{LM}$, $z = \overline{MN}$ (see the figure below), we want to show that $x + y + z \geq 3y$ or, which is the same, $x + z \geq 2y$.

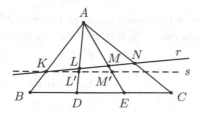

We obviously have $BDEC \overset{A}{\barwedge} KLMN$, so that $(B, D; E, C) = (K, L; M, N)$. This gives

$$\frac{BE}{ED} \cdot \frac{CD}{BC} = \frac{KM}{ML} \cdot \frac{NL}{KN},$$

and hence

$$(-2) \cdot \left(-\frac{2}{3}\right) = \left(-\frac{x+y}{y}\right)\left(-\frac{y+z}{x+y+z}\right).$$

Some algebra shows that this is equivalent to $y^2 + (x+z)y - 3xz = 0$.

Letting $f(y) = y^2 + (x+z)y - 3xz$, in order to get $y \leq \frac{x+z}{2}$ the elementary theory of quadratic functions (cf. [5], for instance) assures that it suffices to show that $f\left(\frac{x+z}{2}\right) \geq 0$. To this end, a simple computation gives

$$f\left(\frac{x+z}{2}\right) = \frac{3}{4}(x-z)^2 \geq 0,$$

so that $y = \frac{x+z}{2}$ if and only if $y = \frac{x+z}{2}$ and $x = z$, i.e., if and only if $x = y = z$.

Now, assume we have equality but r is not parallel to $\overleftrightarrow{BC}$, and let s be the parallel to $\overleftrightarrow{BC}$ through K, and L', M' be its intersections with $\overleftrightarrow{AD}$ and $\overleftrightarrow{AE}$ (see the figure above). Since $\overline{KL} = \overline{LM}$ and (from Thales' theorem) $\overline{KL'} = \overline{L'M'}$, we have $KLL' \sim KMM'$ by SAS. Hence, $L\widehat{L'}K = M\widehat{M'}K$, and this forces $\overleftrightarrow{LL'} \parallel \overleftrightarrow{MM'}$, which is an absurd. $\square$

Back to the development of the theory, we now extend the notion of perspectivity of projective pencils to circles. To be precise, given projective pencils (a_1, b_1, c_1, d_1), (a_2, b_2, c_2, d_2) of centers O_1, O_2, we say that they are **in perspective from a circle** Γ if the six points O_1, O_2, $a_1 \cap a_2$, $b_1 \cap b_2$, $c_1 \cap c_2$, $d_1 \cap d_2$ are pairwise distinct and lie in Γ. Also, in this case we write

$$a_1 b_1 c_1 d_1 \overset{\Gamma}{\barwedge} a_2 b_2 c_2 d_2.$$

In what comes next, we shall show that projective pencils in perspective from a circle also have equal cross ratios. To this end, start by noticing that (9.5) and (9.7) readily give

$$|(a, b; c, d)| = \frac{\sin \widehat{ac}}{\sin \widehat{cb}} \cdot \frac{\sin \widehat{db}}{\sin \widehat{ad}}, \tag{9.8}$$

Fig. 9.25

$$a_1b_1c_1d_1 \stackrel{\Gamma}{\overline{\wedge}} a_2b_2c_2d_2$$

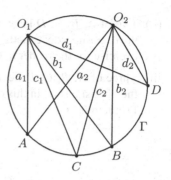

where $\widehat{ac}$ denotes the angle formed by lines a and c, and likewise for $\widehat{cb}$, $\widehat{db}$, $\widehat{ad}$ (note that, since $\sin\theta = \sin(\pi - \theta)$, it does not matter whether we choose $\widehat{ac} = \theta$ or $\pi - \theta$, and accordingly for the other angles).

Theorem 9.33 $a_1b_1c_1d_1 \stackrel{\Gamma}{\overline{\wedge}} a_2b_2c_2d_2 \Rightarrow (a_1, b_1; c_1, d_1) = (a_2, b_2; c_2, d_2).$

Proof Relation (9.8) gives

$$\left|(a_1, b_1; c_1, d_1)\right| = \frac{\sin \widehat{a_1c_1}}{\sin \widehat{c_1b_1}} \cdot \frac{\sin \widehat{d_1b_1}}{\sin \widehat{a_1d_1}}, \quad \left|(a_2, b_2; c_2, d_2)\right| = \frac{\sin \widehat{a_2c_2}}{\sin \widehat{c_2b_2}} \cdot \frac{\sin \widehat{d_2b_2}}{\sin \widehat{a_2d_2}}.$$

Now, let $a_1 \cap a_2 = \{A\}$, $b_1 \cap b_2 = \{B\}$, $c_1 \cap c_2 = \{C\}$, $d_1 \cap d_2 = \{D\}$ (see Fig. 9.25). The inscribed angle theorem assures that if A and C lie at the same side of $\overleftrightarrow{O_1O_2}$ (as in Fig. 9.25), then $\widehat{a_1c_1} = \widehat{a_2c_2}$. Analogously, if A and C lie at opposite sides of $\overleftrightarrow{O_1O_2}$, then $\widehat{a_1c_1} = \pi - \widehat{a_2c_2}$ (draw a picture to convince yourself). In any case, $\sin \widehat{a_1c_1} = \sin \widehat{a_2c_2}$. Since a similar reasoning holds for the remaining angles involved, we conclude from the above that

$$\left|(a_1, b_1; c_1, d_1)\right| = \left|(a_2, b_2; c_2, d_2)\right|.$$

We are left to showing that $(a_1, b_1; c_1, d_1)$ and $(a_2, b_2; c_2, d_2)$ have equal signs, for which we distinguish the following cases:

(i) A, B, C, D lie in a single side of $\overleftrightarrow{O_1O_2}$: this is the situation depicted in Fig. 9.25. It suffices to draw a line r, incident to both (a_1, b_1, c_1, d_1) and (a_2, b_2, c_2, d_2) at the quadruples (A_1, B_1, C_1, D_1) and (A_2, B_2, C_2, D_2), respectively, and to check that the cross ratios $(A_1, B_1; C_1, D_1)$ and $(A_2, B_2; C_2, D_2)$ have equal signs.
In the figure above, the eight points A_1, B_1, C_1, D_1, A_2, B_2, C_2, D_2 are highlighted, but for the sake of clarity of the figure the corresponding labels are omitted.

(ii) Two of A, B, C, D lie at a single side of $\overleftrightarrow{O_1 O_2}$, and the other two points lie at the opposite side: exactly the same method of item (i) works. We leave it as an exercise for the reader to draw a picture and finish the proof (see Problem 6).

(iii) Three of A, B, C, D lie at a single side of $\overleftrightarrow{O_1 O_2}$, and the fourth point lies at the opposite side: in this case, we can also reason as in the previous cases (see Problem 6).

$\square$

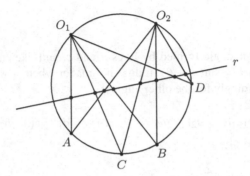

The apparatus above allows us to prove two famous results on Euclidean Geometry, due to Pascal[7] and Brianchon,[8] and we end this section by presenting the usual *projective proofs* of them.

For what comes, we make the following conventions: given pairwise distinct points A, B, C, D, E, F, the **hexagram** $ABCDEF$ is the polygonal line formed by the concatenation of the line segments AB, BC, CD, DE, EF, FA. In particular, two or more of these segments may cross each other, and some three of the points A, B, C, D, E, F may be collinear. The hexagram $ABCDEF$ is **cyclic** if A, B, C, D, E, F are concyclic points, and **tangential** if $\overleftrightarrow{AB}, \overleftrightarrow{BC}, \overleftrightarrow{CD}, \overleftrightarrow{DE}, \overleftrightarrow{EF}, \overleftrightarrow{FA}$ are tangent to a single circle. Finally, in the first case we say that the circle is **circumscribed** to the hexagram, whereas in the second case we say it is **inscribed** in the hexagram (Fig. 9.26).

Theorem 9.34 (Pascal) *Let $ABCDEF$ be a cyclic hexagram. If $\overleftrightarrow{AE} \cap \overleftrightarrow{BD} = \{Z\}$, $\overleftrightarrow{AF} \cap \overleftrightarrow{CD} = \{Y\}$, $\overleftrightarrow{BF} \cap \overleftrightarrow{CE} = \{X\}$, then points X, Y, Z are collinear.*

---

[7]Blaise Pascal, French inventor, mathematician and physicist of the seventeenth century. Apart from his contributions to Mathematics, his work in Physics is recognized through the SI unit for pressure, the *Pascal*. We also owe to Pascal the construction of the first mechanical calculator.
[8]Charles Brianchon, French mathematician of the nineteenth century.

Fig. 9.26 Pascal's theorem

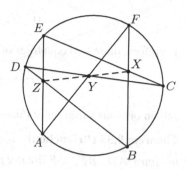

Proof Let Γ denote the circle circumscribed to $ABCDEF$ and (cf. figure below) U, V, Y' be defined by $\overleftrightarrow{AE} \cap \overleftrightarrow{CD} = \{U\}$, $\overleftrightarrow{BD} \cap \overleftrightarrow{CE} = \{F\}$, $\overleftrightarrow{XZ} \cap \overleftrightarrow{CD} = \{Y'\}$. It suffices to show that $Y' = Y$. To this end, start by observing that

$$(\overleftrightarrow{AC}, \overleftrightarrow{AY}, \overleftrightarrow{AU}, \overleftrightarrow{AD}) = (\overleftrightarrow{AC}, \overleftrightarrow{AF}, \overleftrightarrow{AE}, \overleftrightarrow{AD})$$

$$\stackrel{\Gamma}{\overline{\wedge}} (\overleftrightarrow{BC}, \overleftrightarrow{BF}, \overleftrightarrow{BE}, \overleftrightarrow{BD})$$

$$= (\overleftrightarrow{BC}, \overleftrightarrow{BX}; \overleftrightarrow{BE}, \overleftrightarrow{BV}).$$

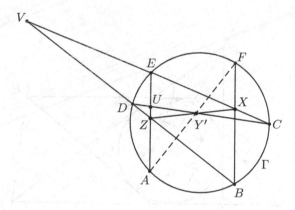

Hence, the definition of cross ratio for projective pencils, together with the previous theorem, give

$$(C, Y; U, D) = (\overleftrightarrow{AC}, \overleftrightarrow{AY}; \overleftrightarrow{AU}, \overleftrightarrow{AD}) = (\overleftrightarrow{BC}, \overleftrightarrow{BX}; \overleftrightarrow{BE}, \overleftrightarrow{BV}) = (C, X; E, V).$$

However, since $CXEV \stackrel{Z}{\overline{\wedge}} CY'UD$, we conclude from (9.6) that

$$(C, X; E, V) = (C, Y'; U, D).   \tag{9.9}$$

Therefore, these two equalities of cross ratios combine to give

$$(C, Y; U, D) = (C, Y'; U, D),$$

and an obvious variation of Problem 1 shows that $Y = Y'$.   □

Theorem 9.35 (Brianchon)   *If $ABCDEF$ is a tangential hexagram and two of the lines $\overleftrightarrow{AD}$, $\overleftrightarrow{BE}$, $\overleftrightarrow{CF}$ are not parallel, then they all concur at a single point.*

Proof   In the notations of Fig. 9.27 assume, without loss of generality, that $\overleftrightarrow{AD}$ and $\overleftrightarrow{BE}$ are not parallel, and let P be their intersection point. We shall show that $\overleftrightarrow{CF}$ also passes through P.

Let Γ be the circle inscribed in $ABCDEF$ and R, S, T, U, V, W be the points where $\overleftrightarrow{AB}$, $\overleftrightarrow{BC}$, $\overleftrightarrow{CD}$, $\overleftrightarrow{DE}$, $\overleftrightarrow{EF}$, $\overleftrightarrow{AF}$ touch Γ, respectively. It is immediate to see that $\overleftrightarrow{RW}$ and $\overleftrightarrow{TU}$ are the polars of A and D with respect to Γ. Therefore, if $\overleftrightarrow{RW}$ $\cap$ $\overleftrightarrow{TU} = \{X\}$, Proposition 9.21 assures that the polar of X with respect to Γ passes through A and D, hence is precisely $\overleftrightarrow{AD}$. Analogously, letting $\overleftrightarrow{RS} \cap \overleftrightarrow{UV} = \{Y\}$ and noting that $\overleftrightarrow{RS}$ and $\overleftrightarrow{UV}$ are the polars of B and E, respectively, we conclude

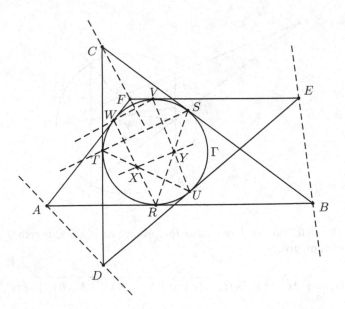

Fig. 9.27   Brianchon's theorem

that $\overleftrightarrow{BE}$ is the polar of Y. Therefore, since P is the intersection point of the polars of X and Y with respect to Γ, we conclude that $\overleftrightarrow{XY}$ is the polar of P.

We now look at the hexagram $RVTUSW$ and distinguish two possibilities:

(i) If $\overleftrightarrow{ST} \parallel \overleftrightarrow{VW}$, then $\overleftrightarrow{CF} \perp \overleftrightarrow{ST}$, so that it passes through the center O of Γ. In turn, Problem 9 assures that $\overleftrightarrow{XY} \parallel \overleftrightarrow{ST}$. Since $\overleftrightarrow{OP} \perp \overleftrightarrow{XY}$, we obtain $\overleftrightarrow{OP} \perp \overleftrightarrow{ST}$, hence $\overleftrightarrow{OP} = \overleftrightarrow{CF}$. In particular, $P \in \overleftrightarrow{CF}$.

(ii) If $\overleftrightarrow{ST} \not\parallel \overleftrightarrow{VW}$, let Z be their point of intersection. Pascal's theorem guarantees that $Z \in \overleftrightarrow{XY}$, which is the polar of P. Thus, invoking Proposition 9.21 again, we obtain that P lies at the polar of Z. However, arguing as above, we conclude that such a polar is precisely $\overleftrightarrow{CF}$.

$\square$

We finish this section with a few remarks on the **duality principle** (which, nevertheless, we shall not state formally).

In the language of Projective Geometry, the theorems of Pascal and Brianchon are prototypes of **dual theorems**. As the reader may suspect, this expression must have something to do with the fact that Pascal's theorem refers to a *cyclic* hexagram and to the *collinearity* of the intersection points of three pairs of diagonals, whereas Brianchon's theorem refers to a *tangential* hexagram and to the *concurrence* of the diagonals determined by three pairs of points.

In order to come a little closer to a precise definition of such a principle, let us rephrase both theorems as below (assuming, for the moment, that all points defined by the intersection of two lines actually exist, i.e., ignoring the fact that the lines could be parallel):

(1) (Pascal's theorem). Let $A_1, A_2, A_3, B_1, B_2, B_3$ be points lying in a circle. Then, the three intersection points of the pairs of lines $\overleftrightarrow{A_1 B_2}$ and $\overleftrightarrow{A_2 B_1}$, $\overleftrightarrow{A_1 B_3}$ and $\overleftrightarrow{A_3 B_1}$, $\overleftrightarrow{A_2 B_3}$ and $\overleftrightarrow{A_3 B_2}$ are collinear.

(2) (Brianchon's theorem). Let $r_1, r_2, r_3, s_1, s_2, s_3$ be lines tangent to a circle. Then, the three lines joining the pairs of points $r_1 \cap s_2$ and $r_2 \cap s_1$, $r_1 \cap s_3$ and $r_3 \cap s_1$, $r_2 \cap s_3$ and $r_3 \cap s_2$ are concurrent.

Although the statements above are somewhat cumbersome, they reveal more clearly the nature of duality: given a *projective theorem* (roughly speaking, a geometric result involving solely *incidence properties* of points, lines and circles— i.e., *passing through, intersecting at, being concurrent, being collinear, being tangent*), it can be shown that one can always manufacture another theorem, called its **dual**, simply by performing the following replacements of expressions: *points* by *lines*; *points lying a circle* by *lines tangent to a circle*; *intersection points of pairs of lines* by *lines joining pairs of points*; *collinearity* by *concurrence*.

Now, what if two lines defining a point were parallel? (As it happens in Problem 9.) In the realm of Euclidean Geometry one has always to deal with such a situation with a separate argument, and the corresponding conclusion will be somewhat different from the one would get if the lines were not parallel. Projective Geometry deals with such a situation by *avoiding parallel lines*. In other words, the *projective plane* extends the euclidean plane by adjoining a **point at infinity** corresponding to each direction in the euclidean plane. Therefore, in the projective setting, the theorems of Pascal and Brianchon are genuine dual results, and can be stated exactly as above.

For a thorough development of Projective Geometry which goes far beyond what we have presented here, and in particular for the proof of the duality principle, we refer the reader to [20].

The theorems of Pascal and Brianchon, as well as some other concepts studied in this chapter, will be extended to conics in Sect. 10.4.

Problems: Sect. 9.4

1. * If A, B, C, D_1, D_2 are pairwise distinct points for which $(A, B; C, D_1) = (A, B; C, D_2)$, show that $D_1 = D_2$.

     For the next two problems, given distinct and collinear points A, B, C, D, we say that AB **separates** CD if $C \in AB$ and $D \notin AB$ or vice-versa.

2. Establish the following properties of the cross ratio of four pairwise distinct collinear points A, B, C, D:

   (a) $(A, B; C, D) < 0$ if and only if AB separates CD.
   (b) $(A, B; C, D) = (B, A; D, C) = (C, D; A, B)$.
   (c) $(A, B; C, D) = 1 - (A, C; B, D)$.
   (d) $(A, B; C, D)(A, B; D, C) = 1$.

3. Let A, B, C, D be pairwise distinct collinear points, with AB separating CD. If $(A, B; C, D) = (B, A; C, D)$, show that (A, B, C, D) is a harmonic quadruple.

     For the next two problems, we say that a **projectivity** between two quadruples of collinear points is a finite number of perspectivities applied in succession.

4. Given a quadruple of collinear points (A, B, C, D), show that there exists a projectivity that applies it to (B, A, D, C) and explain how this relates to the first equality in item (b) of Problem 2.

5. The purpose of this problem is to give another proof of Theorem 9.19. To this end, let $ABCD$ be a quadrilateral in which $\overleftrightarrow{AC} \cap \overleftrightarrow{BD} = \{Q\}$, $\overleftrightarrow{AB} \cap \overleftrightarrow{CD} = \{P\}$ and $\overleftrightarrow{AD} \cap \overleftrightarrow{BC} = \{O\}$.

   (a) If $\overleftrightarrow{AB} \cap \overleftrightarrow{OQ} = \{R\}$, show that there exists a projectivity applying (A, B, P, R) to (B, A, P, R).
   (b) Show that $\left(\overleftrightarrow{OA}, \overleftrightarrow{OB}, \overleftrightarrow{OP}, \overleftrightarrow{OQ} \right)$ is a harmonic pencil.

For item (a), if $\overleftrightarrow{OQ} \cap \overleftrightarrow{CD} = \{S\}$, show that

6. * Complete the proof of Theorem 9.33 by examining cases (i) and (ii).

   The next problem revisits Problem 9, page 317, with the material of this section.

7. Let $ABCD$ be a tangential quadrilateral and M, N, P, Q be the points where sides AB, BC, CD, AD meet its inscribed circle. Prove that lines $\overleftrightarrow{AC}$, $\overleftrightarrow{BD}$, $\overleftrightarrow{MP}$, $\overleftrightarrow{NQ}$ are concurrent.

8. Prove Pappus' theorem (cf. Problem 12, page 137) with the methods developed in this section.

   The next problem establishes a degenerated version of Pascal's theorem.

9. Let $ABCDEF$ be a cyclic hexagram. If $\overleftrightarrow{AF} \cap \overleftrightarrow{CD} = \{Y\}$, $\overleftrightarrow{BF} \cap \overleftrightarrow{CE} = \{X\}$ and $\overleftrightarrow{AE} \parallel \overleftrightarrow{BD}$, prove that $\overleftrightarrow{XY} \parallel \overleftrightarrow{BD}$.

10. We are given a triangle ABC and a point D lying in the angle $\angle BAC$ but outside the triangular region ABC. Points P and Q are the feet of the perpendiculars dropped from D to $\overleftrightarrow{AB}$ and $\overleftrightarrow{AC}$, respectively, whereas points R and S are the feet of the perpendiculars dropped from A to $\overleftrightarrow{CD}$ and $\overleftrightarrow{BD}$, also respectively. Show that $\overleftrightarrow{BC}$, $\overleftrightarrow{PR}$ and $\overleftrightarrow{QS}$ concur.

Chapter 10
Basic Concepts in Solid Geometry

From an informal viewpoint, the main difference between the material of the previous chapters (i.e., Plane Euclidean Geometry), and that of this and the subsequent ones (i.e., *Solid* Euclidean Geometry) lies in the fact that we now have one more *dimension* to play with. In other words, instead of working in a single (*two-dimensional*) plane, we shall now have the entire (*three-dimensional*) space at our disposal.

As we have done so far, we shall generally guide our presentation through the use of the axiomatic method, thus deducing further results from previously stated axioms. Nevertheless, for the sake of simplicity and whenever there is no danger of confusion, in doing so we shall sometimes rely on the reader's intuition, thus not listing an exhaustive set of postulates from which one can build the theory in a strictly axiomatic way. In particular, whenever we state some claim without presenting a proof, the reader is warned to take it as an axiom. As a relevant instance, we postulate that the entire Euclidean Geometry developed in the previous chapters remains valid when we restrict the context to objects situated in some specific plane.

For a rigorous axiomatic presentation of the fundamentals of Solid Euclidean Geometry, we refer the interested reader to elegant book [19].

10.1 Introduction

As it happened in our study of Plane Euclidean Geometry (cf. Sect. 1.1), Solid Euclidean Geometry assumes the concepts of point, line and plane as *primitive* ones. It also takes as primitive the concept of (**euclidean**) **three-dimensional space** (hereafter simply referred to as *space*), which properly contains all planes with which we shall work (and contains at least one plane).

Also as before, points (resp. lines) will be denoted by uppercase (resp. lowercase) Latin letters. Moreover: given in space a point P and a line r, there are only two

© Springer International Publishing AG, part of Springer Nature 2018
A. Caminha Muniz Neto, *An Excursion through Elementary Mathematics, Volume II*,
Problem Books in Mathematics, https://doi.org/10.1007/978-3-319-77974-4_10

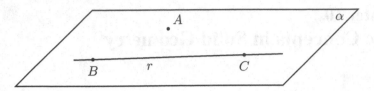

Fig. 10.1 Three non-collinear points determine a single plane

possibilities: $P \in r$ or $P \notin r$; two distinct points A and B in space determine a single line r passing through both of them, which will also be denoted by writing $r = \overleftrightarrow{AB}$ (cf. Fig. 1.3). We shall also assume the concepts of half-line and line segment to be known, exactly as discussed in Sect. 1.1.

A nonempty subset $\mathcal{C}$ of space is **convex** if, for all $A, B \in \mathcal{C}$, we have $AB \subset \mathcal{C}$. We assume that the complement of a plane α consists of two disjoint convex sets, which are called the **half-spaces** determined by α. Moreover, if points A and B belong to distinct half-spaces with respect to α, then we assume that AB always intersects α, and that the intersection is a single point.

As in Plane Geometry, three points lying in a single line will be called **collinear**. Three non-collinear points A, B and C determine a single plane, which will be denoted (ABC) (cf. Fig. 10.1). It is also common usage to denote planes by lowercase Greek letters; in this sense, we can write $\alpha = (ABC)$ to denote plane (ABC) of Fig. 10.1.

If a line r has two points in common with a plane α, then r will be contained in α, and we shall denote this by writing $r \subset \alpha$; in particular, all points of r will be points of α. In Fig. 10.1, we have $r \subset \alpha$, since $B, C \in r \cap \alpha$.

The discussion above guarantees that a line r and a point $A \notin r$ determine a single plane α, which contains r. Indeed, letting B and C be distinct points of r, we conclude that plane (ABC) contains both A and r; on the other hand, any plane containing both A and r will also contain A, B and C, hence will coincide with (ABC). Letting α be the plane determined by A and r, we denote $\alpha = (A, r)$.

Also as in Plane Geometry, two distinct lines r and s are said to be **concurrent** if they have a common point. Lines r and s are **coplanar** if they are contained in a single plane.

In the notations of the previous paragraph, suppose that r and s concur at A and let $B \in r \setminus \{A\}$, $C \in s \setminus \{A\}$. It is immediate to check that (ABC) is the only plane containing both r and s; in particular, we conclude that two concurrent lines are always coplanar.

Two lines r and s are **parallel** if they are coplanar and do not have common points. Nevertheless, a striking difference between Plane and Solid Geometry lies in the fact that parallelism is not the only situation in which two lines in space have no common points. The coming definition clears this point.

Definition 10.1 Lines r and s are said to be **reverse** if they are not coplanar.

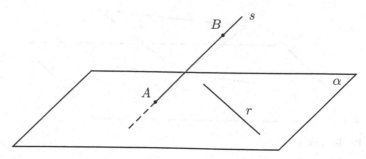

Fig. 10.2 Reverse lines

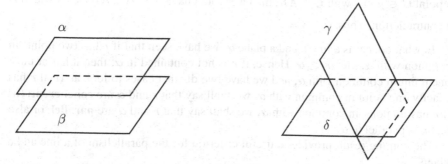

Fig. 10.3 Parallel planes (left) and concurrent planes (right)

In order to see that reverse lines do exist, suppose given a line r and a point A not belonging to r; let $\alpha = (A, r)$ and take a point $B \notin \alpha$ (cf. Fig. 10.2). Setting $s = \overleftrightarrow{AB}$, we assert that lines r and s are reverse. Indeed, if there existed a plane β containing r and s, we should have $\beta = (A, r) = \alpha$. However, since $B \in \beta$, we would then have $B \in \alpha$, which is a contradiction.

Proceeding with the exposition, let us see what can be said on the relative position of two distinct planes, as well as on a line and a plane. To this end, we start by saying that two distinct planes α and β are **parallel** if $\alpha \cap \beta = \emptyset$, and **concurrent** if $\alpha \cap \beta \neq \emptyset$. Moreover, in this last case we *postulate* that α and β have at least two points in common.

Figure 10.3 shows two parallel planes α and β at left and two concurrent planes γ and δ at right. Note that γ and δ intersect along a line; as we shall now see, this is the case with every pair of concurrent planes.

Proposition 10.2 *The intersection of two distinct concurrent planes is a straight-line.*

Proof Let γ and δ be two distinct concurrent planes, and $A, B \in \gamma \cap \delta$. According to our previous discussions, we have $\overleftrightarrow{AB} \subset \gamma, \delta$. On the other hand, if there existed

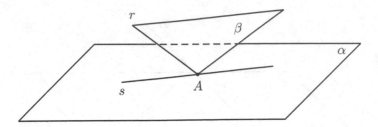

Fig. 10.4 Parallel line and plane

a point $C \in \gamma \cap \delta$ with $C \notin \overleftrightarrow{AB}$, then we would have $\gamma = (C, \overleftrightarrow{AB}) = \delta$, which is a contradiction. Thus, $\gamma \cap \delta = \overleftrightarrow{AB}$. □

In what concerns a line r and a plane α, we have seen that if r has two points in common with α, then $r \subset \alpha$. Hence, if r is not contained in α, then it has at most one point in common with α, and we have two distinct cases to look at: (i) if r has exactly one point in common with α, we shall say that r and α are **concurrent**; (ii) if r has no point in common with α, we shall say that r and α are **parallel**, or also that r is parallel to α.

The coming result provides a useful criterion for the parallelism of a line and a plane.

Proposition 10.3 *A line r and a plane α are parallel if and only if $r \not\subset \alpha$ and there exists $s \subset \alpha$ such that $r \parallel s$.*

Proof Firstly, suppose that r and α are parallel (cf. Fig. 10.4). Then $r \cap \alpha = \emptyset$ and, in particular, $r \not\subset \alpha$. Now, let $A \in \alpha$ and $\beta = (A, r)$. Since $A \in \alpha \cap \beta$, it follows from Proposition 10.2 that $\alpha \cap \beta = s$, for some line s passing through A. Hence, r and s are coplanar and such that $r \cap s \subset r \cap \alpha = \emptyset$, whence r and s are actually parallel.

Conversely, suppose that $r \not\subset \alpha$ and there exists $s \subset \alpha$ for which $r \parallel s$. Then, by definition r and s are coplanar. Letting β be the plane containing them, the condition $r \not\subset \alpha$ guarantees that $\alpha \neq \beta$. Hence, Proposition 10.2 gives $\alpha \cap \beta = s$, so that $r \cap \alpha \subset \beta \cap \alpha = s$. However, since $r \cap s = \emptyset$, we conclude that $r \cap \alpha = \emptyset$, and this proves that r and α are parallel. □

Let us now establish the usual set of sufficient conditions for the parallelism of two distinct planes.

Proposition 10.4 *Distinct planes α and β are parallel if and only if there exist concurrent lines $r, s \subset \alpha$, both parallel to β.*

Proof Suppose first that α and β are parallel planes (cf. Fig. 10.5, left). If $r, s \subset \alpha$ are concurrent, then $r \cap \beta \subset \alpha \cap \beta = \emptyset$, so that r and β are parallel; analogously, s and β are also parallel.

For the converse we argue by contraposition, thus assuming that α and β are non parallel distinct planes, with $\alpha \cap \beta = t$ (cf. Fig. 10.5, right). If $r \subset \alpha$ is a line parallel to β, then $r \cap t = \emptyset$; however, since $r, t \subset \alpha$, we conclude that r and t are parallel lines. Hence, any line s of α, concurrent with r, will also concur with t, and then with β; in particular, $s \not\parallel \beta$. □

Corollary 10.5 *Given a point A outside a plane α, there exists a single plane β, passing through A and parallel to α.*

Proof Let us first prove that such a plane β, parallel to α and passing through A, does exist. To this end (cf. Fig. 10.6), take concurrent lines $r, s \subset \alpha$ and, in the planes (A, r) and (A, s), draw lines r' and s' parallel to r and s, respectively. It follows from Proposition 10.3 that r' and s' are parallel to α. Hence, if $\beta = (r', s')$, then $A \in \beta$ and, by the previous result, $\beta \parallel \alpha$.

Now, suppose that β and γ are distinct planes, both parallel to α and passing through A, with $\beta \cap \gamma = r$ (cf. Fig. 10.7).

Take, in α, a line s not parallel to r, and let $\delta = (A, s)$. If $\beta \cap \delta = u$ and $\gamma \cap \delta = v$, we claim that u and v are both parallel to s and pass through A. Indeed, u is parallel to s for u and s are coplanar, $u \subset \beta, s \subset \alpha$ and $\beta \parallel \alpha$; analogously, $v \parallel s$. Hence, the fifth postulate of Euclid (cf. Postulate 2.17) gives $u = v$. However, since $u, v \neq r$

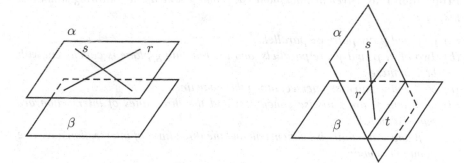

Fig. 10.5 Usual criterion for the parallelism of two distinct planes

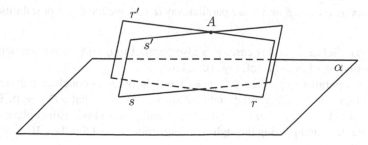

Fig. 10.6 Construction of the parallel plane

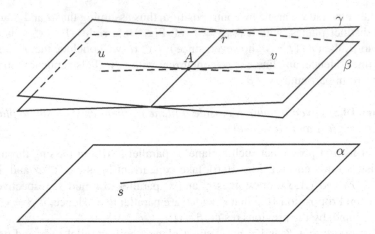

Fig. 10.7 Uniqueness of the parallel plane

(for otherwise we would have $s \parallel r$), we obtain $\beta = (u, r) = (v, r) = \gamma$, which is an absurd.   □

We finish this section by examining the relative positions of three planes in space.

Proposition 10.6 *Given distinct planes α, β and γ, one of the following situations takes place:*

(a) α, β and γ are pairwise parallel.
(b) Two of α, β and γ are parallels, and the remaining plane is concurrent with both of these.
(c) α, β and γ pairwise intersect along the same line.
(d) α, β and γ are pairwise concurrent and the three lines of intersection are pairwise parallel.
(e) α, β and γ are pairwise concurrent and the three lines of intersection have one point in common.

Proof One clearly has two possibilities for α, β and γ: either at least two of them are parallel or any two of them intersect.

If at least two of α, β and γ are parallel, say $\alpha \parallel \beta$, we have two possibilities left for γ:

(i) γ is parallel to β: in this case, γ is also parallel to α and we find ourselves in the situation of item (a) (cf. Fig. 10.8, left).
(ii) γ is not parallel to β: then $\beta \cap \gamma \neq \emptyset$ and, in order to conclude that item (b) must take place (cf. Fig. 10.8, right), it suffices to show that $\alpha \cap \gamma \neq \emptyset$. To this end, take $A \in \beta \cap \gamma$; if $\alpha \cap \gamma = \emptyset$, then β and γ would be distinct planes, both parallel to α and passing through A, which contradicts Corollary 10.5.

Now, suppose that any two of the planes α, β and γ do intersect, with $\alpha \cap \beta = r$, $\beta \cap \gamma = s$ and $\alpha \cap \gamma = t$. There are three possibilities:

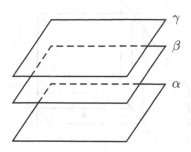
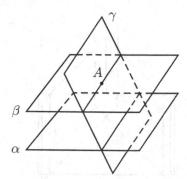

Fig. 10.8 Cases (a) and (b) of the intersection of three planes

Fig. 10.9 Case (c) of the
intersection of three planes

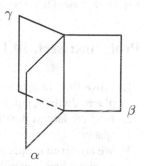

(i) If $r = s$, then r is common to α, β and γ, so that $r = t$ by the uniqueness
of the line of intersection of two concurrent planes. Analogously, by assuming
$r = t$ or $s = t$, we conclude that $r = s = t$, and the three planes intersect
along the same line. This is the situation of item (c) (cf. Fig. 10.9).

(ii) Suppose that r, s and t are pairwise distinct. If $r \cap s \neq \emptyset$, say $r \cap s = \{A\}$, then
$A \in r$ implies $A \in \alpha \cap \beta$ and $A \in s$ implies $A \in \beta \cap \gamma$. Hence, $A \in \alpha \cap \gamma$,
so that $A \in t$ and r, s and t pass through a single point. Likewise, by assuming
that $r \cap t \neq \emptyset$ or $s \cap t \neq \emptyset$ we obtain $r \cap s \cap t \neq \emptyset$, and item (e) occurs (cf.
Fig. 10.10, left).

(iii) Finally, suppose that r, s and t are pairwise distinct, with $r \parallel s$. Then $r \parallel t$
too, for otherwise (namely, if $r \cap t \neq \emptyset$) the previous item would give us
$r \cap s \cap t \neq \emptyset$, which would contradict $r \parallel s$. Analogously, $s \parallel t$ and we are in
the situation of item (d) (cf. Fig. 10.10, right).

$\square$

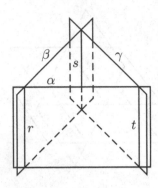 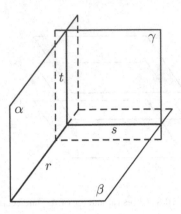

Fig. 10.10 Cases (d) and (e) of the intersection of three planes

Problems: Sect. 10.1

1. Prove that, in space, there exist four noncoplanar points. How many distinct planes do these points determine?

2. Find the largest possible number of regions into which three planes divide the space.

3. We are given in space a point A and a line r, with $A \notin r$. Prove that there exists a unique line s, parallel to r and passing through A.

4. * Let r, s and t be lines in space, such that $r \parallel s$ and $s \parallel t$. Prove that $r \parallel t$.

5. Let α and α' be two distinct planes, and ABC and $A'B'C'$ be triangles in α and α', respectively. If lines $\overleftrightarrow{AB}$ and $\overleftrightarrow{A'B'}$, $\overleftrightarrow{BC}$ and $\overleftrightarrow{B'C'}$, $\overleftrightarrow{AC}$ and $\overleftrightarrow{A'C'}$ are pairwise concurrent, prove that the three points of intersection are collinear.

6. Prove the following version of Thales' theorem in space: let α and α' be parallel planes, let A be a point outside of α and α', and X and Y points in α. If lines $\overleftrightarrow{AX}$ and $\overleftrightarrow{AY}$ intersect α' respectively at points X' and Y', then $\frac{\overline{AX}}{\overline{AX'}} = \frac{\overline{AY}}{\overline{AY'}}$.

7. * Planes α and α' are parallel, A is a point outside of α and α', and XYZ is a triangle in α. If the straightlines $\overleftrightarrow{AX}$, $\overleftrightarrow{AY}$ and $\overleftrightarrow{AZ}$ intersect α' respectively in X', Y' and Z', prove that $X'Y'Z' \sim XYZ$.[1]

---

[1]In consonance with the comments we made right after the proof of Desargues' theorem 4.24, we say that triangles XYZ and $X'Y'Z'$ are **in perspective** from the **vanishing point** A.

10.2 Orthogonality of Lines and Planes

Proceeding with the construction of the elementary concepts and results of Solid Geometry, we now define the angle between two lines r and s, not necessarily coplanar, in the following way: we choose an arbitrary point A in space and, through A, draw the lines u and v, respectively parallel to r and s (cf. Fig. 10.11). Then, we define the **angle** between r and s as the non obtuse angle θ formed by lines u and v.

For such a definition to be meaningful, we must show that θ does not depend on the chosen point A. To this end, choose another point A' in space and draw, through A', the lines u' and v' respectively parallel to r and s. We now consider two cases:

(i) The planes (u, v) and (u', v') coincide: since $u, u' \parallel r$, Problem 4, page 338 assures that $u \parallel u'$, and likewise $v \parallel v'$. Therefore the result follows immediately from Corollary 2.18.

(ii) The planes (u, v) and (u', v') are distinct: as in the previous case, we have $u \parallel u'$ and $v \parallel v'$. Now, letting $\alpha = (u, v)$ and $\alpha' = (u', v')$ (cf. Fig. 10.12), it follows from Proposition 10.4 that $\alpha \parallel \alpha'$. Choose a point $O \in \overleftrightarrow{AA'}$, with $O \notin \alpha \cup \alpha'$. Then, take points $X \in u \setminus \{A\}$, $Y \in v \setminus \{A\}$ and mark the intersections X' and Y' of $\overrightarrow{OX}$ and $\overrightarrow{OY}$, respectively, with α', so that $X' \in u'$ and $Y' \in v'$. Problem 7, page 338 guarantees that $AXY \sim A'X'Y'$; in particular, $X\widehat{A}Y = X'\widehat{A'}Y'$, as we wished to show.

Fig. 10.11 Defining the angle between two lines

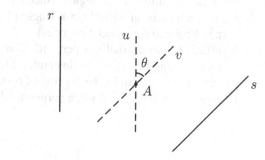

Fig. 10.12 Well defineteness of the angle formed by two reverse lines

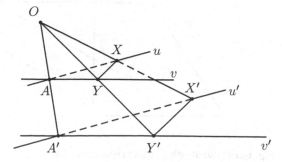

It is pretty clear from the definition above that if θ is the angle formed by lines r and s, then θ is also the angle formed by r' and s, for every line $r' \parallel r$.

Suppose that two lines r and s in space form an angle of $90°$. If they are coplanar, then, as in the planar case, we shall say that r and s are **perpendiculares**, or that r is perpendicular to s (or vice-versa); on the other hand, if r and s are reverse, then we shall say that they are **orthogonal**. In either case, we shall write $r \perp s$ to refer to such a situation, relying in the context to discern whether r and s are actually perpendicular or orthogonal.

In what concerns lines and planes, we say that a line r and a plane α are **perpendicular** if r is orthogonal to every line of α; in this case, we denote $r \perp \alpha$. In Proposition 10.8 we shall show that a line and a plane are perpendicular if and only if the line is orthogonal to two concurrent lines of the plane. However, before we can do that we need a preliminary result.

Lemma 10.7 *Let be given a line r and a plane α.*

(a) *If r is orthogonal to two concurrent lines of α, then r and α are concurrent.*
(b) *If r intersects α in A, then $r \perp \alpha$ if and only if r is perpendicular to every line of α that passes through A.*

Proof

(a) Let s and t be two concurrent lines in α, such that $r \perp s, t$. For the sake of contradiction, suppose that we had $r \parallel \alpha$. Then, it would follow from Proposition 10.3 the existence of a line $r' \subset \alpha$ such that $r \parallel r'$. Therefore, from the definition of the angle formed by two lines in space, we would have $r' \perp s, t$, which is an absurd (in a single plane, one cannot have a single line perpendicular to two concurrent ones).

(b) It suffices to assume that r is perpendicular to every line of α passing through A, then proving that $r \perp \alpha$. To this end, let (cf. Fig. 10.13) s be a line of α not passing through A, and s' be the line of α passing through A and parallel to s. By hypothesis, lines r and s' are perpendicular; however, since $s' \parallel s$, we know

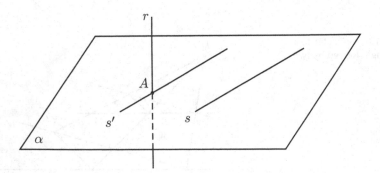

Fig. 10.13 Partial orthogonality criterion for lines and planes

that r and s also form an angle of 90°, so that they are orthogonal. Therefore, r is orthogonal to every line of α, which is the same as saying that $r \perp \alpha$. ☐

We are finally able to prove the desired criterion.

Proposition 10.8 *Given a line r and a plane α, we have $r \perp \alpha$ if and only if r is orthogonal to two concurrent lines of α.*

Proof If $r \perp \alpha$, there is nothing to do (for in this case r is orthogonal to every line in α, by definition). Conversely, assume that r is orthogonal to two concurrent lines s and t of α. Item (a) of the previous lemma assures that r and α concur, say at point A.

Drawing through A lines s' and t', respectively parallel to s and t, we have r perpendicular to s' and t'. Now, from item (b) of Lemma 10.7 (and referring to Fig. 10.14), it suffices to show that r is perpendicular to every line u' of α that passes through A. Let us do this.

Choose points B and B' in r, symmetric with respect to A. Then choose, respectively in s' and t', points C and D such that $\overline{AC} = \overline{AD}$, and let E be the intersection point of lines $\overleftrightarrow{CD}$ and u'. Since $C\widehat{A}B = C\widehat{A}B' = D\widehat{A}B = D\widehat{A}B' = 90°$, we have

$$CAB \equiv CAB' \text{ and } DAB \equiv DAB'$$

by SAS. Hence, $\overline{CB} = \overline{CB'}$ and $\overline{DB} = \overline{DB'}$, so that $CDB \equiv CDB'$ by SSS. Therefore, $B\widehat{C}D = B'\widehat{C}D$, and thus

$$CEB \equiv CEB'$$

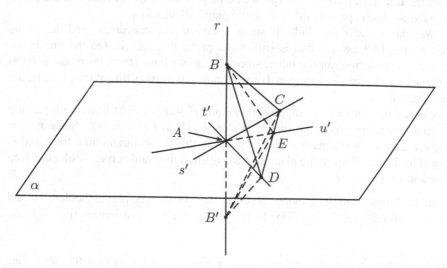

Fig. 10.14 Orthogonality criterion for lines and planes

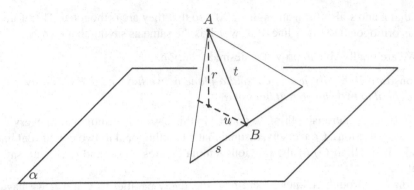

Fig. 10.15 Constructing a line perpendicular to a plano through a point

by SAS again.[2] Hence, $\overline{EB} = \overline{EB'}$, so that EA is the median relative to the basis of the isosceles triangle $BB'E$. As we already know, EA will also be the altitude of $BB'E$ relative to BB', so that $u' \perp r$.   □

In spite of the above, nothing we have done so far guarantees that orthogonal lines and planes actually exist. We remedy this with the coming corollary (in this respect, see also Problem 3, page 348).

Corollary 10.9 *Given a point A and a plane α, with A ∉ α, there exists a single line r, passing through A and perpendicular to α.*

Proof Pick a line $s \subset \alpha$ and, in the plane (A, s), draw line t, passing through A and perpendicular to s (cf. Fig. 10.15). Let B be the point of intersection of s and t, and u be the line in α, passing through B and perpendicular to s. Finally, in the plane (t, u), draw line r, passing through A and perpendicular to u.

We claim that $r \perp \alpha$. Indeed, since s and u are concurrent and lie in α, Proposition 10.8 assures that is suffices to prove that $r \perp s, u$. On the one hand, $r \perp u$ by construction; on the other, since $s \perp t, u$, we have (from Proposition 10.8) $s \perp (t, u)$, and hence s is orthogonal or perpendicular to every line of the plane (t, u); in particular, $s \perp r$.

Concerning uniqueness, suppose that r and r' were distinct lines, both passing through A and perpendicular to α. Let $\beta = (r, r')$ and $s = \alpha \cap \beta$. Since $s \subset \alpha$ and $r, r' \perp \alpha$, we would have $r, r' \perp s$. However, this is a contradiction, for r and r' would be distinct lines in the plane β, passing through A and perpendicular to s (see Problem 17, page 39).   □

In the notations of the previous corollary, let r be the line perpendicular to α and passing through point A (cf. Fig. 10.16). If P is the point of intersection of r and

---

[2]Here and whenever needed in all that follows, for the sake of simplicity we will tacitly assume the validity of the usual cases of congruence and similarity of triangles, as discussed in Sects. 2.1 and 4.2. We shall assume this even if the involved triangles do not lie in the same plane.

Fig. 10.16 Distance from a
point to a plane

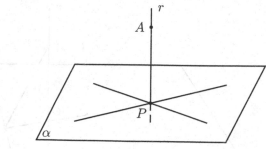

Fig. 10.17 Distance between
two parallel planes

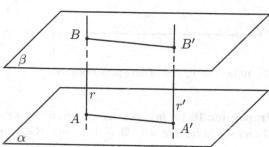

α, we say that P is the **foot of the perpendicular** dropped from A to α, and we
define the **distance** from A to α as being equal to the length of the line segment
AP. Whenever convenient, we denote this by writing

$$\overline{AP} = d(P; \alpha).$$

Also with respect to such a situation, if Q is other point of α, then the triangle
inequality applied to triangle APQ gives

$$d(P; \alpha) = \overline{AP} < \overline{AQ},$$

so that P is the only point of α that is closest to A.

We can also define the distance between two parallel planes α and β, and we
do this in the following way: letting r and r' two distinct lines perpendicular to
α, Problem 2, page 348, assures that r and r' are also perpendicular to β (cf.
Fig. 10.17). On the other hand Problem 4, page 348, shows that r and r' are (coplanar
and) parallel. Hence, if A and B (resp. A' and B') stand for the intersections of r
(resp. r') with α and β, respectively, then $\overleftrightarrow{AB} \parallel \overleftrightarrow{A'B'}$ and $\overleftrightarrow{AA'} \parallel \overleftrightarrow{BB'}$ (for $\alpha \parallel \beta$).
Therefore, $AA'B'B$ is a rectangle, so that $\overline{AB} = \overline{A'B'}$. Thus, we define the
distance between α and β as the common length $\overline{AB} = \overline{A'B'}$.

A small variation of the reasoning employed in the proof of the existence part
of Corollary 10.9 will allow us to prove the coming proposition, which will be a
central tool in several points for what is to come. Such a result is usually referred to
as the **theorem of three perpendiculars**.

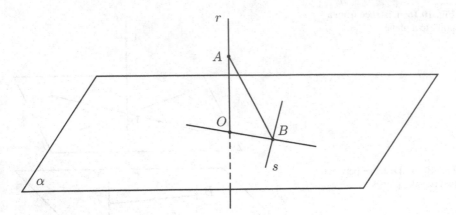

Fig. 10.18 The theorem of three perpendiculars

Proposition 10.10 *In space, we are given a plane α and lines r and s, with $r \perp \alpha$ at O and $s \subset \alpha$ (cf. Fig. 10.18). If $A \in r$ and $B \in s$, then*

$$\overleftrightarrow{AB} \perp s \Leftrightarrow \overleftrightarrow{OB} \perp s.$$

Proof Note first that $r \perp \alpha \Rightarrow r \perp s$. Now, suppose that $\overleftrightarrow{AB} \perp s$. Then, since r and $\overleftrightarrow{AB}$ are concurrent lines, Proposition 10.8 assures that $s \perp (r, \overleftrightarrow{AB})$; in particular, $s \perp OB$.

Conversely, assume that $\overleftrightarrow{OB} \perp s$. Since r and $\overleftrightarrow{OB}$ concur, it follows once more from Proposition 10.8 that $s \perp (r, \overleftrightarrow{OB})$; in particular, $s \perp \overleftrightarrow{AB}$.   □

Continuing with our discussion, let planes α and β be given. We say that α is *perpendicular* to β, and denote $\alpha \perp \beta$, if α contains a line perpendicular to β. Assuming that this is the case (see Fig. 10.19), let $r \subset \alpha$ be a straightline perpendicular to β. Then, letting s be the line of intersection of α and β, we have $r \perp s$. Now, write A for the point of intersection of r and s, and let $t \subset \beta$ be the line passing through A and perpendicular to s. Since $r \perp \beta$, we have $r \perp t$, and thus $t \perp (r, s) = \alpha$. However, since $t \subset \beta$, we are led to the conclusion that $\beta \perp \alpha$. Thanks to the above discussion, the notion of *perpendicularity* of a plane to another one is *symmetric*. Therefore, if planes α and β are such that $\alpha \perp \beta$, from now on we shall simply say that α and β are **perpendicular planes**.

Given a line r and a plane α, with $r \not\perp \alpha$, fix an arbitrary point $A \in r$ and take line s, passing through A and perpendicular to α (cf. Fig. 10.20). Since plane $\beta = (r, s)$ contains s and $s \perp \alpha$, we conclude that $r \subset \beta \perp \alpha$. With a little more work (cf. Problem 7), it is possible to show that β is the only plane containing r and perpendicular to α. Hereafter, we shall assume such a uniqueness result without further comments.

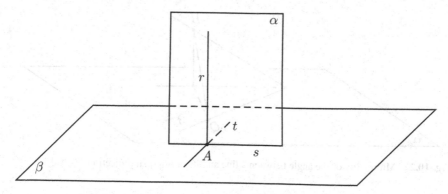

Fig. 10.19 Perpendicularity of planes is a symmetric relation

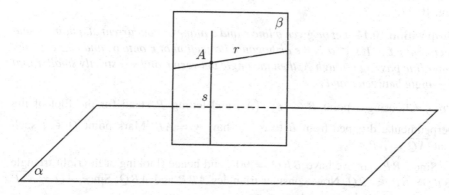

Fig. 10.20 Constructing plane β, perpendicular to α and containing r

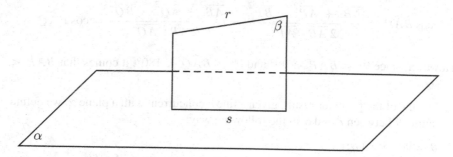

Fig. 10.21 Orthogonal projection of a line onto a plane

In view of Corollary 10.9 and of the constructions above, we can define the *angle* between a line r concurrent with a plane α. To this end, let us first assume that $r \not\perp \alpha$, and define (cf. Fig. 10.21) the **orthogonal projection**, or simply the **projection** of r onto α as the line $s = \alpha \cap \beta$, where β is the single plane perpendicular to α and containing r.

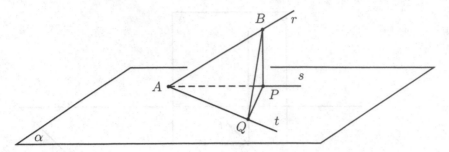

Fig. 10.22 Minimality of the angle between a line and its orthogonal projection

We now have the following result, for whose statement we refer the reader to Fig. 10.22.

Proposition 10.11 *Let be given a line r and a plane α, concurrent at a point A and such that r$\not\perp$α. If s $\subset$ α is the orthogonal projection of r onto α, and t $\subset$ α is any other line passing through A, then the angle between r and s is strictly smaller that the angle between r and t.*

Proof Choosing a point $B \in r \setminus \{A\}$, and letting P stand for the foot of the perpendicular dropped from B to $α$, we have $s = \overleftrightarrow{AP}$. Mark point $Q \in t$ such that $\overline{AQ} = \overline{AP}$.

Since $\overleftrightarrow{BP} \perp α$, we have $B\widehat{P}Q = 90°$, and hence (looking at the right triangle BPQ) $\overline{BP} < \overline{BQ}$. Now, consider triangles ABP and ABQ. Since $\overline{AQ} = \overline{AP}$ and $\overline{BP} < \overline{BQ}$, the cosine law gives

$$\cos B\widehat{A}P = \frac{\overline{AB}^2 + \overline{AP}^2 - \overline{BP}^2}{2\,\overline{AB} \cdot \overline{AP}} > \frac{\overline{AB}^2 + \overline{AQ}^2 - \overline{BQ}^2}{2\,\overline{AB} \cdot \overline{AQ}} = \cos B\widehat{A}Q.$$

However, since $0° < B\widehat{A}P < 90°$ and $0° < B\widehat{A}Q < 180°$, it comes that $B\widehat{A}P < B\widehat{A}Q$.   □

In view of the previous result, given a line r concurrent with a plane $α$, we define the **angle** $θ$ between r and $α$ in the following way:

i. $θ = 90°$, if $r \perp α$;
ii. $θ =$ angle between r and its orthogonal projection onto $α$, if $r \not\perp α$.

We finish this section by studying, in the coming result, the important concept of *common perpendicular* to two reverse lines.

Theorem 10.12 *Given two reverse lines r and s, there exist single points A $\in$ r, B $\in$ s such that $\overleftrightarrow{AB} \perp r, s$. Moreover, if A$'$ $\in$ r and B$'$ $\in$ s are any points, then $\overline{AB} \leq \overline{A'B'}$, with equality holding if and only if A$'$ = A and B$'$ = B.*

Proof Choose a point on r and draw the line s', parallel to s and passing through this point, thus obtaining plane $\alpha = (r, s')$. Analogously, choose a point on s and draw the line r', parallel to r and passing through the point, thus getting the plane $\beta = (r', s)$ (see Fig. 10.23). It follows from Proposition 10.4 that $\alpha \parallel \beta$.

Let γ be the plane containing r and perpendicular to α. Then, γ contains a line perpendicular to α; since $\alpha \parallel \beta$, it is also the case that such a line is also perpendicular to β, so that $\gamma \perp \beta$ as well. Letting $\beta \cap \gamma = r''$, the parallelism of α and β also gives $r'' \parallel r$. However, since r and s are reverse, we conclude that r'' and s concur, say at a point B. Since $B \in \gamma$ and $\gamma \perp \alpha$, line t, passing through B and perpendicular to α, is contained in γ. If A is the point of intersection of r and t, then $\overleftrightarrow{AB} = t$, so that $\overleftrightarrow{AB} \perp \alpha, \beta$. In particular, $\overleftrightarrow{AB} \perp r, s$.

We leave the uniqueness of the common perpendicular as an exercise for the reader (cf. Problem 8).

Concerning the stated inequality, let $A' \in r$ and $B' \in s$ be arbitrary points. If $B'' \in \beta$ is such that $\overleftrightarrow{AB''} \parallel \overleftrightarrow{A'B'}$, Problem 9 guarantees that $\overline{A'B'} = \overline{AB''}$. However, since $\overleftrightarrow{AB} \perp \overleftrightarrow{BB''}$, we obtain

$$\overline{A'B'} = \overline{AB''} \geq \overline{AB},$$

with equality if and only if $B'' = B$, i.e., if and only if $\overleftrightarrow{A'B'} \parallel \overleftrightarrow{AB}$. In particular, if equality holds then $A'B'$ is another common perpendicular of r and s. Therefore,

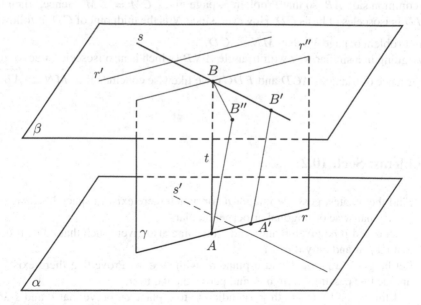

Fig. 10.23 The common perpendicular between two reverse lines

Fig. 10.24 An important
example of common
perpendicular

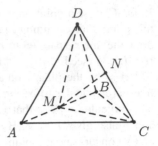

by invoking the uniqueness of the common perpendicular, we conclude that $A = A'$
and $B = B'$.   □

As was anticipated in the proof of the theorem, we say that line $\overleftrightarrow{AB}$ is the
common perpendicular to the reverse lines r and s. Let us take a look at a simple
yet important

Example 10.13 Points A, B, C and D are noncoplanar and such that $\overline{AB} = \overline{CD}$,
$\overline{AC} = \overline{BD}$ and $\overline{AD} = \overline{BC}$. Prove that the common perpendicular to $\overleftrightarrow{AB}$ and the
line segments AB and CD.

Proof Let M and N be the midpoints of AB and CD, respectively (cf. Fig. 10.24).
Since $\overline{AD} = \overline{BC}$, $\overline{BD} = \overline{AC}$, we conclude that $ABD \equiv BAC$ by SSS (AB
is a common side). Now, DM and CM are medians of these triangles relative to
the common side AB, so that Problem 3, page gives $\overline{CM} = \overline{DM}$. Hence, triangle
CMD is isosceles of basis CD. However, since N is the midpoint of CD, it follows
from Problem 6, page 30, that $\overleftrightarrow{MN} \perp \overleftrightarrow{CD}$.

Arguing in a similar way with triangle ANB (which is also isosceles, due to the
congruence of triangles ACD and BDC), we likewise conclude that $\overleftrightarrow{MN} \perp \overleftrightarrow{AB}$.
                                                                                    □

Problems: Sect. 10.2

1. Find the greatest possible integer n for which there exist in space lines r_1, r_2,
   ..., r_n, pairwise orthogonal or perpendicular.
2. * Let α and β be given planes and r be a line also given, such that $r \perp \alpha$. Prove
   that $r \perp \beta$ if and only if $\alpha \parallel \beta$.
3. Let be given a point A and a plane α, with $A \in \alpha$. Prove that there exists a
   unique line passing through A and perpendicular to α.
4. * If lines r and s are both perpendicular to a plane α, prove that r and s are
   parallel.

5. Let α and β be perpendicular planes and $r \subset \alpha$ be a straightline. If $r \perp (\alpha \cap \beta)$, prove that $r \perp \beta$.

6. Prove that, through a given point, there passes only one plane perpendicular to a given line.

7. * Given a line r and a plane α, with $r \not\perp \alpha$, prove that the plane through r and perpendicular to α is unique.

8. * Show that the common perpendicular to two reverse lines is unique.

9. * Planes α and β are parallel, and points A, $A' \in \alpha$ and B, $B' \in \beta$ are such that $\overleftrightarrow{AB} \parallel \overleftrightarrow{A'B'}$. Prove that $\overline{AB} = \overline{A'B'}$.

10. * In space, let be given a plane α and points A and B, not belonging to α and situated in distinct half-spaces with respect to α. Prove that A and B are at equal distances from α if and only if the midpoint of AB lies in α.

11. Four noncoplanar points are given in space. Show that there are exactly seven planes α such that the four points lie at equal distances from α.

12. (Romania) Let ABC be a triangle with $\overline{AB} = c$, $\overline{BC} = a$ and $\overline{AC} = b$. Show that there exists at least one point D in space, such that

$$\overline{AD} = \sqrt{bc}, \quad \overline{BD} = \sqrt{ac}, \quad \overline{CD} = \sqrt{ab}.$$

13. Given reverse lines r and s, show that there are infinitely many points in space which lie at equal distances from r and s.

10.3 Loci in Space

Now that we have at our disposal the basic notions and results concerning lines and planes, we can extend to Space Geometry the important concept of *locus*. As we shall see in what comes next, formally it is pretty much identical to the corresponding concept in the plane, which was presented in Definition 3.1. Nevertheless, by exploring it conveniently, we shall develop a set of quite important and useful tools, which, in conjunction with the material of the first two sections, will allow us to build the more difficult parts of the theory. This being said, we have the following central

Definition 10.14 Given a property $\mathcal{P}$ relative to points in space, its **locus** is the set $\mathcal{L}$ of points in space satisfying the following conditions:

(a) Every point of $\mathcal{L}$ has the property $\mathcal{P}$.
(b) Every point in space that has the property $\mathcal{P}$ belongs to $\mathcal{L}$.

In other words, $\mathcal{L}$ is the locus of property $\mathcal{P}$ if $\mathcal{L}$ is composed by *exactly* those points of space that have property $\mathcal{P}$, no more, no less. As we have done in Sect. 3.1, in what follows we shall study some elementary loci in space, as well as present a few applications of them.

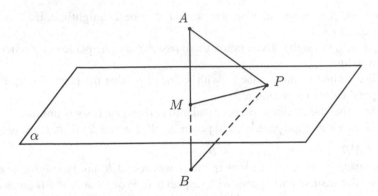

Fig. 10.25 The bisecting plane of line segment AB

The first locus we shall study is the analogous, in space, of the perpendicular bisector of a line segment in a plane.

Proposition 10.15 *Given distinct points A and B, the locus of points in space that are at equal distances from A and B is the plane α, perpendicular to line $\overleftrightarrow{AB}$ and passing through the midpoint of the line segment AB. We say that α is the **bisecting plane** of segment AB.*

Proof Let M be the midpoint of AB and α be defined as in the statement of the proposition. If $P \neq M$ is a point of α (cf. Fig. 10.25), then $\overleftrightarrow{PM} \perp \overleftrightarrow{AB}$, so that $PMA \equiv PMB$ by SAS (PM is a common side of both triangles, $\overline{AM} = \overline{BM}$ and $P\widehat{M}A = P\widehat{M}B = 90°$). Hence, $\overline{PA} = \overline{PB}$.

Conversely, let α be as above and $P \neq M$ be a point in space for which $\overline{PA} = \overline{PB}$. Then, PM is the median of the isosceles triangle PAB relative to the basis AB, so that $\overleftrightarrow{PM} \perp \overleftrightarrow{AB}$. If $\beta = (\overleftrightarrow{AB}, \overleftrightarrow{PM})$ and $\alpha \cap \beta = s$, then s, $\overleftrightarrow{PM} \subset \beta$ and s, $\overleftrightarrow{PM} \perp r$. Hence, $s = \overleftrightarrow{PM}$ and $P \in \alpha$.   □

We now examine the set of points equidistant from three noncollinear given points.

Proposition 10.16 *Given noncollinear points A, B and C, the locus of points in space that are at equal distances from A, B and C is line r, perpendicular to the plane (ABC) and passing through the circumcenter of triangle ABC. We say that r is the **medial line** of triangle ABC.*

Proof Let O be the circumcenter of ABC and r be the line passing through O and perpendicular to the plane (ABC) (cf. Fig. 10.26). If $P \in r \setminus \{O\}$, then triangles AOP, BOP and COP are congruent by SAS, for OP is a common side of them, $\overline{AO} = \overline{BO} = \overline{CO}$ and $A\widehat{O}P = B\widehat{O}P = C\widehat{O}P = 90°$. Hence, $\overline{AP} = \overline{BP} = \overline{CP}$.

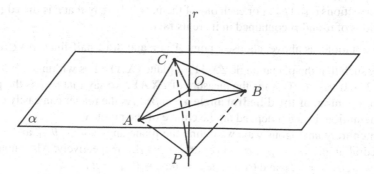

Fig. 10.26 The medial line of triangle ABC

Fig. 10.27 A dihedral angle

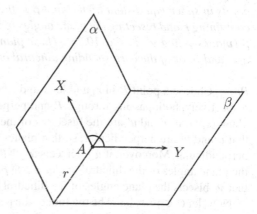

Now, suppose that $\overline{AP} = \overline{BP} = \overline{CP}$. Then, P belongs to the bisecting planes of line segments AB, AC and BC; since the same holds for O, it follows that line $\overleftrightarrow{OP}$ lies in these three planes. However, since the bisecting plane of AB is perpendicular to $\overleftrightarrow{AB}$, we conclude that $\overleftrightarrow{OP} \perp \overleftrightarrow{AB}$; analogously, $\overleftrightarrow{OP} \perp \overleftrightarrow{AC}$ and (which is unnecessary at this point) $\overleftrightarrow{OP} \perp \overleftrightarrow{BC}$. Thus, $\overleftrightarrow{OP} \perp (\overleftrightarrow{AB}, \overleftrightarrow{AC}) = (ABC)$. $\qquad\qquad\square$

Notice that, as a corollary of the proof just presented, we have shown that the medial line of a triangle ABC is the intersection of the bisecting planes of its sides.

In order to continue the discussion of the main loci of Solid Geometry, we need the coming definition.

Definition 10.17 In space, a **dihedral angle** is the intersection of two half-spaces, each of which determined by one of two concurrent planes (cf. Fig. 10.27).

If planes α and β meet at line r, it is immediate to see that they divide the space into four dihedral angles. Letting the half-spaces determined by α (resp. by β) be denoted by α_+ and α_- (resp. β_+ and β_-), these dihedral angles are given by the four intersections $\alpha_\pm \cap \beta_\pm$. For each one of them, we also say that r is the **edge** and the portions of α and β contained in it are its **faces**.

With notations as above, choose a point A in r and draw half-lines $\overrightarrow{AX} \subset \alpha$ and $\overrightarrow{AY} \subset \beta$, such that the planar angle $\angle XAY$ of plane (XAY) lies within $\alpha_+ \cap \beta_+$ (cf. Fig. 10.27). If $\theta_{++} \in (0, \pi)$ is the measure of $\angle XAY$, we say that θ_{++} is the **plane angle**, or **opening, of the dihedral angle** $\alpha_+ \cap \beta_+$. As the reader can easily check, such a definition does not depend on the choice of the point A.

In an entirely analogous way, we define the plane angles θ_{+-}, θ_{-+} and θ_{--} of the dihedral angles $\alpha_+ \cap \beta_-$, $\alpha_- \cap \beta_+$ and $\alpha_- \cap \beta_-$, respectively. Also, note that $\theta_{++} = \theta_{--}$, $\theta_{+-} = \theta_{-+}$ and $\theta_{++} + \theta_{+-} = \theta_{--} + \theta_{-+} = \pi$.

Proposition 10.18 *Given planes α and β concurrent along line r, the locus of points in space equidistant from α and β is the union of two perpendicular planes, containing r and bisecting the plane angles of the dihedral angles formed by α and β (planes γ and γ', in Fig. 10.28). These planes are known as the **bisector planes** of α and β, or of the corresponding dihedral angles.*

Proof Choose a point P in r, and let s and t be straightlines contained in α and β, respectively, both passing through P and perpendicular to r (cf. Fig. 10.28). In the plane (s, t), if u and u' are the bisectors of the angles formed by s and t, we know that u and u' are perpendicular, so that planes $\gamma = (r, u)$ and $\gamma' = (r, u')$ are also perpendicular. Moreover, if u intersects $\alpha_+ \cap \beta_+$, then it is immediate that γ bisects the plane angles of the dihedral angles $\alpha_+ \cap \beta_+$ and $\alpha_- \cap \beta_-$; from this, it follows that γ' bisects the plane angles of the dihedral angles $\alpha_+ \cap \beta_-$ and $\alpha_- \cap \beta_+$.

Now, let $C \in \gamma \backslash r$, let P be the foot of the perpendicular dropped from C to r, and let s and t be the lines passing through P and perpendicular to α and β, respectively. If A and B stand for the feet of the perpendiculars dropped from C to the planes α and β, respectively, we have $A \in s$ and $B \in t$. However, since $C\widehat{P}A = C\widehat{P}B$ and $C\widehat{A}P = C\widehat{B}P = 90°$, it is pretty clear that $CPA \equiv CPB$ by AAS. Therefore,

Fig. 10.28 The bisector planes γ and γ', of α and β

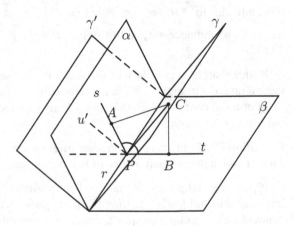

Fig. 10.29 The sphere of
center O and radius R

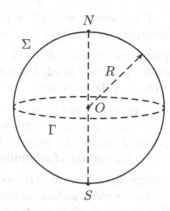

$\overline{CA} = \overline{CB}$, and C is at equal distances from α and β. Analogously, every point of γ' equidists from α and β.

Conversely, if C is equidistant from α and β, it is not difficult to prove (cf. Problem 2) that C lies in one of the planes γ or γ' defined as above. □

We now present one of the most important definitions of all Solid Geometry.

Definition 10.19 We are given a point O in space and a positive real number R. The **sphere** Σ of **center** O and **radius** R, denoted $\Sigma(O; R)$, is the locus of points in space which are at distance R from O (cf. Fig. 10.29).

In the notations of the previous definition, given points $A, B \in \Sigma$, we say that AB is a **chord** of Σ. Also as in Plane Geometry, a chord AB is a **diameter** of Σ if $O \in AB$; in this case, we also say that A and B are **antipodal points** of Σ. In Fig. 10.29, points N and S are antipodal. In general, for every chord AB of Σ, we have $\overline{AB} \leq 2R$, with equality holding if and only if A and B are antipodal. Indeed, if $O \notin AB$, then the triangle inequality applied to triangle AOB, furnishes $\overline{AB} < \overline{AO} + \overline{BO} = 2R$.

The coming example provides another characterization of a sphere as a locus.

Example 10.20 Given in space distinct points A and B, show that the locus of points P such that $A\widehat{P}B = 90°$ is the sphere of diameter AB, except for the points A and B themselves.

Proof Let $\overline{AB} = 2R$ and P be a point in space with $P \notin \overleftrightarrow{AB}$. Letting O be the midpoint of AB, the line segment PO is the median relative to the side AB of triangle PAB. Hence, Corollary 2.44 and its converse give

$$A\widehat{P}B = 90° \Leftrightarrow \overline{PO} = \frac{1}{2}\overline{AB} = R \Leftrightarrow P \in \Sigma(O; R).$$

Thus, the desired locus is $\Sigma(O; R) \setminus \{A, B\}$. □

The previous example explains the shape of Σ, as depicted in Fig. 10.29, according with the following argument: in the notations of the example, and fixed a plane α containing $\overleftrightarrow{AB}$, the locus of the points of α that look at AB under an angle of $90°$ is, as we already know, the union of the two arcs capable of $90°$ on AB. In turn, since the angle is $90°$, such a union coincides with the circle of diameter AB in α, except for the points A and B. Hence, by *turning α around $\overleftrightarrow{AB}$*, we obtain $\Sigma(O; R)$ as the union of all the circles in space with diameter AB, except for the points A and B. Due to this construction, we also say that the sphere of diameter AB is the **surface of revolution** generated by the *revolution* (i.e., rotation) of the semicircle of diameter AB around the straightline $\overleftrightarrow{AB}$. We shall have more to say on more general surfaces of revolution later in these notes.

Our next result clears the nature of the **plane sections** of a sphere, i.e., of the possible intersections of a sphere with a generic plane, not necessarily passing through the center of the sphere.

Proposition 10.21 *Let be given, in space, a plane α and a sphere Σ, of center O and radius R. Let also d be the distance from O to α.*

(a) If $d > R$, then α does not intersect Σ.
(b) If $d = R$, then α intersects Σ at a single point.
(c) If $d < R$, then α intersects Σ along a circle Γ of radius $\sqrt{R^2 - d^2}$, centered at the foot of the perpendicular dropped from O to α.

Proof Let O' be the foot of the perpendicular dropped from O to α, so that $\overline{OO'} = d$.

(a) and (b) Suppose that $d \geq R$. If $P \in \alpha \setminus \{O'\}$, then $P\widehat{O'}O = 90°$ and hence $\overline{PO} > \overline{OO'} = d \geq R$; therefore, $P \notin \Sigma$. However, since $\overline{OO'} = d \geq R$, we have $O' \in \Sigma$ if and only if $d = R$. Hence, α intersects Σ at O' if $d = R$, whereas α does not intersect Σ if $d > R$.

(c) If $P \in \Sigma \cap \alpha$, then $P \neq O'$ and hence $P\widehat{O'}O = 90°$. Therefore, by applying Pythagoras' theorem to triangle POO', we obtain

$$\overline{O'P}^2 = \overline{OP}^2 - \overline{O'O}^2 = R^2 - d^2,$$

so that $\overline{O'P} = \sqrt{R^2 - d^2}$. Thus, P lies in the circle Γ of the statement.

Conversely, if P is a point of Γ, then $\overline{O'P} = \sqrt{R^2 - d^2}$ and we can retrace the steps of the reasoning just presented to conclude that $\overline{OP} = R$, i.e., that $P \in \Sigma$.
$\square$

In the notations of the previous proposition, suppose that $d = R$ and let T be the common point of α and Σ. Then, we shall say that α *is tangent to Σ at T*, or also that α and Σ are **tangent** at T; also in this case, T is the **point of tangency** of α and Σ. Through a point T of a sphere Σ of center O there passes a unique plane α tangent to Σ. Indeed, α must be perpendicular to $\overleftrightarrow{OT}$ and pass through T (Fig. 10.30).

Fig. 10.30 A plane section
of a sphere

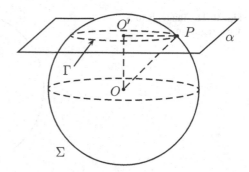

Yet in the notations of Proposition 10.21, assume that α intersects Σ along a circle Γ of center O'. Then, we have just seen that Γ has radius $r = \sqrt{R^2 - d^2}$, with d being equal to the distance from the center O of Σ to α. It follows that $r \leq R$, with equality holding if and only if $d = 0$, i.e., if and only if $O = O'$. If this is the case, we shall say that Γ is a **great circle** or an **equator** of Σ, and that the endpoints of the diameter of Σ perpendicular to the plane of Γ are (in some arbitrarily chosen order) the **North** and **South poles** of Σ with respect to Γ.

If Σ is a sphere of center O and A and B are non antipodal points of Σ, then there is exactly one equator Γ of Σ passing through A and B. Indeed, letting α be the plane of such an equator, the very definition of equator assures that $O \in \alpha$. However, since A and B are non antipodal, it follows that A, O and B are noncollinear. Hence, $\alpha = (AOB)$ and $\Gamma = (AOB) \cap \Sigma$.

We can define the notion of tangency between a line r and a sphere $\Sigma(O; R)$ by imposing that r intersects Σ at a single point T, which will then be called the **point of tangency** of r and Σ. This being the case, if $\alpha = (O, r)$ and $\Gamma(O; R)$ is the equator of Σ contained in α, we have $r \subset \alpha$ and r intersects Γ only at T, so that, in plane α, line r is tangent to the circle Γ at T. Hence, $\overleftrightarrow{OT} \perp r$, which implies that r is contained in the plane tangent to Σ at T. Conversely, it is immediate to check that such a tangent plane is the union of the lines tangent to Σ at T.

Given in space a sphere $\Sigma(O; R)$ and a point P such that $\overline{PO} > R$, we shall show in the coming example that it is possible to draw, through P, infinitely many lines tangent to Σ. To this end, given a circle Γ of center O, we define its **medial line** as the straightline passing through O and perpendicular to the plane of Γ (for a geometric interpretation of the medial line of a circle, see Problem 6).

Example 10.22 We are given a sphere $\Sigma(O; R)$ and a point P with $\overline{PO} = d > R$ (cf. Fig. 10.31).

(a) For $T \in \Sigma$, we have $\overleftrightarrow{PT}$ tangent to Σ if and only if $\overline{PT} = \sqrt{d^2 - R^2}$.

(b) The set of points $T \in \Sigma$ for which $\overleftrightarrow{PT}$ is tangent to Σ is a circle Γ of radius $\frac{R\sqrt{d^2-R^2}}{d}$, contained in Σ and such that $\overleftrightarrow{PO}$ is its medial line.

Fig. 10.31 Tangents to Σ
departing from P

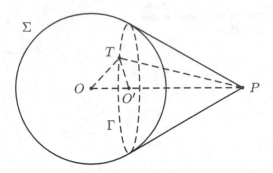

Proof

(a) Since $\overleftrightarrow{PT} \perp \overleftrightarrow{OT}$, triangle POT is right at T. It thus suffices to apply
    Pythagoras' theorem to triangle POT.
(b) From item (a), the set of points T possessing the stated property coincides
    with the intersection of the spheres Σ and $\Sigma'(P; \sqrt{d^2 - R^2})$. Hence, Problem 5
    assures that it is a circle Γ having $\overleftrightarrow{PO}$ as its medial. Finally, if O' is the point
    of intersection of $\overleftrightarrow{PO}$ with the plane of Γ (cf. Fig. 10.31), then O' is the center
    of Γ and $O'T \perp OP$; since triangle OTP is right at T, the metric relations in
    right triangles give

$$\overline{O'T} = \frac{\overline{OT} \cdot \overline{PT}}{\overline{OP}} = \frac{R\sqrt{d^2 - R^2}}{d}.$$

$\square$

   In the notations of the previous example, letting $T \in \Sigma$ be such that $\overleftrightarrow{PT}$ is
tangent to Σ, we shall say that $\overline{PT}$ is the *length of the tangent drawn to Σ through P*.
Item (a) of the example shows that such a length does not depend on the particularly
chosen point of tangency T.

   Another very important locus is the *cylinder of revolution* of a certain *radius* and
with a given *axis*. In order to define it, given in space a line r and a point $A \notin r$, we
define the **distance from A to r** as the distance from A to r in the plane (A, r).

Definition 10.23 Given a positive real number R and a line e, the **cylinder (of
revolution) of axis e** and **radius R**, denoted $\mathcal{C}(e; R)$, is the set of points P in space
lying at distance R from e (cf. Fig. 10.32).

   From now on, whenever there is no danger of confusion, we shall refer to a
cylinder of revolution $\mathcal{C}$ simply as a *cylinder*.

   Given a point A in the cylinder $\mathcal{C}(e; R)$, let α be the plane passing through A
and perpendicular to e. If O is the intersection point of α with the axis e, we have
$\overleftrightarrow{AO} \perp e$ and $\overleftrightarrow{AO} \subset (A, e)$, so that $\overline{AO} = R$; hence, A lies in the circle of center
O and radius R of the plane α. Conversely, one immediately sees that every point

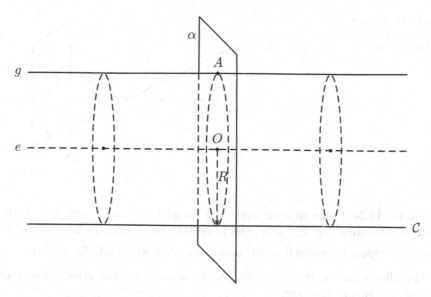

Fig. 10.32 The cylinder of revolution of axis e and radius R

of such a circle lies in C, and thanks to this fact we shall sometimes refer to C as a
right circular cylinder.

If $\beta = (A, e)$ and g is the line contained in β, parallel to e and passing through
A, every point of g is at distance R from e, so that $g \subset C(e; R)$. Line g is said to be
a **generatrix** of the cylinder C.

In view of the previous paragraph, we can easily justify the shape of $C(e; R)$,
as shown in Fig. 10.32: fixed a plane α containing e, the locus of the points of α
which lie at distance R from e is the union of two lines g and g' of α, both parallel
to e. Hence, if we now *turn α around e*, we get $C(e; R)$ as the union of such lines
g (line g' is obtained from g, upon rotating α of 180° from its original position).
Thus, $C(e; R)$ is another example of a *surface of revolution*,[3] this time generated by
revolving a line g around an axis e, provided $g \parallel e$.

The discussion above has also shown that the plane sections of a cylinder by a
plane perpendicular to (resp. containing its) axis is a circle (resp. the union of two
parallel lines). More generally, if the plane is parallel to the axis of the cylinder, it
is not difficult to prove that the corresponding plane section in the cylinder is the
empty set, a generatrix or the union of two parallel generatrices (cf. Problem 14). In
the next section, we shall look at the sections by planes which are neither parallel
nor perpendicular to the axis of the cylinder.

We now define the last locus we shall study in this section.

---

[3]For a general definition, see the beginning of Sect. 12.1.

Fig. 10.33 The cone of
revolution of axis e, vertex V
and opening 2θ

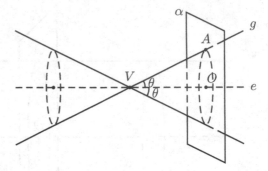

Definition 10.24 Given an acute angle θ, a straightline e and a point $V \in e$, the
cone of revolution $\mathcal{C}(e; V; \theta)$, of **axis** e, **vertex** V and **opening** 2θ is the set of
points A in space for which line $\overleftrightarrow{AV}$ forms an angle θ with e (cf. Fig. 10.33).

Hereafter, whenever there is no danger of confusion we shall refer to a cone of
revolution $\mathcal{C}$ simply as a *cone*.

Let A be a point in a cone $\mathcal{C}(e; V; \theta)$, and α be the plane passing through A and
perpendicular to e. If α intersects e at O, then $\overleftrightarrow{AO} \perp e$ and $\overleftrightarrow{AO} \subset (A, e)$. Since
$V \in e$, we have $\overline{AO} = \overline{VO} \cdot \tan\theta$; hence, A lies in the circle of center O and
radius $\overline{VO} \cdot \tan\theta$ of the plane α. Conversely, it is immediate to check that every
point of such a circle lies in $\mathcal{C}$, and for this reason we shall also sometimes refer to
a cone as a **right circular cone**.

In the notations above, letting $\beta = (A, e)$ and $g = \overleftrightarrow{AV}$, line g forms an angle θ
with e, such that $g \subset \mathcal{C}$; it is called a **generatrix** of cone $\mathcal{C}$.

A plane passing through V and perpendicular to e divides $\mathcal{C} \setminus \{V\}$ in two *pieces*,
which are usually referred to as the **leaves** of the cone. It is pretty clear that each
such leaf is the union of half-lines of the generatrices of the cone, all of origin V.

With the discussion above at our disposal, we can now easily justify Fig. 10.33:
fixed a plane α containing e, the locus of the points P of α for which $\overleftrightarrow{PV}$ forms
an angle θ with e is the union of two lines g and g' of α, both passing through
V. Therefore, *by turning α around e* we obtain the cone $\mathcal{C}(e; V; \theta)$ as the union of
such lines g (g' is got from g, when α turns $180°$ around e). Thus, $\mathcal{C}(e; V; \theta)$ is
yet another example of a *surface of revolution*, this time the one generated by the
revolution of a line g around an axis e, with g not parallel to e and $g \cap e = \{V\}$.

We saw above that the plane section of a cone of revolution by a plane
perpendicular to its axis and not passing through its vertex (resp. containing its
vertex and not perpendicular to its axis) is a circle (resp. the union of two lines
concurrent at the vertex). The coming section will deal with the other plane sections
of a right circular cone, showing that they can be either ellipses, hyperbolas or
parabolas.

Problems: Sect. 10.3

1. We are given a plane α and a positive real number d. Prove that the locus of points in space which are at a distance d from α is the union of two planes parallel to α.

2. * Complete the proof of Proposition 10.18, showing that if a point C lies at equal distances from planes α and β, then (in the notations used there) C belongs to one of the planes γ or γ'.

3. * We consider in space a point V and three half-lines $\overrightarrow{VX}$, $\overrightarrow{VY}$ and $\overrightarrow{VZ}$, not contained in a single plane. We also let $\alpha = (VXY)$ and α_+ the half-space determined by α and containing $\overrightarrow{VZ}$, $\beta = (VXZ)$ and β_+ the half-space determined by β and containing $\overrightarrow{VY}$, $\gamma = (VYZ)$ and γ_+ the half-space determined by γ and containing $\overrightarrow{VZ}$. The **trihedral angle** of **vertex** V and **faces** $\angle XVY$, $\angle XVZ$ and $\angle YVZ$ is the region in space given by $\alpha_+ \cap \beta_+ \cap \gamma_+$ (cf. Fig. 10.34). Prove that the set of points in space which lie at equal distances from the faces of a trihedral angle is a half-line of origin V.

4. We are given two concurrent planes α and β, forming an angle θ with each other, where $0 < \theta < \frac{\pi}{2}$. Given a convex n-gon $\mathcal{P}$ in α, let $\mathcal{Q}$ be its orthogonal projection onto β. Show that:

   (a) $\mathcal{Q}$ is also a convex n-gon.
   (b) $A(\mathcal{Q}) = A(\mathcal{P}) \cos \theta$.

5. * Let $\Sigma_1(O_1; R_1)$ and $\Sigma_2(O_2; R_2)$ be two given spheres. If $|R_1 - R_2| < \overline{O_1 O_2} < R_1 + R_2$, prove that $\Sigma_1 \cap \Sigma_2$ is a circle having $\overleftrightarrow{O_1 O_2}$ as its medial line.

6. * Given a circle Γ of center O, prove that the locus of points in space equidistant from all the points of Γ is its medial line.

7. * In space, we are given a circle Γ and a point A outside the plane of Γ. Prove that there exists a single sphere containing Γ and passing through A.

Fig. 10.34 The trihedral angle of faces $\angle XVY$, $\angle XVZ$ and $\angle YVZ$

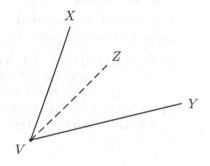

8. (Brazil) $A_1 A_2 \ldots A_n$ is an n-gon inscribed in a circle Γ, and A is a point outside the plane of Γ. For each $1 \le i \le n$ draw the plane α_i, passing through A_i and perpendicular to $\overleftrightarrow{A A_i}$. Prove that $\alpha_1, \ldots, \alpha_n$ all pass through a single point.

9. * Two circles in space pass through two given points but are not contained in a single plane. Prove that there exists a sphere containing both of them.

10. Two circles have a single point A in common and do not lie in a single plane. If the lines tangents to the circles through A coincide,[4] prove that there exists a sphere containing both of them.

11. Lines $\overleftrightarrow{AB}$, $\overleftrightarrow{CD}$, $\overleftrightarrow{EF}$ are noncoplanar and pass through a point P, with $P \in AB, CD, EF$ and $P \ne A, B, C, D, E, F$. If $\overline{AP} \cdot \overline{BP} = \overline{CP} \cdot \overline{DP} = \overline{EP} \cdot \overline{FP}$, prove that there exists a sphere passing through A, B, C, D, E, F.

12. Let A and B be points lying in a single half-space of those determined by a plane α. Find the locus of the points of tangency, with α, of the spheres tangent to α and passing through A and B.

13. (Putnam) We are given three pairwise perpendicular lines r, s and t, all passing through a point O, and another point $C \in t \setminus \{O\}$. Find the locus of points P in space, for which there exist points $A \in r$ and $B \in s$ such that $\overleftrightarrow{AP}$, $\overleftrightarrow{BP}$, $\overleftrightarrow{CP}$ are pairwise perpendicular.

14. * We are given a cylinder $\mathcal{C}$ of axis e, and a plane α parallel to e. If the intersection of α with $\mathcal{C}$ is not empty, prove that it is either a generatrix or the union of two parallel generatrices of the cylinder.

15. * Let $\mathcal{C}$ be a cone of revolution with axis e and vertex V, and α be a plane that intersects $\mathcal{C}$. Prove the following assertions:

   (a) If α is parallel to two distinct generatrices of $\mathcal{C}$, then α intersects both leaves of $\mathcal{C}$.
   (b) If α intersect a single leaf of $\mathcal{C}$ and is parallel to a generatrix g of $\mathcal{C}$, then $\alpha \perp (g, e)$.

   The last three problems elaborate the concept of inversion in euclidean three-dimensional space $\mathcal{E}$.

16. Let O be a point of $\mathcal{E}$ and k be a positive real number. The **inversion** of **center** O and ratio k is the mapping $I : \mathcal{E} \setminus \{O\} \to \mathcal{E} \setminus \{O\}$ that sends $A \ne O$ to $A' \in \overrightarrow{OA} \setminus \{O\}$ such that $\overline{OA} \cdot \overline{OA'} = k^2$. In this case, A' is said to be the **inverse** of A (with respect to I). Do the following items:

   (a) $I \circ I : \mathcal{E} \setminus \{O\} \to \mathcal{E} \setminus \{O\}$ is the identity. In particular I is a bijection from $\mathcal{E} \setminus \{O\}$ to itself, and if $\mathcal{F}_1, \mathcal{F}_2 \subset \mathcal{E} \setminus \{O\}$ are such that $I(\mathcal{F}_1) \subset \mathcal{F}_2$ and $I(\mathcal{F}_2) \subset \mathcal{F}_1$, then $I(\mathcal{F}_1) = \mathcal{F}_2$ and $I(\mathcal{F}_2) = \mathcal{F}_1$.
   (b) The sphere of center O and radius k remains fixed upon I; for this reason, it is called the **inversion sphere** (with respect to I).

---

[4] In such a situation, we say that the two circles are *tangent* to each other.

(c) The inverse of a plane α passing through O is α itself, and the restriction of I to $\alpha \setminus \{O\}$ is the inversion (in the sense of Sect. 9.1) of center O and ratio k.

(d) The inverse of a sphere Σ of center X and passing through O is the plane σ, perpendicular to $\overleftrightarrow{OX}$ and passing through X'.

(e) The inverse of a plane σ not passing through O is a sphere Σ of center X' and passing through O, such that $\overleftrightarrow{OX'} \perp \sigma$ and $\overrightarrow{OX'}$ meets σ at the point X.

(f) The inverse of a sphere Σ of center X and not passing through O is a sphere Σ' of center Y, such that O, X and Y are collinear.

(g) If two spheres, two planes or a sphere and a plane are tangent at a point $T \neq O$, prove that their inverses are tangent at T'. If $T = O$, prove that their inverses are parallel planes.

17. Let Σ be a sphere of center O and radius 1, let Γ be an equator of Σ and α be the plane of Γ. Let also N and S respectively denote the North and South poles of Σ with respect to Γ. If I stands for the inversion of center S and ratio $\sqrt{2}$ in space, prove that:

(a) I applies $\Sigma \setminus \{S\}$ into α, and vice-versa.

(b) Given $A \in \alpha \setminus \{O\}$ with inverse A', if A'' is the intersection of $\overrightarrow{NA'}$ with α, then A'' is the inverse of A' with respect to the inversion of center O and ratio 1 in α.

(c) If Γ_1 is a circle in α, show that $I(\Gamma_1)$ is a circle in Σ.

18. Γ_1 and Γ_2 are spheres of radii 1 and 2, respectively, and are internally tangent at T. Three pairwise tangent identical spheres are also externally tangent to Γ_1 and internally tangent to Γ_2. Compute their radius.

10.4   A Third Look on Conics

This section combines the concepts and results of Solid Geometry we studied so far with the material of Chap. 9 to study yet other properties of conics. In particular, we extend the results of Sects. 9.3 and 9.4 to conics. For more on conics than we will see here, we refer the reader to [1].

As a prelude to the more general result of Theorem 10.26, we start this section by showing that ellipses appear as the plane sections of a cylinder of revolution by planes which are neither parallel nor perpendicular to its axis.

Theorem 10.25 *Let $C(e; R)$ be the cylinder of revolution of axis e and radius R. If plane α is neither parallel nor perpendicular to e, then the corresponding plane section of C (cf. Fig. 10.35) is an ellipse.*

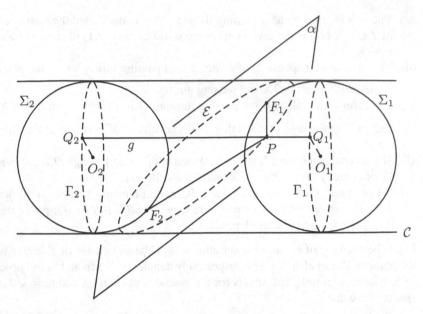

Fig. 10.35 Ellipses as plane sections of a cylinder of revolution

Proof We shall assume without proof the heuristically plausible fact that $\alpha \cap C$ is a *simple closed plane curve*, i.e., a curve contained in a plane, which can be drawn *continuously* (i.e., without taking the pencil out of the paper), is closed (i.e., the drawing ends precisely at the point it has begun) and has no self-intersections.

For $j = 1, 2$, let $\Sigma_j(O_j; R)$ denote a sphere centered at e and tangent to α at F_j (cf. Fig. 10.35). Since α is not perpendicular to e, we have $F_1 \neq F_2$. We claim that F_1 and F_2 are the foci and $\overline{O_1 O_2}$ is the length of the major axis of the ellipse of intersection. To this end, let P be a common point of α and C, and g be the generatrix of C passing through P. For $j = 1, 2$, it is immediate to see that Σ_j intersects C along an equator Γ_j with medial line e, and that g intersects Γ_j at a single point Q_j. Hence, g is tangent to Σ_j at Q_j, and item (a) of Example 10.22 assures that $\overline{PF_j} = \overline{PQ_j}$. Hence,

$$\overline{PF_1} + \overline{PF_2} = \overline{PQ_1} + \overline{PQ_2} = \overline{Q_1 Q_2} = \overline{O_1 O_2},$$

and Theorem 6.22 shows that the plane section is an ellipse.   □

The construction executed in the proof of the theorem above is due to the Belgian mathematician of the nineteenth century Germinal Pierre Dandelin. For this reason, Σ_1 and Σ_2 are usually referred to as the **spheres of Dandelin** of C, relative to the plane α.

Now, we shall see that ellipses, hyperbolas and parabolas appear as plane sections of a cone of revolution. Actually, it is due to this fact that we refer to them as **conics**, or as **conic sections**.

Theorem 10.26 *If C is a cone of revolution and α is a plane not passing through its vertex, then the corresponding plane section of C by α is:*

(a) An ellipse, provided α intersects a single leaf of C and is not parallel to a generatrix.

(b) A hyperbola, in case α intersects both leaves of C.

(c) A parabola, if α intersects only one leaf of C but is parallel to a generatrix.

Proof Let us do the proof of items (b) and (c), leaving the proof of item (a) as an exercise for the reader (cf. Problem 1). For both items, we let V be the vertex and e be the axis of C.

For item (b), as in the proof of Theorem 10.25, we assume without proof the heuristically plausible fact that $\mathcal{H} = C \cap \alpha$ consists of two *simple, open plane curve*,[5] each one contained in a leaf of C (see Fig. 10.36).

Fig. 10.36 Hyperbolas as plane sections of a cone of revolution

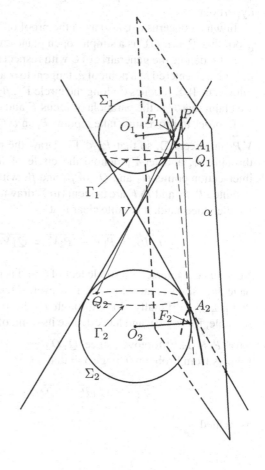

---

[5]Here, *open* stands as the antonym of *closed*, in the sense of the proof of Theorem 10.25.

Let $\Sigma_1(O_1; R_1)$, $\Sigma_2(O_2, R_2)$ denote the spheres centered at points $O_1, O_2 \in e$, tangent to α at F_1, F_2, respectively, and to the leaves of C (this means that Σ_1, Σ_2 are tangent to the generatrices of the cone—cf. Fig. 10.36). Denote by A_1, A_2 the points at which $\overleftrightarrow{F_1F_2}$ intersect C, with A_i lying in the leaf of C to which Σ_i is tangent. We claim that $\mathcal{H}$ is the hyperbola of foci F_1, F_2 and major axis A_1A_2.

For what is left to do, let $\Gamma_i = \Sigma_i \cap C$, so that Γ_1, Γ_2 are circles with medial line e. Take $P \in \mathcal{H}$ and let Q_1, Q_2 be the points of contact of $\overleftrightarrow{PV}$ with Σ_1, Σ_2, respectively. Then, $Q_1 \in \Gamma_1$, $Q_2 \in \Gamma_2$ and, since both PF_i, PQ_i are tangent to Σ_i, item (a) of Example 10.22 gives $\overline{PF_i} = \overline{PQ_i}$ for $i = 1, 2$. Thus,

$$\overline{PF_2} - \overline{PF_1} = \overline{PQ_2} - \overline{PQ_1} = \overline{Q_1Q_2}. \tag{10.1}$$

However, it is pretty clear that $\overline{Q_1Q_2}$ depends only on Σ_1 and Σ_2 (and hence on C and α), thus not on the particular position of P along $\mathcal{H}$. Therefore, $\mathcal{H}$ is indeed a hyperbola.

In what concerns (c), also as in the proof of Theorem 10.25, we assume without proof that $\mathcal{P} = \alpha \cap C$ is a simple, open plane curve (see Fig. 10.37).

Let g denote the generatrix of C with respect to which α is parallel, and $\Sigma(O; R)$ the sphere centered at a point of e, tangent to α at F and to the leaf of the cone that α intersects. If Σ touches C along the circle Γ_2, β_2 is the plane of Γ_2 and $d = \alpha \cap \beta_2$, we claim that $\mathcal{P}$ is the parabola of focus F and directrix d.

In order to prove this, take a point P_1 in $\alpha \cap C$ and mark the point P_2 in which $\overleftrightarrow{VP_1}$ intersects Σ, so that $P_2 \in \Gamma_2$. Draw the plane β_1, parallel to β_2 and passing through P_1, and let Γ_1 denote the circle of intersection of β_1 and C; mark the intersection points Q_1 and Q_2 of β_1 and β_2 with g, respectively.

Since P_1P_2 and P_1F are tangents to Σ drawn through P_1, we have $\overline{P_1F} = \overline{P_1P_2}$. On the other hand, it is quite clear that

$$\overline{P_1P_2} = \overline{P_1V} - \overline{P_2V} = \overline{Q_1V} - \overline{Q_2V} = \overline{Q_1Q_2}.$$

Now, since α intersects a single leaf of C and is parallel to g, item (b) of Problem 15, page 360, shows that $\alpha \perp (g, e)$. However, since we also have $\beta_2 \perp (g, e)$, it follows that $d \perp (g, e)$. Finally, draw through P_1 the parallel to g contained in α (such is possible, for $\alpha \parallel g$), and let R denote its point of intersection with d. Since $d \perp g$, we have $\overleftrightarrow{P_1R} \perp d$. Moreover, since $\overleftrightarrow{Q_1Q_2} = g \parallel \overleftrightarrow{P_1R}$ and $P_1, Q_1 \in \beta_1$, $R, Q_2 \in \beta_2$, it follows from Problem 9, page 349 that $\overline{Q_1Q_2} = \overline{P_1R}$. Thus, we get at last

$$\overline{P_1F} = \overline{P_1P_2} = \overline{Q_1Q_2} = \overline{P_1R} = d(P_1; d),$$

as wished. $\square$

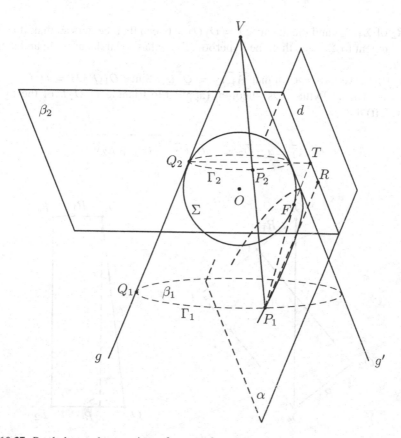

Fig. 10.37 Parabolas as plane sections of a cone of revolution

As in the case of Theorem 10.25, the construction performed in the proof of the previous result is due to Dandelin. For this reason, Σ_1, Σ_2 in item (b) and Σ in item (c) are known as the **spheres of Dandelin** of C relative to α.

At this point, a natural question poses itself: is it true that all ellipses, hyperbolas and parabolas, as defined in Sect. 6.3, appear as plane sections of a cone of revolution? The answer is *yes*, as assured by the next result.

Theorem 10.27 *Every conic can be obtained as the plane section of a cone of revolution by a suitable plane.*

Proof We will prove the theorem again for hyperbolas and parabolas, leaving the case of ellipses to the reader (see Problem 2).

Let $\mathcal{H}$ be a hyperbola of focal distance $2c$ and major axis $2a$. In the notations of the proof of item (b) of Theorem 10.26 and Fig. 10.36, note that C falls completely determined by the choices of R_1, R_2 and $\overline{O_1O_2}$. Moreover, once we have chosen these data, we still have infinitely many possibilities for α, but all of them give equal hyperbolas $C \cap \alpha$. Thus, we only need to show that we are able to choose the radii

R_1, R_2 of Σ_1, Σ_2 and the distance $\ell = \overleftrightarrow{O_1 O_2}$ between their centers so that, if α is a plane tangent to Σ_1, Σ_2, then the hyperbola $C \cap \alpha$ has focal distance $2c$ and major axis $2a$.

In (10.1), we have shown that $\overline{A_1 A_2} = \overline{Q_1 Q_2}$. Since $O_1 \widehat{Q_1} Q_2 = O_2 \widehat{Q_2} Q_1 = 90°$, we get from Pythagoras' theorem (applied to triangle $O_1 O_2 T$ of the figure below, left) that

$$\overline{A_1 A_2} = \overline{Q_1 Q_2} = \sqrt{\overline{O_1 O_2}^2 - (R_1 + R_2)^2}.$$

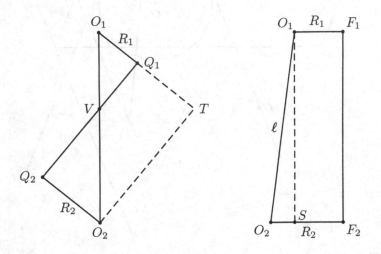

Likewise (see the figure above, right), since $\overleftrightarrow{O_1 F_1}$, $\overleftrightarrow{O_2 F_2} \perp \alpha$, the quadrilateral $O_1 F_1 F_2 O_2$ is a right trapezoid. Then, Pythagoras' theorem (applied to triangle $O_1 O_2 S$ of the figure above, right) gives

$$\overline{F_1 F_2} = \sqrt{\overline{O_1 O_2}^2 - (R_2 - R_1)^2}.$$

Therefore, we only need to show that we can choose R_1, R_2, $\overline{O_1 O_2}$ such that $\overline{O_1 O_2} > R_1 + R_2$ and

$$\sqrt{\overline{O_1 O_2}^2 - (R_1 + R_2)^2} = 2a, \quad \sqrt{\overline{O_1 O_2}^2 - (R_2 - R_1)^2} = 2c.$$

This is immediate: choose any $\ell > 2a$, $2c$, set $\overline{O_1 O_2} = \ell$ and note that the equalities above are equivalent to

$$R_1 + R_2 = \sqrt{\ell^2 - 4a^2}, \quad R_2 - R_1 = \sqrt{\ell^2 - 4c^2};$$

since $c > a$, the solutions of this linear system of equations are positive.

Now, let $\mathcal{P}$ be a parabola of parameter p. In the notations of the proof of item (c) of Theorem 10.26 and Fig. 10.37, note that C falls completely determined by the choices of R and $\overline{VO}$. Moreover, once we have chosen these data, we still have infinitely many possibilities for α, but all of them give equal parabolas $C \cap \alpha$. Thus, we only need to show that we are able to choose the radius R of Σ and the distance $\ell = \overleftrightarrow{VO}$ such that $\ell > R$ and, if α is a plane tangent to Σ and parallel to a generatrix of C, then the parabola $C \cap \alpha$ has parameter p.

To what is left to do, note that $\overleftrightarrow{VO} \perp \overleftrightarrow{Q_2T}$ and $\overleftrightarrow{VQ_2} \perp \overleftrightarrow{Q_2F}$ implies $Q_2\widehat{V}O = F\widehat{Q_2}T$. Since $V\widehat{Q_2}O = Q_2\widehat{F}T = 90°$, the right triangles VQ_2O and Q_2FT are similar by the AA case of similarity.

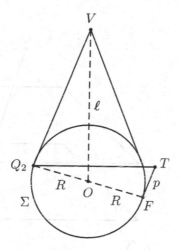

Hence, $\frac{V_2Q}{Q_2O} = \frac{Q_2F}{FT}$ or, which is the same, $\frac{\sqrt{\ell^2 - R^2}}{R} = \frac{2R}{p}$. Solving for ℓ, we find

$$\ell = \sqrt{\frac{4R^4}{p^2} + R^2}.$$

Thus, it suffices to choose any $R > 0$, then let ℓ be given by the last expression above.   □

Apart from the results above, in order to be able to extend the results of Sects. 9.3 and 9.4 to conics we need to digress a bit on the properties of a *central* projection from a plane onto another. We now turn to this concept.

Given in space distinct planes α and β and a point O outside of $\alpha \cup \beta$, there are two possibilities: $\alpha \parallel \beta$ or $\alpha \nparallel \beta$.

In the first case (see the Fig. 10.38), for each point $A \in \alpha$ the line $\overleftrightarrow{OA}$ meets β at a point B, and the map $\pi : \alpha \to \beta$ that sends A to B is a bijection. Such a map is called the **central projection** of center O, from α onto β.

In the second case, Fig. 10.39 shows planes α and β, as well as planes α', β', respectively parallel to α, β and passing through O. Since $\alpha \nparallel \beta$, we also have $\alpha' \nparallel \beta$, $\alpha \nparallel \beta'$, and we let $r_\alpha = \alpha \cap \beta'$, $r_\beta = \beta \cap \alpha'$.

If $A \in r_\alpha$, then $\overleftrightarrow{OA} \subset \beta'$, hence $\overleftrightarrow{OA} \cap \beta = \emptyset$. If $A \in \alpha \setminus r_\alpha$ then $\overleftrightarrow{OA} \not\subset \beta'$, so that $\overleftrightarrow{OA}$ intersects β at a single point B. Thus, we let $\pi : \alpha \setminus r_\alpha \to \beta$ be the map that sends A to B. Note, however, that the image of π is not all of β. Indeed, if $B \in r_\beta$ then $\overleftrightarrow{OB} \parallel \alpha$, thus there exists no $A \in \alpha$ for which $\pi(A) = B$. Nevertheless,

Fig. 10.38 Central projection $\pi : \alpha \to \beta$, for $\alpha \parallel \beta$

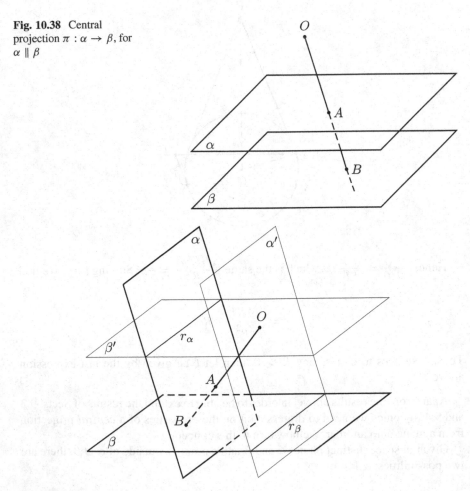

Fig. 10.39 Central projection $\pi : \alpha \to \beta$, for $\alpha \nparallel \beta$

if $B \in \beta \setminus r_\beta$, then $\overleftrightarrow{OB}$ intersects α at a single point A, and $\pi(A) = B$. Therefore, $\pi : \alpha \setminus r_\alpha \to \beta \setminus r_\beta$ is a bijective map. In this case too, this map is the **central projection** of center O, from $\alpha \setminus r_\alpha$ onto $\beta \setminus r_\beta$.

In the above notations, let $\pi : \alpha \setminus r_\alpha \to \beta \setminus r_\beta$ be the central projection of center O. If $\mathcal{F} \subset \alpha$ intersects r_α at a finite number of points, then π maps $\mathcal{F} \setminus r_\alpha$ bijectively onto $\pi(\mathcal{F} \setminus r_\alpha)$.

We are finally in position to state and prove the following consequence of Theorem 10.27.

Corollary 10.28 *Given an ellipse (resp. hyperbola, parabola) in a plane α, there exists another plane β and a point $O \notin \alpha \cup \beta$ such that the central projection $\pi : \alpha \to \beta$ maps the ellipse (resp. hyperbola, parabola) onto a circle (resp. a circle minus two points, a circle minus one point).*

Proof We stick to the notations of the proof of Theorem 10.27, as well as of the discussion that precedes the statement of the corollary. Once more, we only work the proof for hyperbolas and parabolas; the proof for ellipses parallels these and is much easier.

For the case of a hyperbola $\mathcal{H}$, the proof of Theorem 10.27 shows that there exists a cone of revolution C and a plane α such that $\mathcal{H} = C \cap \alpha$. In the notations of Fig. 10.36, let V be the vertex of C, let $\Gamma = \Gamma_1$, β be the plane of Γ and $\pi : \alpha \setminus r_\alpha \to \beta \setminus r_\beta$ be the central projection of center V, from $\alpha \setminus r_\alpha$ onto $\beta \setminus r_\beta$.

Since β' is the plane parallel to β through V and $r_\alpha = \alpha \cap \beta'$, it is clear that r_α does not intersect $\mathcal{H}$. On the other hand, since α' is the plane parallel to α through V and $r_\beta = \alpha' \cap \beta$, it is clear that r_β intersects Γ at two points X, Y.

Now, for each $P \in \mathcal{H}$, let $\overleftrightarrow{VP} \cap \beta = \{Q\}$. Since $\overleftrightarrow{VP}$ is also a generatrix of the cone, we have $Q \in \beta \cap C = \Gamma$, and hence $Q \in \Gamma \setminus \{X, Y\}$. Conversely, if $Q \in \Gamma \setminus \{X, Y\}$, then working the previous argument backwards we prove that there exists a point $P \in \mathcal{H}$ such that $\overleftrightarrow{VP} \cap \beta = \{Q\}$. Therefore, $\pi(\mathcal{H}) = \Gamma \setminus \{X, Y\}$.

The proof in the case of a parabola $\mathcal{P}$ is pretty much the same: the proof of Theorem 10.27 shows that there exists a cone of revolution C and a plane α such that $\mathcal{P} = C \cap \alpha$. In the notations of Fig. 10.37, let V be the vertex of C, $\beta = \beta_2$, $\Gamma = \Gamma_2$ and $\pi : \alpha \setminus r_\alpha \to \beta \setminus r_\beta$ be the central projection of center V, from $\alpha \setminus r_\alpha$ onto $\beta \setminus r_\beta$.

Since β' is the plane parallel to β through V and $r_\alpha = \alpha \cap \beta'$, it is clear that r_α does not intersect $\mathcal{P}$. On the other hand, since α' is the plane parallel to α through V and $r_\beta = \alpha' \cap \beta$, it is clear that r_β intersects Γ at Q_2.

Now, for each $P \in \mathcal{P}$, let $\overleftrightarrow{VP} \cap \beta = \{Q\}$. Since $\overleftrightarrow{VP}$ is also a generatrix of the cone, we have $Q \in \beta \cap C = \Gamma$, and hence $Q \in \Gamma \setminus \{Q_2\}$. Conversely, if $Q \in \Gamma \setminus \{Q_2\}$, then working the previous argument backwards we prove that there exists a point $P \in \mathcal{P}$ such that $\overleftrightarrow{VP} \cap \beta = \{Q\}$. Therefore, $\pi(\mathcal{P}) = \Gamma \setminus \{Q_2\}$. $\square$

We now address the other properties of central projections that, together with the previous corollary, make these maps so important for us.

In the sequel, we only consider central projections between two nonparallel planes, for the case of parallel planes is much easier, generally reducing to similarity arguments. In the statement and proof of the coming result, as well as in the subsequent comments we make, we shall systematically denote the image of a point A (resp. a figure $\mathcal{F}$) with respect to a given central projection by writing A' (resp. $\mathcal{F}'$).

Proposition 10.29 *Let* $\pi : \alpha \setminus r_\alpha \to \beta \setminus r_\beta$ *be the central projection of center O.*

(a) *If s is a line of α distinct from r_α, then s' is a line of β.*

(b) *If lines $s, t \subset \alpha$ are distinct from r_α and from one another, and concur at a point A, then lines s', t' are either parallel or concur at A', according to whether $A \in r_\alpha$ or not.*

(c) *If $A, B, C, D \in \alpha \setminus r_\alpha$, then $(A', B'; C', D') = (A, B; C, D)$.*

Proof For item (a), let γ denote the plane $\gamma = (O, s)$. If $\gamma \parallel \beta$ then, since γ passes through O, we would have $r_\alpha \subset \gamma$. However, since $\alpha \nparallel \beta$, we would also have $\alpha \nparallel \gamma$, so that $s \subset \alpha \cap \gamma = r_\alpha$. This contradicts our choice of s. Now, since $\gamma \nparallel \beta$, we have $s' = \pi(s) = \gamma \cap \beta$, which is a line of β.

For item (b), if $A \notin r_\alpha$ we obviously have $A' = \pi(A) \in \pi(s) = s'$, and likewise $A' \in t'$. Since $s \neq t$, we also have $s' \neq t'$, so that $s' \cap t' = \{A'\}$. On the other hand, if $A \in r_\alpha$, then s', t' continue to be distinct lines of β; however, in this case they cannot have a common point, for otherwise such a point would be the image of A under π (and r_α, whence A, has no image under π). Thus, s', t' are coplanar lines without a common point, whence parallel.

Item (c) follows from Theorem 9.28, together with the fact that quadruples (A, B, C, D) and (A', B', C', D') are in perspective from O. □

It is worth referring to the results of items (a) and (b) of the proposition above by saying that central projections preserve **incidence**, i.e., collinearity of points and concurrence of lines (that concur at a point not lying in r_α). On the other hand, item (c) can be stated in words by saying that *central projections preserve cross ratios*.

We now collect some important consequences of items (a) and (b) of the proposition. We urge the reader to draw (at least mentally) the corresponding pictures, in order to make sure he/she properly understands what is being said.

Corollary 10.30 *Let* $\pi : \alpha \setminus r_\alpha \to \beta \setminus r_\beta$ *be the central projection of center O.*

(a) *If the quadruples (A_1, B_1, C_1, D_1), (A_2, B_2, C_2, D_2) of points of $\alpha \setminus r_\alpha$ are in perspective from a point $P \notin r_\alpha$, then $A_1' B_1' C_1' D_1' \overset{P'}{\overline{\wedge}} A_2' B_2' C_2' D_2'$.*

(b) *If lines a, b, c, d of α are distinct from r_α and form a projective pencil (a, b, c, d) with center $P \notin r_\alpha$, then (a', b', c', d') is a projective pencil of β with center P' and $(a', b'; c', d') = (a, b; c, d)$.*

(c) *Let the lines a_i, b_i, c_i, d_i of α $(i = 1, 2)$ be distinct from r_α and form projective pencils (a_1, b_1, c_1, d_1), (a_2, b_2, c_2, d_2) in α, with centers $P, Q \notin r_\alpha$, respectively. If (a_1, b_1, c_1, d_1), (a_2, b_2, c_2, d_2) are in perspective from a line*

$s \neq r_\alpha$ and $a_1 \cap a_2, b_1 \cap b_2, c_1 \cap c_2, d_1 \cap d_2 \notin r_\alpha$, then $a'_1 b'_1 c'_1 d'_1 \overset{s'}{\overline{\wedge}} a'_2 b'_2 c'_2 d'_2$.
In particular, $(a'_1, b'_1; c'_1, d'_1) = (a'_2, b'_2; c'_2, d'_2)$.

(d) Let the lines a, b, c, d_1, d_2 of α be distinct from r_α and form projective pencils (a, b, c, d_1), (a, b, c, d_2) in α, with center $P \notin r_\alpha$. If $(a', b'; c', d'_1) = (a', b'; c', d'_2)$, then $d_1 = d_2$.

Proof Item (a) and the first part of item (b) follow from the fact that central projections preserve incidence. For the second part of (b), let s be a line of α with $a \cap s = \{A\}, b \cap s = \{B\}, c \cap s = \{C\}, d \cap s = \{D\}$, such that $A, B, C, D \notin r_\alpha$. Then, item (c) of the proposition gives

$$(a', b'; c', d') = (A', B'; C', D') = (A, B; C, D) = (a, b; c, d).$$

The first part of (c) follows again from the incidence-preserving property of central projections. The last part follows from Corollary 9.30. Finally, for (d), Corollary 9.31 gives $d'_1 = d'_2$, so that $d_1 = d_2$. $\square$

In spite of the above the reader is warned that, hereafter, whenever there is no danger of confusion we shall generally adopt the following conventions: we shall generally write $\pi : \alpha \to \beta$, instead of $\pi : \alpha \setminus r_\alpha \to \beta \setminus r_\beta$, to denote the central projection of center O, mapping $\alpha \setminus r_\alpha$ bijectively onto $\beta \setminus r_\beta$. Also, given a figure $\mathcal{F} \subset \alpha$, with $\mathcal{F} \cap r_\alpha$ consisting of at most a finite number of points, we shall write $\mathcal{F}' = \pi(\mathcal{F})$, even though we are thinking of $\pi(\mathcal{F} \setminus r_\alpha)$. In particular, if each of $\mathcal{F} \subset \alpha$ and $\mathcal{G} \subset \beta$ meet r_α and r_β, respectively, at most a finite number of times and satisfy $\pi(\mathcal{F} \setminus r_\alpha) = \mathcal{G} \setminus r_\alpha$, we shall simply write (somewhat imprecisely but rather efficiently) $\mathcal{F}' = \mathcal{G}$.

The discussion above allows us to give an almost complete proof[6] to the following result.

Theorem 10.31 *Let $\mathcal{C}$ be a conic and P be a point not in $\mathcal{C}$. Let r and s be secants to $\mathcal{C}$ passing through P, with $\mathcal{C} \cap r = \{A, B\}, \mathcal{C} \cap s = \{C, D\}$. If $\overleftrightarrow{AC} \cap \overleftrightarrow{BD} = \{R\}$ and $\overleftrightarrow{AD} \cap \overleftrightarrow{BC} = \{Q\}$, then the straightline $\overleftrightarrow{QR}$ does not depend on the choices of r and s. Moreover, if $\overleftrightarrow{QR} \cap \mathcal{C} = \{T_1, T_2\}$, then $\overleftrightarrow{PT_1}$ and $\overleftrightarrow{PT_2}$ are tangent to $\mathcal{C}$.*

Figure 10.40 illustrates the theorem above for an ellipse, and for the sake of simplicity we have chosen to take P outside the bounded region delimited by the ellipse; however, this is completely immaterial. We urge the reader to sketch the cases of a hyperbola and a parabola.

Proof Take a central projection that maps $\mathcal{C}$ onto a circle $\mathcal{C}'$. We assume that, under such a projection, points $A, B, C, D, P, Q, R, T_1, T_2$ map to $A', B', C', D', P', Q', R', T'_1, T'_2$, respectively (this is always the case if $\mathcal{C}$ is an ellipse).

---

[6]The difference from *almost complete* to *complete* lies in the fact that we are not considering *points at infinity*. If we were, then we would get a complete proof.

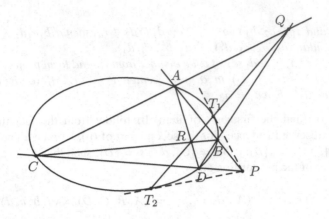

Fig. 10.40 Complete quadrangles inscribed in an ellipse

Theorem 9.23 shows that $\overleftrightarrow{Q'R'}$ is always the polar p of P' with respect to C'. Therefore, $\overleftrightarrow{QR}$ is the inverse image of p with respect to the central projection, and as such does not depend on the choices of r and s. It also shows that $\overleftrightarrow{P'T_1'}$ and $\overleftrightarrow{P'T_2'}$ are tangent to C', and hence (arguing again with the inverse of the central projection) $\overleftrightarrow{PT_1}$ and $\overleftrightarrow{PT_2}$ are tangent to C.   □

In the notations of the statement of the previous result, if $P \notin C$ we say that $\overleftrightarrow{QR}$ is the **polar of P with respect to C**. If $P \in C$, we define the polar of P with respect to C as the tangent to C passing through P.

If $P \notin C$ and we can draw tangents to C passing through P, the previous result shows that the polar of P with respect to C is line $\overleftrightarrow{T_1T_2}$, where T_1, T_2 are the points of contact of such tangents with C. Actually, that result also shows how one can find T_1, T_2 with the aid of a straightedge, only. In this respect, see Problem 4.

We now extend Proposition 9.21 to conics. In this case too, we get an almost complete proof.

Proposition 10.32 *We are given in the plane a conic C and points P and Q. If p is the polar of P and q is that of Q, then $P \in q \Leftrightarrow Q \in p$.*

Figure 10.41 illustrates the result, again for an ellipse.

Proof Take a central projection that maps C onto a circle C'. We assume that, under such a projection, points P, Q map to points P', Q' and lines p, q map to lines p', q' (once more, this is always so if C is an ellipse). The previous discussion assures that p' is the polar of P' and q' is that of Q'. Hence, Proposition 9.21 gives $P' \in q' \Leftrightarrow Q' \in p'$, and applying the inverse of the central projection we obtain $P \in q \Leftrightarrow Q \in p$.   □

A useful way of rephrasing the previous result is as in the following

Fig. 10.41 $P \in q \Leftrightarrow Q \in p$

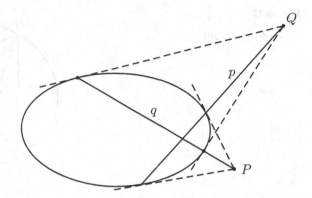

Corollary 10.33 *Let C be a conic, P be a point not in C and RS be a chord of C passing through P. If the tangents to C through R and S are not parallel, then they meet at the polar of P.*

Proof Let r and s be the tangents to C drawn through R and S, respectively, and assume that $r \cap s = \{Q\}$. Since $\overleftrightarrow{QR} = r$ and $\overleftrightarrow{QS} = s$ are tangent to C, the discussion following Theorem 10.31 assures that $\overleftrightarrow{RS}$ is the polar of Q with respect to C. However, since $P \in \overleftrightarrow{RS}$, the previous result shows that Q lies in the polar of P. $\quad\square$

An important particular case of this corollary is isolated in the next one.

Corollary 10.34 *If C is a conic with focus F, then the polar of F is the corresponding directrix.*

Proof This is immediate from the previous corollary, together with Theorem 6.34.
$\quad\square$

We now extend the theorems of Pascal and Brianchon to conics. To this end, we need to extend yet another notion presented in Sect. 9.4, and we do this in the following

Definition 10.35 Given projective pencils (a_1, b_1, c_1, d_1), (a_2, b_2, c_2, d_2) in a plane α, of centers O_1, O_2, respectively, we say that they are **in perspective from a conic** C if the six points O_1, O_2, $a_1 \cap a_2$, $b_1 \cap b_2$, $c_1 \cap c_2$, $d_1 \cap d_2$ are pairwise distinct and lie in C. Also, in this case we write

$$a_1 b_1 c_1 d_1 \overset{C}{\barwedge} a_2 b_2 c_2 d_2.$$

Figure 10.42 illustrates the definition above in the case of an ellipse.

We now extend Theorem 9.33 for projective pencils in perspective from a conic. Here again we present an almost complete proof, which is complete for ellipses.

Fig. 10.42

$$a_1 b_1 c_1 d_1 \overset{\varepsilon}{\wedge} a_2 b_2 c_2 d_2$$

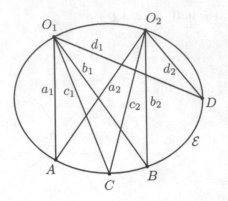

Theorem 10.36 *If the projective pencils* (a_1, b_1, c_1, d_1), (a_2, b_2, c_2, d_2) *of a plane* α *are in perspective from a conic* $\mathcal{C}$*, then* $(a_1, b_1; c_1, d_1) = (a_2, b_2; c_2, d_2)$. *In symbols,*

$$a_1 b_1 c_1 d_1 \overset{\mathcal{C}}{\wedge} a_2 b_2 c_2 d_2 \Rightarrow (a_1, b_1; c_1, d_1) = (a_2, b_2; c_2, d_2).$$

Proof Choose a plane β and a point $O \notin \alpha \cup \beta$ such that the central projection $\pi : \alpha \to \beta$ maps $\mathcal{C}$ onto a circle Γ. Then, (a_1, b_1, c_1, d_1), (a_2, b_2, c_2, d_2) map under π to projective pencils (a_1', b_1', c_1', d_1'), (a_2', b_2', c_2', d_2') of β, which are in perspective from Γ. Theorem 9.33 thus gives $(a_1', b_1'; c_1', d_1') = (a_2', b_2'; c_2', d_2')$, whence Corollary 10.30 shows that $(a_1, b_1; c_1, d_1) = (a_2, b_2; c_2, d_2)$. $\quad\square$

As anticipated above, we now extend Pascal's theorem from circles to conics. The extension of Brianchon's theorem will be the object of Problem 8. We also leave to the reader the (straightforward) task of defining what it means for an hexagram $ABCDEF$ to be inscribed in or tangential to a conic $\mathcal{C}$.

Theorem 10.37 (Pascal) *Let $ABCDEF$ be an hexagram inscribed in a conic. If* $\overleftrightarrow{AE} \cap \overleftrightarrow{BD} = \{Z\}$, $\overleftrightarrow{AF} \cap \overleftrightarrow{CD} = \{Y\}$, $\overleftrightarrow{BF} \cap \overleftrightarrow{CE} = \{X\}$, *then points X, Y, Z are collinear.*

Figure 10.43 illustrates Pascal's theorem in an ellipse, and we suggest the reader to depict the cases of a hyperbola and a parabola.

Proof Choose a plane β and a point $O \notin \alpha \cup \beta$ such that the central projection $\pi : \alpha \to \beta$ maps $\mathcal{C}$ onto a circle Γ. If $ABCDEF$ is mapped onto $A'B'C'D'E'F'$ and X, Y, Z onto X', Y', Z', then $\overleftrightarrow{A'E'} \cap \overleftrightarrow{B'D'} = \{Z'\}$, $\overleftrightarrow{A'F'} \cap \overleftrightarrow{C'D'} = \{Y'\}$, $\overleftrightarrow{B'F'} \cap \overleftrightarrow{C'E'} = \{X'\}$. Pascal's theorem for circles now assures that X', Y', Z' are collinear, and hence X, Y, Z are also collinear. $\quad\square$

Fig. 10.43 Pascal's theorem
for an ellipse

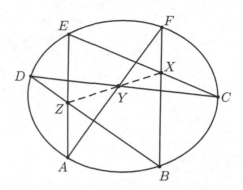

Problems: Sect. 10.4

1. * Complete the proof of Theorem 10.26 by establishing item (a).
2. * Show that every ellipse can be realized as the plane section of a suitable cylinder of revolution and a suitable cone of revolution.
3. Let C be a cone of revolution of vertex V, let α be a plane intersecting both leaves of C and $\mathcal{H} = C \cap \alpha$. Let Σ be one of the Dandelin's spheres of the hyperbola $\mathcal{H}$, let $\Gamma = C \cap \Sigma$ and β be the plane of Γ. If $\pi : \alpha \to \beta$ is the central projection that maps $\mathcal{H}$ to Γ, identify the straightlines of the plane β that are the images of the asymptotes of $\mathcal{H}$ under π.
4. * We are given a conic C and a point $P \notin C$, such that we can draw tangents to C passing through P. Show how such tangents can be drawn with the aid of a straightedge only.
5. Let $\pi : \alpha \backslash r_\alpha \to \beta \backslash r_\beta$ be a central projection. For $X \in \alpha \backslash r_\alpha$, write X' to denote $\pi(X)$. If (A, B, P, Q) is a harmonic quadruple in α, with $A, B, P, Q \notin r_\alpha$, show that (A', B', P', Q') is a harmonic quadruple in β. If $Q \in r_\alpha$, show that P' is the midpoint of $A'B'$.
6. Let C be a conic and P be a point not in C. Let r be a secant to C passing through P, with $C \cap r = \{A, B\}$. If the polar of P intersects r at U, then (A, B, P, U) is a harmonic quadruple.
7. We are given conics C_1 and C_2, and points Q_1 and Q_2. Show how to find, if they exist, all points P with the following property: for $i = 1, 2$, the tangents drawn from P to C_i touch it at the endpoints of a chord passing through Q_i.
8. Extend Brianchon's theorem 9.35 to conics.
9. Let $ABCD$ be a quadrilateral circumscribed to an ellipse (i.e., such that the sides AB, BC, CD, AD are tangent to the ellipse at the points M, N, P, Q. Prove that lines $\overleftrightarrow{AC}$, $\overleftrightarrow{BD}$, $\overleftrightarrow{MP}$, $\overleftrightarrow{NQ}$ are concurrent.

Problems, Sect. 10.1

Chapter 11
Some Simple Solids

In this chapter we apply the concepts and results of the previous one to the study of prisms and pyramids. In particular, we discuss in some detail the geometry of tetrahedra, which, in Solid Geometry, play a role analogous to that of triangles in Plane Geometry.

Along this chapter (and in the coming ones, too) we shall need the following concepts: given in space a point P and a positive real number R, we define the **open ball** of center P and radius R as the set $\mathcal{B}(P; R)$ of points in space whose distance from P is less than R:

$$\mathcal{B}(P; R) = \{Q; \ \overline{PQ} < R\}.$$

A set $\mathcal{A}$ of points in space is **bounded** if $\mathcal{A}$ lies inside an open ball (Fig. 11.1).

11.1 Pyramids and Tetrahedra

Given a convex n-gon $A_1 A_2 \ldots A_n$ and a point V not in the plane of $A_1 A_2 \ldots A_n$, we define the **pyramid** $V A_1 A_2 \ldots A_n$, of **basis** $A_1 A_2 \ldots A_n$ and **vertex** (or **apex**) V, as the bounded portion of space, delimited by $A_1 A_2 \ldots A_n$ and by the triangles $V A_i A_{i+1}$, for $1 \leq i \leq n$, with the convention that $A_{n+1} = A_1$ (cf. Fig. 11.2). We say that the line segments $V A_i$ and $A_i A_{i+1}$ are the **edges** and that triangles $V A_i A_{i+1}$ are the **lateral faces** of the pyramid. Finally, whenever convenient and there is no danger of confusion, we shall collectively refer to the basis and lateral faces of a pyramid as its *faces*. The **interior** of a pyramid is the set of points of a pyramid which does not belong to any of its faces.

Since the basis of the pyramid defined above is an n-gon, we shall say that it is an **n-sided pyramid**. In the cases $n = 3$ and $n = 4$, we shall more frequently call them a **triangular pyramid** and **quadrangular pyramid**, respectively; moreover, we

© Springer International Publishing AG, part of Springer Nature 2018
A. Caminha Muniz Neto, *An Excursion through Elementary Mathematics, Volume II*,
Problem Books in Mathematics, https://doi.org/10.1007/978-3-319-77974-4_11

shall denote their bases as ABC and $ABCD$, also respectively (with the convention, in the quadrangular case, that AC and BD are the diagonals of $ABCD$).

A triangular pyramid $VABC$ (of basis ABC) can also be seen as a triangular pyramid of basis VAB, VAC or VBC (and vertices C, B or A, according to the case). This point of view is mostly important in the cases in which the choice of a particular basis is irrelevant and, if this is so, then we shall refer to $VABC$ simply as the **tetrahedron** of **vertices** V, A, B and C, and **faces** VAB, VAC, VBC and ABC (cf. Fig. 10.24, where we show a tetrahedron $ABCD$).

If P is the foot of the perpendicular dropped from the vertex V to the plane of the basis $A_1A_2 \ldots A_n$ of an n-sided pyramid $VA_1A_2 \ldots A_n$, then we shall say that the line segment VP (or sometimes its length) is the **altitude** or **height** of the pyramid (cf. Fig. 11.2). Such a pyramid is called **regular** (cf. Fig. 11.3) if its basis $A_1A_2 \ldots A_n$ is a regular polygon of center P. In this case, since $\overline{A_iP} = \overline{A_{i+1}P}$, $V\widehat{P}A_i = V\widehat{P}A_{i+1} = 90°$ and VP is a common side, we conclude that $VPA_i \equiv VPA_{i+1}$ by SAS, and hence that $\overline{VA_i} = \overline{VA_{i+1}}$ for $1 \leq i \leq n$.

In the notations of the discussion above, and applying Pythagoras' theorem to triangle VPA_1, we shall promptly get the following result.

Fig. 11.1 The open ball of center P and radius R

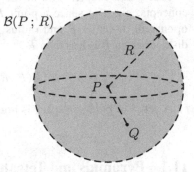

$\mathcal{B}(P\,;\,R)$

Fig. 11.2 Pyramid of basis $A_1A_2 \ldots A_n$ and apex V

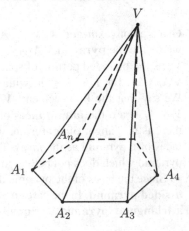

Fig. 11.3 A regular pyramid
of basis $A_1 A_2 \ldots A_n$ and
vertex V

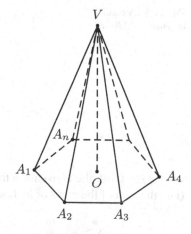

Fig. 11.4 An n-sided regular
pyramid inscribed in a cone
of revolution

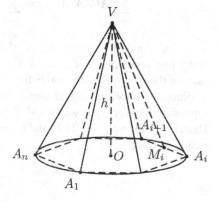

Proposition 11.1 *Let* $V A_1 A_2 \ldots A_n$ *be a regular pyramid of vertex* V *and basis* $A_1 A_2 \ldots A_n$. *If* R *is the circumradius of the basis and* h *is the (measure of the) altitude of the pyramid, then the measure* l *of the lateral edges of it is given by* $l = \sqrt{R^2 + h^2}$.

As an application of the concepts above, let $\mathcal{C}(e; V; \theta)$ be a cone of revolution of axis e, vertex V and opening 2θ, and let α be a plane perpendicular to the axis e of the cone at the point O, with $\overline{VO} = h > 0$. Then, α intersects the cone in a circle $\Gamma(O; R)$ and, hereafter, whenever there is no danger of confusion, we shall also refer to the portion of $\mathcal{C}(e; V; \theta)$ situated between V and α as the **cone of revolution** of radius R, height h and generatrix $g = \sqrt{R^2 + h^2}$; the circle Γ is the **basis** of the cone.

The concept of regular pyramid allows us to define, through the following heuristic argument, the *lateral area* of the cone of revolution described in the previous paragraph. To this end, let $n \geq 3$ be a natural number and $A_1 A_2 \ldots A_n$ be a regular n-gon inscribed in Γ (cf. Fig. 11.4). The n-sided pyramid of basis $A_1 A_2 \ldots A_n$ and vertex V is regular and has height h; letting M_i denote the midpoint

Fig. 11.5 A regular
tetrahedron $ABCD$

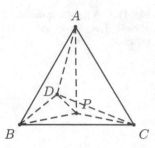

of $A_i A_{i+1}$ (with the convention that $A_{n+1} = A_1$), the *lateral area* of the pyramid
(i.e., the sum of the areas of its lateral faces) is equal to

$$\sum_{i=1}^{n} A(V A_i A_{i+1}) = \sum_{i=1}^{n} \frac{1}{2} \overline{A_i A_{i+1}} \cdot \overline{V M_i} = \frac{1}{2} p_n a_n,$$

where p_n stands for the perimeter of $A_1 A_2 \ldots A_n$ and a_n for the common length
of the line segments $\overline{V M_1}, \ldots, \overline{V M_n}$. Now, as $n \to +\infty$, the union of the lateral
faces of the pyramid better and better approximates the lateral portion of the cone of
revolution. Hence, the lateral area of the pyramid also forms an increasingly better
approximation for what we would like to define as the **lateral area** A of the cone.
Therefore, since $p_n \to 2\pi R$ and $a_n \to g$ as $n \to +\infty$, we define such a lateral
area by setting

$$A = \frac{1}{2} \cdot 2\pi R g = \pi R g. \tag{11.1}$$

Back to tetrahedra, since any face of a tetrahedron can be seen as its basis, the line
segment joining a vertex of a tetrahedron to the foot of the perpendicular dropped
to the opposite face will be called an **altitude** of the tetrahedron. In particular, every
tetrahedron has exactly four altitudes.

Even though every tetrahedron is a triangular pyramid and vice-versa, we shall
follow the standard practice of letting the expression *regular tetrahedron* refer to a
more restrictive notion than *regular triangular pyramid*. More precisely, we say that
$ABCD$ is a **regular tetrahedron** if its six edges have the same length (cf. Fig. 11.5).
The coming example clears the relation between these two notions of regularity.

Example 11.2 Every regular tetrahedron is a regular triangular pyramid (with
respect to any one of its faces taken as basis).

Proof In the notations of Fig. 11.5, let $ABCD$ be a regular tetrahedron and P be
the foot of the altitude dropped from A to the face BCD. Then, triangles ABP,
ACP and ADP are right at P and such that $\overline{AB} = \overline{AC} = \overline{AD}$. Hence, by the
special case of congruence of right triangles, we get $\overline{BP} = \overline{CP} = \overline{DP}$, so that P

is the circumcenter of BCD. However, since BCD is equilateral, we conclude that $ABCD$ is a regular triangular pyramid of basis BCD. □

Proposition 11.1 can be used to justify the existence of regular tetrahedra (nothing that we have done so far assures this) and explain under what conditions the converse of the previous example holds. We do this next.

Example 11.3 Let $ABCD$ be a regular triangular pyramid of vertex A and basis BCD. If a is the length of the edges of the base and h is the altitude of $ABCD$ (also with respect to BCD), then $ABCD$ is a regular tetrahedron if and only if $h = \frac{a\sqrt{6}}{3}$.

Proof Yet in the notations of Fig. 11.5, let P denote the center of face BCD. The definition of regular triangular pyramid (and the previous example) assures that $\overleftrightarrow{AP} \perp (BCD)$, whence $\overline{AP} = h$. On the other hand, letting R denote the circumradius of BCD (which is equilateral), we have $R = \frac{a}{\sqrt{3}}$. Hence, it follows from Proposition 11.1 that

$$\overline{AB} = \overline{AC} = \overline{AD} = \sqrt{\frac{a^2}{3} + h^2},$$

from where we conclude that $ABCD$ is a regular tetrahedron if and only if $\sqrt{\frac{a^2}{3} + h^2} = a$, i.e., if and only if $h = \frac{a\sqrt{6}}{3}$. □

As we shall see in Sect. 12.3, regular tetrahedra form one of the five types of *regular convex polyhedra*.

Continuing with the presentation of the basics facts on pyramids, let be given a pyramid of apex V and basis $A_1 A_2 \dots A_n$. We would like to address the problem of deciding whether there does exist a sphere Σ:

(1) passing through $V, A_1, A_2, \dots, A_n$;
(2) contained in the pyramid and tangent to its basis and lateral faces.

In case (1) holds, we say that Σ is **circumscribed** to the pyramid, or also that the pyramid is *inscribed in* Σ; if (2) holds, then Σ is said to be **inscribed** in the pyramid, and the pyramid is said to be *circumscribed to* Σ (see Fig. 11.6).

It follows from Problems 2 and 3, that if a pyramid possess a circumscribed (resp. inscribed) sphere, then such a sphere is unique. Nevertheless, it is easy to see that a general pyramid need not have a circumscribed or inscribed sphere. In the positive direction, the next result shows that every tetrahedron does have a circumscribed and an inscribed sphere. For its statement, the reader may find it convenient to recall (cf. Problem 3, page 359) that the locus of the points in space equidistant from the faces of a trihedral angle is a half-line departing from its vertex.

Proposition 11.4 *In every tetrahedron:*

(a) *There is a single point O, called the **circumcenter** of the tetrahedron, which lies at equal distances from the vertices of it. In particular, the sphere of center*

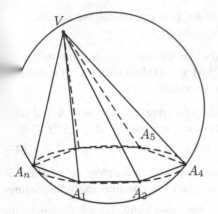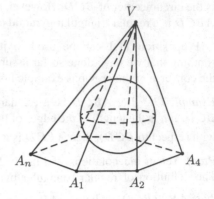

Fig. 11.6 A pyramid admitting a circumscribed (left) and inscribed (right) sphere

Fig. 11.7 The sphere
circumscribed to a
tetrahedron

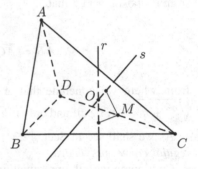

    *O and radius equal to the common distance from O to the vertices of the
tetrahedron is its only circumscribed sphere.*

(b) *There is a single point I, called the **incenter** of the tetrahedron, which lies at
equal distances from the faces of it. In particular, the sphere of center I and
radius equal to the common distance from I to the faces of the tetrahedron is
its only inscribed sphere.*

Proof Let $ABCD$ be a given tetrahedron.

(a) If r and s respectively denote the medial lines of faces BCD and ACD (cf.
Fig. 11.7), then r and s are contained in the medial plane of CD and are not parallel
(for $(BCD) \neq (ACD)$). Hence, r and s intersect each other at a point O, so that O
is equidistant from the vertices B, C, D, as well as from the vertices A, C, D. Thus,
O lies at equal distances from all of the vertices of $ABCD$. Letting R denote the
common distance from O to the vertices of $ABCD$, we conclude that the sphere of
center O and radius R is circumscribed to the tetrahedron.

    Conversely, if O' is the center of a sphere circumscribed to $ABCD$, it follows
from $\overline{O'B} = \overline{O'C} = \overline{O'D}$ that O' lies in the medial line r of BCD. Likewise, O'
also belongs to the medial line s of ACD, whence O' coincides with O.

Fig. 11.8 The sphere
inscribed in a tetrahedron

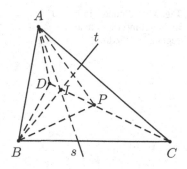

(b) In the trihedral angle of vertex A and faces $\angle BAC$, $\angle BAD$, $\angle CAD$, let (cf. Problem 3, page 359) s be the (half-line) locus of points in space equidistant from its faces; accordingly, we let t be the (half-line) locus of points in space equidistant from the faces of the trihedral angle of vertex B and faces $\angle ABC$, $\angle ABD$, $\angle CBD$ (cf. Fig. 11.8).

The proof of Problem 3, page 359, shows that s and t both lie in the bisector plane α of one of the dihedral angles formed by planes (ABC) and (ABD) (the one that contains the tetrahedron). In order to see that s and t are not parallel, let P denote the intersection of α with edge CD, so that triangle ABP is the planar section of $ABCD$ along α; since s lies within angle $\angle BAP$ and t within angle $\angle ABP$, we conclude that s and t intersect each other at a point I inside triangle ABP. Now, it follows from $I \in s$ that I is at equal distances from the faces (ABC), (ABD) and (ACD) of $ABCD$; analogously, $I \in t$ guarantees that I is equidistant from the faces (BAC), (BAD) and (BCD) of $ABCD$. Hence, I is at equal distances from all of the faces of the tetrahedron, and letting r denote such a common distance we conclude that the sphere of center I and radius r is inscribed in the tetrahedron $ABCD$.

Conversely, if I' is the center of a sphere inscribed in $ABCD$, then the fact that I' lies at equal distances from the faces (ABC), (ABD) and (ACD) assures that belongs to the half-line s above. By the same token, I' belongs to t, whence I' coincides with I. $\qquad \square$

We now examine the previous proposition in the special case of a regular tetrahedron. A more general case will be dealt with in Sect. 11.3.

Example 11.5 If $ABCD$ is a regular tetrahedron, show that its inscribed and circumscribed spheres are concentric. Moreover, letting a stand for the common length of the edges of $ABCD$, show that the respective radii r and R of such spheres are given by

$$r = \frac{a\sqrt{6}}{12} \text{ and } R = \frac{a\sqrt{6}}{4}.$$

Fig. 11.9 Spheres inscribed
and circumscribed to a
regular tetrahedron

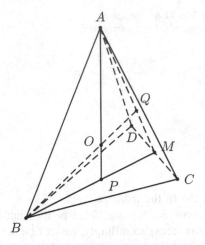

Proof Letting O and P respectively denote the circumcenters of the tetrahedron and of its face BCD (cf. Fig. 11.9), we know from the proof of the previous proposition that $\overleftrightarrow{OP}$ is the medial line of triangle BCD. However, since $\overline{AB} = \overline{AC} = \overline{AD}$, it also follows that $A \in \overleftrightarrow{OP}$. Accordingly, letting Q denote the circumcenter of ACD, we have $B \in \overleftrightarrow{OQ}$.

Now, let M stand for the midpoint of CD. Since triangles BCD and ACD are both equilaterals, of centers respectively equal to P and Q, we have $\overrightarrow{BP} \cap \overrightarrow{AQ} = \{M\}$. On the other hand, since $\overline{MP} = \frac{1}{3}\overline{MB}$ and $\overline{MQ} = \frac{1}{3}\overline{MA}$, triangles MPQ and MBA are similar by SAS. Therefore, $\overleftrightarrow{PQ} \parallel \overleftrightarrow{AB}$ and $\overline{PQ} = \frac{1}{3}\overline{AB}$. In turn, the parallelism of $\overleftrightarrow{PQ}$ and $\overleftrightarrow{AB}$ assures that $PQO \sim ABO$ by AA, so that $\overline{OP} = \frac{1}{3}\overline{AO}$. Hence, it follows from Example 11.3 that

$$\overline{OP} = \frac{1}{4}\overline{AP} = \frac{1}{4} \cdot \frac{a\sqrt{6}}{3} = \frac{a\sqrt{6}}{12}$$

and

$$\overline{AO} = \frac{3}{4}\overline{AP} = \frac{3}{4} \cdot \frac{a\sqrt{6}}{3} = \frac{a\sqrt{6}}{4}.$$

Finally, note that $\overline{AO} = \frac{a\sqrt{6}}{4}$ is the radius of the sphere circumscribed to $ABCD$. On the other hand, since the reasoning above can be replicated to the medial lines of any two faces of $ABCD$, we conclude that is at equal distances from all of the faces of $ABCD$, thus coinciding with I. Therefore, the radius of the sphere inscribed in $ABCD$ is equal to $\overline{OP} = \frac{a\sqrt{6}}{12}$. $\square$

We finish this section by presenting the following beautiful example on a sphere hanging on the vertex of a regular pyramid.

Example 11.6 (Poland) The regular $2n$-gon $A_1 A_2 \ldots A_{2n}$ is the basis of a regular pyramid of vertex V. A sphere Σ, passing through V, intersects the lateral edge $V A_i$ at B_i, for $1 \le i \le 2n$. Prove that

$$\sum_{i=1}^{n} \overline{V B}_{2i-1} = \sum_{i=1}^{n} \overline{V B}_{2i}.$$

Proof The situation described in the statement is depicted in the figure below.

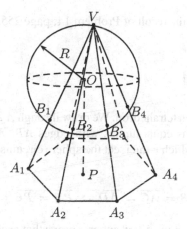

Let O be the center and R be the radius of Σ, and l be the length of the lateral edges of the pyramid. By applying the Intersecting Chords Theorem to $\overrightarrow{A_i V}$ and $\overrightarrow{A_i O}$ we obtain $l(l - \overline{B_i V}) = \overline{A_i O}^2 - R^2$, so that

$$\sum_{k=1}^{n} l(l - \overline{B_{2k} V}) = \sum_{k=1}^{n} (\overline{A_{2k} O}^2 - R^2).$$

Hence,

$$\sum_{k=1}^{n} \overline{B_{2k} V} = \frac{1}{l} \left(n(l^2 + R^2) - \sum_{k=1}^{n} \overline{A_{2k} O}^2 \right),$$

and likewise

$$\sum_{k=1}^{n} \overline{B_{2k-1}V} = \frac{1}{l}\left(n(l^2 + R^2) - \sum_{k=1}^{n}\overline{A_{2k-1}O}^2\right).$$

It thus suffices to prove that $\sum_{k=1}^{n}\overline{A_{2k}O}^2 = \sum_{k=1}^{n}\overline{A_{2k-1}O}^2$. To this end, let P be the foot of the perpendicular dropped from O to the plane of the basis of the pyramid. Pythagoras' theorem applied to triangle OPA_i gives $\overline{A_iO}^2 = \overline{A_iP}^2 + \overline{OP}^2$, so that

$$\sum_{k=1}^{n}\overline{A_{2k}O}^2 = \sum_{k=1}^{n}\overline{A_{2k-1}O}^2 \Leftrightarrow \sum_{k=1}^{n}\overline{A_{2k}P}^2 = \sum_{k=1}^{n}\overline{A_{2k-1}P}^2.$$

It now suffices to invoke the result of Problem 14, page 255.   □

Problems: Sect. 11.1

1. Let $ABCD$ be a given tetrahedron. We draw through A a line r, which intersects face BCD and forms equal angles with edges AB, AC and AD. Letting P denote the point in which r intersect the sphere circumscribed to $ABCD$, prove that

$$\overline{AB} = \overline{AC} = \overline{AD} \Rightarrow \overline{PB} = \overline{PC} = \overline{PD}.$$

2. If a pyramid admits an inscribed sphere, prove that such a sphere is unique.
3. * Prove that a pyramid of vertex V and basis $A_1A_2\ldots A_n$ admits a circumscribed sphere if and only if the n-gon $A_1A_2\ldots A_n$ is cyclic. Moreover, in this case show that the sphere circumscribed to the pyramid is unique.
4. * Let $\mathcal{C}$ be a cone of revolution with vertex V, radius R and altitude h. Draw, at a distance $d < h$ from the plane of the basis of the cone, a parallel plane that intersects it along a circle Γ', of radius R'. The portion of the cone lying between these two planes is called the **frustum of cone** of bases Γ and Γ' (or radii R and R') and altitude d. Prove that the lateral area A of such a solid is given by

$$A = \pi(R + R')d\sqrt{1 + \left(\frac{R - R'}{d}\right)^2}.$$

5. $ABCD$ is a regular tetrahedron and M and N are distinct points, respectively situated in the faces ABD and ACD. Show that $\overline{MN}$, $\overline{BN}$ and $\overline{CM}$ are the lengths of the sides of a triangle.

6. In a tetrahedron $ABCD$, suppose that the foot of the altitude dropped from A coincides with the orthocenter of the face BCD. Prove that this altitude intersects the other three altitudes of the tetrahedron. Moreover, if the foot of the altitude of the tetrahedron dropped from C coincides with the orthocenter of the face ABD too, show that the four altitudes of the tetrahedron are concurrent and that each one of them touches the opposite face in its orthocenter.

7. Let $ABCD$ be a regular tetrahedron of edge length a. Do the following items:

   (a) Prove that the edges AB and CD are orthogonal.
   (b) Among all planar sections of $ABCD$ through a plane parallel to the edges AB and CD, find the one with largest area.

8. (United States) $ABCD$ is a tetrahedron and E, F, G, H, I and J are points lying in the edges AB, BC, CA, AD, BD and CD, respectively. If

$$\overline{AE} \cdot \overline{BE} = \overline{BF} \cdot \overline{CF} = \overline{AG} \cdot \overline{CG} = \overline{AH} \cdot \overline{DH} = \overline{BI} \cdot \overline{DI} = \overline{CJ} \cdot \overline{DJ},$$

   show that E, F, G, H, I and J are all situated on a single sphere.

9. * $ABCD$ is any tetrahedron and M, N, P and Q are the barycenters of faces BCD, ACD, ABD and ABC, respectively. Prove that:

   (a) The line segments AM, BN, CP and DQ intersect at a single point G, called the **barycenter** of the tetrahedron.
   (b) $\dfrac{\overline{AG}}{\overline{GM}} = \dfrac{\overline{BG}}{\overline{GN}} = \dfrac{\overline{CG}}{\overline{GP}} = \dfrac{\overline{DG}}{\overline{GQ}} = 3$.

10. In a regular quadrangular pyramid, the altitude length is h, whereas the basis is a square of side length a. If α is the plane passing through two adjacent vertices of the basis and through the midpoint of the altitude, compute, in terms of a and h, the area of the planar section of the pyramid determined by α.

11. $VA_1A_2 \ldots A_n$ is a regular n-gonal pyramid of basis vertex V, lateral edges of length x and basis edges of length y. If θ is the plane dihedral angle between the planes of two consecutive lateral faces, show that

$$1 - \cos\theta = \frac{2\cos^2 \frac{\pi}{n}}{1 - \left(\frac{y}{2x}\right)^2}.$$

12. $ABDF$ is a regular tetrahedron of edge length y and $ABDC$, $ABFG$ are equal regular triangular pyramids with lateral edges of length x. If the plane dihedral angles formed by the pairs of planes (ABC), (ABG) and (ABC), (ACD) are equal, compute all possible values of $\frac{x}{y}$.

13. (Netherlands) Let $VA_1A_2 \ldots A_n$ be a regular n-gonal pyramid of apex V and basis $A_1A_2 \ldots A_n$. Through a point P lying in the interior of $A_1A_2 \ldots A_n$ we draw a line r, perpendicular to the plane of the basis. For $1 \leq i \leq n$, let B_i denote the point of intersection of line r with the plane (VA_iA_{i+1}) (with the convention that $A_{n+1} = A_1$). Prove that the value of $\overline{PB_1} + \overline{PB_2} + \cdots + \overline{PB_n}$ does not depend on the position of P.

14. (Czechoslovakia) If S stands for the sum of the areas of the faces of a tetrahedron $ABCD$, prove that

$$S \le \frac{1}{2\sqrt{3}}(\overline{AB}^2 + \overline{AC}^2 + \overline{AD}^2 + \overline{BC}^2 + \overline{BD}^2 + \overline{CD}^2).$$

15. Extend Euler's median formula (cf. Problem 6, page 188) to tetrahedra, in the following form: if $ABCD$ is a tetrahedron and M and N are the midpoints of AB and CD, respectively, then

$$\overline{MN}^2 = \frac{1}{4}(\overline{AC}^2 + \overline{AD}^2 + \overline{BC}^2 + \overline{BD}^2 - \overline{AB}^2 - \overline{CD}^2).$$

16. (Czechoslovakia) Prove that every tetrahedron $ABCD$ can be placed in the region of space delimited by two parallel planes, situated at a distance from each other of at most

$$\frac{1}{2\sqrt{3}}\sqrt{\overline{AB}^2 + \overline{AC}^2 + \overline{AD}^2 + \overline{BC}^2 + \overline{BD}^2 + \overline{CD}^2}.$$

17. (Brazil) What is the smallest length a rope loop can have, so that we can pass a regular tetrahedron of edge length l through it?

18. Prove that, in every tetrahedron $ABCD$, one has $\overline{AB} \cdot \overline{CD} + \overline{AD} \cdot \overline{BC} > \overline{AC} \cdot \overline{BD}$.

    For the coming problem, let $\Gamma_i(O_i; R_i)$, for $i = 1, 2$, be two given spheres. We say that Γ_1 and Γ_2 are **externally** (resp. **internally**) **tangent** if $\overline{O_1 O_2} = R_1 + R_2$ (resp. $\overline{O_1 O_2} = |R_1 - R_2|$). In any one of these cases, it is immediate to show that Γ_1 and Γ_2 have a single point in common, which is said to be the **point of tangency** of Γ_1 and Γ_2.

19. $A_1 A_2 A_3 A_4$ is a tetrahedron for which there exist spheres S_1, S_2, S_3, S_4, centered respectively at A_1, A_2, A_3, A_4 and pairwise externally tangent. Suppose that there exist spheres Σ and Σ', centered at a point O in the interior of $A_1 A_2 A_3 A_4$ and such that Σ is tangent to the edges of the tetrahedron, whereas Σ' is tangent to S_1, S_2, S_3, S_4. Prove that $A_1 A_2 A_3 A_4$ is regular.

11.2   Prisms and Parallelepipeds

We are given convex polygons $A_1 A_2 \ldots A_n$ and $A'_1 A'_2 \ldots A'_n$, lying in parallel planes and such that lines $\overleftrightarrow{A_1 A'_1}$, $\overleftrightarrow{A_2 A'_2}$, ..., $\overleftrightarrow{A_n A'_n}$ are parallel to each other. It is immediate to see (cf. Fig. 11.10) that, for $1 \le i \le n$ (with the convention that

Fig. 11.10 A prism of bases
$A_1A_2 \ldots A_n$ and $A'_1A'_2 \ldots A'_n$

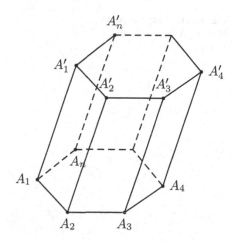

$A_{n+1} = A_1$), the quadrilateral $A_i A_{i+1} A'_{i+1} A'_i$ is a parallelogram and the polygons $A_1A_2 \ldots A_n$ and $A'_1A'_2 \ldots A'_n$ are congruent.[1]

The **prism** of **bases** $A_1A_2 \ldots A_n$ and $A'_1A'_2 \ldots A'_n$ is the bounded portion of space delimited by the polygons $A_1A_2 \ldots A_n$ and $A'_1A'_2 \ldots A'_n$ and by the parallelograms $A_i A_{i+1} A'_{i+1} A'_i$, for $1 \le i \le n$.

We say that points A_1, A_2, ..., A_n, A'_1, A'_2, ..., A'_n are the **vertices** of the prism. Also, the line segments $A_i A_{i+1}$, $A'_i A'_{i+1}$ and $A_i A'_i$ are its **edges** and the line segments $A_i A'_i$, for $1 \le i \le n$, are its **lateral edges**. Parallelograms $A_i A_{i+1} A'_{i+1} A'_i$ are the **lateral faces** of the prism. Whenever will be convenient and there will be no danger of confusion, we shall collectively refer to the bases and lateral faces of a prism as its **faces**. Finally, the **interior** of a prism is the set of the points of a prism which do not belong to any of its faces (Fig. 11.10).

Since the bases of the prism above defined are n-gons, we shall say that it is an **n-gonal prism**. In the cases $n = 3$ and $n = 4$, we shall more frequently use the alternative expressions **triangular** and **quadrangular prisms**, respectively. Also in these case, we shall denote the corresponding bases by ABC and $A'B'C'$, $ABCD$ and $A'B'C'D'$, also respectively (with the convention, in the quadrangular case, that AC and BD are the diagonals of $ABCD$).

The **height** of a prism is the distance between the planes of its bases. A **right prism** is a prism whose lateral edges are perpendicular to the planes of its bases; in particular, if $A_1A_2 \ldots A_n$ and $A'_1A'_2 \ldots A'_n$ are the bases of a right prism, then $A_i A_{i+1} A'_{i+1} A'_i$ is a rectangle for $1 \le i \le n$ (with the convention that $A_{n+1} = A_1$ and $A'_{n+1} = A'_1$) and the common length of its lateral edges is the height of the prism. Finally, a right prism is **regular** (cf. Fig. 11.11) if its bases are regular polygons.

----

[1]Even though we have not explicitly defined what one means by the congruence of polygons, the idea here is that, for $2 \le i \le n - 1$, triangles $A_1 A_i A_{i+1}$ and $A'_1 A'_i A'_{i+1}$ are congruent.

Fig. 11.11 A regular prism
of bases $A_1 A_2 \ldots A_n$ and
$A_1' A_2' \ldots A_n'$

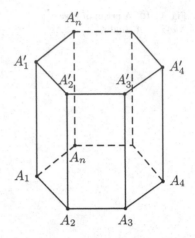

As an application of the concepts above, let $\mathcal{C}(e; R)$ be a cylinder of revolution of axis e and radius R, and α and α' be planes perpendicular to e at points O and O', respectively, and situated at a distance h of each other. Then, α and α' intersect the cylinder along circles $\Gamma(O; R)$ and $\Gamma'(O'; R)$, respectively, such that $\overline{OO'} = h$. Hereafter, whenever there is no danger of confusion, we shall also refer to the portion of $\mathcal{C}(e; R)$ contained between α and α' as the **cylinder of revolution** of radius R and height h. The concept of regular prism allows us to define, by means of the heuristic argument of the next paragraph, the *lateral area* of such a cylinder of revolution.

Let $n \geq 3$ be a natural number, and $A_1 A_2 \ldots A_n$ and $A_1' A_2' \ldots A_n'$ be regular polygons respectively inscribed in Γ and Γ' and such that $\overleftrightarrow{A_i A_i'} \parallel e$ for $1 \leq i \leq n$ (cf. Fig. 11.12). The prism of bases $A_1 A_2 \ldots A_n$ and $A_1' A_2' \ldots A_n'$ is regular of height h; its *lateral area* (i.e., the sum of the areas of its lateral faces) is equal to

$$\sum_{i=1}^{n} A(A_i A_{i+1} A_{i+1}' A_i') = \sum_{i=1}^{n} \overline{A_i A_{i+1}} \cdot h = p_n h,$$

where p_n stands for the perimeter of $A_1 A_2 \ldots A_n$. Now, as $n \to +\infty$, the union of the lateral faces of the prism forms a better and better approximation for the cylinder of revolution. Accordingly, the lateral area of the prism forms a better and better uma approximation for what we would like to define as the **lateral area** A of the cylinder. However, since $p_n \to 2\pi R$ as $n \to +\infty$, we define such a lateral area by setting

$$A = 2\pi R h.$$

An important subclass of prisms is the one formed by **parallelepipeds**, i.e., quadrangular prisms whose bases are parallelograms (cf. Fig. 11.13). A **right**

Fig. 11.12 A regular n-gonal prism inscribed in a cylinder of revolution

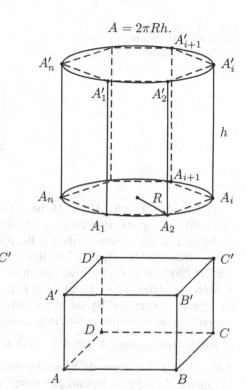

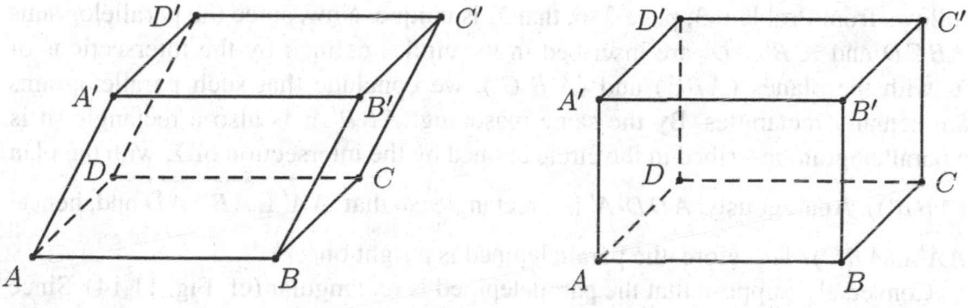

$$A = 2\pi Rh.$$

Fig. 11.13 A parallelepiped (left) and right parallelepiped (right)

parallelepiped is a parallelepiped which is a right prism, and **rectangular parallelepiped** is a right parallelepiped whose bases are rectangles. Notice that the twelve edges of a parallelepiped can be grouped into three sets of four parallel equal edges each; in the notations of Fig. 11.13, these sets are $\{AB, A'B', CD, C'D'\}$, $\{AA', BB', CC', DD'\}$ and $\{AD, A'D', BC, B'C'\}$. Letting a, b and c denote the lengths of the four edges in each of such sets, we shall sometimes say that *the parallelepiped has edge (lengths) a, b and c.*

As with pyramids, a sphere is *circumscribed* to a prism provided it passes through all of its vertices. In general, no such a sphere will exist. Nevertheless, if it does, then it is unique (cf. Problem 1). In the positive direction, we have the following important result.

Proposition 11.7 *A parallelepiped admits a circumscribed sphere if and only if it is rectangular. In this case, if a, b and c are the lengths of its edges and R is the radius of the circumscribed sphere, then*

$$R = \frac{1}{2}\sqrt{a^2 + b^2 + c^2}.$$

Fig. 11.14 Rectangular
parallelepipeds possess
circumscribed speres

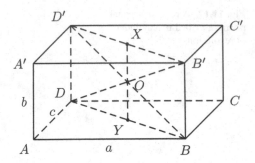

Proof Let $ABCD$ and $A'B'C'D'$ be the bases of the parallelepiped, named as usual. Firstly, suppose that the parallelepiped is inscribed in a sphere Σ of center O. Since Σ is also circumscribed to the pyramid of apex A' and basis $ABCD$, it follows from Problem 3, page 386, that Σ is unique. Now, since the parallelograms $ABCD$ and $A'B'C'D'$ are inscribed in the circles defined by the intersections of Σ with the planes (ABC) and $(A'B'C')$, we conclude that such parallelograms are actually rectangles. By the same reasoning, $ABB'A'$ is also a rectangle (it is a parallelogram inscribed in the circle defined by the intersection of Σ with the plan (ABB')). Analogously, $ADD'A'$ is a rectangle, so that $\overleftrightarrow{AA'} \perp \overleftrightarrow{AB}$, $\overleftrightarrow{AD}$ and, hence, $\overleftrightarrow{AA'} \perp (ABC)$. Therefore, the parallelepiped is a right one.

Conversely, suppose that the parallelepiped is rectangular (cf. Fig. 11.14). Since $BDD'B'$ is a rectangle, its diagonals meet at their common midpoint O; since $\overline{BD'} = \overline{B'D}$, we conclude that O equidists of B, B', D and D'. Letting R denote the common distance from O to such points and applying Pythagoras' theorem to triangles BDD' and ABD, we obtain

$$4R^2 = \overline{BD'}^2 = \overline{BD}^2 + \overline{DD'}^2 = (\overline{AB}^2 + \overline{AD}^2) + \overline{AA'}^2 = a^2 + b^2 + c^2.$$

Now, observe that O is also the midpoint of XY, where X and Y are the midpoints of BD' and BD, respectively. However, since X and Y lie in $A'C'$ and AC (respectively), an argument analogous to the one above allows us to conclude that O is the point of intersection of the diagonals of rectangle $AA'C'C$, and hence is at equal distances from A, A', C and C'. Finally, a computation quite similar to the one we did above shows that the distance from O to such vertices of the parallelepiped is equal to R too. Thus, O is the center and R is the radius of the sphere circumscribed to the parallelepiped.   □

Continuing with the development of the theory, we say that a rectangular parallelepiped is a **cube** (cf. Fig. 11.15) if all of its edges have the same length. In this case, letting $ABCD$ and $A'B'C'D'$ be the bases of the cube, named as usual, and a be the common length of its edges, the proof of the previous proposition assures that all of the line segments AC', $A'C$, BD' and $B'D$ have lengths equal to $a\sqrt{3}$ and concur at the center of the sphere circumscribed to the cube. Such segments (or

Fig. 11.15 A cube of edge
length a and its diagonals

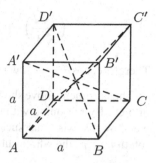

sometimes their common length) are called the **diagonals** of the cube, whereas O is its **center**.

In Sect. 12.3 we shall see that cubes form another one of the five classes of convex regular polyhedras.

Example 11.8 A cube of edge length a has bases $ABCD$ and $A'B'C'D'$, named in the usual way. Show that a $120°$ rotation of space around $\overleftrightarrow{BD'}$ maps the cube into itself.

Proof We shall first compute the opening of the angle formed by the planes $(BC'D')$ and $(BA'D')$ (see figure below).

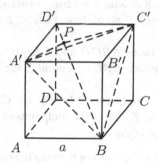

In order to do this, start by noticing that triangles $BA'D'$ and $BC'D'$ are congruent by SSS. Therefore, the feet of the perpendiculars dropped from A' and C' to BD' coincide, say at P. Since $D'\widehat{A'}B = 90°$, the usual metric relations in right triangles applied to $BA'D'$ give $\overline{A'P} \cdot \overline{BD'} = \overline{A'D'} \cdot \overline{A'B}$. By substituting $\overline{A'D'} = a$, $\overline{A'B} = a\sqrt{2}$ and $\overline{BD'} = a\sqrt{3}$, we obtain $\overline{A'P} = \frac{a\sqrt{2}}{\sqrt{3}} = \overline{C'P}$. Now, applying the cosine law to $A'PC'$ with $A'\widehat{P}C' = \theta$, we obtain

$$\overline{A'P}^2 + \overline{C'P}^2 - 2\,\overline{A'P} \cdot \overline{C'P}\cos\theta = \overline{A'C'}^2$$

or, which is the same,

$$\left(\frac{a\sqrt{2}}{\sqrt{3}}\right)^2 + \left(\frac{a\sqrt{2}}{\sqrt{3}}\right)^2 - 2\left(\frac{a\sqrt{2}}{\sqrt{3}}\right)^2 \cos\theta = \left(a\sqrt{2}\right)^2.$$

Therefore, $\cos\theta = -\frac{1}{2}$, whence $\theta = 120°$.

The argument above shows that a $120°$ rotation of the space around $\overleftrightarrow{BD'}$ maps A' into C'. By reasoning with the pairs of triangles $B'BD'$, CBD' and DBD', ABD' we show likewise that such a rotations maps B' into C and D into A. Thus, this rotation maps the whole cube into itself.   □

Problems: Sect. 11.2

1. Prove that a prism of bases $A_1A_2\ldots A_n$ and $A'_1A'_2\ldots A'_n$ admits a circumscribed sphere if and only if it is right and its bases are cyclic polygons. Moreover, in this case, prove that the circumscribed sphere is unique.
2. Let $ABCDEFGH$ be a rectangular parallelepiped of bases $ABCD$ and $EFGH$, with $\overleftrightarrow{AE}$, $\overleftrightarrow{BF}$, $\overleftrightarrow{CG}$ and $\overleftrightarrow{DH}$ parallel. If $D\widehat{H}C = 45°$ and $F\widehat{H}B = 30°$, compute $G\widehat{B}H$.
3. $\mathcal{C}$ is a cube of edge length 2. For each vertex of $\mathcal{C}$, we draw a plane perpendicular to the diagonal of $\mathcal{C}$ incident to that vertex. As the result of these operation, we obtain a solid of 14 faces, 8 of which are congruent triangles and the remaining 6 ones are congruent octagons. If all of these 14 faces have the same area, compute its value.
4. (TT) Consider a cube of bases $ABCD$ and $EFGH$ and lateral edges AE, BF, CG and DH. Take, in the face $ABCD$, a point X such that $\angle AXF = \angle AXH = 90°$. Compute $A\widehat{X}E$.
5. In a cube of edge length a and bases $ABCD$, $A'B'C'D'$ (with vertices labelled in the usual way), X, Y and Z denote the midpoints of edges $A'D'$, AB and CC', respectively. Do the following items:

   (a) Show that the plane (XYZ) passes through the center of the cube.
   (b) Compute, as a function of a, the area of the planar section of the cube through the plane (XYZ).

6. If a plane intersects a rectangular parallelepiped along a regular hexagon, prove that the parallelepiped is actually a cube.
7. We are given a right n-gonal prism with bases of area S. Planes α and β are perpendicular to each other and intersect the prism along the convex n-gons $\mathcal{P}$ and $\mathcal{Q}$, respectively. Find all values of the angle between α and the lateral edges of the prism for which $A(\mathcal{P}) + A(\mathcal{Q})$ is as small as possible.
8. Planes α and β concur at r and form an angle σ, with $0 < \sigma < \frac{\pi}{2}$. A cube $\mathcal{C}$ of edge length 1 has parallel faces $ABCD$, $EFGH$ and lateral edges AE, BF,

CG, DH; also, $ABCD$ is contained in α, with $A \in r$, and $\overrightarrow{AB}$ forms an angle θ with r, where $0 < \theta < \frac{\pi}{2}$.

(a) For $X \in \alpha$, let X' be its orthogonal projection onto β (hence, $A = A'$). Show that the orthogonal projection of C onto β is the convex hexagon $C'B'F'E'H'D'$ such that:

    i. $A'D'C'B'$ and $E'H'G'F'$ are congruent parallelograms contained in $C'B'F'E'H'D'$.

    ii. $H'D'B'F'$ is a rectangle, with $\overleftrightarrow{D'H'},\ \overleftrightarrow{B'F'} \perp r$.

(b) $A(C'B'F'E'H'D') = \cos\sigma + (\sin\theta + \cos\theta)\sin\sigma$.

(c) $A(C'B'F'E'H'D')$ is maximum if and only if the diagonal AG is perpendicular to β.

11.3   More on Tetrahedra

Along this section, we use the material developed so far to discuss in detail two interesting classes of tetrahedra, namely, the *trirectangular* and *isosceles* ones.

A tetrahedron $ABCD$ is **trirectangular** in D if $A\widehat{D}B = A\widehat{D}C = B\widehat{D}C = 90°$ (cf. Fig. 11.16). The coming proposition brings some properties of such tetrahedra.

Proposition 11.9 *If tetrahedron $ABCD$ is trirectangular at D, with $\overline{AD} = a$, $\overline{BD} = b$ and $\overline{CD} = c$, then:*

(a) *Face ABC is an acute triangle.*

(b) *The foot of the altitude of $ABCD$ dropped from D is the orthocenter of face ABC.*

(c) *The radius of the sphere circumscribed to $ABCD$ equals $\frac{1}{2}\sqrt{a^2 + b^2 + c^2}$.*

Proof

(a) Pythagoras' theorem gives $\overline{AB} = \sqrt{a^2 + b^2}$, $\overline{BC} = \sqrt{b^2 + c^2}$ and $\overline{AC} = \sqrt{a^2 + c^2}$. Hence, in order to show that ABC is acute, Corollary 7.26 assures that it suffices to note that

Fig. 11.16 A tetrahedron $ABCD$, trirectangular at D

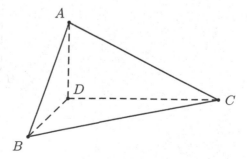

$$\overline{AB}^2 + \overline{BC}^2 = a^2 + 2b^2 + c^2 > a^2 + 2c^2 = \overline{AC}^2,$$

and analogously $\overline{AB}^2 + \overline{AC}^2 > \overline{BC}^2$ and $\overline{AC}^2 + \overline{BC}^2 > \overline{AB}^2$.

(b) If P is the foot of the altitude of $ABCD$ dropped from D (cf. Fig. 11.17), then $\overleftrightarrow{DP} \perp \overleftrightarrow{BC}$. However, since $\overleftrightarrow{AD} \perp \overleftrightarrow{BC}$, it follows that $\overleftrightarrow{BC} \perp (ADP)$. Hence, $\overleftrightarrow{BC} \perp \overleftrightarrow{AP}$, so that $\overleftrightarrow{AP}$ is an altitude of triangle ABC. Analogously, one shows that $\overleftrightarrow{BP}$ and $\overleftrightarrow{CP}$ are the other two altitudes of ABC, whence P is the orthocenter of this face.

(c) Construct points X, Y, Z and W, such that all of the quadrilaterals $ADBZ$, $ADCY$ and $BDCX$ are rectangles. Then, since $ABCD$ is trirectangle, one immediately concludes that $ADBZYCXW$ is a rectangular parallelepiped (cf. Fig. 11.18). If Σ is the sphere circumscribed to such a parallelepiped (cf. Proposition 11.7), then Σ also circumscribes $ABCD$. Letting R denote its radius, it follows from the proposition just referred that $\frac{1}{2}\sqrt{a^2 + b^2 + c^2}$. □

A tetrahedron is **isosceles** if each two opposite edges of it have equal lengths. In the notations of Fig. 11.18, it is straightforward to check that the tetrahedron of vertices A, B, C and W is isosceles, with

Fig. 11.17 The foot of the altitude of a trirectangular tetrahedron, dropped from the vertex of the right angles

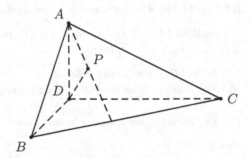

Fig. 11.18 Computing the radius of the sphere circumscribed to a trirectangular tetrahedron

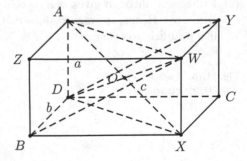

$$\overline{AB} = \overline{CW} = \sqrt{a^2 + b^2}, \quad \overline{AC} = \overline{BW} = \sqrt{a^2 + c^2}, \quad \overline{AW} = \overline{BC} = \sqrt{b^2 + c^2}.$$

Conversely, the following discussion shows, among other interesting properties of isosceles tetrahedra, that each one of these can be constructed as $ABCW$, i.e., departing from an appropriate rectangular parallelepiped. Our discussion follows, essentially, Chapter 9 of [13].

For what comes next, given a tetrahedron $ABCD$, we shall refer to the trihedral angle of faces $\angle BAC$, $\angle BAD$ and $\angle CAD$ (cf. Problem 3, page 359) as the *trihedral angle of vertex A determined by $ABCD$*.

Proposition 11.10 *A tetrahedron is isosceles if and only if the following condition is satisfied: for each vertex X of the tetrahedron, the trihedral angle of vertex X it determines has the sum of the angles of its faces at X equal to $180°$.*

Proof If $ABCD$ is an isosceles tetrahedron, then $ACD \equiv CAB$ and $ABD \equiv BAC$ by SSS, so that $C\widehat{A}D = A\widehat{C}B$ and $B\widehat{A}D = A\widehat{B}C$. Hence,

$$B\widehat{A}C + B\widehat{A}D + C\widehat{A}D = B\widehat{A}C + A\widehat{B}C + A\widehat{C}B = 180°.$$

Conversely, let $ABCD$ be a tetrahedron satisfying the stated condition. In the plane of ABC, construct triangles ABD_2, BCD_3 and ACD_1, respectively congruent to the faces ABD, BCD and ACD of $ABCD$ (cf. Fig. 11.19). Since

$$D_2\widehat{B}A + A\widehat{B}C + C\widehat{B}D_3 = D\widehat{B}A + A\widehat{B}C + C\widehat{B}D = 180°,$$

we have $B \in D_2D_3$; analogously, $C \in D_1D_3$ and $A \in D_1D_2$. Now, since $\overline{AD_1} = \overline{AD_2} = \overline{AD}$, we conclude that A is the midpoint of D_1D_2; analogously, B and C are the midpoints of D_2D_3 and D_1D_3, respectively. Hence, the midsegment theorem assures that

$$\overline{AB} = \frac{1}{2}\overline{D_1D_3} = \overline{CD_1} = \overline{CD},$$

and likewise $\overline{AC} = \overline{BD}$ and $\overline{AD} = \overline{BC}$. Thus, $ABCD$ is isosceles. $\square$

Item (a) of Proposition 11.9 assures that all of the faces of the isosceles tetrahedron $ABCW$ of Fig. 11.18 are acute triangles. As a consequence of the previous result, we shall extend this property, in Corollary 11.12, to all conceivable isosceles tetrahedra. Nevertheless, prior to that we need an auxiliary result which is interesting in itself.

Lemma 11.11 *Given a trihedral angle of faces $\angle XVY$, $\angle XVZ$ and $\angle YVZ$, we have*

$$X\widehat{V}Y < X\widehat{V}Z + Y\widehat{V}Z.$$

Fig. 11.19 *Unfolding tetrahedron ABCD*

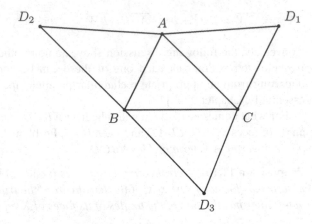

Fig. 11.20 A useful inequality between the angles of the faces of a trihedral angle

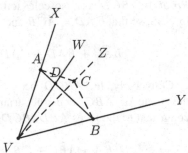

Proof Suppose that $X\widehat{V}Y > Y\widehat{V}Z$ (for otherwise there is nothing to do). In the face $\angle XVY$ of the trihedral angle, draw the half-line $\overrightarrow{VW}$ such that $Y\widehat{V}W = Y\widehat{V}Z$ (cf. Fig. 11.20). Then, plot points $C \in \overrightarrow{VZ}$ and $D \in \overrightarrow{VW}$ with $\overline{VC} = \overline{VD}$, and cut the trihedral with a plane containing $\overleftrightarrow{CD}$, which intersects $\overrightarrow{VX}$ and $\overrightarrow{VY}$ at points A and B, respectively.

The congruence of triangles BVC and BVD (by SAS) assures that $\overline{BC} = \overline{BD}$. On the other hand, by applying the triangle inequality in ABC, we obtain

$$\overline{AD} + \overline{BD} = \overline{AB} < \overline{AC} + \overline{BC} = \overline{AC} + \overline{BD},$$

so that $\overline{AD} < \overline{AC}$. Now, triangles AVC and AVD are such that AV is a common side, $\overline{VC} = \overline{VD}$ and $\overline{AC} > \overline{AD}$; hence, the cosine law guarantees that $A\widehat{V}D < A\widehat{V}C$. Therefore,

$$A\widehat{V}B = A\widehat{V}D + D\widehat{V}B < A\widehat{V}C + C\widehat{V}B = A\widehat{V}C + B\widehat{V}C.$$

$\square$

We are finally in position to prove the coming

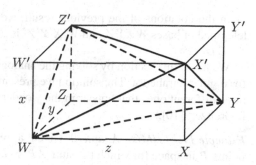

Fig. 11.21 Realization of isosceles tetrahedra

Corollary 11.12 *The faces of an isosceles tetrahedron are acute triangles.*

Proof If $ABCD$ is an isosceles tetrahedron, it follows from the previous lemma and from Proposition 11.10 that

$$2B\widehat{A}C < B\widehat{A}C + B\widehat{A}D + C\widehat{A}D = 180°,$$

i.e., $B\widehat{A}C < 90°$. Analogously, $A\widehat{B}C < 90°$ and $A\widehat{C}B < 90°$, so that ABC is an acute triangle. However, since all the faces of an isosceles tetrahedron are congruent, there is nothing left to do. □

In view of the above result, we state and prove the following *realization theorem* for isosceles tetrahedra.

Theorem 11.13 *If $ABCD$ is an isosceles tetrahedron, then there exists a rectangular parallelepiped of bases $WXYZ$ and $W'X'Y'Z'$ (cf. Fig. 11.21) such that the faces of the isosceles tetrahedron $WZ'X'Y$ are congruent to those of $ABCD$.*

Proof Let $\overline{AB} = \overline{CD} = c$, $\overline{AC} = \overline{BD} = b$ and $\overline{AD} = \overline{BC} = a$. Since ABC is an acute triangle, we have $a^2 + b^2 > c^2$, $a^2 + c^2 > b^2$ and $b^2 + c^2 > a^2$. Therefore, it is easily seen that the system of equations

$$\begin{cases} x^2 + y^2 = c^2 \\ x^2 + z^2 = b^2 \\ y^2 + z^2 = a^2 \end{cases}$$

admits the unique positive solutions

$$x = \sqrt{\frac{b^2 + c^2 - a^2}{2}}, \quad y = \sqrt{\frac{a^2 + c^2 - b^2}{2}} \text{ and } z = \sqrt{\frac{a^2 + b^2 - c^2}{2}}.$$

In turn, if $WXYZW'X'Y'Z'$ is a rectangular parallelepiped with $\overline{WW'} = x$, $\overline{WZ} = y$ and $\overline{WX} = z$, then $\overline{WZ'} = \sqrt{x^2 + y^2} = c$, $\overline{WX'} = \sqrt{x^2 + z^2} = b$ and $\overline{WY} = \sqrt{y^2 + z^2} = a$. □

In the notations of the previous result, we shall say that the rectangular parallelepiped of bases $WXYZ$ and $W'X'Y'Z'$ is *associated* to the isosceles tetrahedron
$ABCD$.

We finish this section by solving the analogue of Steiner's problem (cf. Sect. 7.5)
for regular tetrahedra. The solution we present, up to slight modifications, is one of
three different ones that appear in the wonderful book of the late professor Samuel
L Greitzer [10].

Example 11.14 (IMO—Adapted) Given a regular tetrahedron $ABCD$, find all
points P in space for which the sum $\overline{AP} + \overline{BP} + \overline{CP} + \overline{DP}$ is minimum.

Proof In order to help visualization, we use the rectangular parallelepiped associated to ABC, which in this case is a cube. We claim that the circumcenter O of
$ABCD$ (which coincides with the center of the cube) is the only point in space
which minimizes the sum of the distances from A, B, C, D.

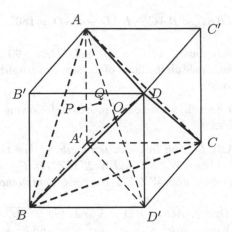

In order to prove this, let P be any other point in space. Since $P \neq O$ and O is
the intersection point of planes $(AA'D'D)$, $(BB'C'C)$, $(ABD'C')$, we can assume,
without any loss of generality, that $P \notin (AA'D'D)$ (for the sake of simplicity, we
plot P inside the cube, but this is completely immaterial). We shall prove that P does
not minimize the sum of the distances to the vertices of $ABCD$; more precisely, if
Q stands for the foot of the perpendicular dropped from P to $(AA'D'D)$, we shall
show that

$$\overline{AP} + \overline{BP} + \overline{CP} + \overline{DP} > \overline{AQ} + \overline{BQ} + \overline{CQ} + \overline{DQ}.$$

For what is left to do, start by looking at the right triangles APQ and DPQ to
obtain

$$\overline{AP} + \overline{DP} > \overline{AQ} + \overline{DQ}. \tag{11.2}$$

Now, note that $\overleftrightarrow{PQ} \parallel \overleftrightarrow{BC}$ and $\overline{BQ} = \overline{CQ}$. Therefore, letting E be the symmetric of C with respect to $\overleftrightarrow{PQ}$ (cf. figure below) we obtain

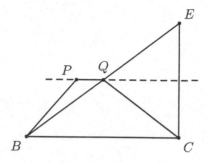

$$\overline{BP} + \overline{CP} = \overline{BP} + \overline{PE} > \overline{BE}$$
$$= \overline{BQ} + \overline{QE} = \overline{BQ} + \overline{QC}. \tag{11.3}$$

Finally, by adding (11.2) and (11.3) we arrive at the desired inequality.   □

Problems: Sect. 11.3

1. (TT) $ABCD$ is a regular tetrahedron with circumscribed sphere Σ. If CC' and DD' are diameters of Σ, compute the angle between planes (ABC') and (ACD').

2. * (France) Let S be a fixed point of a sphere Σ, centered at O. Consider all tetrahedra $SABC$, inscribed in Σ and such that SA, SB and SB are pairwise perpendicular. Prove that all of the planes (ABC) pass through a common point.

3. * Prove that a tetrahedron is isosceles if and only if the common perpendiculars to its pairs of reverse edges join the midpoints of these edges.

4. Prove that, in every isosceles tetrahedron, the common perpendiculars to the pairs of reverse edges intersect at their respective midpoints.

5. Show that a given tetrahedron is isosceles if and only if its incenter and circumcenter coincide.

6. Prove that, in every isosceles tetrahedron, the barycenter (cf. Problem 9, page 387), incenter and circumcenter coincide.

7. (Brazil—adapted) Consider an isosceles tetrahedron with edge lengths equal to a, b and c. Show that:

(a) The distances between the midpoints of the pairs of reverse edges of the tetrahedron are equal to $\sqrt{\frac{b^2+c^2-a^2}{2}}$, $\sqrt{\frac{a^2+c^2-b^2}{2}}$ and $\sqrt{\frac{a^2+b^2-c^2}{2}}$.

(b) The radius of the sphere circumscribed to the tetrahedron is equal to $\sqrt{\frac{a^2+b^2+c^2}{8}}$.

Chapter 12
Convex Polyhedra

We begin this chapter by defining and computing the area of a sphere and establishing a famous result of Girard on the area of a *spherical triangle*. Then, we present the important concept of *convex polyhedron*, which encompasses prisms and pyramids, and apply Girard's theorem to prove the celebrated *theorem of Euler*, which asserts that the *Euler characteristic* of every convex polyhedron is equal to 2. The chapter finishes with using Euler's theorem to obtain the classification of all *regular* polyhedra, and showing that all found possibilities do exist.

12.1 The Area of a Sphere

This section starts with the definition of a *surface of revolution*, as well as with the presentation of a heuristic argument which allows us to define a notion of area for such a surface. To this end, we shall need to use some simple concepts and facts on Calculus, for which we refer the reader to [2] or [5].

Let $a < b$ be given real numbers, and $f : (a, b) \to \mathbb{R}$ be a positive and differentiable function, with continuous derivative $f' : (a, b) \to \mathbb{R}$. In a fixed plane in space, take a cartesian system with axis of abscissas e, and let $\mathcal{G}$ be the graph of f. The **surface of revolution** of **axis** e and **generatrix curve** $\mathcal{G}$ is the set $\mathcal{S}(e; \mathcal{G})$ of points in space obtained by the *rotation* of $\mathcal{G}$ around e, in such a way that, for every $x \in (a, b)$, the point $(x, f(x)) \in \mathcal{G}$ describes the circle of radius $f(x)$, centered at $x \approx (x, 0) \in e$ and contained in the plane passing through x and perpendicular to e (cf. Fig. 12.1).

For real numbers $c, d \in (a, b)$, with $c < d$, let $\mathcal{S}_{[c,d]}$ be the portion of $\mathcal{S}(e; \mathcal{G})$ lying in the strip of the space formed by the planes perpendicular to e at c and d. In order to define the area of $\mathcal{S}_{[c,d]}$, consider real numbers $c = x_0 < x_1 < \cdots < x_n = d$, with $\Delta x_i := x_i - x_{i-1}$ equal to $\frac{d-c}{n}$ for $1 \le i \le n$. For a sufficiently large n, it is pretty reasonable to suppose that the frustum of cone with axis e, radii

© Springer International Publishing AG, part of Springer Nature 2018
A. Caminha Muniz Neto, *An Excursion through Elementary Mathematics, Volume II*,
Problem Books in Mathematics, https://doi.org/10.1007/978-3-319-77974-4_12

Fig. 12.1 The surface of revolution $\mathcal{S}(e; \mathcal{G})$

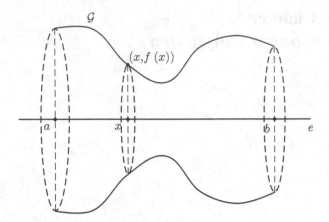

Fig. 12.2 Approximating the area of a surface of revolution

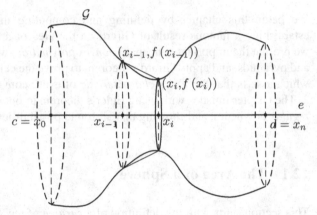

$f(x_{i-1})$ and $f(x_i)$ and height Δx_i forms a good approximation of the portion of $\mathcal{S}_{[c,d]}$ situated between the planes perpendicular to e at x_{i-1} and x_i (cf. Fig. 12.2). Therefore, it is also reasonable to assume that the lateral area A_i of such a frustum of cone constitutes a good approximation for the *area* of the portion of $\mathcal{S}_{[c,d]}$ situated between these planes.

It now follows from Problem 4, page 386, that

$$A_i = \pi(f(x_{i-1}) + f(x_i))\Delta x_i \sqrt{1 + \left(\frac{f(x_i) - f(x_{i-1})}{x_i - x_{i-1}}\right)^2}.$$

However, by the Mean Value Theorem of Lagrange, there exists $\xi_i \in (x_{i-1}, x_i)$ such that $\frac{f(x_i) - f(x_{i-1})}{x_i - x_{i-1}} = f'(\xi_i)$. Moreover, taking into account that f does not vary too much along $[x_{i-1}, x_i]$ when n is sufficiently large, we can assume that $f(x_{i-1}), f(x_i) \cong f(\xi_i)$, and hence

$$A_i \cong 2\pi f(\xi_i)\sqrt{1 + f'(\xi_i)^2}\Delta x_i.$$

By adding the approximations above for the lateral areas of the n frustums of cone thus obtained, we are led to conclude that

$$\sum_{i=1}^{n} 2\pi f(\xi_i)\sqrt{1 + f'(\xi_i)^2}\Delta x_i$$

is a good approximation for what we would like to consider as being the *area* of $S_{[c,d]}$. Also, it is geometrically plausible that such an approximation becomes better and better as $n \to +\infty$.

On the other hand, as $n \to +\infty$ we have that

$$\sum_{i=1}^{n} 2\pi f(\xi_i)\sqrt{1 + f'(\xi_i)^2}\Delta x_i \to 2\pi \int_c^d f(x)\sqrt{1 + f'(x)^2}dx,$$

so that we *define* the **area** A of $S_{[c,d]}$ as

$$A = 2\pi \int_c^d f(x)\sqrt{1 + f'(x)^2}dx. \tag{12.1}$$

Notice that the integral at the right hand side above does make sense, for the hypotheses we made on f guarantee the continuity of the integrand.

Example 12.1 In the previous discussion, let $f : (-R, R) \to \mathbb{R}$ be given by $f(x) = \sqrt{R^2 - x^2}$, so that $f'(x) = \frac{-x}{\sqrt{R^2 - x^2}}$ for $x \in (-R, R)$. Then, $S(e; \mathcal{G})$ is a sphere Σ of radius R, with two antipodal points N and S removed.

However, since

$$\Sigma \setminus \{N, S\} = \bigcup_{0 < \epsilon < R} S_{[-R+\epsilon, R-\epsilon]}$$

and $S_{[-R+\epsilon', R-\epsilon']} \subset S_{[-R+\epsilon, R-\epsilon]}$ for $0 < \epsilon < \epsilon'$, it is reasonable to define the area $A(\Sigma)$ of Σ as

$$A(\Sigma) = \lim_{\epsilon \to 0+} A(S_{[-R+\epsilon, R-\epsilon]}).$$

Thus, with the aid of the Fundamental Theorem of Calculus, we compute

Fig. 12.3 A spherical lune
with opening θ

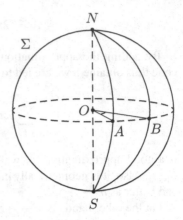

$$A(\Sigma) = \lim_{\epsilon \to 0+} A(\mathcal{S}_{[-R+\epsilon, R-\epsilon]})$$

$$= \lim_{\epsilon \to 0+} 2\pi \int_{-R+\epsilon}^{R-\epsilon} \sqrt{R^2 - x^2} \sqrt{1 + \left(\frac{-x}{\sqrt{R^2 - x^2}}\right)^2} dx$$

$$= \lim_{\epsilon \to 0+} 2\pi R \int_{-R+\epsilon}^{R-\epsilon} dx = 4\pi R^2.$$

In short, we define the **area** A of a sphere of radius R as

$$A = 4\pi R^2. \tag{12.2}$$

Continuing with our discussion, let a sphere $\Sigma(O; R)$, of center O and radius R, be given. For $\theta \in (0, \pi)$, we define a **spherical lune** of *opening* θ in Σ as the intersection of Σ with a dihedral of opening θ, whose edge contains a diameter of Σ (cf. Fig. 12.3, where we show a spherical lune of opening $A\widehat{O}B = \theta$, bounded by the arcs of great circles $\overset{\frown}{NAS}$ and $\overset{\frown}{NBS}$).

For future use, we now need to establish a formula for the area of a spherical lune. Since we have not actually *defined* such an area, the bothered reader can look at the coming lemma as a heuristic argument for deducing only reasonable way of defining it.

Lemma 12.2 *Let be given a sphere $\Sigma(O; R)$ and $\theta \in (0, \pi)$. If $\mathcal{F}$ stands for a spherical lune of opening θ in Σ, then*

$$A(\mathcal{F}) = 2\theta R^2.$$

Proof If $\theta = \frac{2\pi}{n}$ for some natural number $n > 2$, then Σ can be partitioned into n spherical lunes of openings equal to θ, so that

$$A(\mathcal{F}) = \frac{1}{n} A(\Sigma) = \frac{1}{n} \cdot 4\pi R^2 = 2\theta R^2.$$

Now, let $\theta = \frac{2\pi m}{n}$, for some natural numbers $m < n$. Then, $\mathcal{F}$ can be partitioned into m spherical lunes with openings all equal to $\frac{\theta}{m} = \frac{2\pi}{n}$, and by the previous paragraph each one of them has area $\frac{1}{m} \cdot 2\theta R^2$. Hence,

$$A(\mathcal{F}) = m \cdot 2\frac{\theta}{m} R^2 = 2\theta R^2.$$

Finally, if $\alpha = \frac{\theta}{2\pi} \in \left(0, \frac{1}{2}\right)$ is irrational, choose sequences $(r_j)_{j \geq 1}$ and $(s_j)_{j \geq 1}$ of positive rationals such that

$$0 < r_1 < r_2 < \cdots < \alpha < \cdots < s_2 < s_1 < \frac{1}{2}$$

and $r_j, s_j \to \alpha$ as $j \to +\infty$. For each integer $j \geq 1$, pick spherical lunes $\mathcal{F}_j$ and $\mathcal{F}'_j$, with openings respectively equal to $2\pi r_j$ and $2\pi s_j$, such that

$$\mathcal{F}_j \subset \mathcal{F} \subset \mathcal{F}'_j$$

(this is clearly possible from the choices of r_j and s_j). Then

$$A(\mathcal{F}_j) \leq A(\mathcal{F}) \leq A(\mathcal{F}'_j),$$

and it follows from what we did above that

$$4\pi r_j R^2 \leq A(\mathcal{F}) \leq 4\pi s_j R^2.$$

Finally, letting $j \to +\infty$, the squeezing theorem for limits of sequences gives

$$A(\mathcal{F}) = 4\pi \alpha R^2 = 2\theta R^2.$$

$\square$

We consider again a sphere $\Sigma(O; R)$. A **spherical triangle** in Σ is a union $\overset{\frown}{AB} \cup \overset{\frown}{AC} \cup \overset{\frown}{BC}$, with $\overset{\frown}{AB}$, $\overset{\frown}{AC}$, $\overset{\frown}{BC}$ being arcs of great circles in Σ, of lengths less than πR (see Fig. 12.4). In this case, we say that A, B, C are the **vertices** and $\overset{\frown}{AB}$, $\overset{\frown}{AC}$, $\overset{\frown}{BC}$ are the **sides** of the spherical triangle. Whenever there is no danger of confusion, we shall always refer to the spherical triangle above simply by ABC.

A spherical triangle ABC in Σ always divides it into two parts, exactly one of which is contained in a hemisphere (be sure to identify such a part of Σ in Fig. 12.4). We shall also refer to this portion of Σ as the *spherical triangle ABC*.

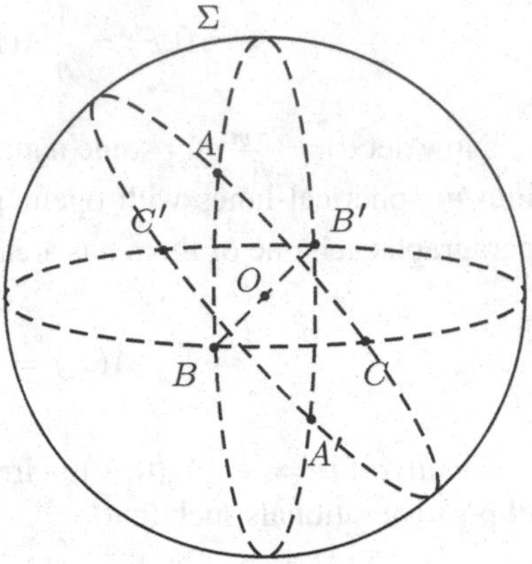

Fig. 12.4 The spherical
triangle ABC

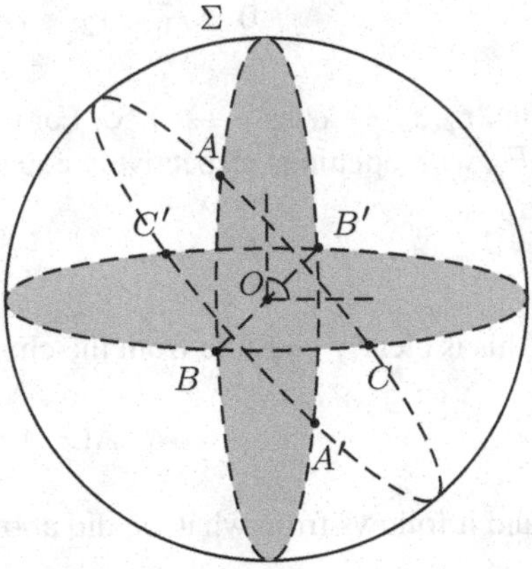

Fig. 12.5 Internal angle at B
of the spherical triangle ABC

Given a spherical triangle ABC in Σ, we define its **internal angle** at B as the
dihedral angle θ formed by the planes (ABO) and (BCO); in particular, $\theta \in (0, \pi)$.
In the notations of Fig. 12.5, planes (ABO) and (BCO) are the same as those of the
gray disks bounded by the great circles passing through A, B and B, C, respectively.
These planes meet at line $\overleftrightarrow{BB'}$, and their dihedral angle is the same as that formed
by the tangents to the corresponding great circles at B.

We have finally arrived at the second central result of this section, which
computes the area of a spherical triangle (the first result was the computation of the
area of a sphere of radius R). It is due to the French mathematician of the sixteenth
century Albert Girard.

Theorem 12.3 (Girard) *Let Σ be a sphere of radius R, and ABC be a spherical
triangle in Σ, with internal angles α, β and γ. Then,*

$$A(ABC) = (\alpha + \beta + \gamma - \pi)R^2. \tag{12.3}$$

Fig. 12.6 Girard's theorem

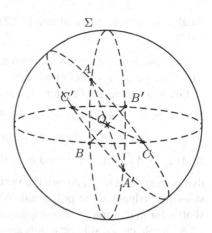

Proof Let O be the center of Σ and A', B', C' be the antipodals of A, B, C, respectively (cf. Fig. 12.6). Let $\mathcal{F}_A$ stand for the spherical lune containing the spherical triangle ABC and defined by the planes (ABB') and (ACC'); let also $\mathcal{F}'_A$ be the spherical lune symmetric to $\mathcal{F}_A$ with respect to $\overleftrightarrow{AA'}$. Define $\mathcal{F}_B$, $\mathcal{F}'_B$, $\mathcal{F}_C$ and $\mathcal{F}'_C$ in analogous ways.

Since α, β and γ are the openings of $\mathcal{F}_A$, $\mathcal{F}_B$, $\mathcal{F}_C$, respectively (and also those of $\mathcal{F}'_A$, $\mathcal{F}'_B$, $\mathcal{F}'_C$, also respectively), Lemma 12.2 gives

$$A(\mathcal{F}_A) = A(\mathcal{F}'_A) = 2\alpha R^2, \quad A(\mathcal{F}_B) = A(\mathcal{F}'_B) = 2\beta R^2$$

and

$$A(\mathcal{F}_C) = A(\mathcal{F}'_C) = 2\gamma R^2.$$

Now, a quick inspection in Fig. 12.6 shows that

$$\Sigma = (\mathcal{F}_A \setminus ABC) \cup (\mathcal{F}'_A \setminus A'B'C') \cup (\mathcal{F}_B \setminus ABC) \cup (\mathcal{F}'_B \setminus A'B'C')$$

$$\cup (\mathcal{F}_C \setminus ABC) \cup (\mathcal{F}'_C \setminus A'B'C') \cup ABC \cup A'B'C'.$$

Since the factors at the right hand side have no interior points in common and the spherical triangles ABC and $A'B'C'$ are congruent, the decomposition above gives

$$A(\Sigma) = A(\mathcal{F}_A \setminus ABC) + A(\mathcal{F}'_A \setminus A'B'C') + A(\mathcal{F}_B \setminus ABC) + A(\mathcal{F}'_B \setminus A'B'C')$$

$$+ A(\mathcal{F}_C \setminus ABC) + A(\mathcal{F}'_C \setminus A'B'C') + A(ABC) + A(A'B'C')$$

$$= A(\mathcal{F}_A) + A(\mathcal{F}'_A) + A(\mathcal{F}_B) + A(\mathcal{F}'_B) + A(\mathcal{F}_C) + A(\mathcal{F}'_C)$$

$$- 6A(ABC) + 2A(ABC)$$

$$= 4\alpha R^2 + 4\beta R^2 + 4\gamma R^2 - 4A(ABC).$$

Finally, it suffices to substitute (12.2) at the left hand side to obtain the desired result.   □

We now extend Girard's theorem to the class of *convex spherical polygons*, for which we need a few preliminaries.

Given a sphere Σ of center O, a *simple closed spherical polygonal* of k sides in Σ is a union $\overset{\frown}{A_1 A_2} \cup \overset{\frown}{A_2 A_3} \cup \ldots \cup \overset{\frown}{A_{k-1} A_k} \cup \overset{\frown}{A_k A_1}$, where, for $1 \leq j \leq k$ (with $A_{k+1} = A_1$), $\overset{\frown}{A_j A_{j+1}}$ is an arc of great circle in Σ, of measure less than π, and $\overset{\frown}{A_i A_{i+1}} \cap \overset{\frown}{A_j A_{j+1}} \neq \emptyset$ if and only if $i = j - 1$, j or $j + 1$. In this case, we say that points $A_1, A_2, \ldots, A_k$ are the **vertices** and arcs $\overset{\frown}{A_i A_{i+1}}$ (for $1 \leq i \leq k$) are the **sides** (or **edges**) of the polygonal. Whenever there is no danger of confusion, we shall refer to the simple closed spherical polygonal above simply as $A_1 A_2 \ldots A_k$.

A simple closed spherical polygonal $A_1 A_2 \ldots A_k$ in Σ divides it into two parts, exactly one of which is contained in a hemisphere of Σ. Such a part is the **spherical polygon** bounded by the polygonal $A_1 A_2 \ldots A_k$. As above, from now on we shall refer to it simply as $A_1 A_2 \ldots A_k$. Also in this case, we define the **internal angle** of the spherical polygon $A_1 A_2 \ldots A_k$ at A_i as the opening of the dihedral angle formed by the planes $(A_{i-1} A_i O)$ and $(A_i A_{i+1} O)$.

With the concepts above at our disposal, we can finally state the coming

Corollary 12.4 *Let Σ be a sphere of radius R. If $A_1 A_2 \ldots A_k$ is a spherical polygon in Σ, then*

$$A(A_1 A_2 \ldots A_k) = \left(\sum_{i=1}^{k} \theta_i - (k-2)\pi \right) R^2,$$

with θ_i standing for the internal angle of $A_1 A_2 \ldots A_k$ at A_i.

Proof Exercise; see Problem 1.   □

Problems: Sect. 12.1

1. * Prove Corollary 12.4.
2. We are given a sphere $\Sigma(O; R)$ and a positive real d, with $d < R$. Cut Σ with a plane α situated at distance d from O. The **spherical cap** of *height* $R - d$ in Σ is the portion of Σ lying in the half-space opposite to O with respect to α. Compute, in terms of R and d, the area of such a spherical cap.
3. An airplane is at distance h from the Earth's surface, which is assumed to be a sphere of radius R. If S stands for the area of the portion of the Earth's surfaces that can be seen by the pilot, compute S in terms of R and h.

4. Let $ABCD$ be a regular tetrahedron inscribed in a sphere of center O and radius R. Compute the measures of the angles of the spherical triangle of vertices A, B and C.

5. (TT—adapted) Let P, P_1, P_2, ..., P_n be points in space, such that each P_i is closer to P than to any other P_j.

   (a) Show that $n \leq 14$.
   (b) Exhibit an allowed configuration with $n = 13$.

12.2 Euler's Theorem

For what comes next we need a few preliminaries.

We start by recalling that, given a point P and a positive real R, the open ball of center P and radius R is the set

$$\mathcal{B}(P; R) = \{Q; \ \overline{PQ} < R\}.$$

Now, if $\mathcal{A}$ denotes a set of points in space, then we say that a point P is an **interior point** of $\mathcal{A}$ if $B(P; R) \subset \mathcal{A}$ for some $R > 0$; in particular, $P \in \mathcal{A}$. The **interior** of $\mathcal{A}$ is the subset $\text{Int}(\mathcal{A})$ of $\mathcal{A}$, formed by its interior points.

The interior of a set can be equal to the set itself, and in this case we shall say that the set is **open**. For instance, the empty set[1] and the whole space are open sets. The open ball $\mathcal{B}(P; R)$ is also an open set (and by the way this justifies the name *open ball*), for if $Q \in \mathcal{B}(P; R)$, then one readily sees that $\mathcal{B}(Q; r) \subset \mathcal{B}(P; R)$ for $r = R - \overline{PQ} > 0$.

A set $\mathcal{A}$ of points in space is **closed** provided its complement $\mathcal{A}^c$ is open. Thus, the whole space and the empty set are also closed. It is also closed

$$\mathcal{B}(P; R)^c = \{Q; \ \overline{PQ} \geq R\}$$

for its complement is the open ball $\mathcal{B}(P; R)$, which we have just seen to be an open set. Another example of closed set is furnished by the **closed ball** $\overline{\mathcal{B}(P; R)}$, where

$$\overline{\mathcal{B}(P; R)} = \{Q; \ \overline{PQ} \leq R\}.$$

Indeed, $\overline{\mathcal{B}(P; R)}^c = \{Q; \ \overline{PQ} > R\}$; therefore, if $Q \in \overline{\mathcal{B}(P; R)}^c$ and $r = \overline{PQ} - R > 0$, then $\mathcal{B}(Q; r) \subset \overline{\mathcal{B}(P; R)}^c$, so that $\overline{\mathcal{B}(P; R)}^c$ is open.

Once more we consider a set $\mathcal{A}$ of points in space. A point P is said to be a **boundary point** of $\mathcal{A}$ if, for every $R > 0$, the open ball $\mathcal{B}(P; R)$ intersects both $\mathcal{A}$ and $\mathcal{A}^c$. The **boundary** of $\mathcal{A}$ is the set $\partial \mathcal{A}$, formed by its boundary points.

---

[1] By the definitions just presented, we are forced to conclude that the interior of the empty set is itself.

The boundary of a set can be disjoint from the set itself. Every open ball is an example of such a situation, provided we show that $\partial \mathcal{B}(P; R) = \Sigma(P; R)$. Indeed, our previous discussions guarantee that no point of $\mathcal{B}(P; R) \cup \overline{\mathcal{B}(P; R)}^c$ lies in the boundary of $\mathcal{B}(P; R)$; on the other hand, for every point $Q \in \Sigma(P; R)$, we have $Q \in \mathcal{B}(P; R)^c$, whereas it is immediate to check that $\mathcal{B}(Q; r)$ intersects $\mathcal{B}(P; R)$ for every $r > 0$.

The coming result brings an important relation between the concepts of closed set and boundary.

Lemma 12.5 *A set $\mathcal{A}$ of points in space is closed if and only if $\partial \mathcal{A} \subset \mathcal{A}$. Moreover, if this the case we have*

$$\mathcal{A} = \text{Int}(\mathcal{A}) \cup \partial \mathcal{A}.$$

Proof Firstly, we assume that $\mathcal{A}$ is closed and prove that $\partial \mathcal{A} \cap \mathcal{A}^c = \emptyset$. Indeed, since $\mathcal{A}^c$ is open, if $P \in \mathcal{A}^c$, then there exists $R > 0$ such that $\mathcal{B}(P; R) \subset \mathcal{A}^c$; in particular, $P \notin \partial \mathcal{A}$.

Conversely, we suppose that $\partial \mathcal{A} \subset \mathcal{A}$ and show that $\mathcal{A}^c$ is open; to this end, if P is a point in $\mathcal{A}^c$, we shall prove that P lies in the interior of $\mathcal{A}^c$. Note first that if the open ball $\mathcal{B}(P; R)$ intersects both $\mathcal{A}$ and $\mathcal{A}^c$ for every $R > 0$, then we will have $P \in \partial \mathcal{A}$, so that $P \in \mathcal{A}$; but this contradicts the fact that $P \in \mathcal{A}^c$. Thus, there exists $R > 0$ for which $\mathcal{B}(P; R)$ intersects at most one of the sets $\mathcal{A}$ and $\mathcal{A}^c$. However, since $P \in \mathcal{B}(P; R) \cap \mathcal{A}^c$, we have that $\mathcal{B}(P; R)$ intersects solely $\mathcal{A}^c$; on the other hand, since $\mathcal{A} \cup \mathcal{A}^c$ is the whole space, it follows that $\mathcal{B}(P; R) \subset \mathcal{A}^c$.

For what is left to do, if $\mathcal{A}$ is a closed set, then the first part gives $\text{Int}(\mathcal{A}) \cup \partial \mathcal{A} \subset \mathcal{A}$. Now, we take $P \in \mathcal{A}$ and show that $P \in \text{Int}(\mathcal{A}) \cup \partial \mathcal{A}$, for which we consider two distinct cases:

- There exists $R > 0$ such that $\mathcal{B}(P; R) \subset \mathcal{A}$: in this case, $P \in \text{Int}(\mathcal{A})$.
- For every $R > 0$, we have $\mathcal{B}(P; R) \not\subset \mathcal{A}$: in this case, since $P \in \mathcal{A}$, it is immediate that $P \in \partial \mathcal{A}$.

$\square$

We are finally in position to define the central object of study of this section.

Definition 12.6 A **polyhedron** is a closed and bounded set in space, with nonempty interior and whose boundary consists of the union of a finite number of polygons satisfying the following conditions:

(a) Any two polygons do not lie in a single plane.
(b) If two polygons intersect, then they have exactly a vertex or an edge in common.
(c) If two such polygons $\mathcal{P}$ and $\mathcal{Q}$ do not intersect, then there exist polygons $\mathcal{P}_1 = \mathcal{P}, \mathcal{P}_2, \ldots, \mathcal{P}_k = \mathcal{Q}$, such that $\mathcal{P}_i$ and $\mathcal{P}_{i+1}$ do intersect, for $1 \leq i < k$.

A **convex polyhedron** is a polyhedron which is also a convex subset of space.

Pyramids and prisms are examples of convex polyhedra. For a general polyhedron $\mathcal{P}$, Lemma 12.5 gives $\partial \mathcal{P} \subset \mathcal{P}$. It can also be proved (though we shall assume it without proof) that if two polygons of the boundary intersect along a common edge, then no other polygon of the boundary contains this edge. Henceforth, we shall sometimes refer to $\partial \mathcal{P}$ as the **surface** of $\mathcal{P}$, and to the polygons that compose it as the **faces** of $\mathcal{P}$. The **edges** (resp. the **vertices**) of $\mathcal{P}$ are the edges (resp. the vertices) of its faces.

We now need the following preliminary result, in whose proof we shall use the concept of *least upper bound* of a set of real numbers bounded from above (cf. Section 7.1 of [5], for instance).

Proposition 12.7 *Let $\mathcal{P}$ be a convex polyhedron, O a point in its interior and Σ a sphere centered at O. For every point $P \in \Sigma$, the half-line $\overrightarrow{OP}$ intersect $\partial \mathcal{P}$ in a single point Q. Moreover, the function*

$$P \in \Sigma \mapsto Q \in \partial \mathcal{P}$$

thus defined is a bijection.[2]

Proof For $P \in \Sigma$, let us firstly show that $\overrightarrow{OP}$ intersects $\partial \mathcal{P}$ at a single point. To this end, start by noticing that the set

$$\{X \in \overrightarrow{OP}; \ X \neq O \ \text{and} \ OX \subset \mathcal{P}\}$$

is nonempty (for O belongs to the interior of $\mathcal{P}$) but does not coincide with $\overrightarrow{OP}$ (for $\mathcal{P}$ is bounded). Writing $\overline{OX} = x$, the discussion above guarantees that

$$d := \sup\{x; \ X \in \overrightarrow{OP} \ \text{and} \ OX \subset \mathcal{P}\}$$

is well defined. If $Q \in \overrightarrow{OP}$ is the only point for which $\overline{OQ} = d$, we claim that $Q \in \partial \mathcal{P}$. In order to establish this fact, Lemma 12.5 assures that it suffices to exclude the possibilities $Q \in \text{Int}(\mathcal{P})$ and $Q \in \mathcal{P}^c$.

- If $Q \in \text{Int}(\mathcal{P})$, take $R > 0$ such that $\mathcal{B}(Q; R) \subset \mathcal{P}$, and then a point $T \in \overrightarrow{OQ} \cap \mathcal{B}(Q; R)$ for which $Q \in OT$. We thus have $T \in \overrightarrow{OP}$, $OT \subset \mathcal{P}$ and $\overline{OT} > \overline{OQ} = d$, which contradicts the definition of d.
- If $Q \in \mathcal{P}^c$ (which is open), take $R > 0$ such that $\mathcal{B}(Q; R) \subset \mathcal{P}^c$, and then a point $T \in \overrightarrow{OQ} \cap \mathcal{B}(Q; R)$ for which $T \in OQ$. This way, $TQ \subset \mathcal{P}^c$, so that $d \leq \overline{OT} < \overline{OQ} = d$. Once more, this is a contradiction.

---

[2] At this point, if we had at our disposal the concept of *continuity* for functions $f : \Sigma \to \partial \mathcal{P}$, we could easily prove that this function f is a *homeomorphism*, i.e., a continuous bijection with continuous inverse. Thanks to this result, the boundary of a convex polyhedron is *homeomorphic* to (i.e., has essentially the *same shape* as) a sphere. Intuitively, such a statement means that one can *continuously deform* a sphere until transform it into the boundary of $\mathcal{P}$.

We are left to showing that, for $P \in \Sigma$, the half-line $\overrightarrow{OP}$ intersects $\partial \mathcal{P}$ at a single point. Arguing by contradiction, assume that $\overrightarrow{OP}$ intersects $\partial \mathcal{P}$ at Q and Q', with $Q \neq Q'$. Suppose $Q' \in OQ$ (the other case can be handled analogously). Since $P \in \text{Int}(\mathcal{P})$ and $Q \in \mathcal{P}$, Problem 1 shows that $Q' \in \text{Int}(\mathcal{P})$, which is a contradiction.

Finally, for $Q \in \partial \mathcal{P}$, it is clear that $\overrightarrow{OQ}$ intersects Σ at a single point; hence, the stated function is also surjective.

Given a (note necessarily convex) polyhedron $\mathcal{P}$, we shall denote by V, E and F its numbers of vertices, edges and faces, respectively. Also,

$$\mathcal{X}(\mathcal{P}) = V - A + F$$

is the **Euler characteristic** of $\mathcal{P}$.

Theorem 12.9 below, due to Euler, is the main result of this section. Nevertheless, before we can prove it we shall need one more auxiliary result, which is also important in itself.

Lemma 12.8 *If a polyhedron has E edges and F_k faces with k edges each, then*

$$2A = 3F_3 + 4F_4 + 5F_5 + \cdots .$$

Proof Since each edge of the polyhedron lies in exactly two faces, it suffices to note that both sides of the equality stated above count each edge of the polyhedron exactly twice. □

We are finally in position to state and prove Euler's theorem.

Theorem 12.9 (Euler) *Every convex polyhedron has Euler characteristic equal to 2.*

Proof Let $\mathcal{P}$ be a convex polyhedron, O be an interior point of $\mathcal{P}$, Σ a sphere centered at O and

$$f : \partial \mathcal{P} \to \Sigma$$
$$Q \mapsto Q'$$

the inverse of the bijection defined in the previous proposition (so that f is also a bijection). If $A_1 A_2 \ldots A_k$ is a face of $\mathcal{P}$, we claim that $A_1' A_2' \ldots A_k' = f(A_1 A_2 \ldots A_k)$ is a spherical polygon in Σ. Indeed (cf. Fig. 12.7), it is immediate to note that $f(A_i A_{i+1})$ is an arc of great circle in Σ, joining A_i' and A_{i+1}'. On the other hand, if α is the plane passing through O and parallel to the plane of the face $A_1 A_2 \ldots A_k$, then $A_1' A_2' \ldots A_k'$ lies in the hemisphere of Σ contained in the same half-space of $A_1 A_2 \ldots A_k$ with respect to α.

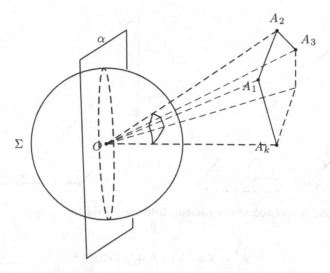

Fig. 12.7 The effect of function f

Letting θ_i stand for the internal angle of $A'_1 A'_2 \ldots A'_k$ at A'_i, Corollary 12.4 gives

$$A(A'_1 A'_2 \ldots A'_k) = \left(\sum_{i=1}^{k} \theta_i - (k-2)\pi \right) R^2, \qquad (12.4)$$

with R denoting the radius of Σ.

We now add both sides of the equality above over all convex spherical polygons obtained from the faces of $\mathcal{P}$:

- for the sum of the left hand sides note that, since f is a bijection and the spherical polygons $A'_1 A'_2 \ldots A'_k$ partition Σ, the sum of their areas equals the area S of Σ.
- concerning the sum of the right hand sides, start by observing that if a vertex A of $\mathcal{P}$ belongs to the faces $\mathcal{F}_1, \ldots, \mathcal{F}_j$ and $\mathcal{F}'_i = f(\mathcal{F}_i)$ for $1 \le i \le j$, then the sum of the internal angles of the convex spherical polygons $\mathcal{F}'_i$ at A' is equal to 2π. On the other hand, if F_k denotes (for $k \ge 3$) the number of faces of $\mathcal{P}$ with exactly k vertices, then $F = F_3 + F_4 + F_5 + \cdots$ and the sum of the right hand sides of (12.4) is successively equal to

$$\left(2\pi V - \sum_{k \ge 3}(k-2)\pi F_k \right) R^2 = \left(2V - \sum_{k \ge 3} k F_k + 2F \right) \pi R^2$$

$$= (2V - 2E + 2F)\pi R^2,$$

where we have used the result of Lemma 12.8 in the last equality.

However, since the sums of equal summands give equal totals, we conclude from the above that

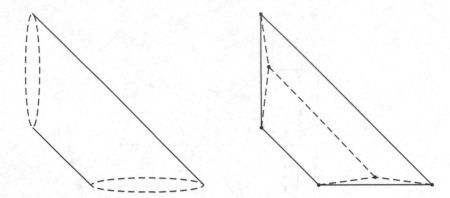

Fig. 12.8 Building a polyhedron with vanishing Euler characteristic

$$4\pi R^2 = S = (V - E + F) \cdot 2\pi R^2,$$

whence $V - E + F = 2$.

Remarks 12.10

i. Euler's theorem is no longer valid for nonconvex polyhedra. In order to see this, we start by cutting a cylinder of revolution with two planes, each one forming an angle of 45° with the axis of the cylinder, thus getting the ellipses of Fig. 12.8, left. Then, we inscribe in each of these ellipses the triangle having as vertices the endpoints of the major axis and one of the endpoints of the minor axis, as shown in Fig. 12.8, right. By joining the vertices of such triangles in pairs, we obtain a convex polyhedron. Finally, by gluing four copies of the polyhedron of Fig. 12.8, we obtain the nonconvex polyhedron $\mathcal{P}$ of Fig. 12.9. For $\mathcal{P}$, one immediately verifies that $V = 12$, $A = 24$ and $F = 12$, whence

$$\mathcal{X}(\mathcal{P}) = V - A + F = 12 - 24 + 12 = 0.$$

ii. It is possible to prove (albeit such a proof involves concepts much deeper than those discussed here) that for every polyhedron $\mathcal{P}$ one has the formula

$$\mathcal{X}(\mathcal{P}) = 2 - 2g, \tag{12.5}$$

where g stands for the *genus* of $\mathcal{P}$, which can be "*defined*" as the " *number of holes*" of $\mathcal{P}$. For the polyhedron of item i. we have $g = 1$, so that $\mathcal{X}(\mathcal{P}) = 2 - 2 = 0$. In this respect, see also Problem 9.

Convex polyhedra form our primordial interest along these notes. Hence, whenever there is no danger of confusion, from now on we shall refer to a convex polyhedron simply as a *polyhedron*.

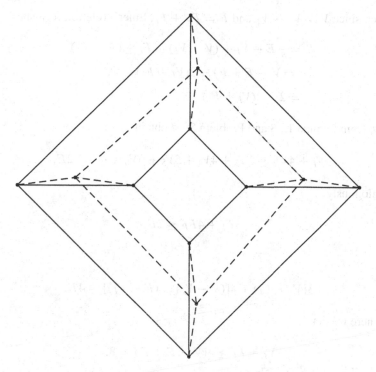

Fig. 12.9 A (nonconvex) polyhedron with vanishing Euler characteristic

As the coming example shows, sometimes Euler's relation is quite a useful tool in the analysis of problems involving convex polyhedra.

Example 12.11 (OCM) We associate the integer -1 to each edge of a convex polyhedron. Then, we associate to each vertex the product of the numbers associated to the edges incident to it. Finally, we associate to each face the product of the numbers associated to its edges. What are the possible values of the sum of all of these numbers?

Proof Let $\mathcal{P}$ be a convex polyhedron with V vertices, E edges and F faces; let also V_i (resp. F_i) stand for the number of vertices (resp. faces) of $\mathcal{P}$ incident with (resp. having) exactly i edges. Setting

$$V_P = V_4 + V_6 + V_8 + \cdots, \quad V_I = V_3 + V_5 + V_7 + \cdots,$$

$$F_P = F_4 + F_6 + F_8 + \cdots \text{ and } F_I = F_3 + F_5 + F_7 + \cdots,$$

it is immediate to see that the sum S of the numbers in the statement is given by

$$S = -A - V_I + V_P - F_I + F_P.$$

However, since $V = V_P + V_I$ and $F = F_P + F_I$, Euler's relation furnishes

$$S = -E - V_I + (V - V_I) - F_I + (F - F_I)$$
$$= (V - E + F) - 2(V_I + F_I)$$
$$= 2 - 2(V_I + F_I).$$

Now, from Lemma 12.8 and Problem 5 we obtain

$$3V_I + 4V_P \leq 3V_3 + 4V_4 + 5V_5 + 6V_6 + \cdots = 2E,$$

and analogously

$$3F_I + 4F_P \leq 2E.$$

Hence,

$$3(V_I + F_I) + 4[(V - V_I) + (F - F_I)] \leq 4E$$

from where we get

$$V_I + F_I \geq 4(V - E + F) = 8.$$

Also from Lemma 12.8 and Problem 5 we get

$$V_I + (2V_3 + 4V_4 + 4V_5 + 6V_6 + \cdots) = 2E$$

and

$$F_I + (2F_3 + 4F_4 + 4F_5 + 6F_6 + \cdots) = 2E,$$

so that V_I and F_I are even numbers.

Writing $V_I + F_I = 2k$, with $k \geq 4$, we conclude that $S = 2 - 4k$, with $k \geq 4$. We claim that all of these values are attained, for which we consider two distinct cases:

(i) $k \geq 4$ is even: we take regular polygons $A_1 A_2 \ldots A_k$ and $A_1' A_2' \ldots A_k'$, centered at points O and O', respectively, and positioned in space in such a way that $\overleftrightarrow{OO'}$ is perpendicular to their planes. We now rotate $A_1' A_2' \ldots A_k'$, with center O' and around $\overleftrightarrow{OO'}$, of $\frac{\pi}{k}$ radians, thus getting a new regular polygon $A_1'' A_2'' \ldots A_k''$. Finally, for $1 \leq i \leq k$, we draw the line segments $A_i A_{i-1}''$ (with $A_0'' = A_k''$), thus obtaining a convex polyhedron with $2k$ vertices, 2 faces which are k-gons and $2k$ triangular faces; also, each vertex of such a polyhedron is adjacent to exactly four others (i.e., $V = V_4$). Therefore, for such a polyhedron we have

$$S = 2 - 2(V_I + F_I) = 2 - 2(0 + 2k) = 2 - 4k.$$

(ii) $k \geq 4$ is odd: take regular polygons $A_1 A_2 \ldots A_k$ and $A_1' A_2' \ldots A_{k-1}'$, centered at points O and O', respectively, and positioned in space in such a way that $\overleftrightarrow{OO'}$ is perpendicular to their planes. Then, impose that $\overleftrightarrow{A_1 A_2} \parallel \overleftrightarrow{A_1' A_2'}$ and the remaining vertices in both polygons lie in the same half-space, of those determined by the trapezoid $A_1 A_2 A_2' A_1'$. Drawing all line segments joining two consecutive points in the list A_2, A_2', A_3, A_3', A_4, A_4', A_5, $\ldots A_k$, A_1', A_1, we obtain a convex polyhedron with $2k - 2$ faces with odd numbers of edges (on face being a k-gon and $2k - 3$ faces being triangular), in which $V_3 = 2$ and $V_4 = 2k - 3$. Therefore, for such a polyhedron one has

$$S = 2 - 2(V_I + F_I) = 2 - 2(2 + (2k - 2)) = 2 - 4k.$$

□

Problems: Sect. 12.2

1. * Let C be a convex set and A and B be points of C, with A being an interior point. Prove that all points in the line segment AB, with the possible exception of B, are also interior points of C.

2. Given a set A of points in space, prove that ∂A is a closed set.

3. (IMO—Shortlist) Prove that there exists no convex polyhedra with exactly 7 edges.

4. (Brazil) Prove that every convex polyhedron has an even number of faces with an odd number of edges.

5. * In a convex polyhedron, let E denote the number of edges and V_k the number of vertices incident to exactly k edges. Prove that $2E = 3V_3 + 4V_4 + 5V_5 + \cdots$.

6. If each pair of vertices of a convex polyhedron is joined by an edge, prove that the polyhedron is a tetrahedron.

7. (IMO—shortlist) Prove that there does not exist a convex polyhedron all of whose plane sections are triangles.

8. (IMO—shortlist) All faces of a convex polyhedron are equilateral triangles. Prove that it has no more than 30 edges.[3]

9. * Show that, for every $n \in \mathbb{N}$, there exists a polyhedron whose Euler characteristic is equal to $2 - 2n$.

---

[3] The maximum value of 30 edges is attained in regular icosahedra (see next section).

10. (Brazil) We are given a convex polyhedron and a point P lying in its interior. Prove that, for at least one of the faces of the polyhedron, the orthogonal projection of P onto that face lies inside it.

11. (IMO—shortlist) A convex polyhedron has 12 faces and satisfies the following conditions:

   (a) All of its faces are isosceles triangles of side lengths x and y, with $x \neq y$.

   (b) Each vertex is incident to either 3 or 6 edges.

   (c) All dihedral angles of the polyhedron are equal.

   Compute $\frac{x}{y}$.

12.3 Regular Polyhedra

In this section we classify the *regular* convex polyhedra, also called **Plato's polyhedra**,[4] in the sense of the following

Definition 12.12 A convex polyhedron is called **regular** if the two following conditions are satisfied:

(a) All of its faces are regular polygons with the same number of edges.

(b) Each vertex is incident with the same number of edges.

We had already been presented to two distinct kinds of regular polyhedra in Sects. 11.1 and 11.2: regular tetrahedra and cubes. These last ones are also known as regular **hexahedra**, thanks to the fact they possess exactly six square faces. On the other hand, if we join the center of each face of a cube to the centers of the four adjacent faces, we obtain a new kind of regular polyhedron, the regular **octahedron**, which has exactly eight triangular faces (cf. Fig. 12.10).

Continuing, we shall now exhibit two other types of regular polyhedra, beginning with the *icosahedron*.

Fig. 12.10 The regular octahedron $ABCDEF$

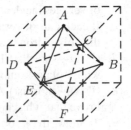

---

[4]In honor of Plato, one of the great philosophers of Classical Greek Antiquity, who lived in the fifth century BC.

Fig. 12.11 Constructing a regular icosahedron

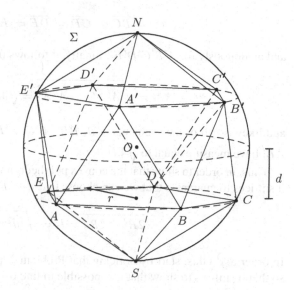

Example 12.13 There exists a regular polyhedron inscribed in (i.e., with all of its vertices lying on the surface of) a sphere $\Sigma(O; R)$, having exactly twelve vertices, twenty triangular faces and thirty edges. Such a polyhedron is called a **regular icosahedron**.

Proof Fix an equator Γ of Σ and let N and S be the corresponding north and south poles (cf. Fig. 12.11). The planes parallel to the plane of Γ, at a distance $d < R$ of O, intersect Σ along two circles of radii r, with $r = \sqrt{R^2 - d^2}$. In one such circle, let $AA''BB''CC''DD''EE''$ be an inscribed regular decagon (note that A'', B'', C'', D'' and E'' are not shown in Fig. 12.11); in the other circle, let $A'B'C'D'E'$ be the inscribed regular pentagon, with $\overleftrightarrow{A'A''}$, $\overleftrightarrow{B'B''}$, $\overleftrightarrow{C'C''}$, $\overleftrightarrow{D'D''}$ and $\overleftrightarrow{E'E''}$ parallel to $\overleftrightarrow{NS}$.

Now, by joining S and N to the vertices of $ABCDE$ and $A'B'C'D'E'$, and drawing line segments AA', $A'B$, BB', $B'C$, $B'D$, $C'D$, DD', $D'E$, EE' and AE', we obtain a convex polyhedron with twelve vertices and twenty triangular faces, in which each vertex is adjacent to five other vertices. Moreover,

$$\overline{SA} = \overline{SB} = \overline{SC} = \overline{SD} = \overline{SE} = \sqrt{(R-d)^2 + r^2} = \sqrt{2R^2 - 2Rd}$$

and

$$\overline{NA'} = \overline{NB'} = \overline{NC'} = \overline{ND'} = \overline{NE'} = \sqrt{2R^2 - 2Rd}.$$

On the other hand, letting l_5 and l_{10} respectively denote the lengths of the sides of the regular pentagon and decagon inscribed in a circle of radius r, we have (cf. Problem 2, page 254)

$$\overline{AB} = \overline{BC} = \overline{CD} = \overline{DE} = \overline{EA} = l_5,$$

and analogously for $A'B'C'D'E'$. Finally, it follows from Pythagoras' theorem that

$$\overline{AA'} = \sqrt{\overline{A'A''}^2 + \overline{A''A}^2} = \sqrt{4d^2 + l_{10}^2},$$

and likewise all of the segments $A'B$, BB', $B'C$, $B'D$, $C'D$, DD', $D'E$, EE' and AE' have measure equal to $\sqrt{4d^2 + l_{10}^2}$.

Thus, in order to show that the convex polyhedron we have constructed is regular, it suffices to guarantee that we can choose $0 < d < R$ such that

$$\sqrt{2R^2 - 2Rd} = l_5 = \sqrt{4d^2 + l_{10}^2}.$$

In order to do this, start by noticing that Problem 2, page 254 gives $l_5^2 = l_{10}^2 + r^2$, so that it suffices to show that it is possible to find $0 < d < R$ for which

$$2R^2 - 2Rd = l_{10}^2 + r^2 = 4d^2 + l_{10}^2.$$

Since $d^2 = R^2 - r^2$, the second equality is equivalent to

$$l_{10}^2 + r^2 = 4(R^2 - r^2) + l_{10}^2,$$

which in turn gives $r = \frac{2R}{\sqrt{5}}$. Finally, substituting this value of r in the equality $2R^2 - 2Rd = l_{10}^2 + r^2$, we ought to have

$$l_{10}^2 = 2R^2 - 2R\sqrt{R^2 - r^2} - r^2$$

$$= 2\left(\frac{r\sqrt{5}}{2}\right)^2 - 2 \cdot \frac{r\sqrt{5}}{2} \cdot \sqrt{\frac{5r^2}{4} - r^2} - r^2$$

$$= r^2\left(\frac{3 - \sqrt{5}}{2}\right) = r^2\left(\frac{\sqrt{5} - 1}{2}\right)^2,$$

which is true (again thanks to Problem 2, page 254).   □

After the proof of Theorem 12.15, the reader will find the (seemingly arbitrary) choices involved in the construction delineated in the previous example to be completely natural. For the time being, let us use such example to present the last type of regular polyhedron.

Example 12.14 There exists a regular polyhedron inscribed (i.e., with all of its vertices) in a sphere $\Sigma(O; R)$, having exactly twenty vertices, twelve pentagonal faces and thirty edges. Such a polyhedron is called a **regular dodecahedron**.

Proof With the previous example at our disposal, let A be a vertex and ABC, ACD, ADE, AEF and AFB be the faces of a regular icosahedron incident at A. If V, W, X, Y and Z stand for the centers of those faces, respectively, it is immediate to verify that $VWXYZ$ is a regular pentagon whose side lengths depend only on the lengths of the edges of the icosahedron. However, since the icosahedron has twelve vertices, if we construct (as above) the regular pentagons corresponding to each one of them we shall obtain a regular dodecahedron. $\qquad\qquad$ $\square$

The coming result guarantees that the examples discussed so far exhaust all possible types of regular polyhedra.

Theorem 12.15 *Every regular polyhedron is a tetrahedron, a hexahedron, an octahedron, a dodecahedron or an icosahedron.*

Proof Let P be a regular polyhedron with V vertices, E edges and F faces, and assume that each face has $n > 2$ edges and each vertex is incident with $m > 2$ edges.

Lemma 12.8 degenerates in the equality $2E = nF$. On the other hand, it follows from Problem 5, page 419, that $mV = 2E$. Now, Euler's formula furnishes

$$2 = V - E + F = \frac{2E}{m} - E + \frac{2E}{n}$$

or, which is the same,

$$\frac{2}{m} + \frac{2}{n} = \frac{2}{E} + 1. \tag{12.6}$$

In order to solve the equation above, note that if $m, n \geq 4$ then

$$\frac{2}{E} + 1 = \frac{2}{m} + \frac{2}{n} \leq \frac{2}{4} + \frac{2}{4} = 1,$$

which is impossible. Therefore, $m = 3$ or $n = 3$. We shall consider these two cases separately:

(i) $m = 3$: then $\frac{2}{n} = \frac{2}{E} + \frac{1}{3} = \frac{E+6}{3E}$, whence

$$n = \frac{6E}{E+6} = 6 - \frac{36}{E+6}.$$

However, since $E > 3$, we conclude that $E+6$ is a divisor of 36 which is larger than 9, so that $E + 6 = 12$, 18 or 36. We then have the possibilities

$$(V, E, F, m, n) = (4, 6, 4, 3, 3), \quad (8, 12, 6, 3, 4), \quad \text{or} \quad (20, 30, 12, 3, 5).$$

(ii) $n = 3$: since (12.6) is symmetric with respect to m and n, we obtain from (i) that

$$(V, E, F, m, n) = (4, 6, 4, 3, 3), \quad (6, 12, 8, 4, 3), \quad \text{or} \quad (12, 30, 20, 5, 3).$$

The above discussion assures that there is at most five distinct possibilities for (V, E, F, m, n). Nevertheless, it could well happen that at least one of them would correspond to more than one type of regular polihedron, so that the list of regular polyhedra we have so far did not exhaust all possibilities. We shall show that this does not happen, and in order to do this we look separately to each one of the possibilities above:

- $(V, E, F, m, n) = (4, 6, 4, 3, 3)$: in this case, since we have four vertices and six edges, each vertex must be joined to the three remaining ones. However, since all faces are triangular, it is immediate that this case corresponds to regular tetrahedra.
- $(V, E, F, m, n) = (6, 12, 8, 4, 3)$: we have six vertices, each of which is joined to four others, and eight triangular faces. Pick a vertex A and let B, C, D, E denote the vertices adjacent to it, corresponding to the faces ABC, ACD, ADE, AEB. Letting F denote the sixth and last vertex, it is immediate to check that F must also be adjacent to B, C, D, E, so that the polyhedron is a regular octahedron.
- $(V, E, F, m, n) = (8, 12, 6, 3, 4)$: an analysis quite similar to that of the previous item assures that this case corresponds to regular hexahedra.
- $(V, E, F, m, n) = (12, 30, 20, 5, 3)$: fix a vertex S of the polyhedron, so that exactly five faces are incident to S, all of which triangular. Let SAB, SBC, SCD, SDE, SEA be such faces (cf. Fig. 12.12). Let $AA'B$ stand for the other face having AB as an edge. If $A'BC$ was a face, we would have only four edges incident to B, which is impossible; hence, the other face having $A'B$ as an edge is $A'BB'$, with $B' \neq C$. Analogously, there exist vertices C', D', E'

Fig. 12.12 The case of an icosahedron

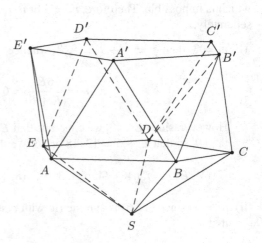

as in Fig. 12.12, i.e., such that $BB'C$, $B'CD$, $B'C'D$, $C'DD'$, $DD'E$, $D'EE'$, AEE', $AA'E'$ are faces of the polyhedron. At this point we have already used eleven vertices, so that, letting N stand for the twelveth one, it must be adjacent to A', B', C', D', E'. Therefore, this case corresponds to that of regular icosahedra.

- $(V, E, F, m, n) = (20, 30, 12, 3, 5)$: this case corresponds to regular dodecahedral and will be left as an exercise for the reader (see Problem 3).

$\square$

Problems: Sect. 12.3

1. Show that it is possible to tile the space with regular tetrahedra and octahedra.
2. * Compute, as a function of R, the length of the edges of a regular icosahedron inscribed in a sphere of radius R.
3. * Complete the proof of Theorem 12.15, showing that the case of the dodecahedron corresponds to $(V, E, F, m, n) = (12, 30, 20, 5, 3)$.
4. * The purpose of this problem is to compute the length of the edges of a regular dodecahedron inscribed in a sphere $\Sigma(X; R)$ (the argument delineated in Example 12.14 makes it clear that such a dodecahedron does exist). To this end, cut a hemisphere of Σ with parallel planes α and α', such that $d(X; \alpha) = d$ and $d(X; \alpha') = d'$, with $d > d'$. Let $\Gamma(O; r)$ (resp. $\Gamma'(O'; r')$) denote the intersection of α (resp. α') with Σ, and $ABCD$ (resp. $A'B'C'D'E'$) be a regular pentagon inscribed in Γ (resp. Γ'), so that $\overleftrightarrow{AB} \parallel \overleftrightarrow{A'B'}$. Finally, adjust d and d' in such a way that, for some points A'', B'', C'', D'' and E'' on Σ, all of $ABB'A''A'$, $BCC'B''B'$, $CDD'C''C'$, $DEE'D''D'$ and $EAA'E''E'$ are regular pentagons. Now, do the following items:

   (a) Use the pentagons $ABCDE$, $ABB'A''A'$, $BCC'B''B'$, $CDD'C''C'$, $DEE'D''D'$ and $EAA'E''E'$ to form a *shell*. Then, gluing such a shell with a copy of it along the polygonal line $A'A''B'B''C'C''D'D''E'E''$, show that one obtains a regular dodecahedron inscribed in Σ.

   (b) If l and l' are the lengths of the edges of the pentagons $ABCDE$ and $A'B'C'D'E'$, show that

$$l' = \left(\frac{\sqrt{5}+1}{2}\right)l \text{ and } r' = \left(\frac{\sqrt{5}+1}{2}\right)r.$$

   (c) Use Pythagoras' theorem to conclude that

$$l^2 = \overline{AA'}^2 = (d - d')^2 + (r' - r)^2.$$

   Then, use the fact that $l = \sqrt{\frac{5-\sqrt{5}}{2}}r$ and $r' = \left(\frac{\sqrt{5}+1}{2}\right)r$ to get $d - d' = r$.

(d) Note that $d = \sqrt{R^2 - r^2}$ and $d' = \sqrt{R^2 - r^2}$, and solve the system of equations

$$\sqrt{R^2 - r^2} - \sqrt{R^2 - r'^2} = r \ \text{ and } \ r' = \Big(\frac{\sqrt{5}+1}{2} \Big) r$$

to get $r = \frac{2\sqrt{2}R}{\sqrt{15+3\sqrt{5}}}$, and hence $l = 2R\sqrt{\frac{3-\sqrt{5}}{6}}$.

Chapter 13
Volume of Solids

In this chapter we present the concept of *volume* of a solid and compute the volumes of various of the solids studied so far. *Cavalieri's principle* will turn to be a central tool for our exposition; in particular, we shall use it to compute the volume of a sphere of a given radius.

13.1 The Concept of Volume

The volume of a solid must be some kind of *measure* of the portion of space it occupies. Then, we hope that two disjoint solids have a combined volume which is equal to the sum of the volumes occupied by each one of them separately. Also, if one of the solids is contained into another, it is reasonable to ask that the volume of the first is less than or equal to that of the second. Finally, in order to numerically express this portion of *occupied space* we need a unit of measure to serve as reference.

It is our purpose in this section to develop the ideas above as rigorously as possible, taking into account our intended scope and audience. To this end, we first need to formulate a general concept of *solid*, in a way that it captures our everyday experience and suffices for our purposes.

Definition 13.1 A **solid** is a set S of points in space satisfying the following conditions:

(a) S is closed, bounded and has nonempty interior.
(b) For all points $A, B \in S$ there exists a polygonal line $A_1 A_2 \ldots A_k$ joining $A = A_1$ to $B = A_k$ and contained in $\text{Int}(S) \cup \{A, B\}$.

Intuitively, item (b) of the above definition guarantees that every solid has *a single piece* (cf. Fig. 13.1).

© Springer International Publishing AG, part of Springer Nature 2018
A. Caminha Muniz Neto, *An Excursion through Elementary Mathematics, Volume II*,
Problem Books in Mathematics, https://doi.org/10.1007/978-3-319-77974-4_13

Fig. 13.1 A solid $\mathcal{S}$

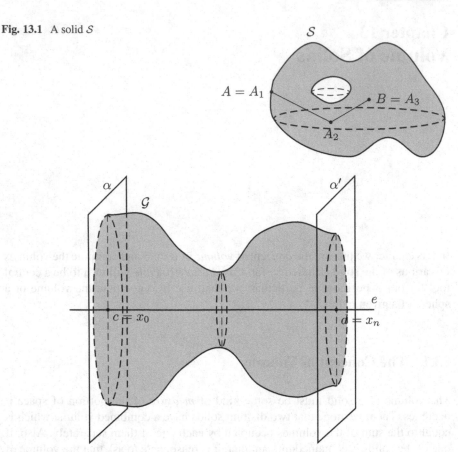

Fig. 13.2 A solid of revolution

Obviously, every convex polyhedron and every closed ball are examples of solids. Other relevant examples are **solid cylinders** (resp. **cones**) **of revolution**, i.e., the closed and bounded portion of space whose boundary is a cylinder (resp. cone) of revolution.

More generally (cf. Fig. 13.2), a **solid of revolution** of *axis e* is the closed and bounded region of space whose boundary is the union of the portion of a surface of revolution of axis e delimited by two planes α, $\alpha' \perp e$, with the closed disks delimited by the intersection circles of α and α' with the surface of revolution. Obviously, every solid of revolution is a solid in the sense of Definition 13.1.

We now need a sufficient condition for the *measurability* of a solid, so that *it is possible* to measure its volume. We isolate such a condition in the definition below, which is a first formulation of *Cavalieri's principle*,[1] alluded to in the introduction to this chapter.

Definition 13.2 (Cavalieri' Principle I) A solid $\mathcal{S}$ is **measurable**[2] if $\mathcal{S} \cap \alpha$ is a set of measurable area, for every plane α intersecting $\mathrm{Int}(\mathcal{S})$.

The definition above is such that all of the solids presented so far are measurable. For instance, since every planar section (that intersects the interior) of a closed ball is a closed disk and every planar section of a convex polyhedron is a convex polygon, we conclude that closed balls and convex polyhedra are measurable solids. On the other hand, Theorems 10.25 and 10.26 guarantee that solid cylinders and cones of revolution are also measurable.

Finally, it is possible to show that general solids of revolution are also measurable, as long as they are generated by the rotation, around the axis of abscissas, of graphs of functions $f : (a, b) \rightarrow \mathbb{R}$ of class C^2 (i.e., twice differentiable and with continuous second derivative.[3]

To each measurable solid $\mathcal{S}$ it is possible to associate a positive real number $\mathcal{V}(\mathcal{S})$, called the **volume** of $\mathcal{S}$, in such a way that the following conditions are satisfied:

1. If $\mathcal{S}$ is a cube of edge length 1, then $\mathcal{V}(\mathcal{S}) = 1$.
2. If $\mathcal{S}_1$ and $\mathcal{S}_2$ are measurable solids with $\mathrm{Int}(\mathcal{S}_1) \cap \mathrm{Int}(\mathcal{S}_2) = \emptyset$ and $\mathcal{S}_1 \cup \mathcal{S}_2$ also measurable, then $\mathcal{V}(\mathcal{S}_1 \cup \mathcal{S}_2) = \mathcal{V}(\mathcal{S}_1) + \mathcal{V}(\mathcal{S}_2)$.
3. If $\mathcal{S}_1$ and $\mathcal{S}_2$ are measurable solids with $\mathcal{S}_1 \subset \mathcal{S}_2$, then $\mathcal{V}(\mathcal{S}_1) \leq \mathcal{V}(\mathcal{S}_2)$.
4. (**Cavalieri's principle II**). If $\mathcal{S}_1$ and $\mathcal{S}_2$ are measurable solids and α is a plane such that $A(\mathcal{S}_1 \cap \alpha') = A(\mathcal{S}_2 \cap \alpha')$ for every plane $\alpha' \parallel \alpha$, then $\mathcal{V}(\mathcal{S}_1) = \mathcal{V}(\mathcal{S}_2)$.
5. If $\mathcal{S}_1$ is a measurable solid and $\mathcal{S}_2$ can be obtained from $\mathcal{S}_1$ by means of a translation along a vector ao longo de um vetor, a rotation along an axis or a reflection along a plane, then $\mathcal{S}_2$ is also measurable and $\mathcal{V}(\mathcal{S}_1) = \mathcal{V}(\mathcal{S}_2)$.

As we commented before, a rigorous proof of the existence of *volume function* $\mathcal{S} \mapsto \mathcal{V}(\mathcal{S})$ with the above properties is a consequence of the modern (and much deeper) theory of measure and integration (cf. Chapter 2 of [9]), and is far beyond the scope of these notes. Hence, we shall consider items 1. to 5. above as our *postulates of volume measurement*.

---

[1] Bonaventura Cavalieri, Italian mathematician of the sixteenth century.

[2] The cornerstone of the usual notion of measurability of solids is the concept of *Jordan content* (cf. [19] or [22]); in this context, one proves that a solid is measurable if and only if its boundary has *null Jordan content*. In the modern theory of measure and integration, it is possible to show (cf. Chapter 2 of [9], for instance) that our definition of measurability *is implied* by the usual one, and that the two versions of Cavalieri's principle we present here, as well as our *Postulates* 2., 3. and 5. for the measurement of volumes (see below), are *theorems*.

[3] What must be proved is that the intersection of the solid with any plane that intersects its interior is a region of measurable area. If the plane is *transversal* to the surface of revolution that defines the solid (in the sense that the plane is not tangent to the surface at any point), then the measurability of the corresponding planar section is a fairly simple consequence of the Implicit Function Theorem (see [22]). The general case follows from this one, through a slightly more complicated argument.

On the other hand, as we shall see in the rest of this section, the above set of postulates will allow us to easily compute the volumes of all of the solids we have met so far. In this sense, the result below is central for all that follows.

Proposition 13.3 *If $\mathcal{P}$ is a parallelepiped of basis $\mathcal{B}$ and height h, then*

$$\mathcal{V}(\mathcal{P}) = A(\mathcal{B})h.$$

Proof Let us do the proof in several steps.

(i) $\mathcal{P}$ is a rectangular parallelepiped whose edges have lengths $a, b, c \in \mathbb{N}$: by partitioning $\mathcal{P}$ into abc cubes, each of which of edge length 1, it follows from postulates 1. and 2. of volume measurement that

$$\mathcal{V}(\mathcal{P}) = abc = A(\mathcal{B})h.$$

(ii) $\mathcal{P}$ is a rectangular parallelepiped whose edges have lengths $a, b, c \in \mathbb{Q}$: let $a = \frac{m}{q}, b = \frac{n}{q}$ and $c = \frac{p}{q}$, with $m, n, p, q \in \mathbb{N}$. Let us pile q^3 copies of $\mathcal{P}$ in a way to obtain a rectangular parallelepiped $\mathcal{Q}$ of edge lengths m, n and p. Item (i), together with postulates 1. and 2. of volume measurement, furnish

$$q^3 \mathcal{V}(\mathcal{P}) = \mathcal{V}(\mathcal{Q}) = mnp,$$

and hence

$$\mathcal{V}(\mathcal{P}) = \frac{mnp}{q^3} = A(\mathcal{B})h.$$

(iii) $\mathcal{P}$ is a rectangular parallelepiped with edge lengths $a, b, c \in \mathbb{R}$: let $(a_n)_{n \geq 1}$, $(b_n)_{n \geq 1}$ and $(c_n)_{n \geq 1}$ be sequences of rational numbers such that $a_n < a, b_n < b, c_n < c$ for every $n \in \mathbb{N}$, and $a_n \to a, b_n \to b, c_n \to c$ as $n \to +\infty$. Also, let $\mathcal{P}_n$ be a rectangular parallelepiped contained in $\mathcal{P}$ and with edges of lengths a_n, b_n, c_n. Item (ii), together with postulate 3. of volume measurement, gives

$$\mathcal{V}(\mathcal{P}) \geq \mathcal{V}(\mathcal{P}_n) = a_n b_n c_n$$

for every $n \in \mathbb{N}$. Letting $n \to +\infty$ in such an inequality and taking into account the arithmetic properties of convergent sequences, we get $\mathcal{V}(\mathcal{P}) \geq abc$.

Now, take sequences $(a_n)_{n \geq 1}, (b_n)_{n \geq 1}, (c_n)_{n \geq 1}$ of rational numbers, with $a_n > a, b_n > b, c_n < c$ for every $n \in \mathbb{N}$ and $a_n \to a, b_n \to b, c_n \to c$ as $n \to +\infty$. By a reasoning similar to the one above, we conclude that $\mathcal{V}(\mathcal{P}) \leq abc$. Therefore, $\mathcal{V}(\mathcal{P}) = abc = A(\mathcal{B})h$.

(iv) $\mathcal{P}$ is any parallelepiped: let $\mathcal{P}'$ be a rectangular parallelepiped with the same altitude h as $\mathcal{P}$, whose basis $\mathcal{B}'$ is a rectangle of area equal to that of the basis $\mathcal{B}$ of $\mathcal{P}$ and such that $\mathcal{B}$ and $\mathcal{B}'$ lie in the same plane α, with $\mathcal{P}$ and $\mathcal{P}'$ situated in the same half-space with respect to α. If α' is a plane parallel to α, the equality of the heights of both parallelepipeds assures that $\mathcal{P} \cap \alpha' \neq \emptyset$ if and only if $\mathcal{P}' \cap \alpha' \neq \emptyset$. Moreover, when this does happen, the corresponding intersections $\mathcal{P} \cap \alpha$ and $\mathcal{P} \cap \alpha'$ are quadrilaterals respectively congruent to $\mathcal{B}$ and $\mathcal{B}'$, thus of equal areas. Hence, item (iii), together with Cavalieri's principle II, guarantees that

$$\mathcal{V}(\mathcal{P}) = \mathcal{V}(\mathcal{P}') = A(\mathcal{B}')h = A(\mathcal{B})h.$$

□

Now that we already know how to compute the volume of parallelepipeds, let us see that the calculation of volumes of solid prisms and cylinders is a trivial task.

Corollary 13.4 *If $\mathcal{P}$ is a prism of basis $\mathcal{B}$ and height h, then*

$$\mathcal{V}(\mathcal{P}) = A(\mathcal{B})h.$$

Proof Let $\mathcal{P}'$ denote a parallelepiped with the same height h as $\mathcal{P}$, whose basis $\mathcal{B}'$ is a parallelogram of area equal to that of $\mathcal{B}$, and such that $\mathcal{B}$ and $\mathcal{B}'$ lie in the same plane α; suppose, further, that $\mathcal{P}$ and $\mathcal{P}'$ are in the same half-space with respect to α.

If α' is a plane parallel to α, the equality of heights for the prism and the parallelepiped guarantees that $\mathcal{P} \cap \alpha' \neq \emptyset$ if and only if $\mathcal{P}' \cap \alpha' \neq \emptyset$. Moreover, when this does happen, such intersections are respectively congruent to $\mathcal{B}$ and $\mathcal{B}'$, hence have equal areas. Therefore, the second form of Cavalieri's principle, together with the previous proposition, gives

$$\mathcal{V}(\mathcal{P}) = \mathcal{V}(\mathcal{P}') = A(\mathcal{B}')h = A(\mathcal{B})h.$$

□

For the coming corollary, we say that the *basis* and *height* of a solid cylinder of revolution $\mathcal{C}$ are respectively the basis and height of the cylinder that constitutes its boundary.

Corollary 13.5 *If $\mathcal{C}$ is a solid cylinder of revolution of radius R and height h, then*

$$\mathcal{V}(\mathcal{C}) = \pi R^2 h.$$

Proof Exercise (cf. Problem 2).   □

In what comes next, we show how to compute the volume of pyramids. To this end we shall need an auxiliary result, which is interesting in itself.

Lemma 13.6

(a) *Let $\mathcal{T}$ be a tetrahedron of basis $\mathcal{B}$ and height h, and α be a plane parallel to $\mathcal{B}$, situated at a distance h' from the vertex of $\mathcal{T}$, with $h' < h$. Then, α intersects $\mathcal{T}$ in a triangle $\mathcal{B}'$, similar to $\mathcal{B}$, such that*

Fig. 13.3 Cutting a
tetrahedron with a plane
parallel to the basis

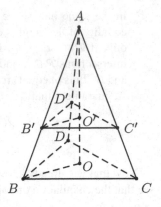

$$\frac{A(\mathcal{B}')}{A(\mathcal{B})} = \left(\frac{h'}{h}\right)^2.$$

(b) Let $\mathcal{T}_1$ and $\mathcal{T}_2$ be two tetrahedra of equal heights, and bases $\mathcal{B}_1$ and $\mathcal{B}_2$, respectively. If $A(\mathcal{B}_1) = A(\mathcal{B}_2)$, then $V(\mathcal{T}_1) = V(\mathcal{T}_2)$.

Proof For item (a), let $\mathcal{T} = ABCD$, $\mathcal{B} = BCD$ and $\mathcal{B} = B'C'D'$, with A, B, B' and A, C, C' being collinear points (cf. Fig. 13.3).

Assume that the feet O and O' of the perpendiculars dropped from A to α and α', respectively, lie in BCD and $B'C'D'$ (the proof in the other case is quite similar to the one we shall present). Since $\alpha' \parallel \alpha$, we have $\overleftrightarrow{BC} \parallel \overleftrightarrow{B'C'}$, $\overleftrightarrow{BD} \parallel \overleftrightarrow{B'D'}$ and $\overleftrightarrow{BO} \parallel \overleftrightarrow{B'O'}$, so that $ABC \sim AB'C'$, $ABD \sim AB'D'$ and $ABO \sim AB'O'$. Therefore,

$$\frac{\overline{B'C'}}{\overline{BC}} = \frac{\overline{AB'}}{\overline{AB}} = \frac{\overline{AO'}}{\overline{AO}} = \frac{h'}{h},$$

and Proposition 5.11 gives

$$\frac{A(\mathcal{B}')}{A(\mathcal{B})} = \frac{A(B'C'D')}{A(BCD)} = \left(\frac{\overline{B'C'}}{\overline{BC}}\right)^2 = \left(\frac{h'}{h}\right)^2.$$

For (b), Postulate 5. of volume measurement allows us to suppose that $\mathcal{B}_1$ and $\mathcal{B}_2$ lie in the same plane α and that $\mathcal{T}_1$ and $\mathcal{T}_2$ are situated in the same half-space with respect to α. Since the tetrahedra have equal heights, for every plane $\alpha' \parallel \alpha$ we have $\mathcal{T}_1 \cap \alpha' \neq \emptyset$ if and only if $\mathcal{T}_2 \cap \alpha' \neq \emptyset$.

Now, let α' be a plane parallel to α, which intersects $\mathcal{T}_1$ and $\mathcal{T}_2$ along triangles $\mathcal{B}'_1$ and $\mathcal{B}'_2$, respectively. If h denotes the common height of $\mathcal{T}_1$ and $\mathcal{T}_2$, and h' the distance from the vertex of $\mathcal{T}_1$ (or of $\mathcal{T}_2$) to α', then the result of item (a) gives

$$\frac{A(\mathcal{B}'_1)}{A(\mathcal{B}_1)} = \left(\frac{d}{h}\right)^2 = \frac{A(\mathcal{B}'_2)}{A(\mathcal{B}_2)}.$$

Mas, como $A(\mathcal{B}_1) = A(\mathcal{B}_2)$, segue daí que $A(\mathcal{B}'_1) = A(\mathcal{B}'_2)$. Therefore, the second form of Cavalieri's principle assures that $\mathcal{T}_1$ and $\mathcal{T}_2$ have equal volumes. □

The previous lemma, together with some ingenuity, suffices for us to compute the volume of a tetrahedron.

Proposition 13.7 *If $\mathcal{T}$ is a tetrahedron of basis $\mathcal{B}$ and height h, then*

$$\mathcal{V}(\mathcal{T}) = \frac{1}{3}A(\mathcal{B})h.$$

Proof Let $UVWX$ be the vertices of $\mathcal{T}$ and UVW be its basis $\mathcal{B}$ (cf. Fig. 13.4). Construct points Y and Z such that $UVXY$ and $VWZX$ are parallelograms. Then, UY and WZ are equal and parallel to VX, so that are also equal and parallel to each other, and $UWZY$ is a parallelogram too. Then, $UVWYXZ$ is a triangular prism of bases UVW and XYZ, and height h; letting $\mathcal{V}$ denote its volume, we have $\mathcal{V} = A(UVW)h = A(\mathcal{B})h$.

Now, split the prism into the three tetrahedra $UVWX$, $UWXY$ and $WXYZ$. If we show that these three tetrahedra have equal volumes, it will follow from what we did above, together with postulate 2. of volume measurement, that

$$3\mathcal{V}(\mathcal{T}) = \mathcal{V}(UVWX) + \mathcal{V}(UWXY) + \mathcal{V}(WXYZ) = \mathcal{V} = A(\mathcal{B})h,$$

and this will finish the proof.

For what is left to do, we have the following:

Fig. 13.4 Partitioning a triangular prism into three tetrahedra of equal volumes

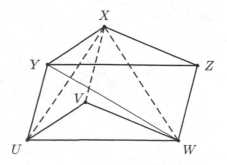

- Since $UVXY$ is a parallelogram, we have $A(UVX) = A(UXY)$. However, since the altitudes of $UVWX$ and $UWXY$ dropped from W are equal, item (b) of the previous lemma assures that $V(UVWX) = V(UWXY)$.
- Since $UWZY$ is a parallelogram, we have $A(UWY) = A(WYZ)$. Also, $UWXY$ and $WXYZ$ have equal altitudes with respect to X, so that a further application of item (b) of the previous lemma gives $V(UWXY) = V(WXYZ)$.

□

We are finally in position of computing the volume of a pyramid.

Corollary 13.8 *If $\mathcal{P}$ is a pyramid of basis $\mathcal{B}$ and height h, then*

$$\mathcal{V}(\mathcal{P}) = \frac{1}{3}A(\mathcal{B})h.$$

Proof Let V be the vertex of the pyramid and $\mathcal{B} = A_1A_2\ldots A_k$ be its basis. For $2 \le k \le n - 1$, let $\mathcal{T}_k$ denote the tetrahedron $VA_1A_kA_{k+1}$ and $\mathcal{B}_k = A_1A_kA_{k+1}$ (cf. Fig. 13.5).

If h stands for the height of $\mathcal{T}$, then h is also the height of $\mathcal{T}_k$, for $2 \le i \le n - 1$. Hence, postulate 2. of volume measurement, together with the result of the previous proposition, give us

$$\mathcal{V}(\mathcal{P}) = \mathcal{V}(\mathcal{T}_2) + \cdots + \mathcal{V}(\mathcal{T}_{n-1})$$

$$= \frac{1}{3}A(\mathcal{B}_2)h + \cdots + \frac{1}{3}A(\mathcal{B}_{n-1})h$$

$$= \frac{1}{3}(A(\mathcal{B}_2) + \cdots + A(\mathcal{B}_{n-1}))h$$

$$= \frac{1}{3}A(\mathcal{B})h.$$

□

Fig. 13.5 Partitioning a pyramid into tetrahedra of equal volumes

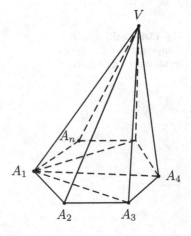

The last result of this section teaches us how to compute the volume of a solid cone of revolution. As the attentive reader can easily notice, its proof is quite similar to that of item (b) of Lemma 13.6.

Corollary 13.9 *If C is a solid cone of revolution of radius R and height h, then*

$$V(C) = \frac{1}{3}\pi R^2 h.$$

Proof Let $\mathcal{T}$ denote a tetrahedron with the same height h as C, and whose basis $\mathcal{B}$ is a triangle of area equal to that of the disk $\mathcal{D}$ that forms the basis of the cone. Assume, without loss of generality, that $\mathcal{B}$ and $\mathcal{D}$ lie in the same plane α, and that C and $\mathcal{T}$ are contained in the same half-space, of those determined by α (cf. Fig. 13.6).

If α' is a plane parallel to α, the equality of the heights of the cone and the tetrahedron assures that $\mathcal{T} \cap \alpha' \neq \emptyset$ if and only if $C \cap \alpha' \neq \emptyset$. For such an α', let h' denote the distance from the vertex of the cone (or of the tetrahedron) to α', let $C \cap \alpha' = \mathcal{D}'$ and $\mathcal{T} \cap \alpha' = \mathcal{B}'$. If R' stands for the radius of $\mathcal{D}'$, then Lemma 13.6 gives

$$\frac{A(\mathcal{B}')}{A(\mathcal{B})} = \left(\frac{h'}{h}\right)^2 = \left(\frac{R'}{R}\right)^2 = \frac{\pi R'^2}{\pi R^2} = \frac{A(\mathcal{D}')}{A(\mathcal{D})}.$$

However, since $A(\mathcal{B}) = A(\mathcal{D})$, it comes that $A(\mathcal{B}') = A(\mathcal{D}')$.

Finally, in view of the last equality above, the second form of Cavalieri's principle furnishes

$$V(C) = V(\mathcal{T}) = \frac{1}{3}A(\mathcal{B})h = \frac{1}{3}A(\mathcal{D})h = \frac{1}{3}\pi R^2 h.$$

□

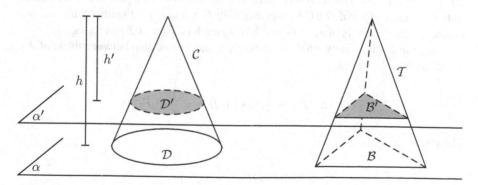

Fig. 13.6 Computing the volume of a cone of revolution

We finish this section by using the material presented here to compute the volume of a *diamond*.

Example 13.10 A convex polyhedron $\mathcal{P}$ has two parallel faces $\mathcal{F}_1$ and $\mathcal{F}_2$. Each other face is a triangle whose vertices are also vertices of $\mathcal{F}_1$ or $\mathcal{F}_2$. Prove that

$$V(\mathcal{P}) = \frac{d}{6}\big(A(\mathcal{F}_1) + A(\mathcal{F}_2) + 4A(\mathcal{S})\big),$$

where d stands for the distance between the planes of $\mathcal{F}_1$ and $\mathcal{F}_2$, and $\mathcal{S}$ for the planar section of $\mathcal{P}$ through a plane parallel to and equidistant from $\mathcal{F}_1$ and $\mathcal{F}_2$.

Solution For the sake of clarity, we shall stick to the case in which both $\mathcal{F}_1$ and $\mathcal{F}_2$ are quadrilaterals, the general case being entirely analogous. Also, in the figure below the quadrilaterals $\mathcal{F}_1 = ABCD$ and $\mathcal{F}_2 = EFGH$ resemble parallelograms, but this is totally immaterial and will not play a role in the proof.

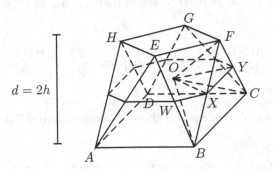

By selecting a point O in the interior of the planar section $\mathcal{S}$, we partition our solid into the eight triangular pyramids of apex O and having each of ABE, BEF, BCF, CFG, CDG, DGH, ADH and AHE as bases (in the figure above we depict only the lateral edges of $OBCF$, together with OX and OY), together with the two quadrangular pyramids of apex O and having each of $\mathcal{F}_1$ and $\mathcal{F}_2$ as bases.

Since the distance between the planes of $\mathcal{F}_1$ and $\mathcal{S}$ (and also between those of $\mathcal{F}_2$ and $\mathcal{S}$) is $h = \frac{d}{2}$, we get

$$V(OABCD) = \frac{1}{3}A(ABCD)h = \frac{1}{6}A(\mathcal{F}_1)d;$$

likewise,

$$V(OEFGH) = \frac{1}{6}A(\mathcal{F}_2)d.$$

In what concerns $OBCF$, observing that $A(BCF) = 4A(XYF)$ we get

$$V(OBCF) = 4V(OXYF) = 4 \cdot \frac{1}{3}A(OXY)h = \frac{2}{3}A(OXY)d.$$

If we perform analogous computations for the remaining triangular pyramids, we will always get $\frac{2}{3}A(\mathcal{T})d$ as result, with $\mathcal{T}$ denoting a triangle whose vertices are O and two consecutive vertices of $\mathcal{S}$. Therefore, by adding these eight expressions, we will obtain

$$\frac{2}{3}\sum_{\mathcal{T}} A(\mathcal{T})d = \frac{2}{3}A(\mathcal{S})d.$$

It now suffices to add the three contributions above to arrive at the desired formula.   □

Problems: Sect. 13.1

1. We are given a rectangular parallelepiped of edge lengths a, b and c. Compute the volume of the octahedron whose vertices are the centers of the faces of it.
2. * Prove Corollary 13.5.
3. Let $ABCD$ denote a tetrahedron trirectangular at A, with $\overline{AD} = a$, $\overline{BD} = b$ and $\overline{CD} = c$. Show that:
   (a) $V(ABCD) = \frac{abc}{6}$.
   (b) The height of $ABCD$ with respect to ABC is equal to $\frac{abc}{2A(ABC)}$.

4. $ABCD$ is an isosceles tetrahedron with edge lengths a, b and c. Show that

$$V(ABCD) = \frac{1}{6\sqrt{2}}\sqrt{(a^2 + b^2 - c^2)(a^2 + c^2 - b^2)(b^2 + c^2 - a^2)}.$$

5. $ABCD$ is a tetrahedron of total area S, and r denotes the radius of its inscribed sphere. Prove that

$$V(ABCD) = \frac{Sr}{3}.$$

Then, use the above formula to show that:

   (a) If $ABCD$ is trirectangular at A, with $\overline{AD} = a$, $\overline{BD} = b$ and $\overline{CD} = c$, then

$$r = \frac{abc}{ab + bc + ca + 2A(ABC)}.$$

   (b) If $ABCD$ is isosceles with edge lengths a, b and c, then

$$r = \frac{1}{2\sqrt{2}A(ABC)}\sqrt{(a^2 + b^2 - c^2)(a^2 + c^2 - b^2)(b^2 + c^2 - a^2)}.$$

6. Point P lies in the interior of an isosceles tetrahedron, and d_1, d_2, d_3, d_4 stand for the distances of P to its faces. Prove that:

   (a) The value of $d_1 + d_2 + d_3 + d_4$ does not depend on the position of P in the interior of the tetrahedron.
   (b) If the tetrahedron is regular with height h, then $d_1 + d_2 + d_3 + d_4 = h$.

7. Let r stand for the radius of the sphere inscribed in a given tetrahedron $ABCD$. Prove that there exists a height h of the tetrahedron for which $h \geq 4r$.

8. Let $ABCD$ be a tetrahedron with barycenter G (cf. Problem 9, page 387). Prove that the four tetrahedra $ABCG$, $ABDG$, $ACDG$ and $BCDG$ have equal volumes.

9. A plane passes through the vertex of a cone of revolution and forms, with the plane of the basis of the cone, an angle of $45°$. It is also known that the plane intersects the basis of the cone along a chord of length $2\sqrt{3}$, and which is seen from the center of the basis through a central angle of $60°$. Compute the volume of the cone.

10. (OCM) A right triangle is rotated around each one of its edges, thus giving us three solids of revolution. Which one of these solids has the greatest volume?

11. In the notations of Example 13.10, do the following items:

    (a) Show that the volume of $\mathcal{P}$ remains unchanged if we translate the faces F_i along the planes containing them.
    (b) Let $\mathcal{F}_1$ be a regular n-gon with edge length a and $\mathcal{F}_2$ be a regular n-gon with edge length b. If the angle between one edge of $\mathcal{F}_1$ and one edge of $\mathcal{F}_2$ is equal to $\frac{\pi}{n}$, compute the volume of $\mathcal{P}$ in terms of a, b, n and d.

12. Compute, in terms of R, the volumes of a regular dodecahedron and icosahedron inscribed in a sphere of radius R.

13. Cut a right n-gonal prism $\mathcal{P}$, of basis $\mathcal{B}_1$ and $\mathcal{B}_2$, by using a plane α, which intersects all of its lateral edges. We say that α divides $\mathcal{P}$ into two *frustums of right prism* say $\mathcal{P}_1$ and $\mathcal{P}_2$, of bases respectively $\mathcal{B}_1$ and $\mathcal{B}_2$; moreover, the portions of the lateral edges of $\mathcal{P}$ contained in $\mathcal{P}_i$ are also called the *lateral edges* of the frustum of right prism $\mathcal{P}_i$ (in Fig. 13.7, we show a frustum of right triangular prism, with basis ABC and lateral edges of lengths a, b and c). Prove that the volume of a frustum of right prism is equal to the product of the area of its basis by the arithmetic mean of the lengths of its lateral edges do tronco.

14. Consider, in the cartesian plane, the parabolas $y = x^2 + 4x + 7$ and $y = x^2 - 2x + 3$. Compute the volume of the solid obtained by the rotation, around the axis of abscissas, of the portion of the plane bounded by such parabolas and by the lines $x = 0$ and $x = 3$.

15. (IMO—adapted)

Fig. 13.7 A frustum of right triangular prism

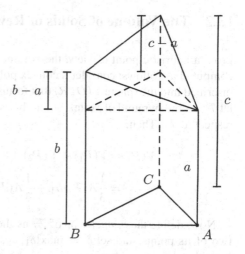

(a) Let LMN be a triangle such that $\overline{LN}, \overline{MN} \le 1$ and $\overline{LM} = x$, with $0 < x < 2$. If H denotes the foot of the altitude relative to LM, prove that $\overline{NH} \le \sqrt{1 - \frac{x^2}{4}}$.

(b) $ABCD$ is a tetrahedron with only one edge of length greater that 1. Prove that $V(ABCD) \le \frac{1}{8}$.

16. Let $ABCD$ be a tetrahedron for which $A(ABC) = A(ABD)$. If α denotes the bisector plane of the dihedral angle formed by the planes (ABC) and (ABD) and which contains $ABCD$, prove that α divides $ABCD$ into two solids of equal volumes.

17. (IMO—shortlist, adapted) $ABCD$ is an isosceles tetrahedron, and M and N are the midpoints of edges AB and CD, respectively. Prove that any plane containing the straightline $\overleftrightarrow{MN}$ divides the tetrahedron into two solids of equal volumes.

18. (IMO shortlist—adapted) Let P be a point in the interior of a regular tetrahedron $A_1 A_2 A_3 A_4$, of volume 1. The four planes passing through P and parallel to the faces of the tetrahedron divide it into 14 convex polyhedra. Let $f(P)$ be the sum of the volumes of those polyhedra which are neither tetrahedra nor parallelepipeds. If h denotes the height of the tetrahedron, show that:

(a) $f(P) = 1 - \frac{1}{h^3}\left(\sum_i d_i^3 + 6 \sum_{i<j<k} d_i d_j d_k\right)$.

(b) $0 < f(P) < \frac{3}{4}$.

13.2   The Volume of Solids of Revolution

From a heuristic point of view, the volume of a closed ball $\overline{B} = \overline{B(O; R)}$ can be computed as follows: consider a convex polyhedron $\mathcal{P}$, with faces $\mathcal{F}_1, \mathcal{F}_2, \ldots, \mathcal{F}_n$, inscribed into the sphere $\Sigma(O; R)$ that forms the boundary of $\overline{B}$. Also, for $1 \leq i \leq n$ let $\mathcal{P}_i$ be the pyramid of vertex O and basis $\mathcal{F}_i$, and let h_i be the height of $\mathcal{P}_i$ with respect to $\mathcal{F}_i$. Then,

$$\mathcal{V}(\mathcal{P}) = \mathcal{V}(\mathcal{P}_1) + \mathcal{V}(\mathcal{P}_2) + \cdots + \mathcal{V}(\mathcal{P}_n)$$

$$= \frac{1}{3} A(\mathcal{F}_1) h_1 + \frac{1}{3} A(\mathcal{F}_2) h_2 + \cdots + \frac{1}{3} A(\mathcal{F}_n) h_n.$$

Now, define the *diameter* δ_i of $\mathcal{F}_i$ as the largest possible distance between any two of its points, and set $\delta = \max\{\delta_1, \delta_2, \ldots, \delta_n\}$. For a sufficiently small δ, we hope that

$$h_i \cong R, \quad \mathcal{V}(\mathcal{P}) \cong \mathcal{V}(\overline{B})$$

and

$$A(\Sigma) \cong A(\mathcal{F}_1) + A(\mathcal{F}_2) + \cdots + A(\mathcal{F}_n).$$

If this is so, then the computations we did in Example 12.1 give

$$\mathcal{V}(\overline{B}) \cong \mathcal{V}(\mathcal{P}) = \frac{1}{3} A(\mathcal{F}_1) R + \frac{1}{3} A(\mathcal{F}_2) R + \cdots + \frac{1}{3} A(\mathcal{F}_n) R$$

$$= \frac{R}{3} \left(A(\mathcal{F}_1) + A(\mathcal{F}_2) + \cdots + A(\mathcal{F}_n) \right)$$

$$\cong \frac{R}{3} A(\Sigma) = \frac{R}{3} \cdot 4\pi R^2 = \frac{4}{3} \pi R^3.$$

In the coming result, we use Cavalieri's principle to show that the argument above does give the correct value for the volume of a closed ball of radius R. The reasoning presented is a variation of that of the Italian mathematician of the sixteenth century Luca Valerio.

Theorem 13.11 *The volume of a closed ball of radius R is equal to $\frac{4}{3}\pi R^3$.*

Proof Let $\mathcal{C}$ be a solid right cylinder of height $2R$, whose bases are two circles of radius R (cf. Fig. 13.8, left).

Inscribe two cones of revolution in $\mathcal{C}$, both with height R and bases equal to those of the cylinder; this way, the vertex of both cones coincides with the midpoint of the axis of the cylinder.

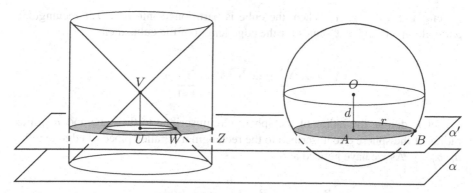

Fig. 13.8 Computing the volume of a closed ball

Let α be the plane containing one of the bases of the cylinder and $\overline{\mathcal{B}} = \overline{\mathcal{B}(O; R)}$ be a closed ball of center O and radius R, tangent to α and lying (with respect to α) in the same half-space as $\mathcal{C}$ (cf. Fig. 13.8, right).

If $\mathcal{S}$ stands for the solid formed by the portion of the cylinder which lies outside both cones, then we shall show that $\mathcal{V}(\overline{\mathcal{B}}) = \mathcal{V}(\mathcal{S})$. Once we have done that, we can compute the volume of $\overline{\mathcal{B}}$ as the difference between the volumes of the cylinder and of the union of the two cones:

$$\mathcal{V}(\overline{\mathcal{B}}) = \mathcal{V}(\mathcal{C}) - 2\mathcal{V}(\text{cone}) = \pi R^2 \cdot 2R - 2 \cdot \frac{1}{3}\pi R^2 \cdot R = \frac{4}{3}\pi R^3.$$

For what is left to do, let α' be a plane parallel to α, contained in the same half-space as $\mathcal{S}$ and $\overline{\mathcal{B}}$ with respect to α, and situated at distance d from O, with $d < R$. Since the height of $\mathcal{C}$ equals the diameter of $\overline{\mathcal{B}}$, it follows that α' intersects $\mathcal{S}$ if and only if it intersects $\overline{\mathcal{B}}$. This being so, we have that α' cuts $\mathcal{S}$ along an annulus of radii $\overline{UW} = d$ and $\overline{UZ} = R$ (look at Fig. 13.8), whereas it cuts $\overline{\mathcal{B}}$ along a disk of center A and radius $\overline{AB} = \sqrt{R^2 - d^2}$. Therefore, the areas of such sections are always equal, and Cavalieri's principle assures that $\mathcal{S}$ and $\overline{\mathcal{B}}$ have equal volumes. □

The coming example applies the formula of the previous theorem. We shall also need to recall that if a rectangular parallelepiped has edge lengths a, b and c, then the radius of its circumscribed sphere is $\frac{1}{2}\sqrt{a^2 + b^2 + c^2}$.

Example 13.12 (Romania) A cube is decomposed in a finite number of rectangular parallelepipeds by means of planes parallel to its faces. If the volume of the sphere circumscribed to the cube is equal to the sum of the volumes of the spheres circumscribed to the parallelepipeds, prove that all of the parallelepipeds are also cubes.

Proof Let AB, AC and AD be three edges of the cube, and suppose that the planes used to partition the cube divide AB into line segments of lengths $a_1, a_2, \ldots, a_r$, divide AC into line segments of lengths $b_1, b_2, \ldots, b_s$, and AD into line segments of lengths $c_1, c_2, \ldots, c_t$. Then, the cube is partitioned into $n = rst$ rectangular parallelepipeds and, if ℓ stands for the edge length of the cube, then

$$\ell = \sum_{i=1}^{r} a_i = \sum_{j=1}^{s} b_j = \sum_{k=1}^{t} c_k.$$

Let R denote the radius of the sphere circumscribed to the cube and R_{ijk} the radius of the sphere circumscribed to the rectangular parallelepiped of edge lengths a_i, b_j, c_k. As we have noticed above,

$$R = \ell\sqrt{3}, \quad R_{ijk} = \sqrt{a_i^2 + b_j^2 + c_k^2}$$

and the stated condition translates into the equality

$$R^3 = \sum_{i,j,k} R_{ijk}^3.$$

Now, by successively applying Jensen's inequality to the function $x \mapsto x^{3/2}$, the inequality between the arithmetic and quadratic means and the arithmetic-geometric means inequality (cf. [5], for instance) we get

$$(2R)^3 = \sum_{i,j,k}(2R_{ijk})^3 \geq n\left(\frac{1}{n}\sum_{i,j,k}(a_i^2 + b_j^2 + c_k^2)\right)^{3/2}$$

$$= \frac{1}{\sqrt{n}}\left(st\sum_{i=1}^{r} a_i^2 + rt\sum_{j=1}^{s} b_j^2 + rs\sum_{k=1}^{t} c_k^2\right)^{3/2}$$

$$\geq \frac{1}{\sqrt{n}}\left(st \cdot r\left(\frac{\ell}{r}\right)^2 + rt \cdot s\left(\frac{\ell}{s}\right)^2 + rs \cdot t\left(\frac{\ell}{t}\right)^2\right)^{3/2}$$

$$= \frac{\ell^3}{\sqrt{n}}\left(\frac{st}{r} + \frac{rt}{s} + \frac{rs}{t}\right)^{3/2} \geq \frac{\ell^3}{\sqrt{n}} \cdot (3\sqrt[3]{rst})^{3/2}$$

$$= \frac{\ell^3}{\sqrt{n}} \cdot (3\sqrt[3]{n})^{3/2} = (\ell\sqrt{3})^3 = (2R)^3.$$

Thus, all of the above inequalities are actually equalities, which gives $a_1 = a_2 = \ldots = a_r, b_1 = b_2 = \ldots = b_s, c_1 = c_2 = \ldots = c_t$ and $r = s = t$.   □

Back to the development of the theory, and conversely to the circle of ideas delineated at the beginning of this section, we shall now present another heuristic reasoning, this time showing how the formula for the volume of a closed ball allows one to envisage the formula for the area of a sphere. To this end, consider two concentric spheres Σ and Σ', of radii R and $R + \epsilon$, respectively (with $\epsilon > 0$), and let S be the solid bounded by $\Sigma \cup \Sigma'$.

On the one hand, the volume of S is also equal to the difference between the volumes of the closed balls with boundaries Σ' and Σ:

$$\mathcal{V}(S) = \frac{4}{3}\pi(R + \epsilon)^3 - \frac{4}{3}\pi R^3 = 4\pi\left(R^2\epsilon + R\epsilon^2 + \frac{\epsilon^3}{3}\right).$$

To link the computation above with the area A of Σ, we partition it into a finite number of spherical triangles. If such triangles are sufficiently small and $\epsilon > 0$ is also small with respect to R, it is reasonable to assume that one can approximate S by the union of a finite number of "*prisms*" of height ϵ, each having the spherical triangles as one of its bases. This being so, we can approximate the volume of S by the sum of the volumes of these "*prisms*"; however, since the sum of the areas of their bases is equal to A, this reasoning gives

$$\epsilon A \cong 4\pi\left(R^2\epsilon + R\epsilon^2 + \frac{\epsilon^3}{3}\right),$$

so that

$$A \cong 4\pi\left(R^2 + R\epsilon + \frac{\epsilon^2}{3}\right).$$

Finally, it is also plausible to assume that the approximation above becomes better and better as we let $\epsilon \to 0$, so that the actual value of A is $A = 4\pi R^2$.

We shall finish this section by deducing a quite useful formula for the computation of the volume of a general solid of revolution. As in Sect. 12.1, we shall need the rudiments of the Integral Calculus, for which we refer the reader to [3] or [5], for instance.

Theorem 13.13 *Let $f : (a, b) \to \mathbb{R}$ be a positive and differentiable function, with continuous derivative. If $[c, d] \subset (a, b)$ and S stands for the solid of revolution generated by the rotation of the graph of $f_{|[c,d]}$ around the axis of abscissas, then*

$$\mathcal{V}(S) = \pi\int_c^d f(x)^2 dx.$$

Proof For $u, v \in [c, d]$ and $r > 0$, let $\mathcal{C}(u, v; r)$ denote the solid right circular cylinder of radius r, with bases centered at $(u, 0)$ and $(v, 0)$ and perpendicular to the axis of abscissas.

Consider real numbers $c = x_0 < x_1 < \cdots < x_n = d$, for which $\Delta x_i := x_i - x_{i-1}$ equals $\frac{d-c}{n}$, for $1 \leq i \leq n$. If

$$m_i = \min\{f(x); \ x \in [x_{i-1}, x_i]\} \quad \text{and} \quad M_i = \max\{f(x); \ x \in [x_{i-1}, x_i]\},$$

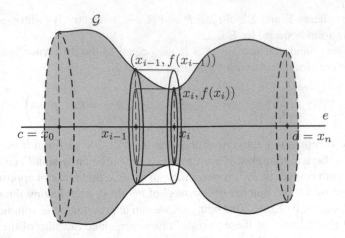

Fig. 13.9 Computing the volume of a solid of revolution

we have (cf. Fig. 13.9)

$$\bigcup_{i=1}^{n} \mathcal{C}(x_{i-1}, x_i; m_i) \subset \mathcal{S} \subset \bigcup_{i=1}^{n} \mathcal{C}(x_{i-1}, x_i; M_i).$$

Hence, postulates 2. and 3. of volume measurement give

$$\sum_{i=1}^{n} \mathcal{V}(\mathcal{C}(x_{i-1}, x_i; m_i)) \leq \mathcal{V}(\mathcal{S}) \leq \sum_{i=1}^{n} \mathcal{V}(\mathcal{C}(x_{i-1}, x_i; M_i)),$$

so that

$$\sum_{i=1}^{n} \pi m_i^2 \Delta x_i \leq \mathcal{V}(\mathcal{S}) \leq \sum_{i=1}^{n} \pi M_i^2 \Delta x_i.$$

Now, note that the first and last expressions in the above inequalities are respectively equal to the lower and upper sums of the restriction of πf^2 to the interval $[c, d]$ and with respect to the partition $\{x_0, x_1, \ldots, x_n\}$. Letting $n \to +\infty$, we obtain

$$\pi \int_c^d f(x)^2 dx \leq \mathcal{V}(\mathcal{S}) \leq \pi \int_c^d f(x)^2 dx,$$

as wished.   □

The coming example shows how to apply the formula deduced in the previous result to compute the volume of a closed ball in another way.

Example 13.14 In the notations of Example 12.1, let $f : (-R, R) \to \mathbb{R}$ be the function given by $f(x) = \sqrt{R^2 - x^2}$, so that the solid of revolution generated by the rotation of the graph of f around the axis of abscissas is a closed ball $\overline{B}$, of radius R, except for the two antipodal points N and S.

Now, given $0 < \epsilon < R$, let $\overline{B}_{[-R+\epsilon, R-\epsilon]}$ be the intersection of $\overline{B}$ with the region of the space bounded by the planes perpendicular to the axis of abscissas at the points $(-R + \epsilon, 0)$ and $(R - \epsilon, 0)$. Since

$$\overline{B} \setminus \{N, S\} = \bigcup_{0 < \epsilon < R} \overline{B}_{[-R+\epsilon, R-\epsilon]}$$

and $\overline{B}_{[-R+\epsilon, R-\epsilon]} \subset \overline{B}_{[-R+\epsilon', R-\epsilon']}$ for $0 < \epsilon < \epsilon'$, it follows from the previous theorem, together with the Fundamental Theorem of Calculus (cf. [5], for instance) that

$$\mathcal{V}(\overline{B}) = \lim_{\epsilon \to 0+} \mathcal{V}(\overline{B}_{[-R+\epsilon, R-\epsilon]})$$

$$= \lim_{\epsilon \to 0+} \pi \int_{-R+\epsilon}^{R-\epsilon} (R^2 - x^2) dx$$

$$= \lim_{\epsilon \to 0+} \pi \left(R^2 x - \frac{x^3}{3} \right) \Big|_{-R+\epsilon}^{R-\epsilon} dx$$

$$= \frac{4}{3} \pi R^3.$$

Problems: Sect. 13.2

1. A **spherical segment** is a solid obtained as the intersection a closed ball with the region in space delimited by two parallel planes. The radii of the planar sections of the ball through such planes are said to be the *radii* of the spherical segment, whereas the distance between the planes is its *height*. Show that the volume $\mathcal{V}$ of a spherical segment of radii r_1 and r_2 and height h is given by

$$\mathcal{V} = \frac{\pi h}{6} \left(3(r_1^2 + r_2^2) + h^2 \right).$$

2. Let ABC be a triangle of area S, and let r be a straightline in the plane of ABC but not intersecting it. If d is the distance from the barycenter of ABC to r, prove that the volume of the solid of revolution obtained by the rotation of ABC around r is $2\pi S d$.

3. Generalize the definition of a solid of revolution to encompass the region of space generated by the rotation, around the axis of abscissas, of the region situated between the graphs of two functions $f, g : [a, b] \to \mathbb{R}$ satisfying the

following conditions: f and g are continuous in $[a, b]$ and such that $0 \le g < f$ in (a, b). More precisely, show that the only reasonable definition for the volume of such a solid is

$$\pi \int_a^b (f(x)^2 - g(x)^2)dx.$$

4. We are given in a plane a line r and a circle $\Gamma(O; R)$, with O situated at distance $d > R$ from r. If $\mathcal{T}$ is the solid of revolution generated by the rotation of Γ around r, compute the volume of $\mathcal{T}$ in terms of R and d (the solid $\mathcal{T}$ is called the **torus of revolution** of radii R and d).

   For Problems 5 to 10, we shall need a small physical digression. Given a simple region $\mathcal{R}$ in the plane, we want to define what one understands by the *barycenter* of $\mathcal{R}$. To this end, let us imagine that $\mathcal{R}$ is a thin plate of metal, of homogeneous density and total mass m; then, the **barycenter** or **center of gravity** G of $\mathcal{R}$ is the point of application of the *weight vector* $m\mathbf{g}$ for purposes of *torque* of $\mathcal{R}$. This way, if we partition $\mathcal{R}$ in a finite, albeit very large number of tiny plates $\mathcal{R}_i$, of masses m_i and barycenters G_i, we ought to have

$$\sum_j m_j \overrightarrow{GG_j} = \mathbf{0}, \tag{13.1}$$

the null vector.

5. Let $f : [a, b] \to \mathbb{R}$ be a function which is continuous in $[a, b]$ and positive in (a, b), and let $\mathcal{R}$ be the region lying under its graph (and above the x-axis). Show that the point $G(x_G, y_G)$ such that

$$x_G = \frac{\int_a^b xf(x)dx}{\int_a^b f(x)dx} \quad \text{and} \quad y_G = \frac{\frac{1}{2}\int_a^b f(x)^2 dx}{\int_a^b f(x)dx}$$

is the only reasonable definition for the barycenter of $\mathcal{R}$.

6. We are given in the plane a triangle ABC, with $\widehat{B}, \widehat{C} < 90°$. Consider a cartesian system of coordinates of origin B, in which $C(a, 0)$, with $a > 0$, and $A(b, h)$, with $0 \le b < a$ and $h > 0$. In such a system, show that the triangular region bounded by ABC coincides with the region $\mathcal{R}$ under the graph of the function $f : [0, a] \to \mathbb{R}$, given by

$$f(x) = \begin{cases} \frac{hx}{b}, & \text{if } 0 \le x \le b \\ \frac{h(x-a)}{b-a}, & \text{if } b \le x \le a. \end{cases}$$

Then, show that the barycenter of ABC, computed in coordinates with the aid of its original definition (cf. Example 6.4), coincides with the barycenter of $\mathcal{R}$, computed with the aid of the formulas of the previous problem.

7. Generalize Problem 5 to the case of the region $\mathcal{R}$ situated between the graphs of the functions $f, g : [a, b] \to \mathbb{R}$, continuous in $[a, b]$ and such that $g < f$ in (a, b).

8. Prove **Pappus' theorem**: in the notations of the previous problem, if $g \geq 0$, e denotes the x-axis and d stands for the distance from the barycenter of $\mathcal{R}$ to e, then the volume of the solid of revolution obtained by the rotation of $\mathcal{R}$ around e is $V = 2\pi Ad$, where A is the area of $\mathcal{R}$.

9. Use Pappus' theorem, together with the formula for the volume of a sphere, to compute the position of the barycenter of a semi-disk of radius R.

10. Use Pappus' theorem to compute again the volume of the torus of revolution with radii R and d (cf. Problem 4).

Chapter 14
Hints and Solutions

Section 1.1

5. Start by drawing a line r and marking a point A on it. Then, mark $B \in r$ so that $\overline{AB} = 5$cm. Finally, point C can be found as one of the intersection points of the circles of center A and radius 6cm, and of center B and radius 4cm.

Section 1.2

6. Review the proof of Proposition 1.9.
7. Apply Proposition 1.9.
9. Use the construction of Example 1.7 to transport angle $\angle A'O'B'$ to an angle with vertex O, and such that one of its sides coincides with $\overrightarrow{OB}$.
10. Our assumptions assure that $A\widehat{O}B = B\widehat{O}C = A\widehat{O}C$. Now, note that the sum of these three angles is equal to $360°$.
11. If α, β and γ are the measures of the angles involved, then $\alpha + \beta + \gamma = 360°$. Assuming that $\alpha \le \beta \le \gamma$, conclude that $\alpha \le 120° \le \gamma$.

Section 1.3

1. For the induction step, let $A_1 A_2 \ldots A_k A_{k+1}$ be a convex polygon of $k + 1$ sides. Diagonal $A_1 A_k$ divides it into two convex polygons: the triangle $A_1 A_k A_{k+1}$ and the convex k-gon $A_1 A_2 \ldots A_k$. Observe that the diagonals of $A_1 A_2 \ldots A_k A_{k+1}$ fall into one of three disjoint sets: (a) $A_1 A_k$; (b) the diagonals of $A_1 A_2 \ldots A_k$; (c) the diagonals $A_i A_{k+1}$, for $2 \le i \le k - 1$. Now, add the numbers of diagonals

of each of these types and apply the induction hypothesis to conclude that there
are exactly $\frac{k(k-3)}{2}$ of the type (b).

2. A convex hexagon has 9 diagonals, and from one vertex of a convex n-gon one
   can trace $n - 3$ diagonals.

3. Let $n, n+1$ and $n+2$ denote the numbers of sides of the polygons. Use the result
   of Proposition 1.12 to get a second degree equation with n as unknown.

Section 2.1

1. Consider scalene triangles ABC and $A'BC$ in the plane, such that $\overline{AB} = \overline{A'B}$
   and $\overline{AC} = \overline{A'C}$ of distinct *orientations*. This intuitive concept can be perceived
   by noting that, if one walks along the perimeter of triangle ABC, from A to B
   and then to C, and then walks along the perimeter of triangle $A'BC$, from A' to
   B' and then to C', then one of these two walks goes clockwise, whereas the other
   one goes counterclockwise.

Section 2.2

1. Follow the construction steps described in Examples 2.9, 2.10 and 2.11.

2. Start by constructing the perpendicular to r passing through A.

3. Start by constructing triangle ABM, where M is the midpoint of the side BC.

4. Let M be the midpoint of side BC and $A' \in \overrightarrow{AM}$ such that $\overline{A'M} = \overline{AM}$. Show
   that $A'MC \equiv AMB$, and then use this fact to construct triangle $AA'C$. After you
   have done that, construct the midpoint M of AA' and get vertex B as the point
   on $\overrightarrow{CM}$ for which $\overline{BM} = \overline{CM}$.

5. Split α in half (constructing its bisector), and then construct triangle ABP,
   where P is the foot of the internal bisector of ABC relative to A. Then, obtain
   vertex C as the intersection of $\overrightarrow{BP}$ and $\overrightarrow{AX}$, where X is chosen in such a way
   that $\overrightarrow{AP}$ is contained in $\angle BAC$ and $B\widehat{A}X = \alpha$.

6. If M is the midpoint of BC, we have shown in Proposition 2.13 that ABM and
   ACM are congruent. Conclude that AM is the internal bisector of $\angle BAC$ and
   that $B\widehat{M}A = C\widehat{M}A$. Finally, use the fact that $B\widehat{M}A + C\widehat{M}A = 180°$ to conclude
   that AM is the altitude of ABC relative to A.

7. If P and H coincide, show that case ASA implies $ABP \equiv ACP$; if M and H
   coincide, use SAS instead of ASA.

8. Since $\overline{OA} = \overline{OB}$, it suffices to apply the results of the two previous problems.

Section 2.3

1. Glue two copies of the first right triangle along the other leg; do the same with two copies of the other right triangle. This way, we get two isosceles triangles, and it suffices to apply the SSS congruence case to them, together with the result of Problem 6, page 30.

2. Start by showing that ADE is isosceles of basis DE, so that $\overline{BD} = \overline{CE}$. Then, apply the SAS congruence case to conclude that triangles DBC and ECB are congruent, and hence that $D\widehat{C}B = E\widehat{B}C$.

3. We consider the case of the medians relative to the sides AB and AC (the analysis of the other two cases is totally analogous). Letting M_b and M_c be the midpoints of the sides AC and AB, respectively, use the SAS congruence case to show that triangles BCM_b and CBM_c are congruent.

4. If E is the foot of the perpendicular dropped from P to the side BC, use de AAS congruence case to show that triangles BAP and BEP are congruent.

5. Apply Corollary 2.18.

6. Mark the angle opposite to β and apply Corollary 2.18.

7. Trace, through the vertex of the angle of measure γ, the line t parallel to r and s. Then, apply Corollary 2.18 twice.

8. Adapt, to the present case, the hint given to the previous problem.

9. Trace half-line $\overrightarrow{BD}$ and apply the exterior angle theorem to triangles ABD and BCD.

10. Let X be the intersection point of AB and CD, and let Y be the intersection point of AE and CD. Apply the exterior angle theorem to triangles BCX and DEY to conclude that the desired sum is equal to $180°$.

11. Use (a) and Theorem 2.19 to prove (b). Now, let $\alpha_1, \ldots, \alpha_n$ be the interior angles of the polygon and $\beta_1, \ldots, \beta_n$ be the corresponding exterior angles, so that $\alpha_i + \beta_i = 180°$ for $1 \le i \le n$. Termwise addition of these equalities, together with (b), give (c).

12. Use the fact that both triangles ABM and ACM are isosceles to conclude that $A\widehat{B}M = B\widehat{A}M = \alpha$ and $A\widehat{C}M = C\widehat{A}M = \beta$. Then, use this to show that the sum of the measures of the interior angles of triangle ABC is twice the measure of $\angle BAC$.

13. Make $\widehat{B} = 2\beta$ and $\widehat{C} = 2\gamma$. Then, use Theorem 2.19 to compute $B\widehat{I}C$ and $B\widehat{A}C$ in terms of β and γ.

14. Apply the result of the previous problem.

15. Adapt, to the present case, the hint given to Problem 13.

16. Make $\widehat{B} = \widehat{C} = \alpha$ and $C\widehat{D}E = \theta$. Then, apply the exterior angle theorem to compute $A\widehat{D}E = A\widehat{E}D$ in terms of α and θ. Finally, use the fact that.

17. Separately consider the cases $A \notin r$ and $A \in r$. For the case $A \notin r$, suppose that there are two distinct lines s and t, both perpendicular to r and passing through A. Letting B and C be the intersection points of s and t with r, respectively, compute the sum of the measures of the interior angles of triangle ABC to reach a contradiction.

18. Make $B\widehat{A}D = C\widehat{A}D = \alpha$. Then, use the exterior angle theorem to compute $A\widehat{D}C$ and $A\widehat{D}B$ in terms of α, $\widehat{B}$ and $\widehat{C}$.

19. Let $\widehat{A} = \alpha$. Use Proposition 2.13 and the exterior angle theorem several times to compute the angles of triangles AEF, DEF, CDE and BCD in terms of α. Then, apply Theorem 2.19 to triangle ABC.

20. Start by showing that the pairs of triangles AMB and DME, AMF and DMC, BCM and EMF are congruent. Then, use these congruences to show that $\widehat{A} = \widehat{D}$, $\widehat{B} = \widehat{E}$ and $\widehat{C} = \widehat{F}$. Finally, apply the result of Problem 11.

21. For item (a), trace line r, passing through B and parallel to AC. If R and S are the feet of the perpendiculars dropped from P to the lines r and $\overleftrightarrow{AC}$, respectively, start by showing that the desired sum is equal to $\overline{RS}$. Do the same for item (b), showing first that what we want to compute is exactly $|\overline{QU} - \overline{QV}|$, with U and V being the feet of the perpendicular dropped from Q to r and $\overleftrightarrow{AC}$, respectively.

22. Trace $DG \parallel BC$, with $G \in AB$, and then mark point F as the intersection of CG and BD. Compute $B\widehat{E}C$ and conclude, with the aid of Problem 2, that $\overline{BE} = \overline{BC} = \overline{BF}$. Then, conclude that $E\widehat{F}G = 40° = E\widehat{G}F$. Finally, use these facts to show that triangles EGD and EFD are congruent.

Section 2.4

1. Use triangle inequality to show that the length of the third side cannot be 14cm.

2. Apply the triangle inequality. It is also worth observing that $a < b + c$, $b < a + c$, $c < a + b$ automatically imply $a, b, c > 0$; for instance, $2a = (a + b - c) + (a + c - b) > 0$.

3. Apply the triangle inequality, together with the result of Proposition 2.23.

4. Apply the triangle inequality to sides QR of AQR, PR of BPR and PQ of CPQ. Then, add the inequalities thus obtained.

6. Factorise $a^3 + b^3$ and use twice the fact that $a + b > c$.

7. Apply the triangle inequality to triangles PAC and PBD.

8. Argue by induction on $n \geq 3$. The initial case is given by the triangle inequality. For the induction step, let a convex polygon $A_1 A_2 \ldots A_k A_{k+1}$ be given, with $k \geq 3$; apply the induction hypothesis to $A_1 A_2 \ldots A_k$ and the triangle inequality to $A_1 A_k A_{k+1}$ to show that $\overline{A_1 A_{k+1}} < \sum_{j=1}^{k} \overline{A_j A_{j+1}}$.

9. We illustrate the proof with the situation depicted in the figure below, the general case being pretty much the same.

   In order to show that the perimeter of $ABCDEF$ is less than that of $UVWXY$, apply the generalization of the triangle inequality, as given by the previous problem, to each one of the polygons AUG, $BGVH$, CHI, DWJ, $EJXY$, FKL and add the results. For instance, in AUG write $\overline{AB} + \overline{BG} < \overline{AU} + \overline{UG}$; in $BGVH$ write $\overline{BC} + \overline{CH} < \overline{BG} + \overline{GV} + \overline{VH}$, and so on.

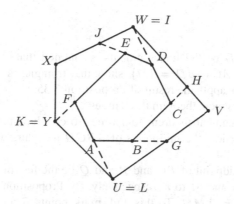

10. If A' and D' denote the symmetrics of A and D with respect to r and s, respectively, let B and C be the intersections of segment $A'D'$ with r and s, also respectively. Choose $B' \in r$ and $C' \in s$ are such that $B' \neq B$ or $C' \neq C$. Arguing in a way analogous to that of the solution of Example 2.28 to obtain $\overline{AB} + \overline{BC} + \overline{CD} < \overline{AB'} + \overline{B'C'} + \overline{C'D}$. To this end, apply the result of the previous problem.

11. Let B' be the symmetric of B with respect to AC and C' the symmetric of C with respect to AB'. If P' is the symmetric of P with respect to AC, we have $\overline{PQ} = \overline{P'Q}$ and $\overline{CP} = \overline{CP'} = \overline{C'P'}$, so that

$$\overline{BQ} + \overline{PQ} + \overline{CP} = \overline{BQ} + \overline{P'Q} + \overline{C'P'} > \overline{BC'} = 2l.$$

12. Suppose, without loss of generality, that $\overline{PA} < \overline{AR}$. Through P, draw the parallel $\overleftrightarrow{QR}$ to $\overleftrightarrow{BC}$, with $Q \in AB$ and $R \in AC$. Then, use Proposition 2.23 and the triangle inequality to show that $\overline{AP} < \overline{AR}$ and

$$\overline{BP} + \overline{PC} < \overline{BQ} + \overline{QR} + \overline{CR} < \overline{BQ} + \overline{AQ} + \overline{CR}.$$

13. If airplanes coming from cities B and C landed in city A, use Proposition 2.23 to conclude that $B\widehat{A}C > 60°$. Then, use this fact to show, by contradiction, that we cannot have six airplanes landing at a single city.

14. Arguing by contradiction if the foot of the perpendicular from P to $\overleftrightarrow{A_1A_2}$ does not lie inside this segment, then $P\widehat{A_1}A_2 \geq 90°$ or $P\widehat{A_2}A_1 \geq 90°$. Assume, without loss of generality, that $P\widehat{A_2}A_1 \geq 90°$. Then, since the feet of the perpendiculars dropped from P to the remaining sides of the n-gon do not lie inside them too, we successively conclude that $P\widehat{A_3}A_2 \geq 90°, \ldots,$ $P\widehat{A_n}A_{n-1} \geq 90°, P\widehat{A_1}A_n \geq 90°$. Therefore, $\overline{PA_1} > \overline{PA_2} > \ldots > \overline{PA_n} > \overline{PA_1}$, which is an absurd.

Section 2.5

1. Let AB and CD be the line segments, so named that $ABCD$ is a convex quadrilateral. If $AC \cap BD = \{M\}$, show that triangles ABM and CDM are congruent. Then, apply the result of Proposition 2.35.

2. Apply the midsegment theorem three times.

3. Apply the midsegment theorem four times to conclude that the quadrilateral having as its vertices the midpoints of $ABCD$ has pairs of opposite sides of equal lengths.

4. Let M be the midpoint of BC and P and Q be the feet of the perpendiculars dropped from A and M to r, respectively. By Proposition 2.40, it suffices to show that $\overline{AP} = 2\overline{MQ}$. To this end, mark points R and S such that R is the midpoint of AG and S is the foot of the perpendicular dropped from R to r; then, apply Proposition 2.38 to show that triangles RSG and MQG are congruent.

5. Let M be the midpoint of the side BC and G the barycenter of ABC. Proposition 2.38 and Example 2.6 guarantee that we can construct triangle BGM. From this, it is immediate to obtain vertex C. Finally, mark on the half-line $\overrightarrow{MG}$ the point A such that $G \in AM$ and $\overline{AG} = 2\overline{GM}$.

6. Let G be the barycenter and M_a and M_b be the midpoints of the sides BC and AC, respectively. Let also $\overline{AB} = c$, $\overline{AC} = b$ and $\overline{BC} = a$. For the first part, apply item (b) of Example 2.27, together with Proposition 2.38. For the second, start by applying the triangle inequality to M_aGM_b, together with Proposition 2.38, to conclude that $\frac{2}{3}(m_a+m_b) > \frac{c}{2}$; then, argue in an analogous way to obtain $\frac{2}{3}(m_b + m_c) > \frac{a}{2}$ and $\frac{2}{3}(m_a + m_c) > \frac{b}{2}$. By termwise adding these three inequalities, we obtain the desired one.

7. Note that R is the barycenter of triangle AQB and apply the result of Proposition 2.38.

8. Let M and N be the midpoints of CD and AB, respectively. Draw through M the parallels to the legs and mark points P and Q, in which such parallel lines intersect AB. Assuming, without loss of generality, that $\overleftrightarrow{MP} \parallel \overleftrightarrow{AD}$, conclude that $APMD$ and $BCMQ$ are parallelograms. Then, use this fact to show that $M\widehat{M}Q = 90°$, to compute PQ and show that MN is the median relative to the hypotenuse of the right triangle PMQ. Finally, apply the result of Corollary 2.44.

9. Letting M be the point of intersection of the diagonals of $ABCD$, use the formula for the midsegment of a trapezoid twice to show that the sum of the distances of A and C to r is equal to the sum of the distances of B and D to r.

10. Trace $CE \parallel AD$, with $E \in AB$. Then, conclude that $AECD$ is a parallelogram, and use this fact to show that triangle BCE is isosceles of basis CE.

11. Mark point E on AB, such that $\overline{BE} = \overline{BC}$. Then, show that $AECD$ is a parallelogram.

12. Let $ABCD$ be a trapezoid of bases AB and CD and legs BC and AD, such that $\overline{AB} = a$, $\overline{BC} = b$, $\overline{BC} = c$ and $\overline{AD} = d$. Draw, through C, the parallel to AD and suppose that such a line intersects basis AB at E. Then, construct triangle EBC with the aid of Example 2.6.

13. If M is the midpoint of BC, use Corollary 2.44 to conclude that triangle ABM is equilateral.

14. Apply Corollary 2.44 to triangles BCH_c and BCH_b.

15. Mark points G and H, with G being the midpoint of BC and H the intersection point of half-lines $\overrightarrow{EG}$ and $\overrightarrow{AB}$. Establish the congruence of triangles BGH and CGE, and then use the stated condition to conclude that triangle AEH is isosceles of basis EH. Finally, apply the result of Problem 6, page 6, together with the congruence of triangles ABG and ADF.

16. Successively construct rectangles $BCD'A'$ and $CD'A''B'$, with $\overline{A'B} = \overline{AB}$ and $\overline{B'C} = \overline{BC}$. Then, let Q' and Q'' be the symmetrics of Q with respect to the straightlines $\overleftrightarrow{AB}$ and $\overleftrightarrow{BC}$, respectively, so that $Q' \in \overleftrightarrow{AD}$ and $Q'' \in A'D'$; let also P' be the symmetric of P with respect to $\overleftrightarrow{BC}$ and R the symmetric of Q'' with respect to $\overleftrightarrow{CD'}$, so that $P' \in CD'$ and $R \in A''D'$. Show that the perimeter of $MNPQ$ is equal to $\overline{Q'M} + \overline{MN} + \overline{NP'} + \overline{P'R}$ and that $\overline{AQ'} = \overline{A''R}$. Finally, apply the result of Problem 8, page 8.

17. Suppose $\overline{AB} < \overline{AC}$ (the case $\overline{AB} = \overline{AC}$ is trivial and the case $\overline{AB} > \overline{AC}$ is entirely analogous to the one we are looking at). Let Q be the point of intersection of side AC with half-line $\overrightarrow{BP}$; show that P is the midpoint of BQ and, then, apply the midsegment theorem to triangle BQC.

Section 3.1

1. The center of the desires circle must be at a distance r from the point A and belong to the perpendicular bisector of the line segment AB. Show that there is a solution if $\overline{AB} = 2r$ and two solutions if $\overline{AB} < 2r$.

3. After we have drawn a segment AB of length c, vertex C must belong to the circle centered at B and with radius a, as well as to the line $\overleftrightarrow{AX}$ such that $B\widehat{A}X = \alpha$.

4. A lies on the circle centered at the midpoint of BC and of radius m_a, but cannot be on $\overleftrightarrow{AB}$.

6. Once we have constructed the side BC, the vertex A is found as the intersection of the parallel to $\overleftrightarrow{BC}$ at distance h_a with the circle of center B and radius c.

7. Obviously, vertex A is the point of intersection of r and s. Then, B belongs to r and to the parallel to s at distance h_b. Vertex C is constructed analogously.

8. B lies in two loci: the circle of center A and radius a, and the locus of the points of the plane which are at a distance b from r (cf. Example 3.3).

9. Start by drawing a line r (which will contain side AC) and marking a point C on it. Then, get B as the intersection of these two loci: the circle of center C and radius a, and the parallels to r at a distance h_b r.

10. Letting B_1 and B_2 be two distinct points on r, and M_1 and M_2 be the midpoints of segments AB_1 and AB_2, respectively, M_1M_2 is a midsegment of triangle AM_1M_2. Now, apply the result of Problem 2, page 58, to conclude that the height of AM_1M_2 relative to A is half the length of the height of AB_1B_2 relative to the same vertex.

11. Apply the result of Problem 8, page 31, together with Proposition 3.5.

12. Use the result of Problem 8, page 31 to show that the desired locus is the diameter of Γ perpendicular to r, except for its endpoints.

13. Assuming that the problem has been solved, mark on the line $\overleftrightarrow{BC}$ and out of the side BC points B' and C' such that $B \in B'C$, $C \in BC'$ and $\overline{BB'} = c$, $\overline{CC'} = b$. Then, note that $\overline{B'C'} = 2p$ and, by the exterior angle theorem, $A\widehat{B'}C = \frac{\beta}{2}$ and $A\widehat{C'}B = \frac{\gamma}{2}$.

14. Start by showing that, at an isosceles right triangle, the height relative to the hypotenuse has length equal to half that of the hypotenuse. Then, apply the result of Proposition 2.40 twice to show that M is at a constant distance from $\overleftrightarrow{AB}$.

Section 3.2

1. Draw the circle Γ of center O and radius $\overline{OB} = \overline{OC}$. Apart from the fact that $A \in \Gamma$, we cannot say anything else about it. Indeed, whichever position along Γ we choose for A (except those occupied by B and C), we will automatically have $\overline{OA} = \overline{OB}$ and $\overline{OA} = \overline{OC}$, so that O will indeed be the circumcenter of ABC.

2. Find A as the intersection of two loci: half-line $\overrightarrow{BX}$, such that $X\widehat{B}I = I\widehat{B}C$ and $\overrightarrow{BI}$ is contained in $\angle XBC$, and half-line $\overrightarrow{CY}$, such that $Y\widehat{C}I = I\widehat{C}B$ and $\overrightarrow{CI}$ is contained in $\angle YCB$.

3. A is the intersection of two loci: the perpendicular to $\overleftrightarrow{BC}$ passing through H and the perpendicular to $\overleftrightarrow{BH}$ passing through C.

4. Look at P as the orthocenter of a triangle having A as one of its vertices.

5. For the converse, assume first that H and I coincide. If H_a stands for the foot of the altitude relative to BC, note that $H_a \in BC$; the sum of the angles of ABH_a then gives $\frac{1}{2}\widehat{A} + \widehat{B} = 90°$; analogously, show that $\frac{1}{2}\widehat{A} + \widehat{C} = 90°$ and $\frac{1}{2}\widehat{B} + \widehat{C} = 90°$, so that $\widehat{A} = \widehat{B} = \widehat{C} = 60°$. Now, suppose that H and O coincide. Use the facts that $\overleftrightarrow{AO} \perp \overleftrightarrow{BC}$ and $\overline{BO} = \overline{CO}$ to conclude that que $\overleftrightarrow{AO}$ is the perpendicular bisector of BC, and hence that $\overline{AB} = \overline{AC}$; accordingly, show

that $\overline{AB} = \overline{BC}$. Finally, assume that I and O coincide. Then, $\overline{AI} = \overline{BI}$, so that $\frac{1}{2}\widehat{A} = \frac{1}{2}\widehat{B}$; likewise, show that $\frac{1}{2}\widehat{A} = \frac{1}{2}\widehat{C}$.

6. Let M and N stand for the midpoints of AC and AB, respectively, and H for the orthocenter. If H' denotes the symmetric of H with respect to N, show that $H'\widehat{A}M = 90°$, and then construct A as the intersection of a semicircle of diameter $H'M$ with the perpendicular to $\overleftrightarrow{MN}$ through H.

Section 3.3

1. Let s be a line passing through P, distinct from the tangent t constructed as in Proposition 3.16. Let O be the center of Γ and α be the measure of the acute angle formed by s and t. Mark point $Q \in s$, lying in the same half-plane as O with respect to t and such that $P\widehat{O}Q = 2\alpha$. Show that $Q \in \Gamma$.

2. Assuming that the problem has been solved, let O be the center of one such circle. Then, $\overleftrightarrow{AO} \perp r$ at A, so that O belongs to the line s, perpendicular to r at A. Note that the point A itself does not belong to the locus.

3. Let O be the center of one of the desired circles (we shall show that there are two of them). The previous problem assures that O lies in the perpendicular to r through P. On the other hand, since O is at the same distance from r and s, it also belongs to the bisector of one of the angles formed by them.

4. If O is the center of such a circle, then the distance from O to r is equal to R.

5. Assume the problem has been solved and let O be the center of one of the desired circle. Then, O belongs to the bisector of one of the angles formed by r and s, and is at distance R from r. There are four solution circles.

6. Letting O be the center of one such circle, we know that O is at equal distances from r and s, as well as from r and t. Hence O is found as the intersection point of the parallel to r and s situated at equal distances from them, and the internal bisector of one of the angles formed by r and t. There are two solutions.

7. First note that $P \in \Gamma_1 \cap \Gamma_2$ if and only if $\overline{PO_1} \le R_1$ and $\overline{PO_2} \le R_2$; in this case, use the triangle inequality to conclude that $|R_1 - R_2| \le \overline{O_1O_2} \le R_1 + R_2$. Now, look separately at each of the items above.

8. Let Γ' be a circle of center O' and radius R'. If $O' \in \overleftrightarrow{AO} \setminus \{A\}$ and $R' = \overline{AO'}$, show that Γ' is tangent to Γ at A.

9. If Γ' has center O', radius r and is tangent to Γ, then Problem 7 gives $\overline{OO'} = R \pm r$.

10. Letting R be the point of tangency of $\overleftrightarrow{BC}$ and Γ, item (a) of Proposition 3.28 gives $\overline{BR} = \overline{BP}$ and $\overline{CR} = \overline{CQ}$. Use these relations to conclude that the perimeter of ABC is equal to $\overline{AP} + \overline{AQ}$.

11. Letting P be the tangency point, Proposition 3.28 gives $M\widehat{A}P = \frac{1}{2}B\widehat{A}P$ and $N\widehat{A}P = \frac{1}{2}D\widehat{A}P$.

12. Let B, C and R denote the points of tangency of $\overleftrightarrow{AP}$, $\overleftrightarrow{AQ}$ and $\overleftrightarrow{PQ}$ with Γ, respectively. By Proposition 3.28 we have $P\widehat{O}R = \frac{1}{2}B\widehat{O}R$ and $Q\widehat{O}R = \frac{1}{2}C\widehat{O}R$. Now, use the fact that the sum of the angles of $ABOC$ is equal to $360°$.

13. Note first that $X\widehat{B}Y = 180° - B\widehat{X}Y - B\widehat{Y}X$. Then, use the inscribed angle theorem to show that the measures of the angles $\angle AXB$ and $\angle AYB$ do not depend on the direction of $\overleftrightarrow{XY}$.

14. Apply the inscribed angle theorem to show that triangles DEF and BEF are both isosceles.

15. If O is the point on AP such that $\overline{PQ} = \overline{BP}$, show that triangle BPQ is equilateral, and hence that $A\widehat{Q}B = 120°$. Then, use the hypotheses, together with the facts already deduced and the inscribed angle theorem to show that triangles ABQ and CBP are congruent by AAS.

16. In both cases, adapt the argument in the proof of item (a) of Proposition 3.21.

17. Show that the center of these arcs are the symmetric, with respect to AB, of the centers of the arcs capable of $180° - \alpha$ on AB.

18. After having constructed a line segment BC of length a, obtain vertex A as the point of intersection of two loci: the arcs capable of α on BC and the parallels to $\overleftrightarrow{BC}$ at distance h_a. You will find four solutions, symmetric in pairs with respect to $\overleftrightarrow{AB}$ and to the perpendicular bisector of BC.

19. If $\widehat{A} = \alpha$, look at A as the intersection of two arcs capable of $\frac{\alpha}{2}$, respectively on BM and CM. Then, use the fact that $\overline{BM} = \overline{CM}$ to get $\overline{AB} = \overline{AC}$.

20. Use the fact that $M\widehat{A}P = N\widehat{A}P = 45°$ to obtain A as the intersection of two arcs capable of $45°$, one on MP and the other on NP. Then, mark $B \in \overrightarrow{AM}$ and $D \in \overrightarrow{AN}$ such that $\overline{AB} = \overline{AD} = l$.

21. Use the result of Problem 13, page 39.

22. Firstly, note that there are four such common tangents: two *external* common tangents, which leave both circles in a single half-plane, and two *internal* common tangents, which leave the circles in opposite half-planes. Let us look at the construction of an external common tangent r (for the construction of the internal ones, just adapt what we are about to describe). Let O and O' be the centers and R and R' the radii of Γ and Γ', respectively, and T and T' the tangency points of r with Γ and Γ', also respectively. Assuming, without loss of generality, that $R > R'$, draw line s, parallel to r and passing through O', and mark the point S of intersection of the radius OT with s. Triangle $OO'S$ is right at S and such that $\overline{OS} = R - R'$; construct it to obtain the point S and, then, mark the intersection T of the half-line $\overrightarrow{OS}$ with Γ; finally, draw the desired common tangent r as being the line parallel to s and passing through T.

23. Assuming that the problem has been solved, let M and N be the midpoints of line segments AX and AY, respectively, so that $\overline{MN} = \frac{l}{2}$. Construct triangle

O_1O_2P, right at P and such that $\overleftrightarrow{O_1P} \parallel \overleftrightarrow{MN}$. Show that there will be a solution if and only if $\overline{O_1O_2} \geq \frac{l}{2}$.

24. Look at the hint given to the previous problem.

25. First of all, consider the three arcs capable of $120°$, on the sides of ABC and located outside the triangle. Then, apply the construction of Problem 23.

26. Carefully analyse the hints given to the two previous problems, trying to adapt them to the condition of greatest possible side length for MNP.

27. Assuming that the problem has been solved, let A' be the point on the half-line $\overrightarrow{BA}$ such that $\overline{BA'} = l$. With the aid of the exterior angle theorem, show that $B\widehat{A'}C = \frac{\alpha}{2}$. Now, construct A' as the intersection of the two following loci: the circle of center B and radius l and the arcs capable of $\frac{\alpha}{2}$ on BC. Finally, obtain A as the intersection of $A'B$ with the perpendicular bisector of segment $A'C$.

28. Mark point $A' \in \overrightarrow{BA} \setminus AB$, such that $\overline{AA'} = \overline{AC}$. Then, apply the exterior angle theorem to show that $B\widehat{A'}C = \frac{1}{2}B\widehat{A}C$. Now, apply the result of Example 3.25 to show that the circle of center M and radius $\overline{MB} = \overline{MC}$ passes through A'. Finally, use this fact, together with $\overleftrightarrow{MN} \perp \overleftrightarrow{A'B}$, to conclude that $\overline{BN} = \overline{A'N}$.

Section 3.4

1. Let R be the length of the circumradius, and a and b those of $\overline{BC}$ and $\overline{AC}$. Draw a circle of radius R, choose a point C on it and mark chords AC and BC on it, such that $\overline{AC} = b$ and $\overline{BC} = a$.

2. Use the inscribed angle theorem to show that $M\widehat{A}N = 90°$.

3. In order to show that $AM \perp NP$, compute the interior angle formed by those chords. Argue analogously to show that $BN \perp MP$ and $CP \perp MN$.

4. If ABC is the triangle formed by the points of intersection of the given lines, the desired points are the incenter and the excenters of ABC.

5. We look at the case in which ABC is an acute triangle (the other cases are entirely analogous). We have

$$O\widehat{A}C = \frac{1}{2}(180° - A\widehat{O}C) = \frac{1}{2}(180° - 2\widehat{B}) = 90° - \widehat{B} = B\widehat{A}H.$$

Now, letting I be the incenter of ABC, we have

$$H\widehat{A}I = B\widehat{A}I - B\widehat{A}H = C\widehat{A}I - O\widehat{A}C = O\widehat{A}I.$$

6. Let ABC be an acute triangle (the proof in the remaining cases is quite similar), H be its orthocenter, H_a the foot of the altitude relative to A and P the other point of intersection of $\overrightarrow{AH}$ with the circumscribed circle. Since $\overleftrightarrow{HH_a} \perp \overleftrightarrow{BC}$,

O_1O_2P, right at P and such that $\overleftrightarrow{O_1P} \parallel \overleftrightarrow{MN}$. Show that there will be a solution if and only if $\overline{O_1O_2} \geq \frac{l}{2}$.

24. Look at the hint given to the previous problem.

25. First of all, consider the three arcs capable of 120°, on the sides of ABC and located outside the triangle. Then, apply the construction of Problem 23.

26. Carefully analyse the hints given to the two previous problems, trying to adapt them to the condition of greatest possible side length for MNP.

27. Assuming that the problem has been solved, let A' be the point on the half-line $\overrightarrow{BA}$ such that $BA' = l$. With the aid of the exterior angle theorem, show that $B\widehat{A'}C = \frac{\alpha}{2}$. Now, construct A' as the intersection of the two following loci: the circle of center B and radius l and the arcs capable of $\frac{\alpha}{2}$ on BC. Finally, obtain A as the intersection of $A'B$ with the perpendicular bisector of segment $A'C$.

28. Mark point $A' \in \overrightarrow{BA} \setminus AB$, such that $\overline{AA'} = \overline{AC}$. Then, apply the exterior angle theorem to show that $B\widehat{A'}C = \frac{1}{2}B\widehat{A}C$. Now, apply the result of Example 3.25 to show that the circle of center M and radius $\overline{MB} = \overline{MC}$ passes through A'. Finally, use this fact, together with $\overleftrightarrow{MN} \perp \overleftrightarrow{A'B}$, to conclude that $\overline{BN} = \overline{A'N}$.

Section 3.4

1. Let R be the length of the circumradius, and a and b those of $\overline{BC}$ and $\overline{AC}$. Draw a circle of radius R, choose a point C on it and mark chords AC and BC on it, such that $\overline{AC} = b$ and $\overline{BC} = a$.

2. Use the inscribed angle theorem to show that $M\widehat{A}N = 90°$.

3. In order to show that $AM \perp NP$, compute the interior angle formed by those chords. Argue analogously to show that $BN \perp MP$ and $CP \perp MN$.

4. If ABC is the triangle formed by the points of intersection of the given lines, the desired points are the incenter and the excenters of ABC.

5. We look at the case in which ABC is an acute triangle (the other cases are entirely analogous). We have

$$O\widehat{A}C = \frac{1}{2}(180° - A\widehat{O}C) = \frac{1}{2}(180° - 2\widehat{B}) = 90° - \widehat{B} = B\widehat{A}H.$$

Now, letting I be the incenter of ABC, we have

$$H\widehat{A}I = B\widehat{A}I - B\widehat{A}H = C\widehat{A}I - O\widehat{A}C = O\widehat{A}I.$$

6. Let ABC be an acute triangle (the proof in the remaining cases is quite similar), H be its orthocenter, H_a the foot of the altitude relative to A and P the other point of intersection of $\overleftrightarrow{AH}$ with the circumscribed circle. Since $\overleftrightarrow{HH_a} \perp \overleftrightarrow{BC}$,

it suffices to show that $\overline{HH_a} = \overline{H_aP}$. To this end, use the inscribed angle theorem to establish the congruence, by ASA, of triangles BH_aP and BH_aH.

7. Let Γ be the portion of the circumscribed circle lying below $\overleftrightarrow{BC}$. Use the result of the previous problem to show that the desired locus is the arc symmetric of Γ with respect to $\overleftrightarrow{BC}$.

8. Use the result of Problem 15, page 39, to show that I_a varies along one of the arcs capable of $90° - \frac{\alpha}{2}$ on BC.

9. Start by showing that AH is the bisector of $\angle I_1HI_2$; then, show that $I_1\widehat{A}I_2 = 90° - \frac{1}{2}I_1\widehat{H}I_2$ and apply the result of the previous problem.

10. Adapt the hint given to Problem 20, page 83, applying the result of Proposition 3.37 to obtain two points on the diagonal AC, for instance.

11. Letting M be the midpoint of BC, use the result of Problem 17, page 59, to compute $\overline{PM}$ in terms of $\overline{AB} = c$ and $\overline{AC} = b$; then, compute $\overline{QR}$ in terms of a and b with the aid of Proposition 3.36 and use the result of Problem 12, page 39.

12. In the notations of Fig. 3.31, note that triangle ANI_a is right at N and such that $\overline{AN} = p$, $\overline{NI_a} = r_a$; hence, we can thus construct it. After that trace the excircle relative to BC (of center I_a and radius r_a), as well as the other tangent to it passing through A. Now, note that we can mark on AN the tangency point E of the incircle with side AC, since $\overline{EN} = a$. After that, mark the incenter I of ABC as the point of intersection of AI_a with the perpendicular to AN passing through E. Finally, trace the incircle and one of the common internal tangents to the incircle and the excircle relative to A (cf. Problem 22, page 83) to get points B and C.

13. Let M be the other intersection of $\overleftrightarrow{AI}$ with the circumcircle of ABC. Use the inscribed angle theorem to show that both ABO and BEM are equilateral, and hence that $\overline{OM} = \overline{DB}$ and (by Proposition 3.37) $\overline{IM} = \overline{BM} = \overline{BE}$. Finally, note that $\overrightarrow{BE} \perp \overrightarrow{MI}$ and $\overrightarrow{BD} \perp \overrightarrow{MO}$ to conclude that $D\widehat{B}E = O\widehat{M}I$ and, thus, that $DBE \equiv OMI$.

Section 3.5

1. Letting Γ be the circle of diameter AB and P its intersection point with $\overleftrightarrow{BC}$, use the inscribed angle theorem to show that $A\widehat{P}C = 90°$.

2. For item (a), start by showing that BCH_bH_c is a cyclic quadrilateral; then, conclude that $H_b\widehat{H}_cC = H_b\widehat{B}C$ and hence that $A\widehat{H}_cH_b = H_b\widehat{C}B$. For (b), use (a), together with the fact that $H_c\widehat{A}O = B\widehat{A}O = 90° - \frac{1}{2}\cdot A\widehat{O}B = 90° - A\widehat{C}B$.

3. Let A, B, C, D, E and F be the six points of intersection of the four lines, such that A, B and C are collinear, with $E \in BF$, $D \in AF$ and $E \in CD$. If $P \neq A$ is the other point of intersection of the circumcircles of triangles ACD and ABF, it suffices to show that P belongs to the circumcircles of triangles BCE

and DEF. To this end, use the fact that quadrilaterals $ACPD$ and $PBAF$ are cyclic to obtain

$$P\widehat{C}E = P\widehat{A}D = P\widehat{A}F = P\widehat{B}F = P\widehat{B}E.$$

4. Firstly, note that $B\widehat{C}D = 90°$. Then, successively show that $M\widehat{B}Q = M\widehat{N}Q$ and $MNBQ$ is cyclic; from this, conclude that $M\widehat{N}B = 90°$. Following, successively show that $M\widehat{D}P = M\widehat{N}P$ and $MDNP$ is cyclic; then, show that $M\widehat{P}D = 90°$.

5. If E is the foot of the perpendicular dropped from P to $\overleftrightarrow{AC}$, show that $P\widehat{Q}A = P\widehat{C}A = P\widehat{C}E = P\widehat{D}E$.

6. Use the result of the previous problem to reduce the present one to the computation of $\angle QAQ'$, with Q and Q' being the points of Γ obtained from P and P', respectively, as there.

7. Assuming that the polygon is cyclic, argue as in the proof of Proposition 3.9 to conclude that the perpendicular bisectors of its sides all pass through a single point.

8. Use Pitot's theorem, together with the computations of Proposition 3.36.

9. First assume that the polygon does possess an inscribed circle Γ, of center O. If A is one of its vertices and B and C are the feet of the perpendiculars dropped from O to the sides incident with A, show that triangles OAB and OAC are congruent, so that $B\widehat{A}O = C\widehat{A}O$. For the converse, argue in an analogous way.

10. Use that $EPCN$, $ABCD$ and $PEQD$ are cyclic to show that $N\widehat{P}E = Q\widehat{P}E$; argue in a similar way for the remaining vertices of $MNPQ$ and, then, apply the result of the previous problem.

11. Argue in a way analogous to that of the first part of the proof of Pitot's theorem.

12. Let $\overline{AB} = a$ and $\overline{BC} = b$. Assuming the problem has been solved, mark point D on the arc $\overset{\frown}{AC}$ of Γ not containing B, and let $\overline{CD} = x$ and $\overline{AD} = y$. Firstly, show that we can assume $a \neq b$, say $a < b$, so that we must have $x - y = b - a > 0$. Now, if $E \in CD$ satisfies $\overline{CE} = b - a$, then E lies in the circle of center C and radius $b - a$; use the fact that ADE is isosceles and $ABCD$ is cyclic to show that E also lies in one of the arcs capable of $180° - \frac{1}{2}A\widehat{B}C$ on AC.

Section 4.1

1. Apply Thales' theorem.
2. Set $\overline{AB} = a$, $\overline{AP_1} = x$ and $\overline{AP_2} = y$; then, use the given equality to show that $x = y$.
3. Adapt the construction steps listed in Example 4.4.

4. Let P be the foot of the internal angle bisector relative to BC. Draw through B the parallel to $\overleftrightarrow{AP}$ and mark its intersection point B' with $\overleftrightarrow{AC}$. From this point on, proceed as in the case of the external angle bisector in the text.

5. Verify that the argument in the proof of the angle bisector theorem, delineated in the hint given to the previous problem, remains valid in the present case to show that $\overline{AB} = \overline{AC}$.

6. Apply the angle bisector theorem and the construction of the fourth proportional to construct a line segment of length $\overline{AC}$.

7. Apply the internal bisector theorem to show that $\frac{BM}{AB} = \frac{CN}{AC}$. Then, apply the partial converse of Thales' theorem.

8. Let G be the barycenter and M_a, M_b and M_c be the midpoints of the sides BC, AC and AB, respectively. Mark point $P \in \overrightarrow{GM_a}$ such that $\overline{GM_a} = \overline{M_aP}$. Since the diagonals of $GCPB$ intersect at their midpoints, quadrilateral $GCPB$ is a parallelogram, and hence $\overline{BP} = \overline{GC}$. Now, Proposition 2.38 guarantees that $\overline{GP} = \frac{2}{3}m_a$, $\overline{GB} = \frac{2}{3}m_b$ and $\overline{BP} = \frac{2}{3}m_c$. By reasoning as in Example 4.3, we can construct line segments of lengths $\frac{2}{3}m_a$, $\frac{2}{3}m_b$ and $\frac{2}{3}m_c$. Then, with the aid of Example 2.6, we can construct triangle BGP. Finally, we mark $A \in \overrightarrow{PG}$ such that G is the midpoint of AP, and $C \in \overrightarrow{BM_a}$ such that M_a is the midpoint of BC.

Section 4.2

1. In each case, adapt the reasoning of the proof of Proposition 4.7.

2. Let A be the point in which the upper horizontal side of the middle square encounters the right vertical side of the largest one; likewise, let B be the point in which the upper horizontal side of the smaller square encounters the right vertical side of the middle one. Use the similarity of triangles XAY and YBZ to compute the desired length.

3. Let M_a and M'_a stand for the midpoints of sides BC and $B'C'$, respectively. Use SAS to show that ABM_a and $A'B'M'_a$ are similar with similitude ratio k, and then conclude that $\frac{m_a}{m'_a} = k$. Now, if H_a and H'_a (resp. P_a and P'_a) stand for the feet of the altitudes (resp. internal angle bisectors) relative to A and A', respectively, argue in an analogous way for the pairs of triangles ABH_a and $A'B'H'_a$ (resp. ABP_a and $A'B'P'_a$).

4. If $Q \in AB$ is such that $PQ \perp AB$, then $\overline{AQ} = \overline{PQ}$ and $PQB \sim CAB$.

5. Let X and Y denote the intersection points of r with $\overleftrightarrow{AB}$ and $\overleftrightarrow{AC}$, respectively. Use the result of the previous problem, together with the fact that XAY is a right isosceles triangle.

6. Start by showing that $FAD \sim EAB$.

7. Let O and O' denote the centers of the given circles and P and P' the points of tangency of them with one of the sides of the angle. Use the similarity of triangles AOP and $AO'P'$ to compute the desired length.

8. For item (a), use the fact that $CNP \sim CBA$. For item (b), recall that the diagonals of a rhombus bisect its internal angles.

9. Draw, through C, the parallel to $\overleftrightarrow{AB}$ and mark its intersection point F with $\overline{DE}$. Then, use the fact that $CFD \sim BMD$ and $CFE \sim AME$.

10. Write h and h' to denote the lengths of the altitudes of the trapezoids $MNCD$ and $ABCD$. If $\overline{MP} = x$, use the similarities $MPD \sim ABD$ and $MPA \sim DCA$, together with the result of Problem 3, to get $\frac{x}{a} = \frac{h}{h'}$ and $\frac{x}{b} = \frac{h'-h}{h'}$. From these, conclude that $x = \frac{ab}{a+b}$. Now, argue in a similar way to compute $\overline{NP}$.

11. Apply the result of the previous problem to the trapezoid $BCYX$.

12. In order to show that $\overline{EF} = \overline{FG}$, use the similarity $AEG \sim ADC$, together with the result of Problem 3. With this equality at hand, show that in order to obtain $\overline{FG} = \overline{GH}$ it suffices to prove that $\overline{EG} = \overline{FH}$. For what is left to do, let h and h' respectively denote the altitudes of trapezoids $ABHE$ and $ABCD$; use the pairs of similarities $AEG \sim ADC$ and $BHF \sim BCD$, together with the result of Problem 3, to get $\overline{EG} = \overline{FH} = \frac{h}{h'} \cdot \overline{CD}$.

13. If D is the foot of the internal angle bisector relative to B, show that $ABC \sim ADB$.

14. Use the result of the previous problem, together with the converse of Phytagoras' theorem.

15. The inscribed angle theorem gives $A\widehat{A'}C = A\widehat{B}H_a$. Therefore, triangles $AA'C$ and ABH_a are similar by AA.

16. Proposition 3.20 gives $A\widehat{B}D = A\widehat{C}F$ and $A\widehat{C}D = A\widehat{B}E$. Use this to conclude that $ABD \sim ACF$ and $ACD \sim ABE$. Finally, use these two similarities to derive the desired relation.

17. Use the metric relations on right triangles.

18. Apply the metric relations in right triangles.

19. For item (a), use the metric relations in right triangles.

20. Use item (a) of the previous problem.

21. Check that the triangles BMP, CNM and APN are congruent by SAS, so that MNP is equilateral. Now, letting M_a be the midpoint of BC, show that $BMP \sim BM_aA$ by SAS, and hence $P\widehat{M}B = A\widehat{M_a}B = 90°$.

22. If $x = \sqrt{a^2 + b^2}$ and $y = \sqrt{a^2 + b^2 - c^2}$, we have $y = \sqrt{x^2 - c^2}$. Now, Pythagoras' theorem gives x as the length of the hypotenuse of a right triangle of legs a and b, whereas y is found, again from Pythagoras, as the length of the other leg of a right of hypotenuse x and leg c.

23. Assume, without loss of generality, that $\overline{BD} = \overline{DE} = \overline{EF} = \overline{FC} = 5$. Now, letting $\overline{AB} = x$ and $\overline{AC} = y$, Pythagoras' theorem applied to ABC gives $x^2 + y^2 = 400$. On the other hand, letting P be the foot of the perpendicular dropped from D to the side AB, use similarity of triangles to conclude that $\overline{BP} = \frac{x}{4}$ and $\overline{DP} = \frac{y}{4}$, whence, again from Pythagoras' theorem, $\overline{AD}^2 =$

$(\frac{3x}{4})^2 + (\frac{y}{4})^2$. Now, reason likewise to find expressions for $\overline{AE}^2$ and $\overline{AF}^2$ in terms of x and y, and use them to compute $\overline{AD}^2 + \overline{AE}^2 + \overline{AF}^2$ in terms of $x^2 + y^2$.

24. The problem only makes sense for $l \leq 2R$, and if $l = 2R$ the locus reduces to the center O of Γ. Assume, then, that $l < 2R$. Letting AB be a chord of length l and M be its midpoint, the fact that AOB is isosceles of basis AB gives $OM \perp AB$. Hence, Pythagoras' theorem gives $\overline{OM} = \sqrt{R^2 - (\frac{l}{2})^2}$ and, letting r denote such a length, we conclude that M lies in the circle of center O and radius r. Conversely, it is not difficult to reverse the steps above to conclude that every point of such a circle is the midpoint of a chord of length l of $\Gamma(O; R)$.

25. Letting P be such a point and PT one of the tangents drawn from P to Γ, Pythagoras' theorem gives $\overline{PO}^2 = \overline{PT}^2 + \overline{TO}^2 = l^2 + R^2$. Hence, P lies in the circle of center O and radius $\sqrt{R^2 + l^2}$. Conversely, reverse the steps of this reasoning to show that, for every point of such a circle, one can draw tangents of length l to $\Gamma(O; R)$.

26. Write A for the top and B for the basis of the building, as well as C and D for the initial and final positions of the pedestrian. Let $A\widehat{C}B = \alpha$ and $A\widehat{D}B = \frac{\alpha}{2}$. The exterior angle theorem gives $D\widehat{A}C = A\widehat{C}B - A\widehat{D}B = \frac{\alpha}{2}$, so that triangle ACD is isosceles of basis AD. Hence, $\overline{AC} = \overline{CD} = 50$, and applying Pythagoras' theorem to triangle ABC we obtain $\overline{AB} = \sqrt{50^2 - 25^2} = 25\sqrt{3}$.

27. Assume, without loss of generality, that $A \in r$, $B \in s$ and $C \in t$, and let D and E denote the feet of the perpendiculars dropped from A and B to t, respectively. Now, letting $\overline{AB} = \overline{AC} = \overline{BC} = l$, we have $\overline{CE} = \sqrt{l^2 - 1}$, $\overline{CD} = \sqrt{l^2 - 16}$ and $\overline{DE} = \sqrt{l^2 - 9}$. On the other hand, it is not difficult to check that $D \in CE$, so that $\sqrt{l^2 - 1} = \sqrt{l^2 - 16} + \sqrt{l^2 - 9}$. Solve this equation to find the value of l.

28. Since the birds had equal velocities, departed at the same moment and arrived together, we conclude that the fountain is at equal distances from the top of both towers. Letting d denote such a common distance, use Pythagoras' theorem to show that the distances from the fountain to the bases of the towers are equal to $\sqrt{d^2 - 900}$ and $\sqrt{d^2 - 1600}$. Solve the equation $\sqrt{d^2 - 900} + \sqrt{d^2 - 1600} = 50$ to find the value of d; then, compute $\sqrt{d^2 - 900}$ and $\sqrt{d^2 - 1600}$.

29. Let ABC be the triangle, with $\overline{AB} = 5\sqrt{2}$, $\overline{BC} = 7$ and $A\widehat{B}C = 135°$. If H stands for the foot of the altitude dropped from A, we have $A\widehat{B}H = 45°$, so that ABH is a right and isosceles triangle. Apply Pythagoras' theorem to it to get $\overline{AH} = \overline{BH} = 5$. Then, apply Pythagoras' theorem to triangle AHC to obtain $\overline{AC} = 13$.

30. The parallels to the sides of $ABCD$ through P divide $ABCD$ into four rectangles $\mathcal{R}_1$, $\mathcal{R}_2$, $\mathcal{R}_3$ and $\mathcal{R}_4$. Label them in such a way that $\mathcal{R}_1$ and $\mathcal{R}_2$ have only vertex P in common, and let a_1, b_1 be the lengths of the sides de $\mathcal{R}_1$ and a_2, b_2 be those of $\mathcal{R}_2$. Apply Pythagoras' theorem four times to show that $\overline{AP}^2 + \overline{CP}^2 = \overline{BP}^2 + \overline{DP}^2 = a_1^2 + b_1^2 + a_2^2 + b_2^2$.

31. Assume, without loss of generality, that P lies in the smaller arc $\overset{\frown}{AD}$. From the external angle theorem, we obtain $A\widehat{P}C = B\widehat{P}D = 90°$. Hence, Pythagoras' theorem gives $\overline{PA}^2 + \overline{PC}^2 = \overline{AC}^2 = 200$ and, analogously, $\overline{PB}^2 + \overline{PD}^2 = 200$.

32. The diagonals divide the quadrilateral into four right triangles. Apply Pythagoras' theorem to each one of them.

33. If r stands for the radius of the circle inscribed in $ABCD$ and $l = \overline{BC}$, then $\overline{AD} = 2r$ and, from Pitot's theorem, $2r + l = 16$. On the other hand, if E is the foot of the perpendicular dropped from C to the basis AB, then the fact that $ABCD$ is right at A guarantees that $C\widehat{E}B = 90°$, so that Pythagoras' theorem furnishes $(2r)^2 + 8^2 = l^2$. Solve the system of equations thus obtained to find the values of r and l. Then, letting O denote the center of the inscribed circle, observe that

$$O\widehat{C}B + O\widehat{B}C = \frac{1}{2}D\widehat{C}B + \frac{1}{2}A\widehat{B}C = 90°,$$

whence OBC is right at O. Finally, apply the metric relations in rights triangles to compute $\overline{OB}$ and $\overline{OC}$.

34. Let O_1, O_2 and O_3 respectively denote the centers of the circles of radii R, r and x. For $1 \leq i \leq 3$, let A_i be the foot of the perpendicular dropped from A_i to s. Use Pythagoras' theorem to conclude that $\overline{A_1 A_2} = \sqrt{(R+x)^2 - (R-x)^2} = 2\sqrt{Rx}$ and, analogously, $\overline{A_1 A_3} = 2\sqrt{Rr}$ and $\overline{A_2 A_3} = 2\sqrt{rx}$. Then, use the fact that $\overline{A_1 A_3} = \overline{A_1 A_2} + \overline{A_2 A_3}$ to obtain the desired relation.

35. Apply the result of the previous problem.

36. Let r be a line passing through two of the n given points, say A and B, and which is at a minimum positive distance from one of the remaining points, say C. By the sake of contradiction, assume that r contains at least another one of the n points, say D. Let H denote the foot of the perpendicular dropped from C to r and assume, without loss of generality, that A and D lie in a single half-line, of those H determines on r. If $E \in AH$, show that $d(E; \overleftrightarrow{AC}) < d(C; r)$, which is a contradiction.

Section 4.3

1. Let $P \in BC$ be the foot of the internal angle bisector relative to A. Start by constructing point $Q \in \overleftrightarrow{BC} \setminus BC$, such that $\frac{BQ}{QC} = \frac{BP}{PC}$—for an easy way of constructing Q, review the construction steps listed in Example 4.17. Then, obtain vertex A as the intersection of the circle of diameter PQ with that of center C and radius b.

2. The solution of this problem is a slight variation of that of the previous one.

3. Draw an arbitrary line r and mark on it points B and C such that $\overline{BC} = a$. Then, obtain vertex A as the intersection of the two following loci: the Apollonius' circle relative to (B, C) in the ratio $\frac{3}{4}$, and the circle centered at the midpoint of BC and with radius m_a.

4. The desired radius is equal to $\frac{1}{2}\overline{PQ}$, with $P \in BC$ and $Q \in \overleftrightarrow{BC} \setminus BC$ being such that $\frac{\overline{BP}}{\overline{PC}} = \frac{\overline{BQ}}{\overline{QC}} = k$. Use such relations to compute $\overline{BP}$ and $\overline{BQ}$ in terms of k and $\overline{BC}$; then, notice that $\overline{PQ} = \overline{BQ} \pm \overline{BP}$, according to whether $0 < k < 1$ or $k > 1$.

5. Let $\overline{AB} = a$, $\overline{BC} = b$, $\overline{CD} = c$ and $\overline{AD} = d$; assume, without loss of generality, that $a + d \geq b + c$ and $c + d \geq a + b$. Use Ptolemy's theorem to show that a point P with the stated properties can be found in the arc $\overset{\frown}{AD}$ not containing B, as the intersection of Γ with the Apollonius' circle relative to (A, C) in the ratio $\frac{a+d-p}{c+p-b}$, with $p = \overline{AC}$.

6. For items (a) and (b), use the inscribed angle theorem to conclude that, if $P \neq B$ is the intersection point of the circumcircles of triangles BCD and ABF, then $B\widehat{P}D = D\widehat{P}C = 60°$ and $A\widehat{P}F = F\widehat{P}B = 60°$. Then, successively conclude that (i) $A\widehat{P}C = 120°$; (ii) P lies in the circumcircle of ACE; (iii) A, P, D and C, P, F are two triples of collinear points. For item (c), use the results of items (a) and (b), together with Corollary 4.19.

7. Use the identity $\left(\frac{1-t^2}{1+t^2}\right)^2 + \left(\frac{2t}{1+t^2}\right)^2 = 1$, together with the converse of Pythagoras' theorem.

8. Use the construction of the previous problem, together with Ptolemy's theorem, to show that one can find a circle Γ' and points $A_1', A_2', \ldots, A_n' \in \Gamma'$ such that $A_1' A_2'$ is a diameter of Γ' and $\overline{A_i' A_j'} \in \mathbb{Q}$ for $1 \leq i < j \leq n$. If O is the center and R is the radius of Γ', let Γ be the circle of center O and radius kR. For $1 \leq i \leq n$, let $A_i \in \overrightarrow{OA_i'}$ be such that $\overline{OA_i} = k\,\overline{OA_i'}$, so that $A_i \in \Gamma$. Show that we can choose k in such a way that $\overline{A_i A_j} \in \mathbb{N}$ for $1 \leq i < j \leq n$.

9. If O' is the circumcenter of ABD and M and M' are the midpoints of sides BC and AD, respectively, use Theorem 4.20 to conclude that $\overline{BH} = 2\,\overline{O'M'} = 2\,\overline{OM}$. Then, notice that $\overline{BC} = 2\,\overline{MC}$ to conclude that triangles BHC and MOC are similar.

10. Theorem 4.20 allows us to successively mark the positions of the barycenter G and vertex A of ABC. Once this has been done, points B and C are obtained as the intersection points of the circle of center O and radius AO with the straightline passing through M and perpendicular to $\overleftrightarrow{OM}$.

11. Let P be the point in which the internal angle bisector relative to BC intersects the circumcircle of triangle AHO. The inscribed angle theorem guarantees that $H\widehat{A}P = H\widehat{O}P$ and $O\widehat{A}P = O\widehat{H}P$. On the other hand, it follows from Problem 5, page 92, that $H\widehat{A}P = O\widehat{A}P$, and hence $H\widehat{O}P = O\widehat{H}P$. Therefore, $\overline{HP} = \overline{OP}$, and thus P belongs to the perpendicular bisector of HO.

12. First note that $\overleftrightarrow{AH_1} \parallel \overleftrightarrow{BH_2}$. Now, Theorem 4.20 furnishes $\overline{AH_1} = \overline{BH_2} = 2\overline{OM}$, with O being the circumcenter of $ABCD$ and M the midpoint of CD. Hence, Problem 1, page 58, guarantees that AH_1H_2B is a parallelogram.

13. Let M be the midpoint of BC. Since $\overline{AO} = \overline{CO}$ and (from Theorem 4.20) $\overline{AH} = 2\overline{OM}$, we have $\overline{CO} = 2\overline{OM}$. Now, use the result of Problem 13, page 58.

14. Let H denote the orthocenter of ABC, let H_a, H_b, H_c denote the feet of the altitudes dropped from A, B, C, respectively, and P_a, P_b, P_c the other points of intersection of $\overrightarrow{AH}$, $\overrightarrow{BH}$, $\overrightarrow{CH}$ with the circumcircle Γ of ABC. Use the result of Problem 6, page 92, together with item (a) of Theorem 4.20, to show that $\overline{AP_b} = \overline{AP_c} = \overline{AH} = \sqrt{4R^2 - a^2}$. Again from Problem 6, page 92, conclude that $d(H, \overleftrightarrow{P_b P_c}) = 2r$. The reasoning above gives the following construction: draw a circle $\Gamma(O; R)$ and pick a point $A \in \Gamma$. Draw the circle $\Sigma(A; \ell)$, with $\ell = \sqrt{4R^2 - a^2}$, so that $\Gamma \cap \Sigma = \{P_b, P_c\}$. Find H as the intersection of Σ with the parallel u to $\overleftrightarrow{P_b P_c}$ at a distance $2r$, with u and A lying in opposite sides with respect to $\overleftrightarrow{P_b P_c}$. Finally, construct P_a as the other intersection of $\overrightarrow{AH}$ with Γ, and then B and C as the intersections of the perpendicular bisector of HP_a with Γ.

Section 4.4

1. By Menelaus' theorem, it suffices to show that $\frac{AZ}{ZB} \cdot \frac{BX}{XC} \cdot \frac{CY}{YA} = -1$. To this end, assume that $\overleftrightarrow{AA'}$, $\overleftrightarrow{BB'}$ and $\overleftrightarrow{CC'}$ concur at O (the case in which $\overleftrightarrow{AA'}$, $\overleftrightarrow{BB'}$ and $\overleftrightarrow{CC'}$ are parallel is much simpler) and apply Menelaus' theorem to triangles $BB'Z$ (with the triple of collinear points A, A', O), $AA'Y$ (with the triple of collinear points C, C', O) and $CC'X$ (with the triple of collinear points B, B', O).

2. Use the similarities $BB'C \sim A'AC$, $BCC' \sim BA'A$ and $BB'A \sim C'CA$.

3. Adapt the discussion of the case in which the triangle is acute, noticing that exactly two of the altitudes of the triangle are external to it.

4. Let $P \in \overleftrightarrow{BC} \setminus BC$ and $Q \in \overleftrightarrow{AC} \setminus AC$ be the feet of the external angle bisectors relative to A and B, respectively, and $R \in AB$ be the foot of the internal angle bisector relative to C. Use the angle bisector theorem to show that $\frac{AR}{RB} \cdot \frac{BP}{PC} \cdot \frac{CQ}{QA} = 1$.

5. For item (c), apply the angle bisector theorem. For item (d), apply Ceva and Menelaus theorems.

6. Use Ceva's theorem in conjunction with item (a) of Proposition 3.36.

7. Use Ceva's theorem in conjunction with item (c) of Proposition 3.36.

8. It suffices to show that if $\overleftrightarrow{AA''}$ and $\overleftrightarrow{CC''}$ concur at a point P, then $P \in \overleftrightarrow{BB''}$. To this end assume, without loss of generality, that $A' \in BC$ and $B' \in \overrightarrow{CA} \setminus AC$, so that $C' \in \overrightarrow{BA} \setminus AB$ (the remaining cases are entirely analogous). If $B\widehat{A}A'' = \alpha$, the isogonality of $\overrightarrow{AA'}$ and $\overrightarrow{AA''}$ guarantees that $C\widehat{A}A' = \alpha$, whereas the parallelism of $\overleftrightarrow{AA'}$ and $\overleftrightarrow{CC'}$, together with the isogonality of $\overrightarrow{CC'}$ and $\overrightarrow{CC''}$, furnishes $C''\widehat{C}B = A\widehat{C}C' = C\widehat{A}A' = \alpha$. Hence, we have $B\widehat{C}P = B\widehat{C}C'' = \alpha = B\widehat{A}A'' = B\widehat{A}P$, so that $BACP$ is a cyclic quadrilateral; it thus follows that $P\widehat{B}C = P\widehat{A}C$. Finally, this last equality, in conjunction with the isogonality of $\overrightarrow{AA'}$ and $\overrightarrow{AA''}$ and the parallelism of $\overleftrightarrow{AA'}$ and $\overleftrightarrow{BB'}$, give us $P\widehat{B}C = A''\widehat{A}C = A'\widehat{A}B = A\widehat{B}B'$. It comes that $\overrightarrow{BB'}$ and $\overrightarrow{BP}$ are isogonal with respect to B, and this shows that $P \in \overleftrightarrow{BB''}$.

9. Apply three times the result of Problem 5, page 92.

10. Let $P' \in \overleftrightarrow{QR}$ be such that $\overrightarrow{AP'} \perp \overleftrightarrow{QR}$. Use the fact that $AQPR$ is a cyclic quadrilateral to show that the half-lines $\overrightarrow{AP}$ and $\overrightarrow{AP'}$ form equal angles with the internal bisector of ABC relative to A, and lie in opposite half-planes with respect to that internal bisector.

11. Apply the theorems of Ceva and Menelaus.

12. Firstly, assume that $\overleftrightarrow{AB'} \cap \overleftrightarrow{BC'} = \{X\}$, $\overleftrightarrow{AB'} \cap \overleftrightarrow{A'C} = \{Y\}$ and $\overleftrightarrow{A'C} \cap \overleftrightarrow{BC'} = \{Z\}$. Apply Menelaus' theorem to triangle XYZ, with respect to each of the following triples of collinear points: A, C, E; B', C, D; A', B, F; A, B, C; A', B', C'. Now, consider the case in which at least one of the pairs of lines $\overleftrightarrow{AB'}$ and $\overleftrightarrow{BC'}$, $\overleftrightarrow{AB'}$ and $\overleftrightarrow{A'C}$, $\overleftrightarrow{A'C}$ and $\overleftrightarrow{BC'}$ are parallel.

13. Suppose, without loss of generality, that $\overline{AB} < \overline{BC} < \overline{AC}$, so that $P \in \overrightarrow{CB} \setminus BC$, $Q \in \overrightarrow{CA} \setminus AC$ and $R \in \overrightarrow{AB} \setminus AB$. By Menelaus' theorem, it suffices to prove that $\frac{BP}{PC} \cdot \frac{CQ}{QA} \cdot \frac{AR}{RB} = -1$. To this end, observe that Proposition 3.20 gives $PAB \sim PCA$ by AA. Therefore, $\frac{\overline{BP}}{\overline{AP}} = \frac{\overline{AP}}{\overline{CP}} = \frac{\overline{AB}}{\overline{AC}}$, whence $\frac{BP}{PC} = -(\frac{\overline{AB}}{\overline{AC}})^2$. Finally, argue in a similar way to obtain $\frac{CQ}{QA} = -(\frac{\overline{BC}}{\overline{AB}})^2$ and $\frac{AR}{RB} = -(\frac{\overline{AC}}{\overline{BC}})^2$.

14. Assume ABC acute (the remaining cases can be dealt with analogously). Applying Menelaus' theorem to triangle ABC, with respect to the triple of collinear points D, E and P, we obtain $\frac{BP}{PC} \cdot \frac{CE}{EA} \cdot \frac{AD}{DB} = -1$. Now, the triangle similarities $CEH_a \sim CH_aA$ and $AEH_a \sim AH_aC$ furnish the equalities $\overline{CE} = \frac{\overline{CH_a}^2}{\overline{AC}}$ and $\overline{AE} = \frac{\overline{AH_a}^2}{\overline{AC}}$, so that $\frac{CE}{EA} = \frac{\overline{CH_a}^2}{\overline{AH_a}^2}$; analogously, we have $\frac{AD}{DB} = \frac{\overline{AH_a}^2}{\overline{BH_a}^2}$. Now, argue in a similar way with respect to the other two sides of ABC. Finally, multiplying the relations thus obtained, we arrive at the equality $\frac{BP}{PC} \cdot \frac{CQ}{QA} \cdot \frac{AR}{RB} = -1$, whence Menelaus' theorem guarantees the collinearity of points P, Q and R.

15. Let A_1, A_2, A_3 be such that $\overleftrightarrow{A_i A_j}$ is a common external tangent to Γ_i and Γ_j, for $1 \le i < j \le 3$, and Γ_1, Γ_2, Γ_3 lie inside triangle $A_1 A_2 A_3$. If O_i is the center of Γ_i, show that triangles $A_1 A_2 A_3$ and $O_1 O_2 O_3$ are in perspective with respect to the incenter of $A_1 A_2 A_3$, then apply Desargues' theorem.

Section 4.5

1. Show that $PBC \sim PCA$.
2. Letting E be the other endpoint of the diameter of Γ passing through D, apply the intersecting chords theorem to AB and DE. Alternatively, apply (4.5), with C in place of P.
3. Apply the intersecting chords theorem.
4. Start by using the intersecting chords theorem to compute $\overline{CE}$. Now, let O be the center of the circle and F and G be the feet of the perpendiculars dropped from O to AB and CD, respectively; successively compute $\overline{OF}$, $\overline{OG}$ and $\overline{OE}$, this last one with the aid of Pythagoras' theorem. Finally, apply (4.5) to compute the desired radius.
5. Let D be the other endpoint of the diameter of the circumcircle of ABC passing through A. Apply the intersecting chords theorem to the chords AD and BC of such a circle.
6. Let ABC be a triangle right at A. Draw the circle of center C and radius $\overline{AC}$, and let D and E be the intersection points of such a circle with line $\overleftrightarrow{BC}$. Now, apply the version of the intersecting chords theorem of Proposition 4.30.
7. Letting T be one of such points, start by using the intersecting chords theorem to compute $\overline{PT}$ in terms of $\overline{PA}$ and $\overline{PB}$. Then, apply the construction of Problem 20, page 117.
8. If we construct the point T of contact of the desired circle with r, then the circle itself will be the circumcircle of ABT. For what is left to do, there are two distinct cases to consider: (i) $\overleftrightarrow{AB} \parallel r$: in this case, T is the intersection point of the perpendicular bisector of AB with r; (ii) $\overleftrightarrow{AB} \not\parallel r$: letting P be the intersection of $\overleftrightarrow{AB}$ and r, the intersecting chords theorem gives $\overline{PT}^2 = \overline{PA} \cdot \overline{PB}$; now, use Problem 20, page 117, to construct a line segment of length $\overline{PT}$, and then construct T.
9. Use the fact that $\overline{PA_1} = \frac{1}{2} \overline{A_1 A_2} = \overline{PA_2}$ to conclude that P lies in the radical axis of Γ_1 and Γ_2; then, argue analogously for point Q.
10. Since CD is a diameter of Σ, triangle CDE is right at E; therefore, the metric relations in right triangles furnish $\overline{ME}^2 = \overline{CM} \cdot \overline{DM}$. Now, apply the intersecting chords theorem to the chords AB and CD of Γ to conclude that $\overline{AM} = \overline{BM} = \overline{EM}$.
11. Firstly, show that both circles pass through the foot H_a of the altitude of ABC relative to BC. Then, show that the pair of chords MN and AH_a of

the circle of diameter AC, as well as the pair of chords PQ and AH_a of the circle of diameter AB, intersect at the orthocenter H of ABC. Apply the intersecting chords theorem to each one of such pairs of chords to conclude that $\overline{MH} \cdot \overline{NH} = \overline{PH} \cdot \overline{QH}$. Then, apply the intersecting chords theorem again to conclude that M, N, P and Q are concyclic.

12. Assume, without loss of generality, that $X \in BY$. Proposition 4.29 assures that it suffices to prove that $\overline{BC} \cdot \overline{BO} = \overline{BX} \cdot \overline{BY}$. To this end, let $P \in BC$ and $Q \in \overrightarrow{BC} \setminus BC$ be such that PQ is a diameter of Γ; use the fact that $\overline{BP} \cdot \overline{BQ} = \overline{BX} \cdot \overline{BY}$, together with the result of Problem 4, page 126, to obtain the desired equality.

13. For $i = 1, 2$, let $\Gamma_i(O_i; R_i)$. Let also P be the center of a circle intersecting Γ_i along a diameter $A_i B_i$ of it, for $i = 1, 2$. Use the fact that $\overline{PA_1} = \overline{PA_2}$ to get $\overline{PO_1}^2 - \overline{PO_2}^2 = R_2^2 - R_1^2$. Then, compare such a relation with (4.9) to conclude that P belongs to the line symmetric to the radical axis of Γ_1 and Γ_2 with respect to the midpoint of $O_1 O_2$.

14. Letting N be the foot of the external angle bisector relative to C, Example 4.23 assures that the points N, E and D are collinear. Hence, both C and D lie in the Apollonius' circle relative to (A, B), in the ratio $\frac{AF}{FB}$. Letting M denote the center of such a circle, it follows from the previous problem that quadrilateral $AMCD$ is cyclic, so that $C\widehat{M}D = \frac{1}{2}\widehat{A}$ and $D\widehat{M}A = \widehat{C}$. Finally, apply the external angle theorem to obtain $M\widehat{C}F = M\widehat{F}C = \frac{1}{2}\widehat{C} + \widehat{B}$, and add the angles of the isosceles triangle MCF to find $\widehat{A} = 120°$.

15. If PQ is the diameter of the circumcircle of ABC passing through I and O, show that the intersecting chords theorem applied to chords FG and PQ, together with Euler's theorem 4.32, gives

$$\overline{DF} \cdot \overline{EG} + r(\overline{DF} + \overline{EG}) + r^2 = 2Rr,$$

where R is the circumradius of ABC. Then, use the fact that $\overline{FG} \le 2R$ to obtain $\overline{DF} + \overline{EG} \le 2(R - r)$. The desired inequality easily follows from this. By reviewing the arguments above, it is pretty easy to conclude that the equality holds if and only if FG is a diameter of the circumcircle of ABC.

16. Let $\overline{AB} = c$, $\overline{AC} = b$ and $\overline{BC} = a$. Apply the bisector theorem and the intersecting chords theorem to get $\overline{AD} \cdot \overline{DK} = \frac{a^2 bc}{(b+c)^2}$; then, use the similarity of triangles $ABD \sim AKC$ to obtain $\overline{AD} \cdot \overline{AK} = bc$ and, hence,

$$\overline{AD}^2 = bc - \overline{AD} \cdot \overline{DK} = bc - \frac{a^2 bc}{(b+c)^2}$$

and

$$\frac{\overline{AD}}{\overline{DK}} = \frac{\overline{AD}^2}{\overline{AD} \cdot \overline{DK}} = \left(\frac{b+c}{a}\right)^2 - 1.$$

Finally, from the above computations (and analogous ones for $\frac{BE}{EL}$ and $\frac{CF}{FM}$), conclude that it suffices to establish the inequality

$$\left(\frac{b+c}{a}\right)^2 + \left(\frac{a+c}{b}\right)^2 + \left(\frac{a+b}{c}\right)^2 \geq 12,$$

with equality if and only if $a = b = c$. In turn, obtain such an inequality by means of a suitable application of the inequality between the arithmetic and geometric means (cf. [5]), for instance.

17. Assume, without loss of generality, that Γ_1 is the circle interior to α, and let P (resp. Q) be the point of tangency of α and Γ_1 (resp. α and Γ_2); let also S and T be the intersection points of segment $O_1 O_2$ respectively with Γ_1 and Γ_2, C the intersection point of segments PQ and ST, B the other intersection of line $\overleftrightarrow{AC}$ with α and R and U the other intersections of half-lines $\overrightarrow{PQ}$ and $\overrightarrow{ST}$ with Γ_2. Adapt the arguments presented in the solution of Example 4.40 to successively show that: $\overleftrightarrow{PS} \parallel \overleftrightarrow{RT}$; quadrilateral $PSQU$ is cyclic; $\overline{AC} \cdot \overline{CB} = \overline{SC} \cdot \overline{CU}$; quadrilateral $ASBU$ is cyclic. Finally, construct point B as being the other intersection of line $\overleftrightarrow{AC}$ with the circumcircle of triangle ASU, thus again reducing the problem to that of Example 4.39.

19. Let $\Gamma_1(O_1; R_1)$, $\Gamma_2(O_2; R_2)$ and $\Gamma_3(O_3; R_3)$ be the three given circles, and $\Gamma(O; R)$ be one of the circles we wish to construct. For the sake of simplicity, assume that Γ_1, Γ_2 and Γ_3 are pairwise exterior, and that Γ is exteriorly tangent to all of them; suppose (also without loss of generality) that $R_1 < R_2, R_3$ (the other cases can be dealt with in pretty much the same way). If $\Gamma_2'(O_2; R_2 + R_1)$, $\Gamma_3'(O_3; R_3 + R_1)$ and $\Gamma'(O_1; R + R_1)$, then Γ' passes through O_1 and is tangent to Γ_2' and Γ_3'. Therefore, we have reduced the present problem to the one considered in Example 4.40 and in the two previous problems. Generically, there are eight solutions.

20. Let K and L (resp. M and N) be the feet of the perpendiculars dropped from P (resp. Q) to the sides AB and AC, respectively. According to Problem 10, page 137, we have $\overleftrightarrow{AP} \perp \overleftrightarrow{MN}$. Letting R be the intersection point of $\overleftrightarrow{AP}$ and $\overleftrightarrow{MN}$, conclude that the quadrilaterals $LRPN$ and $PRMK$ are cyclic. Then, use this fact, together with the intersecting chords theorem, to show that $\overline{AM} \cdot \overline{AK} = \overline{AL} \cdot \overline{AN}$ and, hence, that the quadrilateral $LMKN$ is cyclic too. Finally, note that the center of the circle circumscribed to such a quadrilateral is the intersection point of the perpendicular bisectors of line segments LM and KN, thus coinciding with the midpoint of segment PQ. Argue in an analogous way with the feet of the perpendiculars dropped from P and Q to the side BC.

21. If P and Q denote the points of contact of α with $\overleftrightarrow{BC}$ and $\overleftrightarrow{CD}$, and R and S denote the intersections of $\overleftrightarrow{PQ}$ with $\overleftrightarrow{AB}$ and $\overleftrightarrow{AD}$, respectively, show that $\overline{AR} = \overline{AS}$. Letting O stand for the intersection point of the perpendiculars

through R and S to $\overleftrightarrow{AB}$ and $\overleftrightarrow{AD}$, respectively, show that $\overline{OR} = \overline{OS}$. Finally, if β is the circle of center O and radius $\overline{OR} = \overline{OS}$, show that points B and D lie on the radical axis of α and β, so that $E, F \in \beta$.

22. Firstly, suppose that D, E and F are collinear. Applying the intersecting chords theorem to the circumcircles of ABC and AEF, with respect to the pairs of chords AM and BC, AM and EF, we conclude that $\overline{BD} \cdot \overline{CD} = \overline{ED} \cdot \overline{EF}$. Hence, again from the intersecting chords theorem, $BFCE$ is cyclic. However, since M belongs to the perpendicular bisectors of segments BC and EF, it follows that M is the center of the circle circumscribed to this quadrilateral; in particular, we have $\overline{BM} = \overline{FM}$. The coincidence of the incenters of triangles ABC and AEF now follows from Proposition 3.37. Finally, in order to establish the converse, it suffices to reverse the steps of the argument above.

Section 5.1

1. Write $A(AECF) = A(ABCD) - A(BCE) - A(CDF)$ and show that $A(BCE) = A(CDF) = \frac{1}{4}A(ABCD)$.

2. Since $\overleftrightarrow{AD} \| \overleftrightarrow{BC}$, triangle CDE has height 5 with respect to DE. Also, $\overline{DE} = \overline{AD} - \overline{AE}$.

3. Start by invoking the congruence of triangles ADF and DCE to conclude that $D\widehat{G}F = 90°$. Then, apply the metric relations in right triangles—cf. Theorem 4.10—to compute $\overline{DG}$ and $\overline{GF}$.

4. For item (a), apply Pythagoras' theorem to triangle ABM, where M is the midpoint of BC.

5. Item (a) of the previous problem assures that the altitude of CDE with respect to CD is equal to $1 - \frac{\sqrt{3}}{2}$. Hence, $A(CDE) = \frac{1}{2}\left(1 - \frac{\sqrt{3}}{2}\right)$. Item (b) of the previous problema gives $A(ABE) = \frac{\sqrt{3}}{4}$. Now, show that $AED \equiv BEC$.

6. Firstly, show that $ABE \equiv ADF$. Set $\overline{BE} = x$ to obtain $\overline{CE} = \overline{CF} = 1 - x$; then, apply Pythagoras' theorem to ABE and CEF to get $\sqrt{1 + x^2} = \overline{AE} = \overline{EF} = (1 - x)\sqrt{2}$. Compute x and apply the result of item (b) of Problem 4.

7. Let P be a point in the interior of ABC and x, y and z be the distances of P to the sides BC, AC and AB. Equality $A(ABC) = A(BCP) + A(ACP) + A(ABP)$ gives $x + y + z = \frac{2S}{a}$, where a is the length of the sides of ABC.

8. Letting S denote the area of ABC, we have $h_a = \frac{2S}{a}$, $h_b = \frac{2S}{b}$ and $h_c = \frac{2S}{c}$. Hence, $a + h_a = b + h_b = c + h_c$ if and only if $a + \frac{2S}{a} = b + \frac{2S}{b} = c + \frac{2S}{c}$. Now, $a + \frac{2S}{a} = b + \frac{2S}{b}$ is equivalent to $(a - b)(2S - ab) = 0$, and hence to $a = b$ or $ab = 2S$. Analogously, $a + \frac{2S}{a} = c + \frac{2S}{c}$ if and only if $a = c$ or $ac = 2S$. If $c \neq a, b$, then $ac = 2S = bc$, so that $a = b$. Thus, $\frac{ac}{2} = S = \frac{ah_a}{2}$, and hence $c = h_a$, which assures that ABC is right at B; therefore, $b = \overline{AC} > \overline{BC} = a$, which is a contradiction.

9. We can assume, without loss of generality, that ABC is right at A. Since BB' and CC' intersect at their respective midpoints, $BCB'C'$ is a parallelogram. Since $\overleftrightarrow{AA'} \perp \overleftrightarrow{BC}$ and $\overleftrightarrow{B'C'} \parallel \overleftrightarrow{BC}$, we have $\overleftrightarrow{AA'} \perp \overleftrightarrow{B'C'}$. Mark points H and H', intersections of $\overleftrightarrow{AA'}$ with BC and $B'C'$, respectively. Since $ABC \equiv A'BC \equiv AB'C'$, we have $\overline{A'H'} = \overline{A'H} + \overline{AH} + \overline{AH'} = 3\overline{AH}$. Hence, $A(A'B'C') = \frac{1}{2}\overline{B'C'} \cdot \overline{A'H'} = \frac{3}{2}\overline{BC} \cdot \overline{AH} = 3A(ABC) = 3\text{m}^2$.

10. If M, N and P stand for the midpoints of sides BC, AC and AB, respectively, use the midsegment theorem to conclude that triangles ANP, BMP, CMN and MNP are congruent.

11. By the hint given to the previous problem, we have $A(DPQ) = \frac{1}{4}A(ACD)$ and $A(BMN) = \frac{1}{4}A(ABC)$; hence, $A(DPQ) + A(BMN) = \frac{1}{4}A(ACD) + \frac{1}{4}A(ABC) = \frac{1}{4}A(ABCD)$; analogously, $A(AMQ) + A(CNP) = \frac{1}{4}A(ABCD)$. Now, compute $A(MNPQ)$ as the difference between the area of $ABCD$ and the sum of the areas of triangles AMQ, BMN, CNP and DPQ.

12. Let $ABCD$ and $OMNP$ be the given squares, and O be the center of $ABCD$. Firstly, note that $\overline{OC} = \frac{1}{2}\overline{AB}\sqrt{2} = \frac{\sqrt{2}}{2} < 1 = \overline{OM}$. Now, there are two essentially distinct cases to consider: (i) OM intersects BC at a point $Q \neq B, C$: then OP intersects CD at a point $R \neq C, D$, and the portion of the plane common to the squares is the convex quadrilateral $OQCR$. It is immediate to check that $D\widehat{O}R = C\widehat{O}Q$, and hence $OQC \equiv ORD$ by ASA. Therefore,

$$A(OQCR) = A(OQC) + A(OCR) = A(ODR) + A(OCR)$$

$$= A(OCD) = \frac{1}{4}A(ABCD) = \frac{1}{4}.$$

(ii) OM passes through C: then OP passes through D, and the portion of the plane common to both squares is triangle OCD.

13. Let K, L, M and N denote the intersection points of $\overleftrightarrow{BJ}$ and DE, $\overleftrightarrow{CI}$ and FG, $\overleftrightarrow{AH}$ and BC, $\overleftrightarrow{AH}$ and IJ, respectively. It is immediate that $BDK \equiv AEH$ and $CFL \equiv AGH$, so that $A(ABDE) = A(ABKH)$ and $A(ACFG) = A(ACLH)$. Now, notice that parallelograms $ABKH$ and $MBJN$ have equal bases AH and MN and altitudes relative to such bases also equal, whence $A(ABKH) = A(MBJN)$; analogously, $A(ACLG) = A(MCIN)$. Therefore, $A(ABDE) + A(ACFG) = A(ABKH) + A(ACLH) = A(MBJN) + A(MCIN) = A(BCIJ)$.

14. Let $ABCD$ be a convex quadrilateral of diagonals AC and BD concurrent at O. Let also P and Q denote the feet of the perpendiculars dropped from vertices B and D, respectively, to the diagonal AC. Condition $A(ABC) = A(ACD)$ furnishes the equality $\overline{BP} = \overline{DQ}$. However, since $BOP \sim DOQ$ by AA, we have $\frac{BO}{DO} = \frac{BP}{DQ} = 1$, so that O is the midpoint of BD. Analogously, condition $A(ABD) = A(CBD)$ assures that O is the midpoint of AC. Hence,

the diagonals of $ABCD$ intersect at their midpoints, and it follows that $ABCD$ is a parallelogram.

15. For item (b), let $A(BFP) = x$ and $A(AEP) = y$. It follows from (a) that $\frac{A(ABD)}{A(ACD)} = \frac{\overline{BD}}{\overline{CD}} = \frac{A(BPD)}{A(CPD)}$, and hence $\frac{84+x+40}{y+35+30} = \frac{40}{30}$. Conclude, in a likewise manner, that $\frac{84+y+35}{x+40+30} = \frac{84}{x}$. Then, solve the linear system of equations thus obtained.

16. For the first part, item (a) of the previous problem gives

$$\frac{\overline{BA'}}{\overline{A'C}} = \frac{A(ABA')}{A(ACA')} = \frac{A(BPA')}{A(CPA')}.$$

Now, use the fact that $\frac{a}{b} = \frac{c}{d} \Rightarrow \frac{a}{b} = \frac{c}{d} = \frac{a-c}{b-d}$.

17. Let $\mathcal{P} = A_1A_2\ldots, A_n$, $\Gamma(O; R)$ be the inscribed circle and d the distance from O to r. Moreover, suppose that r intersects A_1A_2 at X and A_iA_{i+1} at Y, and let

$$\overline{XA_2} + \overline{A_2A_3} + \cdots + \overline{A_iY} = \overline{YA_{i+1}} + \overline{A_{i+1}A_{i+2}} + \cdots + \overline{A_nA_1} + \overline{A_1X} = a.$$

Finally, suppose that $O \in YA_{i+1}A_{i+2}\ldots A_nA_1X$. Then,

$$A(XA_2\ldots A_iY) = A(XA_2\ldots A_iYO) - A(XOY)$$

$$= A(XOA_2) + A(A_2OA_3) + \cdots + A(A_iOY) - A(XOY)$$

$$= \frac{R}{2}(\overline{XA_2} + \overline{A_2A_3} + \cdots + \overline{A_iY}) - \frac{d}{2}\,\overline{XY}$$

$$= \frac{1}{2}(aR - d\,\overline{XY})$$

and, analogously,

$$A(YA_{i+1}A_{i+2}\ldots A_nA_1X) = A(YA_{i+1}A_{i+2}\ldots A_nA_1XO) + A(XOY)$$

$$= \frac{1}{2}(aR + d\,\overline{XY}).$$

Hence, $A(XA_2\ldots A_iY) = A(YA_{i+1}A_{i+2}\ldots A_nA_1X)$ if and only if $aR - d\,\overline{XY} = aR + d\,\overline{XY}$, i.e., if and only if $d = 0$. But this is the same as saying that $O \in \overleftrightarrow{XY} = r$.

18. Let $ABCD$ be the quadrilateral, of diagonals AC and BD, let $\overline{AB} = a$, $\overline{CD} = b$, $\overline{AC} = d$ and $a+b+d = 16$. If h_1 and h_2 stand for the altitudes of ABC and ACD relative to AC, then $d(h_1 + h_2) = 64$. On the other hand, the inequality between the arithmetic and geometric means gives

$$d(h_1 + h_2) \le d(a + b) \le \frac{1}{4}(d + a + b)^2.$$

19. To prove the inequality of item (a) is the same as to show that $(\overline{AP} + x)a \geq ax + by + cz = 2A(ABC) = ah_a$, where h_a the height of ABC with respect to BC. For the first inequality of item (b), apply the inequality of item (a) to Q, the symmetric of P with respect to the internal bisector of $\angle BAC$; for the other two inequalities, argue in the same way. For (c), add the inequalities of (b) termwise and apply the inequality between the arithmetic and geometric means to show that $\frac{a}{b} + \frac{b}{a}, \frac{a}{c} + \frac{c}{a}, \frac{b}{c} + \frac{c}{b} \geq 2$. Finally, for (d), note that equality in (c) implies that ABC is equilateral; then, equality in (b) implies that P coincides with its symmetrics with respect to the internal bisectors of ABC.

Section 5.2

1. Let l and a be the lengths of the sides of the equilateral triangle and the square. The result of Problem 4, page 156, assures that we must have $\frac{l^2\sqrt{3}}{4} = a^2$ or, which is the same, $l^2 = \frac{4a^2}{\sqrt{3}} = \frac{2a}{\sqrt{3}} \cdot 2a$. Hence, l is the geometric mean between line segments of lengths $\frac{2a}{\sqrt{3}}$ and $2a$. You might find it useful to take a look at Problem 19, page 117.

2. No! To see why, note first that the rules of movement maintain the area of the triangle whose vertices are the positions of the ants. Then, observe that the area of the triangle formed by the initial positions of the ants is twice as that of a triangle having the midpoints of three sides of the original triangle as vertices.

3. Suppose, with no loss of generality, that AC and BD are the diagonals of $ABCD$. There are three essentially distinct cases to consider: (i) $E \in AB$ and $F, G \in BC$: then $A(EFG) \leq A(ABC) = \frac{1}{2}A(ABCD)$; (ii) $E \in AD$ and $F, G \in BC$: then $A(EFG) \leq A(EBC) = A(ABC) = \frac{1}{2}A(ABCD)$; (iii) $E \in AB$, $F \in BC$ and $G \in CD$: letting H and L be the intersections of EG and AD with the parallel to $\overleftrightarrow{AB}$ drawn through F, we get, by case (ii),

$$A(EFG) = A(EFH) + A(HFG) \leq \frac{1}{2}A(ABFL) + \frac{1}{2}A(LFCD) = \frac{1}{2}A(ABCD).$$

4. For item (a), let N and P be the midpoints of AC and AB, respectively. Use the midsegment theorem to conclude that the altitude of ANM relative to MN is equal to half the altitude of ABC relative to BC; then, show that $A(AMN) = \frac{1}{4}A(ABC)$. For item (b), let G be the barycenter of ABC and M be the midpoint of BC. Mark, on $\overrightarrow{AM}$, the point D such that M is the midpoint DG. Then, successively conclude that $BDCG$ is a parallelogram and that CDG is a triangle whose sides are equal to $\frac{2}{3}$ of the lengths of the medians of ABC. Finally, show that $A(CDG) = A(BCG) = \frac{1}{3}A(ABC)$ and invoke the result of Proposition 5.11 to obtain $A(CDG) = \frac{4}{9}A(DEF)$.

5. Note that $A(ABC) = A(ABF)$, $A(CDE) = A(BCD)$ and $A(AEF) = A(DEF)$.

6. Letting E denote the foot of the perpendicular dropped from C to $\overleftrightarrow{AB}$, we have $\overline{CE} = 5$cm. Apply Pythagoras' theorem to triangle BCE to obtain $\overline{CE} = 12$cm, and hence $A(ABCD) = 186$cm^2. For what is left to do, denoting by x the distance from vertex A to line $\overleftrightarrow{BC}$, we have $186 = A(ABCD) = A(ACD) + A(ABC) = 78 + \frac{13x}{2}$, so that $x = 12$cm.

7. First note that if $ABCD$ is such a trapezoid, with bases AB and CD and legs AD and BC, we must have $\overline{AB} = 2$ and $\overline{CD} = 1$ (or vice-versa), and $\overline{AD} = \overline{BC} = 1$. Now, letting E be the foot of the perpendicular dropped from D to $\overleftrightarrow{AB}$, we have $\overline{AE} = \frac{1}{2}$ and, from Pythagoras' theorem, $\overline{DE} = \frac{\sqrt{3}}{2}$. Hence, $A(ABCD) = \frac{1}{2}(2+1)\frac{\sqrt{3}}{2} = \frac{3\sqrt{3}}{4}$. Now, Problem 4, page 156 guarantees that the area of an equilateral triangle of side length n is equal to $\frac{n^2\sqrt{3}}{4}$. Thus, for the partition to be feasible, the quotient of the areas of the equilateral triangle and the trapezoid, $\frac{n^2}{3}$, must be an integer; in turn, this implies that 3 must divide n. In order to see that such a condition is also sufficient, let $n = 3k$, with $k \in \mathbb{N}$. Start by partitioning the equilateral triangle of side length n into equilateral triangles of side length 3; then, take one of these small equilateral triangles, say ABC, and mark its center O, together with points $A' \in BC$, $B' \in AC$ and $C' \in AB$ for which $\overline{A'C} = \overline{B'A} = \overline{C'B} = 2$; trapezoids $A'CB'O$, $B'AC'O$ and $C'BA'O$ have side lengths 1, 1, 1, 2 and partition ABC.

8. Letting F be the intersection point of $\overleftrightarrow{AD}$ and $\overleftrightarrow{BE}$, we have $BCE \equiv FDE$ by ASA, so that E is the midpoint of BF. Hence, ABF has twice the area of ABE, and it follows that

$$A(ABCD) = A(ABED) + A(BCE) = A(ABED) + A(DEF)$$

$$= A(ABF) = 2A(ABE) = 720\text{cm}^2.$$

Alternatively, let $\overline{AD} = a$, $\overline{BC} = b$ and h be the altitude of $ABCD$. Denoting by G the midpoint of AB, triangles BEG and AEG have height $\frac{h}{2}$ with respect to EG, so that

$$A(ABE) = A(AEG) + A(BEG) = \frac{1}{2}\overline{EG} \cdot \frac{h}{2} + \frac{1}{2}\overline{EG} \cdot \frac{h}{2}$$

$$= \overline{EG} \cdot \frac{h}{2} = \frac{(a+b)h}{4} = \frac{1}{2}A(ABCD).$$

9. Let a and b denote the lengths of the bases AB and CD, respectively, and let h be the altitude of $ABCD$. Also, let h_1 and h_2 denote the altitudes of triangles ABE and CDE with respect to the sides AB and CD, respectively. Then, $h_1 + h_2 = h$ and, since $ABE \sim CDE$, we have $\frac{h_1}{h_2} = \frac{a}{b}$. Therefore, $h_1 = \frac{ah}{a+b}$ and $h_2 = \frac{bh}{a+b}$, whence

$$\sqrt{A(ABE)} + \sqrt{A(CDE)} = \sqrt{\frac{a^2h}{2(a+b)}} + \sqrt{\frac{b^2h}{2(a+b)}}$$

$$= \sqrt{\frac{(a+b)h}{2}} = \sqrt{A(ABCD)}.$$

10. Let $\overleftrightarrow{DE}$, $\overleftrightarrow{FG}$ and $\overleftrightarrow{HI}$ be the drawn parallels, with $D, F \in AB$, $G, H \in BC$ and $E, I \in AC$. Since $BHPD$ and $CEPG$ are parallelograms, we have $\overline{DP} = \overline{BH}$ and $\overline{EP} = \overline{CG}$. Hence, Proposition 5.11 gives

$$1 = \frac{\overline{BH}}{\overline{BC}} + \frac{\overline{FG}}{\overline{BC}} + \frac{\overline{CG}}{\overline{BC}}$$

$$= \frac{\overline{DP}}{\overline{BC}} + \frac{\overline{FG}}{\overline{BC}} + \frac{\overline{EP}}{\overline{BC}}$$

$$= \sqrt{\frac{A(DFP)}{A(ABC)}} + \sqrt{\frac{A(GHP)}{A(ABC)}} + \sqrt{\frac{A(EIP)}{A(ABC)}}$$

$$= \frac{1}{\sqrt{A(ABC)}} + \frac{2}{\sqrt{A(ABC)}} + \frac{3}{\sqrt{A(ABC)}},$$

so that $A(ABC) = 36\,\text{cm}^2$.

11. Proposition 5.13 gives $\frac{1}{r_a} = \frac{p-a}{pr}$. Write the analogous relations involving r_b and r_c and, then, add the three resulting equalities.

12. Let ABC be an acute triangle and H_a, H_b and H_c be the feet of the altitudes dropped from A, B and C, respectively. Show that $AH_bH_c \sim ABC$ and, hence, that the median of AH_bH_c relative to H_bH_c is the symmedian of ABC relative to BC.

13. Partially adapt the hint given to the previous problem, noticing that, in every right triangle, the altitude relative to the hypothenuse is the symmedian relative to it.

14. Let $\overline{AB} = c$, $\overline{AC} = b$ and x, y be the distances from P to $\overleftrightarrow{AB}$ and $\overleftrightarrow{AC}$, respectively. Proposition 5.16 guarantees that AP is symmedian if and only if $\frac{x}{y} = \frac{c}{b}$. Now, observe that

$$\frac{\overline{BP}}{\overline{CP}} = \frac{A(ABP)}{A(ACP)} = \frac{cx}{by},$$

and this relation, together with the previous one, furnishes the desired result.

15. Use the fact that the barycenter divides each median in the ratio $2 : 1$ to show that the distance from G to BC is equal to $1/3$ of the altitude of ABC relative

to A. Now, use the equality $\frac{ah}{2} = pr$ to conclude that such a distance coincides with that from I to BC.

16. Adapt, to the present situation, the proof of Carnot's theorem for acute triangles, presented in the text.

17. Apply Carnot's theorem and its generalization, established in the previous problem, to each one of the three triangles that partition $ABCDE$.

18. Letting $\overleftrightarrow{AD} \cap \overleftrightarrow{BC} = \{T\}$, apply Desargues' theorem 4.24 to the pairs of triangles DMN and KLB, CLM and AKN, to show that $\overleftrightarrow{QT} \parallel \overleftrightarrow{AB}$ and $\overleftrightarrow{PT} \parallel \overleftrightarrow{AB}$, and hence that P, Q and T are collinear points. Then, use similarity of triangles to show that, if h is the altitude of the trapezoid and h' the distance from O to $\overleftrightarrow{PQ}$, we have $h' = \frac{2hk}{k^2-1}$. Also by means of a similarity argument, conclude that

$$\overline{QT} = \frac{\overline{AB}(k \cdot \overline{DO} + \overline{AO})}{(k-1)(\overline{AO} + \overline{BO})} \quad \text{and} \quad \overline{PT} = \frac{\overline{AB}(k \cdot \overline{CO} + \overline{BO})}{(k-1)(\overline{AO} + \overline{BO})},$$

so that $\overline{PQ} = \frac{2 \cdot \overline{AB}}{k-1}$.

Section 5.3

1. Concerning $A_1 A_2 \ldots A_n$, let O be its center, l be the length of its sides and a be the common distance from O to its sides. Let also C_i be the foot of the perpendicular dropped from P to $\overleftrightarrow{A_i A_{i+1}}$ (with $A_{n+1} = A_1$). Equalities

$$\sum_{i=1}^{n} A(O A_i A_{i+1}) = A(A_1 A_2 \ldots A_n) = \sum_{i=1}^{n} A(P A_i A_{i+1})$$

give $n \cdot \frac{la}{2} = \sum_{i=1}^{n} \frac{l \cdot \overline{PC_i}}{2}$, so that $\overline{PC_1} + \overline{PC_2} + \cdots + \overline{PC_n} = na$.

2. Let $\mathcal{P} = A_1 A_2 \ldots A_n$, $\mathcal{Q} = A'_1 A'_2 \ldots A'_n$ and O and O' be the centers of $\mathcal{P}$ and $\mathcal{Q}$, respectively. Apply the result of Proposition 5.11 to each one of the pairs of triangles $O A_i A_{i+1}$ and $O' A'_i A'_{i+1}$, with the convention that $A_{n+1} = A_1$ e $A'_{n+1} = A'_1$.

3. Letting l be the side length of the desired 20-gon, apply the result of the previous problem to conclude that $\frac{5^2}{l^2} + \frac{12^2}{l^2} = 1$.

4. Letting $\mathcal{P}_R$ and $\mathcal{Q}_R$ be regular n-gons, the first of which inscribed and the second circumscribed to a circle Γ of radius R, we have $A(\mathcal{P}_R) \le A(\Gamma) \le A(\mathcal{Q}_R)$ and $A(\mathcal{P}_1) \le \pi \le A(\mathcal{Q}_1)$. Now, apply the result of Problem 2 to conclude that $A(\mathcal{P}_R) = A(\mathcal{P}_1)R^2$ and $A(\mathcal{Q}_R) = A(\mathcal{Q}_1)R^2$.

5. In the notations of Fig. 5.15, we have

$$l_{2n}^2 = \overline{A_i D_i}^2 = \overline{A_i C_i}^2 + \overline{C_i D_i}^2 = \left(\frac{l_n}{2}\right)^2 + (R - a_n)^2$$

$$= \frac{l_n^2}{4} + \left(R - \sqrt{R^2 - \frac{l_n^2}{4}}\right)^2 = 2R^2 - R\sqrt{4R^2 - l_n^2}.$$

The rest follows from this, together with the fact that $l_4 = R\sqrt{2}$.

7. Follow steps analogous to those that led to (5.13).

8. Denote by R, R_1 and R_2 the radii of Γ, Γ_1 and Γ_2, respectively. It follows from $\overline{AC} = \overline{AB} + \overline{BC}$ that $R = R_1 + R_2$, and hence $S = \frac{\pi}{2}(R^2 - R_1^2 - R_2^2) = \pi R_1 R_2$. On the other hand, since $A\widehat{D}C = 90°$, the usual metric relations in right triangles furnish $\overline{BD}^2 = \overline{AB} \cdot \overline{BC} = 4R_1 R_2$, so that $\frac{\overline{BD}^2}{S} = \frac{4}{\pi}$.

9. Let O be the midpoint of hypothenuse BC of ABC or, which is the same, the center of Γ. Let also $A\widehat{O}B = \alpha$, so that $A\widehat{O}C = 180° - \alpha$, and $\overline{AB} = 2R_1$, $\overline{AC} = 2R_2$, $\overline{BC} = 2R$. The area of the portion of Γ_1 exterior to Γ is equal to $\frac{1}{2}\pi R_1^2 - \left(\frac{\alpha}{360°} \cdot \pi R^2 - A(AOB)\right)$, whereas the area of the portion of Γ_2 exterior to Γ is equal to $\frac{1}{2}\pi R_2^2 - \left(\frac{180° - \alpha}{360°} \cdot \pi R^2 - A(AOC)\right)$. Hence, the sum of the areas of these portions of Γ_1 and Γ_2 equals

$$\frac{1}{2} \cdot \pi(R_1^2 + R_2^2) - \frac{1}{2} \cdot \pi R^2 + A(AOB) + A(AOC) = A(ABC),$$

for Pythagoras' theorem gives $R_1^2 + R_2^2 = R^2$.

10. If ABC is right or obtuse, there is nothing to do, for in these cases ABC lies within a semicircle of Γ. If ABC is acute, let H be its orthocenter and A', B' and C' be the other points of intersection of the half-lines semirretas $\overrightarrow{AH}$, $\overrightarrow{BH}$ and $\overrightarrow{CH}$ with Γ; the result of Problem 6, page 92, guarantees that

$$A(\Gamma) > A(AC'BA'CB'A)$$

$$= (A(ABH) + A(ABC')) + (A(ACH) + A(ACB'))$$

$$+ (A(BCH) + A(BCA'))$$

$$= 2A(ABH) + 2A(ACH) + 2A(BCH) = 2A(ABC).)$$

11. If $\alpha = 360° \cdot \frac{m}{n}$, with $m, n \in \mathbb{N}$ and $m < n$, define the length of $\widehat{AB}$ as being equal to the supremum of the lengths of the polygonals $A_1 A_2 \ldots A_{km} A_{km+1}$, with $k \in \mathbb{N}$ and $A_1 A_2 \ldots A_{kn}$ being a regular kn-gon inscribed in Γ and such that $A_1 = A$, $A_{km+1} = B$. Show that

$$\frac{\ell(A_1 A_2 \ldots A_{km} A_{km+1})}{\ell(A_1 A_2 \ldots A_{kn-1} A_{kn})} = \frac{m}{n}$$

and conclude from here that $\ell(\overset{\frown}{AB}) = \frac{m}{n} \cdot 2\pi R$. If $\alpha = 360° \cdot x$, with $x \in (0, 1)$ being irrational, take a sequence $0 < r_1 < r_2 < \cdots < x$ of rationals such that $\sup\{r_1, r_2, \ldots\} = x$. Then, define the length of $\overset{\frown}{AB}$ as the supremum of the lengths of the arcs $\overset{\frown}{AB_n}$, with $B_n \in \overset{\frown}{AB}$ and $A\hat{O}B_n = \alpha_n$ such that $\alpha_n = 360° \cdot r_n$. Finally, conclude that $\ell(\overset{\frown}{AB}) = x \cdot 2\pi R$.

12. First of all, note that the error will be equal to $2\sqrt{2} + 2\sqrt{3} - 2\pi \cong 0.00933$. Then, construct line segments with lengths equal to the sides of an equilateral triangle and of a square inscribed in the circle.

13. Let $\mathcal{P}_n$ and $\mathcal{Q}_n$ be as in the text, so that $\ell(\Gamma) = \sup \ell(\mathcal{P}_n) = \inf \ell(\mathcal{Q}_n)$. Show that it suffices to prove that $\ell(\mathcal{P}_n) < \ell(\mathcal{Q})$ and $\ell(\mathcal{P}) < \ell(\mathcal{Q}_n)$ for every $n \in \mathbb{N}$. Finally, in order to establish these inequalities, invoke the result of Problem 9, page 44.

14. Firstly, note that one may suppose that not all of the given segments are parallel. Now, let $s_1, s_2, \ldots, s_n$ denote the n given segments, fix a point O in the plane and construct a convex polygon $\mathcal{Q} = A_1A_2 \ldots A_{2n-1}A_{2n}$ in the following way (ℓ stands for length): $\overleftrightarrow{A_1A_2} \parallel s_1$ and $\overline{A_1A_2} = \ell(s_1)$, $\overleftrightarrow{A_2A_3} \parallel s_2$ and $\overline{A_2A_3} = \ell(s_2)$, $\ldots$, $\overleftrightarrow{A_{n-1}A_n} \parallel s_n$ and $\overline{A_{n-1}A_n} = \ell(s_n)$, $\mathcal{Q}$ is symmetric with respect to O. Let R be the minimum distance from O to one of the sides of $\mathcal{Q}$ and Γ be the circle of center O and radius R. Apply the result of the previous problem to Γ and $\mathcal{Q}$ to get $R < \frac{1}{\pi}$, then choose r.

Section 6.1

1. Letting $O(0, 0)$, $A(a, b)$, $A'(-a, -b)$, $B(a, 0)$ and $B'(-a, 0)$, show that triangles AOB and $A'OB'$ are congruent.

3. If $A(a, b)$ and $B(c, d)$ are points for which $\overline{AP} = \overline{BP}$, use formula (6.4) to show that $(a - c)\left(a + c - \frac{2}{3}\right) + b^2 - d^2 = 2(b - d)\sqrt{2}$. Now, assuming that $a, b, c, d \in \mathbb{Z}$, use the irrationality of $\sqrt{2}$ to conclude that $b = d$, and hence $a = c$. For what is left to do, first observe that a disk of center P and sufficiently small radius will not contain any point of integer coordinates; then, use the first part of the problem to conclude that, as we continuously increase the radius of such a disk, points of integer coordinates are "swallowed" one at a time.

4. According to the previous problem, there exists a circle $\Gamma(O; R)$ passing through exactly one point of integer coordinates and such that the corresponding disk D contains exactly 2018 points of integer coordinates. Let P be the point of integer coordinates in D for which $r = \overline{OP}$ is as large as possible. Show that, for infinitely many values of $n \in \mathbb{N}$, there exists a regular n-gon centered at O and contained in the region of the plane bounded by the circles Γ and $\Sigma(O; r)$.

5. Suppose, without loss of generality, that $A(0, y)$, $B(x_1, 0)$ and $C(x_2, 0)$, so that formula (6.4) gives $\overline{AB} = \sqrt{x_1^2 + y^2}$, $\overline{AC} = \sqrt{x_2^2 + y^2}$ and $\overline{BC} = |x_1 - x_2|$. Then, write the coordinates of the midpoint M_a of BC and apply (6.4) once more to compute $m_a = \overline{AM_a}$.

6. Apply the formula of the previous problem to suitable medians of each one of the triangles BDM, ABC and ACD. Then, adequately combine the relations thus obtained to get the stated formula.

7. Choose a cartesian system in which $O(0, 0)$, $A(x_1, y_1)$, $B(x_2, y_2)$ and $C(x_3, y_3)$, so that $x_i^2 + y_i^2 = R^2$ for $1 \leq i \leq 3$. Then, use the result of Example 6.4 to write the coordinates of G in terms of the x_i's and y_i's, and compute both sides of the desired equality with the aid of the formula for the distance between two points.

8. Choose a cartesian system of origin O, in which $A\left(\frac{2}{\sqrt{3}}, 0\right)$, $B\left(-\frac{1}{\sqrt{3}}, 1\right)$ and $C\left(\frac{-1}{\sqrt{3}}, -1\right)$. If $P(x, y)$, show that $x^2 + y^2 = \frac{1}{3}$ and use (6.4) to compute $\overline{AP}^2 + \overline{BP}^2 + \overline{CP}^2$. A proof that AP, BP, CP are the sides of a triangle was presented in Problem 1, page 268. For the computation of its area S, write $a = \overline{AP}$, $b = \overline{BP}$, $c = \overline{CP}$ and use Heron's formula to obtain

$$16S^2 = 4a^2b^2 - (a^2 + b^2 - c^2)^2.$$

Then, substitute the expressions for a^2, b^2 and c^2, obtained with the aid of (6.4).

9. For fixed $1 \leq j < k \leq m$, we have $\overline{A_i B_j}^2 - \overline{A_i B_k}^2 = j - k$ for $1 \leq i \leq n$. Hence, Proposition 6.8 assures that all of A_1, A_2, ..., A_n belong to a line r, which is perpendicular to $\overleftrightarrow{B_j B_k}$. However, since the indices j and k in the above argument were arbitrarily chosen, it follows that B_1, B_2, ..., B_m belong to a line s, which is perpendicular to r.

10. Let $\Gamma_i(O_i; R_i)$ for $i = 1, 2$, and let O be the center of a circle of radius R, intersecting both Γ_1 and Γ_2 along diameters. Show that $\overline{OO_i}^2 = R^2 - R_i^2$ and, hence, that $\overline{OO_1}^2 - \overline{OO_2}^2 = R_2^2 - R_1^2$. Subsequently, use Proposition 6.8 to conclude that the desired locus is a line perpendicular to $\overleftrightarrow{O_1 O_2}$, and rework the proof of Theorem 4.35 to conclude that such a line is symmetric to the radical axis of Γ_1 and Γ_2 with respect to the midpoint of $O_1 O_2$.

11. Draw a rectangle $ABCD$, of sides parallel to the coordinate axes and having points A_1, A_2 and A_3 on its sides. Then, compute the area of $A_1 A_2 A_3$ as the difference between the area of $ABCD$ and the sum of the areas of at most three right triangles, all of them with legs parallel to the axes.

12. Show that points $A_k(k, k^2)$, for $1 \leq k \leq n$, satisfy the stated conditions.

13. Argue by contradiction, applying the result of Problem 11.

15. For item (a), once we have partitioned $\mathcal{P}$ into $n - 2$ triangles with diagonals intersecting only at vertices, the same proof presented to the convex case goes through also in this more general case. For item (b), by overlapping two distinct partitions of $\mathcal{P}$ into triangles, we split $\mathcal{P}$ into a finite number of convex polygons

with disjoint interiors. Show that adding the areas of these polygons in two distinct ways give us the sums of the areas of the $n-2$ triangles obtained in the two original partitions of $\mathcal{P}$.

16. For item (a), argue by induction in the number of lattice points contained in the interior or along the sides of the triangle. For item i., use the result of Problem 14 and (a). For ii., start by observing that the sum of the interior angles of the fundamental triangles has three different types of summands: those relative to the vertices of $\mathcal{P}$; those relative to the $B-n$ points of integer coordinates lying on the sides of $\mathcal{P}$ but which are not vertices of $\mathcal{P}$; those relative to the I points of integer coordinates situated in the interior of $\mathcal{P}$. Then, compute the total contribution of each of these types of summands to obtain $180°(n-2)$, $180°(B-n)$ and $360°I$, respectively. For iii., note that the sum of the angles of the k fundamental triangles is $180°k$.

17. For (b), start by noticing that $\mathcal{R}$ can be partitioned into at most 4 simple polygons, one of which is ABC. Then, apply item i. of the previous problem to each such polygon but ABC. Now, note that since the vertices of $\mathcal{R}$ are lattice points, S must be a natural number and $\mathcal{R}$ can be partitioned into S squares 1×1, hence in $2S$ fundamental (right isosceles) triangles. Finally, apply item iii. of the previous problem. For (c), let $\mathcal{R}$ be partitioned into the $2S$ fundamental triangles $\Delta_1, \Delta_2, \ldots, \Delta_{2S}$, with $\Delta_1 = ABC$. Use the result of Problem 11 to write

$$S = A(\mathcal{R}) = \sum_{i=1}^{2S} A(\Delta_i) \geq 2S \cdot \frac{1}{2} = S.$$

18. Item iii. of Problem 16 guarantees that $\mathcal{P}$ can be partitioned into $2I + B - 2$ fundamental triangles. Then, apply item (c) of the previous problem.

19. Let $\mathcal{Q}$ denote the square and $\mathcal{P}$ denote the smallest convex polygon containing all lattice points covered by $\mathcal{Q}$. Use Pick's theorem, together with the fact that $\mathcal{P} \subset \mathcal{Q}$, to get $\frac{B}{2} + I \leq n^2 + 1$, where B and I stand for the number of lattice points lying along the perimeter and in the interior of $\mathcal{P}$, respectively. Now, note that (according to Problem 9, page 44) the perimeter of $\mathcal{P}$ is less than or equal to $4n$, and in turn that this gives $B \leq 4n$. Finally, use the two inequalities above to get $B + I \leq (n+1)^2$.

Section 6.2

1. It is easy to see that the bisector of odd quadrants has equation $y = x$. On the other hand, the line passing through A and B has equation $x + y = a + b$, and hence is perpendicular to that bisector. Moreover, it intersects the bisector at point $C \left(\frac{a+b}{2}, \frac{a+b}{2} \right)$, and it is immediate to check that $\overline{AC} = \overline{BC}$.

4. One possibility is to use the synthetic method to obtain the coordinates of B from the congruence of triangles OAA_x and OBB_y, where A_x and B_y denote the feet of the perpendiculars dropped from A and B to the horizontal and vertical axes, respectively. Alternatively, first note that $\overleftrightarrow{OA}$ has equation $y = \frac{b}{a}x$, so that $\overleftrightarrow{OB}$ has equation $y = -\frac{a}{b}x$; then, by imposing that $\overline{OB} = \overline{OA}$, obtain $x = \pm b$ and, hence, $y = \mp a$. However, since B belongs to the second quadrant, we must have $x = -b$ and $y = a$.

5. Apply the results of Propositions 6.13 and 6.14, as well as Remark 6.10.

6. Since $x_0, y_0 \neq 0$, such a line does not pass through the origin of the cartesian system. Hence, Remark 6.10 (with $k = -\frac{1}{c}$) guarantees that we can write the desired equation in the form $ax + by - 1 = 0$. Now, by imposing that $(x_0, 0)$ and $(0, y_0)$ belong to it, obtain $a = \frac{1}{x_0}$ and $b = \frac{1}{y_0}$.

7. Separately consider the cases $x \geq a$ and $y \geq b$, $x \geq a$ and $y < b$, $x < a$ and $y \geq b$, $x < a$ and $y < b$ to conclude that the region under analysis is a square centered at (a, b), with diagonals parallel to the axes and side lengths equal to $c\sqrt{2}$.

8. Adapt, to the present case, the proof of Proposition 6.1.

9. For item (b)—the proof of item (a) is pretty much analogous—assume first that P, Q and R are collinear; then, by the previous problem, there exists $u \in \mathbb{R}$ for which $R = (1 - u)P + uQ$. Substitute expressions (6.9) for P, Q and R and analyse the equality thus obtained. Conversely, suppose that the equality of item (b) is satisfied, and let $R' \in \overleftrightarrow{AB}$ be the point for which P, Q and R' are collinear, with $R' = (1-u')A+u'B$. It follows from (b) and the first part above that

$$\frac{u'}{1-u'} = -\frac{(1-s)(1-t)}{st} = \frac{u}{1-u}$$

whence, $u = u'$, and then $R = R'$.

10. Write the condition $\frac{BP}{PC} = -\frac{BQ}{QC}$ in terms of the parametric descriptions of P and Q.

11. If $P(x, y)$, show that the problem is equivalent to minimizing $f(x) + g(y)$, for certain second degree functions f and g. Then, apply the theory of maxima and minima of second degree functions (cf. Section 6.2 of [5], for instance).

12. When c attains its largest possible value, line r will be tangent to the circle Γ, so that the solution of the system of equations

$$\begin{cases} ax + by = c \\ x^2 + y^2 = 1 \end{cases}$$

will consist of a single ordered pair. On the other hand, by substituting $x = \frac{c-by}{a}$ in the second equation, we obtain the second degree equation $(a^2 + b^2)y^2 - 2bcy + (c^2 - a^2) = 0$, and the give system of equations will have a solution if and only if $4b^2c^2 - 4(a^2 + b^2)(c^2 - a^2) \geq 0$. In turn, such

4. One possibility is to use the synthetic method to obtain the coordinates of B from the congruence of triangles OAA_x and OBB_y, where A_x and B_y denote the feet of the perpendiculars dropped from A and B to the horizontal and vertical axes, respectively. Alternatively, first note that $\overleftrightarrow{OA}$ has equation $y = \frac{b}{a}x$, so that $\overleftrightarrow{OB}$ has equation $y = -\frac{a}{b}x$; then, by imposing that $\overline{OB} = \overline{OA}$, obtain $x = \pm b$ and, hence, $y = \mp a$. However, since B belongs to the second quadrant, we must have $x = -b$ and $y = a$.

5. Apply the results of Propositions 6.13 and 6.14, as well as Remark 6.10.

6. Since $x_0, y_0 \neq 0$, such a line does not pass through the origin of the cartesian system. Hence, Remark 6.10 (with $k = -\frac{1}{c}$) guarantees that we can write the desired equation in the form $ax + by - 1 = 0$. Now, by imposing that $(x_0, 0)$ and $(0, y_0)$ belong to it, obtain $a = \frac{1}{x_0}$ and $b = \frac{1}{y_0}$.

7. Separately consider the cases $x \geq a$ and $y \geq b$, $x \geq a$ and $y < b$, $x < a$ and $y \geq b$, $x < a$ and $y < b$ to conclude that the region under analysis is a square centered at (a, b), with diagonals parallel to the axes and side lengths equal to $c\sqrt{2}$.

8. Adapt, to the present case, the proof of Proposition 6.1.

9. For item (b)—the proof of item (a) is pretty much analogous—assume first that P, Q and R are collinear; then, by the previous problem, there exists $u \in \mathbb{R}$ for which $R = (1 - u)P + uQ$. Substitute expressions (6.9) for P, Q and R and analyse the equality thus obtained. Conversely, suppose that the equality of item (b) is satisfied, and let $R' \in \overleftrightarrow{AB}$ be the point for which P, Q and R' are collinear, with $R' = (1 - u')A + u'B$. It follows from (b) and the first part above that

$$\frac{u'}{1 - u'} = -\frac{(1 - s)(1 - t)}{st} = \frac{u}{1 - u}$$

whence, $u = u'$, and then $R = R'$.

10. Write the condition $\frac{BP}{PC} = -\frac{BQ}{QC}$ in terms of the parametric descriptions of P and Q.

11. If $P(x, y)$, show that the problem is equivalent to minimizing $f(x) + g(y)$, for certain second degree functions f and g. Then, apply the theory of maxima and minima of second degree functions (cf. Section 6.2 of [5], for instance).

12. When c attains its largest possible value, line r will be tangent to the circle Γ, so that the solution of the system of equations

$$\begin{cases} ax + by = c \\ x^2 + y^2 = 1 \end{cases}$$

will consist of a single ordered pair. On the other hand, by substituting $x = \frac{c-by}{a}$ in the second equation, we obtain the second degree equation $(a^2 + b^2)y^2 - 2bcy + (c^2 - a^2) = 0$, and the give system of equations will have a solution if and only if $4b^2c^2 - 4(a^2 + b^2)(c^2 - a^2) \geq 0$. In turn, such a condition is equivalent to $|c| \leq \sqrt{a^2 + b^2}$, so that the largest possible value of c is $c = \sqrt{a^2 + b^2}$. For what is left to do, notice that if (x_0, y_0) is a common point of r and Γ (for some value of c) and $u = kx_0$, $v = ky_0$, we have

$$|au + bv| = |k||ax_0 + by_0| = |k|c \leq |k|\sqrt{a^2 + b^2}$$
$$= \sqrt{u^2 + v^2}\sqrt{a^2 + b^2},$$

since $x_0^2 + y_0^2 = 1$.

13. Choose a cartesian coordinate system in which $\Gamma(O; R)$ and $A(a, 0)$, with $O(0, 0)$ and $a > R$. Then, show that the equation of line r is $y = m(x - a)$, for some $m \in \mathbb{R}$, and impose that the solution of the system of equations formed by Γ and r is $B(x_0, y_0)$ and $P(x_1, y_1)$, with $x_0 + a = 2x_1$ and $y_0 = 2y_1$.

14. Adapt the proof of Proposition 6.8 to the present case to conclude that the desired locus is the empty set, if $k\sqrt{2} < \overline{AB}$, the midpoint of AB, if $k\sqrt{2} = \overline{AB}$, or a circle centered at the midpoint of AB and radius $\frac{1}{2}\sqrt{2k^2 - \overline{AB}^2}$, if $k\sqrt{2} > \overline{AB}$.

15. In the notations of the statement of Theorem 4.16, choose a cartesian system in which $B(0, 0)$ and $C(a, 0)$. Let $A(x, y)$ and write the condition $\overline{AB} = k \cdot \overline{AC}$ in coordinates to obtain the equation

$$(k^2 - 1)x^2 + (k^2 - 1)y^2 - 2ak^2x + a^2k^2 = 0.$$

Then, divide both sides by $k^2 - 1$ and complete squares.

Section 6.3

2. Adapt, to the present case, the proof of Theorem 6.22.

3. If $\overleftrightarrow{A_1A_2} \cap d_2 = \{P\}$, use metric relations in right triangles to compute OP in terms of a and c.

4. For the first part, let A_1A_2 and B_1B_2 denote the major and minor axes of the hyperbola, respectively, let F_1 and F_2 be its foci and O its center, with $A_i \in OF_i$ and $\overline{F_1F_2} = 2c$. Let C be the intersection point of lines $x = a$, $y = b$. Since $\overline{OA_1} = a$, $\overline{OB_1} = b$, Pythagoras' theorem gives $\overline{OC} = c$. Also, the equation of $\overleftrightarrow{OC}$ is readily seen to be $y = \frac{b}{a}x$, so that $\overleftrightarrow{OC}$ is one of the asymptotes of the hyperbola. For (a), it now suffices to recall that the diagonals of a rectangle are perpendicular if and only if the rectangle is a square. For (b), given O, A_1 and F_1 the following construction works: find C as the intersection of the perpendicular to $\overleftrightarrow{OF}$ through A with the circle of center O and radius c.

Then, find B_1 as the intersection of the perpendicular to $\overleftrightarrow{OF}$ through O with the parallel to $\overleftrightarrow{OF}$ through C.

5. Perform the change of coordinates $(x, y) \mapsto (x' - y', x' + y')$. What is its geometrical effect?

6. Letting $A(a, \frac{1}{a})$, $B(b, \frac{1}{b})$ and $C(c, \frac{1}{c})$, prove that the equations of the altitudes of ABC passing through A and B are $\frac{ay-1}{x-a} = abc$ and $\frac{by-1}{x-b} = abc$, respectively. Then, solve the system of equations thus obtained to get $H(-\frac{1}{abc}, -abc)$, where H stands for the orthocenter of ABC.

7. We can assume, without loss of generality, that $a > 0$. We write (6.14) with γ in place of c and translate the y-axis to the position of the line $x = \alpha$. This way we get a cartesian system in which the new coordinates (x', y') relate to the old ones by means of the formulas $x' = x - \alpha$, $y' = y$. In this new system, the equation of the parabola is obtained from (6.14) changing x by $x' + \alpha$. Since the name we give to the variable is irrelevant, we shall simply write $x + \alpha$ in place of x in (6.14), thus arriving at the equation

$$
y = \frac{1}{2p}(x + \alpha)^2 - \left(\frac{p}{2} + \gamma\right)
$$

$$
= \frac{1}{2p}x^2 + \frac{\alpha}{p}x + \left(\frac{\alpha^2}{2p} - \frac{p}{2} - \gamma\right).
$$

It now suffices to notice that we can adjust the values of α, γ and p so that

$$
\begin{cases}
\frac{1}{2p} = a \\
\frac{\alpha}{p} = b \\
\frac{\alpha^2}{2p} - \frac{p}{2} - \gamma = c
\end{cases}.
$$

8. Letting O denote the origin of the cartesian plane, apply the theorem of intersecting chords to find the point $D(0, y)$ such that $\overline{OA} \cdot \overline{OB} = \overline{OC} \cdot \overline{OD}$.

9. If $P(x, y)$, show that $\overline{PQ_1}$ and $\overline{PQ_2}$ have lengths $x - \frac{ay}{b}$ and $x + \frac{ay}{b}$ in some order. Then, compute

$$
\overline{PQ_1} \cdot \overline{PQ_2} = \left(x - \frac{ay}{b}\right)\left(x + \frac{ay}{b}\right) = a^2\left(\frac{x^2}{a^2} - \frac{y^2}{b^2}\right) = a^2.
$$

Argue in a likewise manner to show that $\overline{PR_1} \cdot \overline{PR_2} = b^2$.

10. Assume, without loss of generality, that $\mathcal{E}$ has equation $\frac{x^2}{a^2} + \frac{y^2}{b^2} = 1$ and that r is not parallel to the axes of $\mathcal{E}$. Then, Proposition 6.13 shows that there exists a fixed real number $\alpha \neq 0$ such that the equation of s has the form $y = \alpha x + \beta$, for some (varying) real number β. Set $P(x_1, y_1)$ and $Q(x_2, y_2)$, so that $M\left(\frac{x_1 + x_2}{2}, \frac{y_1 + y_2}{2}\right)$. Then, note that the ordered pairs (x_1, y_1) and (x_2, y_2) solve the system of equations

$$\begin{cases} y = \alpha x + \beta \\ b^2x^2 + a^2y^2 = a^2b^2 \end{cases}$$

and eliminate y to obtain

$$(a^2\alpha^2 + b^2)x^2 + 2a^2\alpha\beta x + a^2(\beta^2 - b^2) = 0,$$

and hence $\frac{x_1+x_2}{2} = -\frac{a^2\alpha\beta}{a^2\alpha^2+\beta^2}$. Now, compute

$$\frac{y_1 + y_2}{2} = \alpha\left(\frac{x_1 + x_2}{2}\right) + \beta = -\frac{b^2\beta}{a^2\alpha^2 + \beta^2}.$$

Finally, show that M lies in the straightline of equation $b^2x + a^2\alpha y = 0$.

11. Letting R_n denote the radius and $(0, y_n)$ the center of Γ_n, one has $y_2 = y_1 + R_2$ and $y_n = R_1 + 2R_2 + \cdots + 2R_{n-1} + R_n$ for $n > 2$. Show that the condition of tangency between Γ_n and the hyperbola translates into the algebraic fact that the system of equations

$$\begin{cases} x^2 + (y - y_n)^2 = R_n^2 \\ x^2 - y^2 = 1 \end{cases}$$

must have exactly two solutions, hence that $R_n^2 = \frac{y_n^2}{2} + 1$. Then, show that $R_{n+1}^2 - R_n^2 = \frac{1}{2}(y_{n+1}^2 - y_n^2)$ and use the fact that $y_{n+1} = R_n + R_{n+1} + y_n$ to successively get $2(R_{n+1} - R_n) = y_{n+1} + y_n$ and $2y_{n+1} = 3R_{n+1} - R_n$. Finally, deduce that $R_{n+1} - 6R_n + R_{n-1} = 0$, $R_1 = 1$, $R_2 = 3$ and, hence (with the aid of the material of Section 3.2 of [5], for instance) that $R_n = \frac{1}{2}((3 + 2\sqrt{2})^{n-1} + (3 - 2\sqrt{2})^{n-1})$, for every $n \geq 1$.

Section 6.4

1. We have to show that if C is an ellipse or a hyperbola in canonical form and P is a point of C, then there exists a single line r, passing through P and tangent to C. If C is an ellipse and P is one of the endpoints of its major axis, the problem is trivial: the tangent to C through P is unique and coincides with the parallel to the minor axis of C passing through P. Otherwise, letting $P(x_0, y_0)$, we need to show that there exists a single value of $m \in \mathbb{R}$ for which the straightline r of equation $y = m(x - x_0) + y_0$ is tangent to C. As in the case of a parabola, we do this by imposing that the system of equations

$$\begin{cases} b^2x^2 \pm a^2y^2 = a^2b^2 \\ y = m(x - x_0) + y_0 \end{cases}.$$

has (x_0, y_0) as its single solution (of course, the sign $+$ or $-$ is chosen according to whether C is an ellipse or a hyperbola). To this end, adapt the reasoning presented in the text for the case of a parabola. (As in that case, the argument just delineated also shows that the value of m coincides with the one found by computing $y'(x_0)$, where $y = y(x)$ is such that $\frac{x^2}{a^2} \pm \frac{y(x)^2}{b^2} = 1$ and (x_0, y_0) lies in the graph of $x \mapsto y(x)$.)

2. For item (a) of Theorem 6.26, if $Q \in \mathcal{P}$ is distinct from P and U is the foot of the perpendicular dropped from Q to d, check that $\overline{FQ} = \overline{TQ} > \overline{UQ}$, so that $Q \notin \mathcal{P}$. Then, use the result of Theorem 6.25. For item (a) of Proposition 6.28, suppose first that r is tangent to $\mathcal{P}$ at P, and let T be the foot of the perpendicular dropped from P to d. Item (a) of Theorem 6.26 assures that r is the internal angle bisector of the isosceles triangle FPT. Therefore, r is also perpendicular bisector and altitude relative to FT, and this is the same as saying that T is the symmetric of F with respect to r. For the converse, argue analogously.

3. This is immediate from Theorem 6.26, since the internal and external angle bisectors of triangle PF_1F_2 at P are perpendicular.

4. As in the hint given to the previous problem, let T be the foot of the perpendicular dropped from P to d. If $\overrightarrow{FY}$ is the half-line perpendicular to but not intersecting d, then $F\widehat{P}T = X\widehat{F}Y$; thus, the internal bisector of $\angle XFY$ is parallel to that of $\angle FPT$, hence perpendicular to $\overleftrightarrow{FT}$. We now have the following construction: draw $\overrightarrow{FY}$, and then the internal bisector $\overrightarrow{FZ}$ of $\angle XFY$. The perpendicular to $\overleftrightarrow{FZ}$ passing through F intersects d at T, and the perpendicular to d passing through T intersects $\overrightarrow{FX}$ at P.

5. Adapt the analysis of that example to the present case, with the aid of item (b) of Proposition 6.28.

6. Let F be the given focus and r, s and t be the given tangents. Proposition 6.28 assures that if P, Q and R are the symmetrics of F with respect to r, s and t, then the other focus is the circumcenter of triangle PQR, while the major axis is the circumradius of PQR.

7. If r is one of the tangents we are looking for, and P is the foot of the perpendicular dropped from F_2 to r, then item (b) of Corollary 6.31 assures that P lies in the auxiliary circle of $\mathcal{H}$. Since $F_2\widehat{P}Q = 90°$, the point P can be constructed as the intersection of the auxiliary circle with the circle of diameter F_2Q. In general, there are two distinct solutions.

8. Let A_1A_2 be the major axis of C, with $\overline{A_1A_2} = 2a$, and let $\overline{F_1F_2} = 2c$. If Γ is the auxiliary circle of C, Corollary 6.31 assures that $P_1, P_2 \in \Gamma$. Also, if Q is the other intersection of $\overleftrightarrow{F_1P_1}$ with Γ, then $\overline{F_1Q} = \overline{F_2P_2}$. Since for an ellipse we have $\overline{A_1F_1} \cdot \overline{A_2F_1} = (a - c)(a + c) = b^2$, the intersecting chords theorem applied to A_1A_2 and QP_1 gives the result. In the case of a hyperbola, $\overline{A_1F_1} \cdot \overline{A_2F_1} = (c - a)(c + a) = b^2$, and we are also done.

9. Use item (a) of Corollary 6.31 to show that the feet of the perpendiculars dropped from the focus of the parabola to the three tangents are collinear. Then, apply the theorem of Simson-Wallace.

10. Note that no two of r, s, t, u can be parallel, and no three pass through the same point (for there are at most two tangents to $\mathcal{P}$ passing through any given point). If ABC is the triangle determined by r, s, t, CDE is the one determined by r, s, u and AEF is that determined by s, t, u, then the previous problem assures that the focus of $\mathcal{P}$ lies in the circumcircles of ABC, CDE and AEF. Thus, the position of F is entirely determined, and there is only one solution.

11. Let r and s be perpendicular tangents to $\mathcal{E}$, intersecting each other at a point P. If R and S are the symmetrics of F_2 with respect to r and s, respectively, then Proposition 6.28 shows that R and S lie in the director circle Σ centered at F_1. Show that P is the midpoint of RS and $R\widehat{F_2}S = 90°$. Then, conclude that $F_1P \perp RS$ and $\overline{PF_2} = \overline{PR} = \overline{PS}$, so that $\overline{PF_1}^2 + \overline{PF_2}^2 = \overline{F_1R}^2 = 4a^2$. Finally, use the result of Problem 5, page 188, to compute

$$\overline{PO}^2 = \frac{1}{4}\left(2(\overline{PF_1}^2 + \overline{PF_2}^2) - \overline{F_1F_2}^2\right) = 2a^2 - c^2 = a^2 + b^2,$$

thus concluding that P lies in the circle of Monge of $\mathcal{E}$. For the converse, rework the steps above in reverse order.

12. Adapt, to the present case, the hint given to the previous problem.

13. The figure below illustrates the situation, and we want to show that $X\widehat{P}T_1 = F\widehat{P}T_2$ and $T_1\widehat{F}P = T_2\widehat{F}P$.

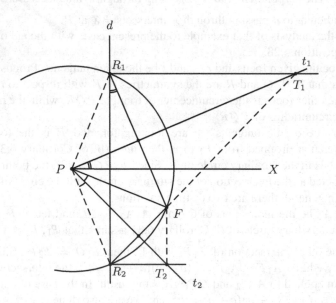

Arguing as in the case of the ellipse, we let R_1 and R_2 be the symmetrics of F with respect to t_1 and t_2, respectively, so that $R_1, R_2 \in d$ and $\overline{PF} = \overline{PR_1} = \overline{PR_2}$. For the first part, since $\overleftrightarrow{PT_1} \perp \overleftrightarrow{FR_1}$ and $\overleftrightarrow{PX} \perp \overleftrightarrow{R_1R_2}$, we get $X\widehat{P}T_1 = R_2\widehat{R_1}F$. Since t_2 is the perpendicular bisector of FR_2, we obtain $F\widehat{P}T_2 = R_2\widehat{P}T_2$. Finally, the inscribed angle theorem gives us $R_2\widehat{R_1}F = \frac{1}{2}R_2\widehat{P}F$, and this finished the proof. For the second part, note that $P\widehat{F}T_2 = P\widehat{R_2}T_2$ and $P\widehat{F}T_1 = P\widehat{R_1}T_1$, and we are left to proving that $P\widehat{R_2}T_2 = P\widehat{R_1}T_1$. This is clear, for $R_2\widehat{R_1}T_1 = R_1\widehat{R_2}T_2 = 90°$ and $P\widehat{R_1}R_2 = P\widehat{R_2}R_1$.

Section 7.1

5. From $\frac{x^2}{a^2} + \frac{y^2}{b^2} = 1$ we get $\left|\frac{x}{a}\right| \leq 1$. Then, there exists $\alpha \in [0, 2\pi)$ such that $\frac{x}{a} = \cos\alpha$, or $x = a\cos\alpha$. By substituting $x = a\cos\alpha$ in the equation of $\mathcal{E}$, we obtain

$$\frac{y^2}{b^2} = 1 - \frac{(a\cos\alpha)^2}{a^2} = \sin^2\alpha,$$

so that $y = \pm b\sin\alpha$. If $y = b\sin\alpha$, we can take $\theta = \alpha$; if $y = b\sin(-\alpha)$, it follows from $\cos(-\alpha) = \cos\alpha$ and $\sin(-\alpha) = -\sin\alpha$ that $x = a\cos(-\alpha)$ and $y = b\sin(-\alpha)$, so that we can take $\theta = -\alpha$. In order to show uniqueness, suppose $\theta' \in [0, 2\pi)$ is such that $x = a\cos\theta'$ and $y = a\sin\theta'$. Then $\cos\theta = \cos\theta'$, $\sin\theta = \sin\theta'$ and, since $\theta, \theta' \in [0, 2\pi)$, we have $\theta = \theta'$. What is left to do is immediate (Fig. 14.1).

6. For item (a), letting $A(a\cos\alpha, b\sin\alpha)$, $B(a\cos\beta, b\sin\beta)$, $C(a\cos\gamma, b\sin\gamma)$, we have $A'(a\cos\alpha, a\sin\alpha)$, $B'(a\cos\beta, a\sin\beta)$, $C'(a\cos\gamma, a\sin\gamma)$. Use the result of Problem 11, page 189, to show that $A(ABC) = \frac{b}{a}A(A'B'C')$. For (b), item (a) assures that it suffices to maximize the area of $A'B'C'$, so that it must be equilateral. Now, note that if $\overleftrightarrow{BC} \parallel r$, then the slope of r is $m = \frac{b(\sin\beta - \sin\gamma)}{a(\cos\beta - \cos\gamma)}$, whereas that of $\overleftrightarrow{B'C'}$ is $m' = \frac{\sin\beta - \sin\gamma}{\cos\beta - \cos\gamma}$. Hence, $m' = \frac{a}{b}m$, so that we can find the direction corresponding to m'.

7. Choose a cartesian coordinate system in the plane with respect to which $\mathcal{E}$ has equation $\frac{x^2}{a^2} + \frac{y^2}{b^2} = 1$, and let Γ be the auxiliary circle of $\mathcal{E}$, of radius a. If $A_1A_2\ldots A_n$ is a convex polygon inscribed in $\mathcal{E}$ and $A'_1, A'_2, \ldots, A'_n$ in Γ are defined as in item (a) of the previous problem, show that $A(A'_1A'_2\ldots A'_n) = \frac{a}{b}A(A_1A_2\ldots A_n)$, then $A(A_1A_2\ldots A_n) \leq \pi ab$. Now, given $\epsilon > 0$, *lift* to Γ to show that $A_1A_2\ldots A_n$ can be chosen in such a way that $A(A_1A_2\ldots A_n) \geq \pi ab - \epsilon$.

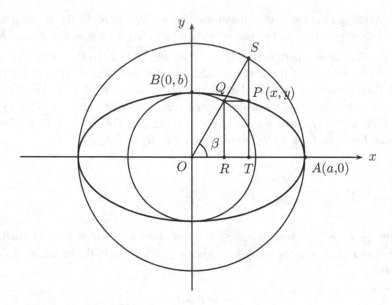

Fig. 14.1 Parametric equations of the ellipse $\mathcal{E}$

Section 7.2

1. Expand $\sin(a - b) = \sin(a + (-b))$ with the aid of the already proved part of item (b) of Proposition 7.18. Argue analogously for $\tan(a - b)$.
2. Write $15° = 45° - 30°$ and apply the formulas of Proposition 7.18. Alternatively, let $a = 15°$ in Corollary 7.19.
3. Apply the fundamental relation of trigonometry to item (a) of Corollary 7.19.
4. Note that $45° = 2 \cdot 22°30'$.
5. The answers are: (a) $\{x \in \mathbb{R}; \ x = k\pi, \ \exists k \in \mathbb{Z}\}$; (b) $\{x \in \mathbb{R}; \ x = \frac{\pi}{2} + k\pi, \ \exists k \in \mathbb{Z}\}$; (c) $\{x \in \mathbb{R}; \ x = k\pi, \ \exists k \in \mathbb{Z}\}$.
6. Use the results of the previous problem in conjunction with those of Proposition 7.22. For instance,

$$\sin x = \sin \alpha \ \Leftrightarrow \ \sin x - \sin \alpha = 0$$

$$\Leftrightarrow 2 \sin\left(\frac{x - \alpha}{2}\right) \cos\left(\frac{x + \alpha}{2}\right) = 0.$$

Thus, $x = \alpha + 2k\pi$ or $x = \pi - \alpha + 2k\pi$, for some $k \in \mathbb{Z}$.
7. In each item, write A as a sine or cosine and apply the results of the previous problem.
8. Adapt, to the present case, the hint given to the previous problem.
9. The fundamental relation of Trigonometry assures that it suffices to compute $\cos(x - y)$. To this end, note that

$$b^2 + a^2 = 2 + 2 \sin x \sin y + 2 \cos x \cos y = 2 + 2 \cos(x - y).$$

10. For the other formula in item (a), write $\sin a - \sin b = \sin a + \sin(-b)$; argue in a likewise manner to obtain the formula for $\tan a - \tan b$. In what concerns item (b), write $\cos a \pm \cos b = \sin \left(\frac{\pi}{2} - a \right) \pm \sin \left(\frac{\pi}{2} - b \right)$ and apply the formulas of item (a).

11. For the first formula, multiply the left hand side by $-2 \sin \frac{b}{2}$, unfold the product $-2 \sin \frac{b}{2} \sin(a + kb)$ (with the aid of Proposition 7.22) and apply the formula for telescoping sums (cf. Section 3.3 of [5], for instance). For the second equality, use the fact that $\cos x = \sin \left(\frac{\pi}{2} - x \right)$.

12. Set $a = 0$ and $b = \frac{2\pi}{n}$ in the formulas of the previous problem.

13. Letting G be the foot of the perpendicular dropped from E to AB, apply Pythagoras' theorem to triangle AEG to compute $\overline{CE}$. Then, compute $\tan E\widehat{A}G$, $\tan D\widehat{A}F$ and use the formula for $\tan(2\theta)$ in terms of $\tan \theta$ to conclude.

14. Let θ denote the acute angle between AP and r, and successively compute $\overline{AP}$, $\overline{BP}$ and the area of ABP in terms of θ to find $1 / \sin(2\theta)$.

15. Show that $\tan(A\widehat{D}B + A\widehat{E}B) = 1$ by using the formula for the tangent of a sum; you will need to compute the lengths of BC and of the altitude relative to it.

16. Let O be the center of the polygon (which coincides with the center of the circle), let A and B be two consecutive vertices and M the midpoint of AB. Since AOB is isosceles of basis AB, we have $A\widehat{O}M = \frac{\pi}{n}$ and $OM \perp AB$; hence, $l_n = 2 \overline{AM} = 2R \sin \frac{\pi}{n}$ and $a_n = \overline{OM} = R \cos \frac{\pi}{n}$. It now suffices to let n be successively equal to 3, 4, 6 and 8, substituting, in each case, the (already computed) values for the sine and cosine of $\frac{\pi}{3}$, $\frac{\pi}{4}$, $\frac{\pi}{6}$ and $\frac{\pi}{8}$, respectively.

17. Let O be the center of the polygon, let A and B be two consecutive vertices of it and M be the midpoint of AB. Let also $A' \in OA$ and $B' \in OB$ be the centers of two of the n circles, and M' be the intersection of OM with $A'B'$. Then, $OM' \perp A'B'$, $\overline{A'M'} = r$ and $\overline{OM'} = a_n - r$, with a_n standing for the apothem of the polygon (cf. previous problem). Apply some Trigonometry to the right triangle $A'OM'$ to compute r as a function of a_n and $A'\widehat{O}M' = \frac{\pi}{n}$; then do the same to the right triangle AOM to compute a_n as a function of l and $A\widehat{O}M = \frac{\pi}{n}$.

18. Let θ_1, θ_2, θ_3, θ_4 and θ_5 denote the central angles corresponding to the edges AB, BC, CD, DE and EA, respectively. If R is the circumradius of the pentagon and d the desired distance, show that

$$a = 2R \sin \frac{\theta_1}{2} \sin \left(\frac{\theta_1 + \theta_2}{2} \right), \quad b = 2R \sin \left(\frac{\theta_1 + \theta_2}{2} \right) \sin \left(\frac{\theta_4 + \theta_5}{2} \right),$$

$$c = 2R \sin \frac{\theta_5}{2} \sin \left(\frac{\theta_4 + \theta_5}{2} \right) \quad \text{and} \quad d = 2R \sin \frac{\theta_5}{2} \sin \frac{\theta_1}{2}.$$

Then, note that $d = \frac{ac}{b}$.

19. Letting H_a stand for the foot of the altitude relative to A and $\overline{HH_a} = h_a$, start by showing that $\overleftrightarrow{HG} \parallel \overleftrightarrow{BC} \Leftrightarrow \overline{HH_a} = d(G; \overleftrightarrow{BC}) = \frac{h_a}{3}$. Then, compute

$$\overline{HH_a} = \overline{AB} \cdot \cos \widehat{B} \cdot \tan H\widehat{B}D = \overline{AB} \cdot \cos \widehat{B} \cdot \cot \widehat{C}.$$

20. Show that $MCN \equiv MAN$, then use this to conclude that $N\widehat{M}L$ is right and isosceles. Now, use the given area relation to obtain $\overline{AB} = 2\overline{LN}$. Finally, if $\overline{AC} = \ell$ and $C\widehat{A}P = \alpha$, compute $\overline{LN} = \overline{AN} - \overline{AL} = \ell\cos\alpha - \ell\sin\alpha$ and $\overline{AB} = \ell\sqrt{2}$.

21. Firstly, note that $\frac{PB}{BH} = \frac{BA}{BH} = \frac{BC}{BA} = \frac{BC}{BP}$, so that $BPH \sim BPC$. Now, if $\widehat{B} = 60°$, then $\frac{PH}{PC} = \frac{BP}{BC} = \frac{1}{2} = \frac{HM}{MC}$, whence PM bisects $\angle HPC$ by the internal bisector theorem. Conversely, if PM bisects $\angle HPC$, then $BPH \sim BPC$ implies $\frac{PH}{PC} = \frac{PB}{BC} = \cos \widehat{B}$. However, the internal bisector theorem gives

$$\frac{PH}{PC} = \frac{HM}{MC} = \frac{BM - BH}{MC} = 1 - \frac{BH}{MC}$$

$$= 1 - \frac{BH}{\overline{AB}^2/(2\,\overline{BH})} = 1 - 2\left(\frac{\overline{BH}}{\overline{AB}}\right)^2$$

$$= 1 - 2\cos^2 \widehat{B}.$$

Hence, $\cos \widehat{B} = 1 - 2\cos^2 \widehat{B}$, so that $\cos \widehat{B} = \frac{1}{2}$ and $\widehat{B} = 60°$.

22. Choosing a cartesian system in which the coordinate axes coincide with the axes of the given parabolas, we can assume that they have equations $y = ax^2 + b$ and $x = cy^2 + d$, with $a, c > 0$. Letting $A(x_1, y_1)$, $B(x_2, y_2)$, $C(x_3, y_3)$, $D(x_4, y_4)$, start by noticing that x_1, x_2, x_3, x_4 (resp. y_1, y_2, y_3, y_4) are pairwise distinct. If m_1, m_2, m_3, m_4 denote the angular coefficients of $\overleftrightarrow{AB}$, $\overleftrightarrow{BC}$, $\overleftrightarrow{CD}$, $\overleftrightarrow{AD}$, compute $m_1 = \frac{y_2 - y_1}{x_2 - x_1} = a(x_2 + x_1)$ and, likewise, $m_2 = a(x_3 + x_2)$, $m_3 = a(x_3 + x_4)$, $m_4 = a(x_1 + x_4)$. Letting α and β denote the trigonometric angles from the horizontal axis to lines $\overleftrightarrow{AB}$ and $\overleftrightarrow{BC}$, respectively, we have $m_1 = \tan\alpha$, $m_2 = \tan\beta$. Let $\delta = A\widehat{B}C$ and $\epsilon = A\widehat{D}A$. If $\delta, \epsilon \neq 90°$, show that $\delta = \alpha + \pi - \beta$ and use item (c) of Proposition 7.18 to compute

$$\tan\delta = \frac{a(x_1 - x_3)}{1 + a^2(x_1 + x_2)(x_2 + x_3)}, \quad \tan\epsilon = \frac{a(x_3 - x_1)}{1 + a^2(x_1 + x_4)(x_3 + x_4)}.$$

Then, use Trigonometry again to show that

$$\delta + \epsilon = \pi \Leftrightarrow \tan\delta = -\tan\epsilon \Leftrightarrow (x_1 + x_2)(x_2 + x_3) = (x_1 + x_4)(x_3 + x_4).$$

Now, show that x_1, x_2, x_3, x_4 are the roots of the polynomial equation $a^2cx^3 + 2abcx^2 - x + (b^2 + d) = 0$ and use this fact to conclude that $x_1 + x_2 + x_3 + x_4 = 0$; finally, use this last relation to conclude the proof. If $\delta = 90°$ (resp. $\epsilon = 90°$), use an argument similar to the above to show that $\epsilon = 90°$ (resp. $\delta = 90°$).

Section 7.3

1. Apply the cosine law to compute $a = \overline{BC}$ in terms of b, c and $B\widehat{A}C$; then, compare the expression thus obtained with the one given in the problem.

2. In the notations of Problem 16, apply the result of Problem 13, page 116, together with the cosine law, to a triangle of sides R, l_{10} and l_{10}—and, obviously, interior angles of $36°$, $72°$ and $72°$.

3. Apply the cosine law to a triangle of sides l_{10}, R and R.

4. At instant t after the fight, apply the cosine law to the three triangles formed by the pub's location and by the positions of two of the musketeers. This will give the distances between each two of them in terms of t, so that it will suffice to use the converse to Pythagoras' theorem.

5. Apply the internal bisector theorem twice to compute $\overline{BD}$ and $\overline{AE}$ in terms of the lengths of the sides of ABC. Then, show that the given equality amounts to $a^2 + b^2 - ab = c^2$ and apply the cosine law.

6. Take point D such that $ABCD$ is a parallelogram, and let Q be the point inside ACD for which APQ is equilateral. Prove that $\overline{CQ} = \overline{BP}$, and hence that $Q\widehat{P}C = 60°$ or $30°$. To conclude, apply the cosine law to triangle APC.

7. Start by considering the convex hexagon inscribed in the same circle but with consecutive sides of lengths a, b, a, b, a, b. Show that this second hexagon has area equal to that of the first one, and that its internal angles all measure $120°$. Then, apply the cosine law and the result of Problem 16, page 247, for the computation of l_3.

8. Apply, in triangles BDM, ABC and ACD, the formula for the length of a median.

9. The first part follows from the previous problem. For the second part, start by constructing, from a given triangle ABC, parallelogram $ABDC$; then, note that $\overline{AD}$ is twice the length of the median of ABC relative to BC.

10. Let $ABCD$ be the trapezoid, with basis AB and CD and diagonals AC and BD, such that $\overline{AB} > \overline{CD}$. Assume also that $B\widehat{A}D = \alpha$ and $A\widehat{B}C = \beta$, with $\alpha > \beta$. Finally, let E be the intersection point of $\overrightarrow{AD}$ and $\overrightarrow{CD}$, with $\overline{AE} = a$, $\overline{BE} = b$, $\overline{CE} = x$ and $\overline{DE} = y$. Apply the cosine law to triangles AEC and BDE, noticing that $\frac{a}{y} = \frac{b}{x}$.

11. Let O and O' be the centers of Γ and Σ, respectively, let $E \neq A$ be the intersection point of $\overrightarrow{AO'}$ and Γ, and E that of AD and BC. Start by observing that AE is the internal bisector of $\angle BAC$; then, use this fact to compute AE, BE and CE. Now, use the intersecting chords theorem to compute DE and, finally, apply Stewart's relation to triangle AOD and the cevian OO'.

12. The internal bisector theorem gives $\overline{BP} = \frac{ac}{b+c}$ and $\overline{CP} = \frac{ab}{b+c}$. Apply Stewart's theorem, together with some elementary algebra.

13. Let P denote the foot of the internal bisector of $\angle BAC$. Apply the internal bisector theorem to triangle ABP, together with the result of the previous problem, to get

$$\overline{AI} = \frac{b+c}{2p} \cdot \overline{AP} = \frac{\sqrt{bcp(p-a)}}{p}.$$

Then, note that this last expression is always less than $\sqrt{bc}$.

14. Let O be the center and R the radius of the circle circumscribed to the polygon. Assume, with no loss of generality, that P lies in the angular region $\angle A_1 O A_2$. If $P\widehat{O}A_2 = \alpha$ then, by walking from A_1 to A_2, then to A_3 and henceforth, we have $P\widehat{O}A_{2k} = \alpha + \frac{(k-1)\pi}{n}$. Therefore, the cosine law gives

$$\overline{A_{2k}P}^2 = \overline{A_{2k}O}^2 + \overline{OP}^2 - 2\,\overline{A_{2k}O} \cdot \overline{OP}\cos\left(\alpha + \frac{(k-1)\pi}{n}\right)$$

$$= R^2 + \overline{OP}^2 - 2R \cdot \overline{OP}\cos\left(\alpha + \frac{(k-1)\pi}{n}\right).$$

Analogously, $\overline{A_{2k-1}P}^2 = R^2 + \overline{OP}^2 - 2R \cdot \overline{OP}\cos\left(\frac{k\pi}{n} - \alpha\right)$, so that

$$\sum_{k=1}^{n} \overline{A_{2k}P}^2 = \sum_{k=1}^{n} \overline{A_{2k-1}P}^2$$

if and only if

$$\sum_{k=1}^{n} \cos\left(\alpha + \frac{(k-1)\pi}{n}\right) = \sum_{k=1}^{n} \cos\left(\frac{k\pi}{n} - \alpha\right).$$

Finally, in order to establish the equality above, apply the result of Problem 11, page 247. For an alternative solution, see the problems of Section 13.2 of [6].

15. Let M be the midpoint of BC and $A_1 \neq A$ be the point in which $\overrightarrow{AM}$ intersects the circumcircle of ABC. If $\overline{MA_1} = x$, use the intersecting chords theorem, together with the fact that $\overline{AA_1} \leq 2R$, to conclude that $m_a^2 + \frac{a^2}{4m_a} \leq 2R$. Finally, use the formula for the length of m_a to obtain $\frac{b^2+c^2}{2m_a} \leq 2R$.

16. Assume that the circles have different radii (the other case can be dealt with analogously), and let r and s denote the drawn inner and outer tangents, respectively. Let N and M (resp. P and Q) denote the points of tangency of r and s, respectively, with the larger and smaller circles, also respectively. Finally, let C denote the intersection point of r and s, and B and A the points of intersection of r and s with the other common outer tangent, respectively. It suffices to prove that the point R of intersection of the two chords lies in the bisector of $\angle BAC$. To this end, compute $M\widehat{R}Q = 90°$ and set $B\widehat{A}C = 2\alpha$, $B\widehat{C}A = 2\theta$, $\overline{AB} = c$, $\overline{AC} = b$, $\overline{BC} = a$, $\overline{MN} = x$, $\overline{NR} = y$, with $b > c$. Compute $x = 2(p-b)\cos\theta$, $y = (b-c)\cos\theta$, $\overline{MR} = x + y = a\cos\theta$. Now, let S denote the foot of the perpendicular dropped from R to AC, set $\overline{RS} = t$,

$\overline{AS} = z$ and compute $t = a \cos\theta \sin\theta = \frac{a\sin(2\theta)}{2}$ and $z = \overline{AM} - a\cos^2\theta = p - a\cos^2\theta$, where p is the semiperimeter of ABC. We need to prove that $t = z \tan\alpha$. To this end, note that

$$t = z\tan\alpha \Leftrightarrow \frac{a\sin(2\theta)}{2} = (p - a\cos^2\theta)\tan\alpha$$

$$\Leftrightarrow \frac{c\sin(2\alpha)}{2} = \left(a(1 - 2\cos^2\theta) + b + c\right)\frac{\tan\alpha}{2}$$

$$\Leftrightarrow 2c\sin\alpha\cos\alpha = \left(b + c - a\cos(2\theta)\right)\tan\alpha$$

$$\Leftrightarrow 2c\cos^2\alpha = b + c - a\left(\frac{a^2 + b^2 - c^2}{2ab}\right)$$

$$\Leftrightarrow c\left(\cos(2\alpha) + 1\right) = \frac{b^2 + c^2 + 2bc - a^2}{2b}$$

$$\Leftrightarrow \cos(2\alpha) = \frac{b^2 + c^2 - a^2}{2bc},$$

which is true.

17. Let M and m denote the lenghts of the greatest and smallest of the $\binom{6}{2} = 15$ line segments connecting two of the given points, and let $\mathcal{P}$ denote the smallest convex polygon containing the six points. There are two possibilities: (i) $\mathcal{P}$ is not an hexagon: considering the possible numbers of sides of P, we conclude that there are three of the vertices, say A, B, C, such that ABC contains a fourth point D of the six given ones. Since

$$A\widehat{D}B + B\widehat{D}C + C\widehat{D}A = 360°,$$

the measure of at least one of these angles is at least $120°$. Suppose, without loss of generality, that $A\widehat{D}B \geq 120°$. By applying the cosine law to triangle ADB, we obtain

$$m^2 \geq AB^2 = AD^2 + BD^2 - 2AB \cdot BD\cos\angle ADB$$

$$\geq m^2 + m^2 - 2m^2\cos 120° = 3m^2.$$

(ii) $\mathcal{P}$ is a hexagon: the sum of the internal angles of $\mathcal{P}$ is $720°$, we conclude that the measure of at least one of these six angles is at least $120°$. Let AD and BD be the edges of such an angle. Then, $\angle ADB \geq 120°$ and it suffices to reason as in the end of (i).

18. If M is the intersection point of EF and DX, show that $\frac{EM}{FM} = \frac{EX \cdot ED}{FX \cdot DF}$ (*). Let k be the power of P with respect to Γ and $P' \neq X$ be the other intersection of $\overleftrightarrow{AP}$ and Γ, so that $\overline{PP'} = \frac{k}{PX}$. Now, compute the power of A with respect

to Γ to find $\overline{AE}^2 = \frac{AX}{PX}(\overline{AP} \cdot \overline{PX} + k)$ (**). Apply Stewart's relation to triangle AEP', together with the similarity of triangles AXE and AEP' and (**) to find $\frac{AP}{AX} \cdot \overline{EX}^2 = k + \overline{EP}^2$. Use an analogous reasoning to obtain $\frac{AP}{AX} \cdot \overline{FX}^2 = k + \overline{FP}^2$. Then, use (*) to get

$$\frac{\overline{EM}}{\overline{FM}} = \left(\frac{k + \overline{EP}^2}{k + \overline{FP}^2}\right)^{1/2} \cdot \frac{\overline{DE}}{\overline{DF}}.$$

Letting N and Q be the points of intersection of BP and CP with DF and DE, respectively, we deduce as above that

$$\frac{\overline{FN}}{\overline{DN}} = \left(\frac{k + \overline{FP}^2}{k + \overline{DP}^2}\right)^{1/2} \cdot \frac{\overline{EF}}{\overline{DE}} \quad \text{and} \quad \frac{\overline{DQ}}{\overline{EQ}} = \left(\frac{k + \overline{DP}^2}{k + \overline{EP}^2}\right)^{1/2} \cdot \frac{\overline{DF}}{\overline{EF}}.$$

Finally, note that $\frac{EM}{MF} \cdot \frac{FN}{ND} \cdot \frac{DQ}{QE} = 1$ and apply Ceva's theorem.

Section 7.4

1. Apply the sine law to each summand at the left hand side.
2. Apply the sine law to ABC, together with the formula for the sine of double arcs (cf. Corollary 7.19) to compute $\cos A\widehat{C}B$ in terms of a, b and c; then, substitute the result into the cosine law.
3. Let r be the common difference of the progression, and α, β and γ be the measures of the angles opposite the sides of lengths $l - r$, l and $l + r$, respectively, so that $\alpha < \beta < \gamma$. Then, $\gamma = \alpha + 90°$ and, since $\alpha + \beta + \gamma = 180°$, we obtain $\beta = 90° - 2\alpha$. Now, it follows from the sine law that

$$\frac{l - r}{\sin \alpha} = \frac{l}{\sin(90° - 2\alpha)} = \frac{l + r}{\sin(90° + \alpha)}$$

or, which is the same,

$$\frac{l - r}{\sin \alpha} = \frac{l}{\cos(2\alpha)} = \frac{l + r}{\cos \alpha}.$$

Finally, look at the equalities above as a system of equations in r and α to conclude the analysis of the problem.
4. Take point D for which $ABDC$ is a parallelogram and apply sine law to ADC, together with the formula for the sine of the double arc.
5. Firstly, compute the area of ABC with the aid of Heron's formula. Then, let O be the center of the semicircle and adapt the idea of the proof of Proposition 5.13 to triangles ABO and ACO to conclude that $A(ABC) = \frac{27R}{2}$.

6. Look separately at the cases $D \in AB$, $D \in AC$ and $D \in BC$. For example, if $D \in BC$, then apply the sine law to triangle DPQ and note that $\overline{AD} \geq \overline{AH_a}$ (with $H_a \in BC$ being the foot of the altitude relative to BC).

7. Set $\overline{AB} = c$, $\overline{AC} = b$ and $\overline{BC} = a$ and use the sine formula for the area of a triangle and the cosine law to show that the given inequality is equivalent to

$$a^2 + b^2 \geq ab(\sqrt{3}\sin\widehat{C} + \cos\widehat{C}).$$

Then, apply the inequality between the arithmetic and geometric means for two positive reals (cf. Section 5.1 of [5], for instance), together with (7.10).

8. Set $\overline{AC} = b$, $\overline{AB} = c$, $\overline{BC} = a$ and $\overline{BD} = \overline{CD} = t$. Apply Ptolemy's theorem to the quadrilateral $ABDC$ to get $(c + b)t = a \cdot \overline{AD}$. From this, use the sine law to arrive at

$$\overline{BK} + \overline{CL} = t(\sin\widehat{B} + \sin\widehat{C}) = \overline{AD} \cdot \sin\widehat{A} \leq \overline{AD}.$$

9. Letting $\overline{AB} = c$, $\overline{AC} = b$ and $\overline{BC} = a$, use Corollary 7.31 and the sine law to show that $A(AB_2C_1) = \frac{a^3}{4R}$ and

$$A(BA_1A_2C) = \frac{(b+c)^2\sin\widehat{A}}{2} - \frac{abc}{4R} = \frac{a}{4R}(b^2 + bc + c^2),$$

where R stands for the circumradius of ABC. Get analogous formulas for the areas of A_1BC_2, B_1CAB_2, A_2B_1C and ABC_2C_1, and add all of them to obtain

$$A(A_1A_2B_1B_2C_1C_2) = \frac{1}{4R}((a^2 + b^2 + c^2)(a + b + c) + 4abc).$$

Finally, apply the inequality between the arithmetic and geometric means for three positive reals (cf. Section 5.1 of [5], for instance), once more in conjunction with Corollary 7.31, to conclude.

10. Letting $\theta = A\widehat{B}M$, first show that $\theta = \widehat{C} - \widehat{B}$; then, compute $\overline{AN}$, $\overline{BN}$ and $\overline{AC}$ in terms of $\overline{MN} = x$, $\overline{AB} = c$ and the involved angles, use the sine law and Trigonometry in right triangles.

11. Apply formula (7.17) to all summands in the equality $A(ABC) = A(ABP) + A(ACP)$.

12. Apply the sine formula for the area of a triangle together with the inequality between the arithmetic and geometric means for two positive reals (cf. Section 5.1 of [5], for instance).

13. Letting M be the intersection point of AC and BD, one has $A(ABCD) = A(ABM) + A(BCM) + A(CDM) + A(DAM)$. Then, apply the sine formula for the area of a triangle to compute the area of each one of the four triangles at the right hand side.

14. Apply the formula of Problem 13, together with the inequality between the arithmetic and geometric means for two positive reals (cf. Section 5.1 of [5], for instance).

15. Apply the sine formula for the area of a triangle to compute S, S_1 and S_2, thus transforming the desired inequality into one involving the lengths of AE, BE, CE and DE. Then, apply the inequality between the arithmetic and geometric means for two positive reals (cf. Section 5.1 of [5], for instance).

16. If $A\widehat{D}C = \theta$, then $A\widehat{B}C = \pi - \theta$, so that

$$A(ADM) = A(ABCM) \Rightarrow A(ABCD) = 2A(ADM)$$

$$\Rightarrow (x + c)y \sin\theta + ab\sin(\pi - \theta) = 2xy \sin\theta$$

$$\Rightarrow xy = cy + ab.$$

$$(14.1)$$

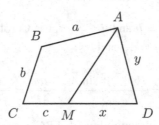

On the other hand, the equality between the perimeters gives $x + y = a + b + c$. Hence, it follows from (14.1) that

$$xy + y^2 = (x + y)y = (a + b)y + cy = (a + b)y + xy - ab,$$

so that $y^2 - by = ay - ab$, or $(y - a)(y - b) = 0$. Therefore, either $y = a$ or $y = b$.

17. Apply the cosine law to triangles ABC and CDA to find an expression for $\cos A\widehat{B}C$ in terms of a, b, c and d. Then, use such a formula to compute $\sin A\widehat{B}C$, and thus the area of $ABCD$ as $A(ABC) + A(CDA)$.

18. Letting $\alpha = B\widehat{A}D$ and $\beta = A\widehat{B}C$, use the sine law to show that $\frac{x}{y} = \frac{\sin\alpha}{\sin\beta}$. Then, use the sine formula for the area of a triangle to compute the area of $ABCD$ in two distinct ways, thus showing that $(ad + bc)\sin\alpha = (ab + cd)\sin\beta$.

19. Apply Ptolemy's and Hypparchus' theorems (see the previous problem) to compute the lengths of the diagonals of $ABCD$ in terms of a, b, c and d. Then, use compass and straightedge to construct the corresponding line segments.

20. If such a sequence did exist, the square and the triangle would have equal areas and perimeters. Analyse the feasibility of this fact in light of Example 7.34.

21. For the first part, apply the sine law to triangles ABP, ACP and ABC, noticing that $\sin A\widehat{P}B = \sin A\widehat{P}C$.

22. For item (b), use the result of (a), together with the fact that $R \geq 2r$ (cf. Theorem 4.32). For (a), start by showing that, if H and O denote the orthocenter and the circumcenter of ABC and H_1 the symmetric of H with respect to O, then $\overleftrightarrow{A_1 H_1} \perp \overleftrightarrow{AH}$, $\overleftrightarrow{B_1 H_1} \perp \overleftrightarrow{BH}$ and $\overleftrightarrow{C_1 H_1} \perp \overleftrightarrow{CH}$. Subsequently, if AA' is a diameter of the circumcircle of ABC, use the theorem on the Euler median and the similarity $AOM \sim AA'A_1$ to obtain $\overline{A_1 H_1} = 2(R + x)$, where x denotes the distance from O to BC. Make analogous computations for $\overline{B_1 H_1}$ and $\overline{C_1 H_1}$, then showing that

$$
\begin{aligned}
A(A_1 B_1 C_1) &= A(A_1 B_1 H_1) + A(A_1 C_1 H_1) + A(B_1 C_1 H_1) \\
&= 2R^2(\sin \widehat{A} + \sin \widehat{B} + \sin \widehat{C}) \\
&\quad + 2(xy \sin \widehat{C} + xz \sin \widehat{B} + yz \sin \widehat{A}) \\
&\quad + 2R((x + y) \sin \widehat{C} + (x + z) \sin \widehat{B} + (y + z) \sin \widehat{A}).
\end{aligned}
$$

Finally, apply the sine law and Carnot's theorem to reach the desired result.

Section 7.5

1. If $D \neq C$ is chosen in such a way that ABD is also equilateral, use Ptolemy's inequality to show that $\overline{AP} + \overline{BP} \geq \overline{DP}$. Then, notice that $\overline{DP} > \overline{CP}$.
2. In the cartesian plane, if $A(-1, 1)$, $B(1, -1)$, $C(-2, -2)$ and $P(x, y)$, then ABC is isosceles and we want to minimize the sum $\overline{AP} + \overline{BP} + \overline{CP}$. To this end, apply the solution for Steiner's problem to compute the coordinates of the desired point P, and then the minimum of the given expression.
3. Set $\overline{AC} = a$, $\overline{CE} = b$, $\overline{AE} = c$ and apply Ptolemy's inequality to quadrilateral $ABCE$ to find $\frac{BC}{BE} \geq \frac{a}{b+c}$. Analogously, show that $\frac{DE}{DA} \geq \frac{b}{a+c}$ and $\frac{FA}{FC} \geq \frac{c}{a+b}$. Thus, it suffices to show that $\geq \frac{a}{b+c} + \frac{b}{a+c} + \frac{c}{a+b} \geq \frac{3}{2}$, which follows from Chebyshev's inequality (cf. Example 5.21 of [5], for instance).

Section 8.1

1. Firstly, noting that AC and $A'C'$ are equal and parallel, conclude that $AA'C'C$ is a parallelogram, so that AA' and CC' are also equal and parallel. Then, use the fact that the pairs of line segments AB, CD and $A'B'$, $C'D'$ are equal and parallel to conclude that CD and $C'D'$ satisfy these conditions too. This will show that $CC'D'D$ is a parallelogram, and hence that CC' and DD' are equal and parallel. Finally, conclude that AA' and DD' are also equal and parallel, so that the same holds for AD and $A'D'$.

2. Write $\mathbf{v} = k\mathbf{w}$ and take moduli in both sides of this equality to obtain $|k| = \frac{\|v\|}{\|w\|}$.

3. Item (a) is immediate if $k_1 = 0$ or $k_2 = 0$; if $k_1 k_2 \neq 0$, it follows at once from the definition of $k\mathbf{v}$, by looking at the possible signs of the product $k_1 k_2$. For item (b), separately consider the cases $k_1 + k_2 < 0$, $k_1 + k_2 = 0$ and $k_1 + k_2 > 0$. For item (c), apply the parallelogram rule to both sides, and use similarity of triangles to compare the results.

4. Write the given condition as $\overrightarrow{OC} - \overrightarrow{OA} = k(\overrightarrow{OB} - \overrightarrow{OA})$ and expand.

5. Fix a point O in the plane and use the result of the previous problem to write $\overrightarrow{NM} = \overrightarrow{OM} - \overrightarrow{ON} = \frac{1}{2}(\overrightarrow{OA} - \overrightarrow{OC})$. Then, conclude that $\overrightarrow{NM} = \overrightarrow{PQ}$.

6. If P stands for the foot of the internal bisector relative to BC, use the internal bisector theorem to conclude that $\overrightarrow{BP} = \frac{c}{b+c} \overrightarrow{BC}$. Then, apply the result of Problem 4 to find $\overrightarrow{OP} = \frac{b\overrightarrow{OB} + c\overrightarrow{OC}}{b+c}$. Finally, repeat the argument above to get $\overrightarrow{AI} = \frac{b+c}{a+b+c} \overrightarrow{AP}$ and, then, the stated formula.

7. Let M_a, M_b, M_c denote the orthocenters of AH_bH_c, BH_aH_c, CH_aH_b, respectively, note that $P_aH_cHH_b$ is a parallelogram. Then, letting H be the origin of vectors and writing simply X instead of $\overrightarrow{OX}$, show that $P_a = H_b + H_c$. Then, setting $X := \frac{1}{2}(H_a + H_b + H_c)$, show that $\frac{P_a + H_a}{2} = \frac{P_b + H_b}{2} = \frac{P_c + H_c}{2}$, whence $P_a P_b P_c$ and $H_a H_b H_c$ are symmetric with respect to X, thus congruent.

8. Letting M be the midpoint of BC, use the fact that $\overleftrightarrow{AH} \| \overleftrightarrow{OM}$ and $\overline{AH} = 2\overline{OM}$ (cf. Theorem 4.20) to conclude that $\overrightarrow{AH} = 2\overrightarrow{OM}$. Then, write $\overrightarrow{AH} = \overrightarrow{OH} - \overrightarrow{OA}$ and apply the result of Problem 4 to find $\overrightarrow{OM} = \frac{1}{2}(\overrightarrow{OB} + \overrightarrow{OC})$.

9. Letting O be the center of the polygon, write $\overrightarrow{PA_i} = \overrightarrow{OA_i} - \overrightarrow{OP}$ and use the fact that $\overrightarrow{OA_1} + \cdots + \overrightarrow{OA_n} = \mathbf{0}$.

10. Given a point O, let $\mathbf{a}_i = \overrightarrow{OA_i}$, $\mathbf{m}_i = \overrightarrow{OM_i}$ and $\mathbf{b}_i = \overrightarrow{OB_i}$. Show that

$$\mathbf{b}_{i+1} + \mathbf{b}_i = 2\mathbf{m}_i = \mathbf{a}_{i+1} + \mathbf{a}_i;$$

then, add the first and last expressions above for $1 \le i \le 2n$ in order to reach the desired result.

11. Fix an arbitrary point O and let $\overrightarrow{OA} = \mathbf{a}$, $\overrightarrow{OA'} = \mathbf{a}'$ and likewise for $\overrightarrow{OB}$, $\overrightarrow{OB'}$ etc. Use the result of Problem 4 to obtain $2\mathbf{b} - \mathbf{a} = \mathbf{a}'$, $2\mathbf{c} - \mathbf{b} = \mathbf{b}'$, $2\mathbf{d} - \mathbf{c} = \mathbf{c}'$ and $2\mathbf{a} - \mathbf{d} = \mathbf{d}'$. Then, look at these four equalities as a linear system of equations in $\mathbf{a}$, $\mathbf{b}$, $\mathbf{c}$ and $\mathbf{d}$ that, once solved, gives $\mathbf{a}$ in terms of $\mathbf{a}'$, $\mathbf{b}'$, $\mathbf{c}'$ and $\mathbf{d}'$.

12. Start by arguing in a way similar to the hints given to the two previous problems to conclude that $Q_i Q_{i+1} = -\frac{1}{4} P_i P_{i+1}$ for $1 \le i \le 5$.

Section 8.2

1. Use the result of Problem 4, page 200.
2. We can assume, without loss of generality, that $\overline{AD} = 1$. Choosing a cartesian system of origin D in which $\overrightarrow{OA} = \mathbf{j}$, we have $\overrightarrow{DC} = -\mathbf{i}$ and $\overrightarrow{DG} = -\sqrt{3}\mathbf{j}$. Apply the result of the previous problem several times to show that $\overrightarrow{DL} = -(1 + \sqrt{3})\mathbf{i} + 3\mathbf{j}$. Finally, write $\overrightarrow{GC}$ and $\overrightarrow{GL}$ in terms of the canonical basis $\mathbf{i}$ and $\mathbf{j}$.
4. Use the previous problem, noticing that if $\overrightarrow{OQ}$ is obtained from $\overrightarrow{OP}$ through a clockwise rotation of θ radians, then the coordinates of P in system $x'Oy'$ coincide with those of Q in system xOy.
5. For item (c), show that the suggested rotation of coordinates transforms the equation $ax^2 + 2bxy + cy^2 + f = 0$ for $\mathcal{C}$ into $a'(x')^2 + 2bx'y' + c'(y')^2 + f' = 0$, with $2b' = 2b\cos 2\theta - (a - c)\sin 2\theta$. Then, show that this last equation admits a solution θ when $a \neq c$. For (d), look separately at the cases $a'c' = 0$, $a'c' < 0$ and $a'c' > 0$, taking the sign of f' into account in each of them.
6. Choose a cartesian system having the center O of one of the hexagons as origin, and such that one of its vertices, say A, is the point $(1, 0)$. Show that the vertices of the hexagons are the points P in the plane for which $\overrightarrow{OP} = a\mathbf{i} + b\mathbf{e}_{\frac{\pi}{3}}$, with $a, b \in \mathbb{Z}$. Now, suppose that there exists a square $P_1 P_2 P_3 P_4$ with all of the P_i's of the above form; conclude that $\overrightarrow{P_2 P_3}$ is obtained from $\overrightarrow{P_1 P_2}$ by means of a trigonometric rotation of $\pm\frac{\pi}{2}$, and apply the result of Problem 1 to reach a contradiction.
7. For item (a), start by writing $\overrightarrow{PD} = \overrightarrow{PC} + \overrightarrow{CD}$, and noticing that $\overrightarrow{CD}$ is obtained from $\overrightarrow{CB}$ by means of a trigonometric rotation of $\frac{\pi}{3}$ radians. For (b), use the fact that $\beta - \alpha = \gamma - \beta = \frac{2\pi}{3}$ and $\gamma - \alpha = \frac{4\pi}{3}$.

Section 8.3

1. Two representatives of a vector are opposite sides of a parallelogram. Hence, the sides of the angles formed by the canonical representatives, in two distinct cartesian systems, of given nonzero vectors $\mathbf{v}$ and $\mathbf{w}$ are parallel in pairs.
2. Show that $\theta(\mathbf{v}, \mathbf{w}) + \theta(-\mathbf{v}, \mathbf{w}) = \pi$.
3. For Corollary 6.11, recall that if $ax + by + c = 0$ is the equation of a given line r, then $\mathbf{v} = (a, b)$ is orthogonal to r, so that the line we wish to find has parametric equations $(x, y) = t(a, b)$. Adapt such kinds of arguments to prove the propositions referred to.
4. First of all, note that $\mathbf{v} = \left(\frac{a}{\sqrt{a^2 + b^2}}, \frac{b}{\sqrt{a^2 + b^2}} \right)$ is a unit vector orthogonal to r. Now, choose an arbitrary point $Q(x, y)$ on r and compute, on the one hand,

$$\overrightarrow{PQ} \cdot \mathbf{v} = (x - x_0, y - y_0) \cdot \left(\frac{a}{\sqrt{a^2 + b^2}}, \frac{b}{\sqrt{a^2 + b^2}} \right)$$

$$= \frac{a(x - x_0) + b(y - y_0)}{\sqrt{a^2 + b^2}} = \frac{-(ax_0 + by_0 + c)}{\sqrt{a^2 + b^2}};$$

on the other, letting θ denote the angle between $\overrightarrow{PQ}$ and $\mathbf{v}$, we have

$$| \overrightarrow{PQ} \cdot \mathbf{v}| = || \overrightarrow{PQ} || \cdot ||\mathbf{v}|| \cdot |\cos \theta| = || \overrightarrow{PQ} || \cdot |\cos \theta| = d(P; r).$$

5. Letting $\overrightarrow{OP} = \mathbf{p}$ and $\overrightarrow{OA_i} = \mathbf{a}_i$ for $1 \le i \le n$, show that

$$\sum_{i=1}^{n} ||\mathbf{a}_i - \mathbf{p}||^2 = \sum_{i=1}^{n} ||\mathbf{a}_i||^2 - 2\mathbf{p} \cdot \sum_{i=1}^{n} \mathbf{a}_i + n||\mathbf{p}||^2.$$

Then, apply the result of Proposition 8.6.)
6. Apply the result of the previous problem, noticing that, in a regular $2n$-gon $A_1 A_2 \ldots A_{2n}$, the n-gons $A_1 A_3 \ldots A_{2n-1}$ and $A_2 A_4 \ldots A_{2n}$ are regular and congruent.
7. The result of Problem 8, page 274, shows that

$$\overrightarrow{OH}^2 = \langle \overrightarrow{OH}, \overrightarrow{OH} \rangle = \langle \overrightarrow{OA} + \overrightarrow{OB} + \overrightarrow{OC}, \overrightarrow{OA} + \overrightarrow{OB} + \overrightarrow{OC} \rangle$$

$$= \overrightarrow{OA}^2 + \overrightarrow{OB}^2 + \overrightarrow{OC}^2 + 2\langle \overrightarrow{OA}, \overrightarrow{OB} \rangle + 2\langle \overrightarrow{OA}, \overrightarrow{OC} \rangle + 2\langle \overrightarrow{OB}, \overrightarrow{OC} \rangle.$$

Now, setting $A\widehat{C}B = \widehat{C}$, use the fact that $A\widehat{O}B = 2\widehat{C}$ to compute

$$2\langle \overrightarrow{OA}, \overrightarrow{OB} \rangle = 2\overrightarrow{OA} \cdot \overrightarrow{OB} \cos 2\widehat{C} = 2R^2(1 - 2\sin^2 \widehat{C})$$

$$= 2R^2 - (2R \sin \widehat{C})^2 = 2R^2 - a^2,$$

and likewise for $2\langle \overrightarrow{OA}, \overrightarrow{OC} \rangle$ and $2\langle \overrightarrow{OB}, \overrightarrow{OC} \rangle$.
8. Write $a + b + c = 2p$ and use the result of Problem 6, page 274, together with the sine law and formulas (5.3) and (7.17) to compute

$$4p^2 \overline{OI}^2 = 4p^2\| \overrightarrow{OI}\|^2 = \|a\ \overrightarrow{OA} + b\ \overrightarrow{OB} + c\ \overrightarrow{OC}\|^2$$
$$= R^2(a^2 + b^2 + c^2) + 2abR^2(1 - 2\sin^2\widehat{C}) +$$
$$+ 2acR^2(1 - 2\sin^2\widehat{B}) + 2bcR^2(1 - 2\sin^2\widehat{A})$$
$$= R^2(a + b + c)^2 - ab(2R\sin\widehat{C})^2 - ac(2R\sin\widehat{B})^2 - bc(2R\sin\widehat{A})^2$$
$$= 4p^2 R^2 - abc(a + b + c) = 4p^2 R^2 - 2pabc$$
$$= 4p^2(R^2 - 2Rr).$$

9. First note that

$$\overline{AP}^2 = |\overrightarrow{IP} - \overrightarrow{IA}|^2 = \langle \overrightarrow{IP} - \overrightarrow{IA}, \overrightarrow{IP} - \overrightarrow{IA} \rangle$$
$$= \overline{IP}^2 + \overline{IA}^2 - 2\langle \overrightarrow{IP}, \overrightarrow{IA} \rangle.$$

Now, write $\overline{BP}^2$ and $\overline{CP}^2$ similarly, and use the result of Problem 6, page 274, to conclude that

$$a\,\overline{AP}^2 + b\,\overline{BP}^2 + c\,\overline{CP}^2 = (a + b + c)\overline{IP}^2 + a\,\overline{IA}^2 + b\,\overline{IB}^2 + c\,\overline{IC}^2.$$

10. Letting $\overline{AC} = b$ and $\overline{AB} = c$, use the sine and cosine laws to show that

$$\overrightarrow{CD} \cdot \overrightarrow{OE} = \left(- \overrightarrow{BC} + \frac{1}{2}\ \overrightarrow{BA}\right) \cdot \left(- \overrightarrow{BO} + \frac{1}{2}\ \overrightarrow{BA} + \frac{1}{3}\ \overrightarrow{BC}\right)$$
$$= \frac{1}{6}(b^2 - c^2).$$

11. For item (e), set $q = 2\sin\frac{\pi}{n}$, so that $l_k = l_0 q^k$. Since $n > 6 \Rightarrow 0 < q < 1$, basic Real Analysis (cf. Section 7.2 of [5], for instance) gives $l_k \to 0$ as $k \to +\infty$. Now, observe that the distance between two points with integer coordinates is at least 1, so that it is impossible for $\mathcal{P}_k$ to have vertices of integer coordinates if k is sufficiently large.

Section 9.1

1. Apply the inversion of center A and ratio $\overline{AC}^2$, so that α_C and β_C remain unchanged. If the inverse r of Γ forms and angle $\theta \in \left(0, \frac{\pi}{2}\right)$ with AB, show that $r_C = \overline{B'C} \cdot \tan\frac{\theta}{2}$ and $R_C = \overline{B'C} \cdot \cot\frac{\theta}{2}$, where B' stands for the inverse of B.

2. Apply the inversion of center A and ratio $\overline{AC}$ to show that the desired equality is equivalent to $\overline{B'D'}^2 = \overline{B'C'}^2 + \overline{C'D'}^2$. Then, use the orthogonality condition to show that this relation does hold.

3. Apply an inversion centered at a point of Γ to reduce the present problem to Problem 8, 148.

4. Apply an inversion of center A to reduce the problem to that of constructing a line tangent to two nonconcentric given circles.

5. Let $\Gamma_1(O_1; R_1)$ and $\Gamma_2(O_2; R_2)$, with $R_1 < R_2$ (the case $R_1 = R_2$ can be dealt with analogously). If $\Gamma(O; R)$, then the circle $\tilde{\Gamma}(O; R_1 + R)$ passes through O_1, is tangent to $\tilde{\Gamma}_2(O_2; R_2 + R)$ and to the line s, parallel to r at a distance R_1 and lying in the half-plane opposite to that of Γ_1 and Γ_2. This reduces the present problem to the previous one.

6. The inversion of pole N and ratio $\overline{NC} = \overline{ND}$ leaves α and β fixed, maps $\overleftrightarrow{AB}$ into the circumcircle of NPQ, $\overleftrightarrow{CP}$ into the circumcircle of ACN and $\overleftrightarrow{DQ}$ into the circumcircle of BDN.

7. The inversion of center B and ratio $\overline{AB}$ transforms α and β in two perpendicular lines passing through A and the circumcircle of BCD in $\overleftrightarrow{C'D'}$, which passes through the center of the circumcircle of $AC'D'$.

8. Choose one of the given points, say O, and apply an inversion of center O. Show that the set of the inverses of the $n-1$ remaining points satisfies the following condition: any line passing through two of them contains a third one. Then, apply the result of Problem 36, page 119.

9. If Σ is orthogonal to Γ and $\Sigma \cap \Gamma = \{P, Q\}$, then $P' = P$, $Q' = Q$ and (from the orthogonality) $\overleftrightarrow{OP}$ and $\overleftrightarrow{OQ}$ are tangent to Σ. Hence, Σ' passes trough P, Q and is orthogonal to $\Gamma' = \Gamma$, whence $\Sigma' = \Sigma$. The converse is analogous.

10. Let $\overline{AP} = \overline{AQ} = r$ and α be the circumcircle of APQ. The inversion of center A and ratio r maps α into $\overleftrightarrow{PQ}$ and (from the result of the previous problem) leaves Σ fixed. If M_a is the midpoint of BC, then $A\widehat{P}M_a = A\widehat{Q}M_a = 90°$, so that AM_a is a diameter of α. Letting H_a be the foot of the perpendicular dropped from A to BC, we also have $A\widehat{H_a}M_a = 90°$, whence $H_a \in \alpha$. Therefore, $H_a' \in \overleftrightarrow{PQ}$. It now suffices to show that $H_a' = H$. To this end, let H_c be the foot of the perpendicular dropped from C to AB, and note that $H_c = B'$. Now, use the fact that $BH_cH_a'H_a = BB'H_a'H_a$ and BH_cHH_a are cyclic to conclude that $H_a' = H$.

11. Let I be the inversion with center A, with respect to which Γ_n remains fixed. Then, I transforms α and β into the tangents to Γ_n perpendicular to $\overleftrightarrow{AC}$. In turn, show that this forces $\Gamma_{n-1}, \ldots, \Gamma_0$ to be transformed into circles congruent to Γ_n all lying in the strip of the plane bounded by α' and β'. The result now follows at once.

12. Along this hint, given an inscribed polygon $\mathcal{P}$ we write $(\mathcal{P})$ for its circumcircle. Apply the inversion of pole A and modulus $\overline{AF} \cdot \overline{AB}$, so that $F' = B$, $E' = C$

and $(AFHE)' = \overleftrightarrow{BC}$. Then, $H' = D'$ and $\overleftrightarrow{FE}' = (ABC)$. This gives G' as the other point of intersection of $\overleftrightarrow{AD}$ and (ABC) (namely, $G' \neq A$). On the other hand, $(DEF)' = (BHC)$, which shows that $N' = M$. Thus, since $S \in (DEF)$, we have S' is the other point of intersection of $\overleftrightarrow{AD}$ and (BHC) (i.e., $S' \neq H$). By symmetry with respect to $\overleftrightarrow{BC}$, we also have $HMA \cong G'MS'$. Now, use the fact that both $SNMS'$ and $GNMG'$ are cyclic.

13. Essentially all of the results we have studied remain true, with the same proofs; it suffices for the reader to rework them. A notable exception is the inversion circle $\Gamma(O; k)$: although we have $I(\Gamma) = \Gamma$, for $A \in \Gamma$ we now have $I(A) = B$, with AB being a diameter of Γ.

14. Suppose first that A lies outside the disk bounded by Γ. The inversion of center A and ratio $k = \sqrt{\overline{AE} \cdot \overline{AF}}$ (which is constant, by the intersecting chords theorem) maps E to D and F to C, hence maps the circumcircle of AEF to the line $r = \overleftrightarrow{EF}$. If B is the inverse of B, show that B depends only on A and B. Then conclude that the circumcenter of AEF lies in the perpendicular bisector of AB. Now, assume that A lies in the disk bounded by Γ. Argue as above, this time considering the opposite inversion of ratio k.

14. Rework the proofs of the results of this section in this setting.

15. Firstly, show that P, M and N are collinear (an elementary argument works). Then, note that the opposite inversion of center P and ratio k, with $k^2 = \overline{PA} \cdot \overline{PA}$ (which is constant, by the intersecting chords theorem) transforms the circumcircle of BPM into the line passing through B and parallel to AA. Then, use the collinearity of M, N and P to conclude that the inverse of M is the point M such that $PAMB$ a rectangle. Finally, note that

$$\overline{PM} \cdot \overline{PN} = \frac{1}{2}\overline{PM} \cdot \overline{PM} = \frac{1}{2}PA \cdot PA.$$

16. For item (a), use Proposition 3.36 to show that the inversion circle has center M and passes through D and J, hence is orthogonal to Γ and Σ. For item (b), use the angle bisector theorem to calculate $\overline{MP}$; for $\overline{MH_a}$, use some Trigonometry in the right triangle ABH_a and the cosine law to compute $\overline{MH_a} = \frac{a}{2} - c\cos\widehat{B} = \frac{b^2-c^2}{2a}$. For (c), start by observing that $\overleftrightarrow{RS}$ is also tangent to Γ and Σ. Then, note that $\overline{MN} = \frac{c}{2}$, and that $\overleftrightarrow{MN} \| \overleftrightarrow{AB}$ implies $MPQ \sim BPR$; in turn, show that such a similarity allows us to compute $\overline{MQ}$. Finally, for (d), note that items (b) and (c) imply that $\overleftrightarrow{RS}$ is the inverse of the nine-point circle, and that inversions preserve tangency.

Section 9.2

1. For (a) $\Rightarrow$ (b), let I be the inversion of center O and some modulus $k > 0$, and let r be the inverse of Γ. Since a is not tangent to Γ, we conclude that a meets r, say at A_1. Analogously, b, c, d meet r, say at B_1, C_1, D_1, and (A_1, B_1, C_1, D_1) is a harmonic quadruple. Then, exactly one of C_1, D_1 lies in $A_1 B_1$, and this implies $AB \cap CD \neq \emptyset$. Also,

$$\frac{\overline{A_1 C_1}}{\overline{C_1 B_1}} = \frac{k^2 \cdot \overline{AC}}{\overline{OA} \cdot \overline{OC}} \cdot \frac{\overline{OC} \cdot \overline{OB}}{k^2 \cdot \overline{CB}} = \frac{\overline{AC}}{\overline{CB}} \cdot \frac{\overline{OB}}{\overline{OA}},$$

and likewise $\frac{\overline{A_1 D_1}}{\overline{D_1 B_1}} = \frac{\overline{AD}}{\overline{DB}} \cdot \frac{\overline{OB}}{\overline{OA}}$. Hence, $\frac{\overline{AC}}{\overline{CB}} = \frac{\overline{AD}}{\overline{DB}}$ follows from $\frac{\overline{A_1 C_1}}{\overline{C_1 B_1}} = \frac{\overline{A_1 D_1}}{\overline{D_1 B_1}}$. The converse can be established quite similarly.

2. Apply the result of the previous problem twice, with Γ equal to the circle of diameter OO_1.

3. Start by choosing a points $O \notin \overleftrightarrow{AB}$, with $\overline{OA} \neq \overline{OB}$, and an arbitrary point $Q \in \overleftrightarrow{OC} \setminus \{O, C\}$. Then, plot U and V as the intersections of $\overleftrightarrow{AQ}$ and $\overleftrightarrow{BO}$, $\overleftrightarrow{AO}$ and $\overleftrightarrow{BQ}$. Finally, plot D as the intersection of $\overleftrightarrow{AB}$ and $\overleftrightarrow{UV}$.

Section 9.3

1. If p_i is the polar of P with respect to Γ_i, show that Q is the intersection of p_1 and p_2.

2. The result can read like this: "*Let ABCD be an isosceles trapezoid with bases AD, BC and circumcircle* Γ. *If* $AC \cap BD = \{R\}$ *and* $\overleftrightarrow{AB} \cap \overleftrightarrow{CD} = \{P\}$, *then the perpendicular to* $\overleftrightarrow{PR}$ *at R is the polar of P with respect to* Γ". For the proof, adapt that of Theorem 9.23.

3. Use the straightedge to draw two secants $\overleftrightarrow{AB}$ and $\overleftrightarrow{CD}$ to Γ, both passing through P. If the convex quadrilateral of vertices A, B, C, D has diagonals AC and BD, show that $\overleftrightarrow{AB}$ and $\overleftrightarrow{CD}$ can be chosen so that $\overleftrightarrow{AD}$ and $\overleftrightarrow{BC}$ are not parallel. Then, use the straightedge again to obtain the intersection points Q of $\overleftrightarrow{AD}$, $\overleftrightarrow{BC}$ and R of $\overleftrightarrow{AC}$, $\overleftrightarrow{BD}$. Since $\overleftrightarrow{QR}$ is the polar of P with respect to Γ, we can plot the intersection points T_1, T_2 of $\overleftrightarrow{QR}$ and Γ. Then, $\overleftrightarrow{PT_1}$ and $\overleftrightarrow{PT_2}$ are tangent to Γ.

4. If such tangents meet at P, use Theorem 9.23 to show that $\overleftrightarrow{XY}$ is the polar of P with respect to Γ. Now, recall that if H is the orthocenter of ABC, then it is the incenter of its orthic triangle. Finally, use the result of Example 4.23.

5. Let H_b and H_c stand for the feet of the altitudes relative to B and C, respectively. Since $\overleftrightarrow{PQ}$ is the polar of A with respect to Σ and $A \in \overleftrightarrow{BH_c}, \overleftrightarrow{CH_b}$, we know that $\overleftrightarrow{PQ}, \overleftrightarrow{BH_b}$ and $\overleftrightarrow{CH_c}$ are concurrent.

6. Use the fact that N belongs to the polars of A and D to conclude that $\overleftrightarrow{AD}$ is the polar of N; accordingly, show that $\overleftrightarrow{BE}$ is the polar of M. Finally, use these two facts to show that $\overleftrightarrow{MN}$ is the polar of P.

7. Since $ABCD$ is not a rectangle, we have $\widehat{B} \neq 90°$; thus, we let $\overleftrightarrow{AD} \cap \overleftrightarrow{BC} = \{P\}$. Now, let Γ denote the circumcircle of $ABCD$ and $\overleftrightarrow{QT}$ be one of the tangents to Γ drawn from Q ($T \in \Gamma$). Point T can be constructed as the intersection of the perpendicular to $\overleftrightarrow{OQ}$ through R and the circle of diameter OQ, and after that we can also draw Γ. Since R is the orthocenter of OPQ (cf. Brocard's theorem), P can be found as the intersection of $\overleftrightarrow{RT}$ with the perpendicular to $\overleftrightarrow{OR}$ through Q. Finally, A is one of the intersection points of Γ and the circle of diameter PQ.

8. If $MP \cap NQ = \{R\}$, then $\overleftrightarrow{UV}$ is the polar of R and $\overleftrightarrow{MP}$ is the polar of W. Since $R \in \overleftrightarrow{MP}$, we conclude that $W \in \overleftrightarrow{UV}$. Now, argue in a similar way to see that $X \in \overleftrightarrow{UV}$.

9. Let $MP \cap NQ = \{R\}$ and Γ be the circle inscribed in $ABCD$. First assume that $\overleftrightarrow{MN} \not\parallel \overleftrightarrow{PQ}, \overleftrightarrow{MQ} \not\parallel \overleftrightarrow{NP}$, and set $\overleftrightarrow{MN} \cap \overleftrightarrow{PQ} = \{U\}, \overleftrightarrow{MQ} \cap \overleftrightarrow{NP} = \{U\}$. Then, $\overleftrightarrow{RU}$ is the polar of V and $\overleftrightarrow{RV}$ is the polar of U. Moreover, since $\overleftrightarrow{AM}$, $\overleftrightarrow{AQ}$ are tangent to Γ and $U \in \overleftrightarrow{MQ}$, we conclude that A belongs to the polar of U, i.e., $A \in \overleftrightarrow{RV}$. Accordingly, $C \in \overleftrightarrow{RV}$, whence, A, C, R are collinear. Likewise, B, D, R are collinear. Now, if $\overleftrightarrow{MN} \parallel \overleftrightarrow{PQ}$ but $\overleftrightarrow{MQ} \not\parallel \overleftrightarrow{NP}$, then a similar reason shows that A, C, R are collinear, whereas an elementary argument established the collinearity of B, D, R. Finally, if $\overleftrightarrow{MN} \parallel \overleftrightarrow{PQ}$ and $\overleftrightarrow{MQ} \parallel \overleftrightarrow{NP}$, then $MNPQ$ is a rectangle and $ABCD$ is a rhombus, and there is nothing left to do.

10. Let α be the circumcircle of BKN and β be that of ABC. Let I be the inversion of center B that leaves γ fixed, so that $K' = A, A' = K, N' = C, C' = N$, $\alpha' = \overleftrightarrow{AC}$ and $\beta' = \overleftrightarrow{KN}$. Use the fact that M lies in both α and β to show that $\overleftrightarrow{BM}, \overleftrightarrow{AC}$ and $\overleftrightarrow{KN}$ concur at M'. Now, let P and Q be the feet of the tangents drawn from B to γ, so that $\overleftrightarrow{PQ}$ is the polar of B with respect to γ. Since $\overleftrightarrow{AK} \cap \overleftrightarrow{CN} = \{B\}$ and $\overleftrightarrow{KN} \cap \overleftrightarrow{AC} = \{M'\}$, we conclude that M', P and Q are collinear. Use the fact that $\overleftrightarrow{BO} \cap \overleftrightarrow{PQ} = \{O'\}$ and $M = M''$ to show that $M'OO'M$ is cyclic, whence $M'\widehat{M}O = M'\widehat{O'}O = 90°$.

Section 9.4

1. The stated condition amounts to $\frac{AC}{CB} \cdot \frac{D_1 B}{AD_1} = \frac{AC}{CB} \cdot \frac{D_2 B}{AD_2}$. This in turn gives $\frac{D_1 B}{AD_1} = \frac{D_2 B}{AD_2}$, so that D_1 and D_2 divide AB, as an oriented segment, in the same ratio. But we already know that this implies $D_1 = D_2$.

2. Check that all items are straightforward consequences of the definition of cross ratio.

3. Apply items (b) and (d) of the previous problem to get $(B, A; C, D) = \frac{1}{(A,B;C,D)}$. Then, conclude that $(A, B; C, D) = \pm 1$, and exclude the possibility 1 by applying item (a) of the previous problem.

4. For the first part, let r be the line passing through A, B, C, D and take a point $O \notin r$. Let $s \neq r$, $\overleftrightarrow{AO}$ be a line passing through A and intersecting $\overleftrightarrow{OB}$, $\overleftrightarrow{OC}$, $\overleftrightarrow{OD}$ respectively at points B_1, C_1, D_1. Show that we can choose s in such a way that $\overleftrightarrow{OB}$ and $\overleftrightarrow{C_1 D}$ meet at a point E, and then note that

$$ABCD \; \overset{O}{\barwedge} \; AB_1 C_1 D_1 \; \overset{D}{\barwedge} \; BB_1 EO \; \overset{C_1}{\barwedge} \; BADC.$$

For the second part, note that projectivities preserve cross ratios, so that the geometric construction delineated above assures that $(A, B; C, D) = (B, A; D, C)$.

5.

$$ABPR \; \overset{D}{\barwedge} \; OQSR \; \overset{C}{\barwedge} \; BAPR.$$

For item (b), since projectivities clearly preserve cross ratios, we have from (a) that $(A, B; P, R) = (B, A; P, R)$. Now, use the result of Problem 3 to conclude that $(A, B; P, R) = -1$, and note that this finishes the proof.

6. Argue as in the proof of case (i).

7. Apply Brianchon's theorem to the hexagrams $AMBCPD$ and $ABNCDQ$.

8. Check that the proof presented for Pascal's theorem also works in this case, verbatim.

9. As in the proof of the ordinary version of Pascal's theorem, let Γ be the circle circumscribed to $ABCDEF$, let $\overleftrightarrow{CD} \cap \overleftrightarrow{AE} = \{U\}$ and $\overleftrightarrow{BD} \cap \overleftrightarrow{CE} = \{V\}$. Also as there,

$$\left(\overleftrightarrow{AC}, \; \overleftrightarrow{AY}, \; \overleftrightarrow{AU}, \; \overleftrightarrow{AD} \right) = \left(\overleftrightarrow{AC}, \; \overleftrightarrow{AF}, \; \overleftrightarrow{AE}, \; \overleftrightarrow{AD} \right)$$

$$\overset{\Gamma}{\barwedge} \left(\overleftrightarrow{BC}, \; \overleftrightarrow{BF}, \; \overleftrightarrow{BE}, \; \overleftrightarrow{BD} \right)$$

$$= \left(\overleftrightarrow{BC}, \; \overleftrightarrow{BX}; \; \overleftrightarrow{BE}, \; \overleftrightarrow{BV} \right),$$

whence $(C, Y; U, D) = (C, X; E, V)$. Now, use the definition of cross ratio to get $\frac{CD}{DY} \cdot \frac{UY}{CU} = \frac{CV}{VX} \cdot \frac{EX}{CE}$. Use this and the given parallelism (via Thales' theorem) to arrive at $\frac{DY}{UY} = \frac{VX}{EX}$, and hence at $\frac{DY}{UY} = \frac{VX}{EX}$. Finally, obtain $\frac{DY}{DU} = \frac{VX}{VE}$ and use the parallelism of $\overleftrightarrow{DV}$ and $\overleftrightarrow{VE}$ to get $\overleftrightarrow{XY} \parallel \overleftrightarrow{DV} = \overleftrightarrow{BD}$.

10. First of all, note that points P, Q, R, S lie in the circle of diameter AD. Then, let $\overleftrightarrow{PR} \cap \overleftrightarrow{QS} = \{E\}$ and apply Pascal's theorem to the hexagram $ARSTQP$ to conclude that B, C, E are collinear.

Section 10.1

1. Pick a plane α in space. Since α is properly contained in space, we can choose a point $A \notin \alpha$. Now, by choosing three noncollinear points $B, C, D \in \alpha$, we obtain the noncoplanar points A, B, C and D. It is immediate to check that the planes (ABC), (ABD), (ACD) and (BCD) are pairwise distinct, and are all of the planes containing three of these four points.
2. In each of the cases listed in Proposition 10.6, compute the number of regions into which the space gets divided by the planes α, β and γ.
3. Apply the fifth postulate of Euclid in the plane (A, r).
4. If r, s and t all lie in a single plane, there is nothing to do. Otherwise, let $\alpha = (r, s)$, $\beta = (s, t)$ and $\gamma = (A, r)$, where A is a point of t. If $\beta \cap \gamma = t'$, then Proposition 10.6 assures that r, s and t' are parallel. However, since $A \in t, t' \subset \beta$ and $t, t' \parallel s$, we have $t = t'$; in particular, $r \parallel t$.
5. Show that the points of intersection all lie in the line of intersection of α and α'.
6. Apply the usual version of Thales' theorem to the plane (AXY).
7. Since $\overleftrightarrow{XY} \parallel \overleftrightarrow{X'Y'}$, we have $AXY \sim AX'Y'$; analogously, $AYZ \sim AY'Z'$. It thus follows that $\frac{XY}{X'Y'} = \frac{AY}{AY'} = \frac{YZ}{Y'Z'}$. Likewise, $\frac{XY}{X'Y'} = \frac{XZ}{X'Z'}$, so that $XYZ \sim X'Y'Z'$ by the SSS case.

Section 10.2

1. Fix a point A in space and draw, through A, line s_i parallel to r_i. Then, s_1, $s_2, \ldots, s_n$ all pass through A and are pairwise perpendicular. Now, show that $n = 3$.
2. Suppose first that $r \perp \beta$. Draw distinct planes γ and δ containing r, and let s and s' (resp. t and t') be the intersections of γ (resp. of δ) with α and β, respectively. Then $r \perp \alpha \Rightarrow r \perp s, t$ and $r \perp \beta \Rightarrow r \perp s', t'$. However, since $s, s' \subset \gamma$ and $t, t' \subset \delta$, it follows that $s \parallel s'$ and $t \parallel t'$. Hence, Proposition 10.4 assures that α and β are parallel planes. Conversely, if $\alpha \parallel \beta$ then we can choose concurrent

lines $s, t \subset \alpha$ and $s', t' \subset \beta$ such that $s \parallel s'$ and $t \parallel t'$. Since $r \perp \alpha$, we have $r \perp s, t$, and thus $r \perp s', t'$. Proposition 10.8 now assures that $r \perp \beta$.

3. Pick a plane $\beta \parallel \alpha$ and, with the aid of Corollary 10.9, draw the line r, passing through A and perpendicular to β. Now, use the result of the previous problem.

4. We have to prove that r and s are coplanar and have no common points. To this end, let A be the point of intersection of r with α, and r' be the parallel to s passing through A, so that $r' \perp \alpha$. If $r' \neq r$, we would have two lines passing through A and perpendicular to α, which is an absurd; thus $r' = r$. Use this fact to finish the proof.

5. Let $\alpha \cap \beta = s$, so that $r \perp s$. Since $\beta \perp \alpha$, plane β contains a line t perpendicular to α, and hence orthogonal or perpendicular to every line of α; in particular, $t \perp r$. Thus, s and t are concurrent lines of β with $r \perp s, t$, so that (from Proposition 10.8) $r \perp \beta$.

6. For the existence part, first consider the case in which the given point A does not lie in the given line r. Draw, in the plane (A, r), line s, passing through A and perpendicular to r at B. Then, draw another plane β containing r and, in β, draw line t, passing through B and perpendicular to r. Finally, since s and t are concurrent and such that $r \perp s, t$, we have $r \perp (s, t)$.

7. Let β be a plane containing r and perpendicular to α. By definition, β contains a line s such that $s \perp \alpha$. Since $r \not\perp \alpha$, we have $r \not\parallel s$ (otherwise, since s is orthogonal to two concurrent lines of α, the same would happen with r, and we would have $r \perp \alpha$). Now, let A be the intersection point of r and s, let β' be another plane containing r and perpendicular to α, and pick a line $s' \subset \beta'$ with $s' \perp \alpha$. Then, $s' \neq s$ (for otherwise $\beta' = (r, s') = (r, s) = \beta$) and we can assume, without loss of generality, that $A \in s'$ (for, if this is not the case, exchange s' by the parallel to it passing through A). Letting B and B' be the intersections of s and s' with α, respectively, it follows from $s, s' \perp \alpha$ that $s, s' \perp \overleftrightarrow{BB'}$. Hence, in the plane (s, s'), the sum of the angles of the triangle ABB' would be larger than $180°$, which is impossible.

8. Let r and s be given reverse lines. In the notations of the proof of Theorem 10.12, if $A'B'$ is a common perpendicular to r and s, then $\gamma = (r, \overleftrightarrow{A'B'})$ is perpendicular to α and, hence, also to β. Now, apply the uniqueness of the plane perpendicular to α and containing r, established in the previous problem, to obtain the desired uniqueness.

9. Let C and C' be the feet of the perpendicular dropped from A and A' to β, respectively. The argument that guaranteed the consistency of the definition of distance between two parallel planes assures that $\overline{AC} = \overline{A'C'}$. Moreover, since $B\widehat{A}C = B'\widehat{A'}C'$ and $A\widehat{C}B = A'\widehat{C'}B' = 90°$, triangles ABC and $A'B'C'$ are congruent by ASA.[1] Hence, $\overline{AB} = \overline{A'B'}$.

---

[1]Here, we are implicitly assuming that one can apply the usual cases of congruence of triangles, even though the triangles under consideration do not lie in a single plane.

10. Let C and C' be the feet of the perpendicular dropped from A and A' to β, respectively. The argument that guaranteed the consistency of the definition of distance between two parallel planes assures that $\overline{AC} = \overline{A'C'}$. Moreover, since $B\widehat{A}C = B'\widehat{A'}C'$ and $A\widehat{C}B = A'\widehat{C'}B' = 90°$, triangles ABC and $A'B'C'$ are congruent by ASA.[2] Hence, $\overline{AB} = \overline{A'B'}$.

11. Let A, B, C and D be the given points and α be a plane such that A, B, C and D are at equal distances from α. There are two distinct possibilities: (i) A and B lie in one of the half-spaces determined by α, whereas C and D lie in the other: if M and N (resp. P and Q) denote the midpoints of AC and BC (resp. of AD and BD), respectively, then, $M, N, P, Q \in \alpha$, by the previous problem. On the other hand, by applying the midsegment theorem to ABC and ABD, we conclude that $\overleftrightarrow{MN}, \overleftrightarrow{PQ} \| \overleftrightarrow{AB}$; hence, $\overleftrightarrow{MN} \| \overleftrightarrow{PQ}$, so that M, N, P and Q are indeed coplanar and $\alpha = (MNP) = (MNQ)$. (ii) A, B and C all lie in one of the half-spaces determined by α, while D lies in the other: it follows from the previous problem that α passes through the midpoints P, Q and R of the line segments AD, BD and CD, respectively. On the other hand, since A, B and C are noncollinear, the same happens with P, Q and R, so that $\alpha = (PQR)$.

12. If I denotes the incenter of ABC and t is the straightline perpendicular to (ABC) at I, we shall show that D can be taken in t. To this end, let $M \in BC$, $N \in AC$ and $P \in AB$ be the points where the incircle of ABC touches its sides. If r is the inradius and $2p$ is the perimeter of ABC, then Pythagoras' theorem gives

$$\overline{DI}^2 = \overline{AD}^2 - \overline{AI}^2 = \overline{AD}^2 - \overline{AN}^2 - r^2$$
$$= \overline{AD}^2 - (p - a)^2 - r^2.$$

By the same token, $\overline{DI}^2 = \overline{BD}^2 - (p-b)^2 - r^2 = \overline{CD}^2 - (p-c)^2 - r^2$, so that one must have

$$\overline{AD}^2 - (p-a)^2 = \overline{BD}^2 - (p-b)^2 = \overline{CD}^2 - (p-c)^2.$$

However, if $AD = \sqrt{bc}$, $BD = \sqrt{ac}$ and $CD = \sqrt{ab}$, the equalities above are indeed satisfied. In order to finish the problem, one needs also to prove that $bc \geq (p-a)^2 + r^2$ (for $\overline{DI}^2 \geq 0$), and likewise $ac \geq (p-b)^2 + r^2$, $ab \geq (p-c)^2 + r^2$. In turn, this is the same as proving that $bc \geq \overline{AI}^2$, $ac \geq \overline{BI}^2$, $ab \geq \overline{CI}^2$, and this was done in Problem 13, page 255.

13. Let $\overleftrightarrow{AB}$ be the common perpendicular of r and s, with $A \in r$ and $B \in s$, and let M be the midpoint of AB. If $r' \| r$ and $s' \| s$ pass through M and $\alpha = (r', s')$,

---

[2]Here, we are implicitly assuming that one can apply the usual cases of congruence of triangles, even though the triangles under consideration do not lie in a single plane.

then $\overleftrightarrow{AB} \parallel r'$, s'. If t is the angle bisector of one of the angles formed by r' and s' and $P \in t$, show that P lies at equal distances from r and s.

Section 10.3

1. Fix a point $A \in \alpha$ and draw line r, perpendicular to α through A. Mark on r points B and B' such that $\overline{AB} = \overline{AB'} = d$. If β and β' respectively denote the planes passing through B and B' and parallel to α, show that the desired locus is the union of β and β'.

3. Look at the intersections of the bisector planes of the dihedral angles $\alpha_+ \cap \beta_+$, $\alpha_+ \cap \gamma_+$ and $\beta_+ \cap \gamma_+$.

4. For (a), let $\mathcal{P} = A_1 A_2 \ldots A_n$ and γ_i be the plane perpendicular to β and containing $\overleftrightarrow{A_i A_{i+1}}$, with $A_{n+1} = A_1$ (in each case there is only one such plane, for $\overleftrightarrow{A_i A_{i+1}}$ is not perpendicular to β). For $1 \leq i \leq n$ we cannot have $\gamma_i = \gamma_{i+1}$, (with $\gamma_{n+1} = \gamma_1$), for otherwise $\overleftrightarrow{A_i A_{i+1}} = \gamma_i \cap \alpha = \gamma_{i+1} \cap \alpha = \overleftrightarrow{A_{i+1} A_{i+2}}$ and A_i, A_{i+1}, A_{i+2} (with $A_{n+2} = A_2$) would be collinear. Also, since $\mathcal{P}$ is contained in a single half-space of those determined by γ_i, then so is $\mathcal{Q}$. Therefore, $\mathcal{Q}$ is the intersection of β (which is a convex set) with n of the half-spaces determined by the γ_i's (one half-space for each plane). Thus, $\mathcal{Q}$ is also a convex set, hence a convex n-gon. For item (b) note firstly that (by tiling $\mathcal{P}$ in triangles) it suffices to establish the formula for triangles. Secondly, let $\alpha \cap \beta = r$ and look at the case of a triangle ABC with $\overleftrightarrow{BC} \parallel r$. Thirdly, consider a triangle ABC such that $\overleftrightarrow{AB} \cap r = \{F\}$, $\overleftrightarrow{AC} \cap r = \{E\}$, $\overleftrightarrow{BC} \cap r = \{D\}$. If $ABC \cap r = \emptyset$, write $A(ABC) = A(AEF) - A(CDE) + A(BDF)$ and apply the previous case to each such triangle; if $ABC \cap r \neq \emptyset$, adapt this argument to establish the formula.

5. Let α be a plane containing $\overleftrightarrow{O_1 O_2}$, and $\Gamma_j = \Sigma_j \cap \alpha$ for $j = 1, 2$. Then Γ_j is an equator of Σ_j, and Problem 7, page 82, shows that $\Gamma_1 \cap \Gamma_2$ consists of two points A and A', symmetric with respect to $\overleftrightarrow{O_1 O_2}$. By letting α rotate around $\overleftrightarrow{O_1 O_2}$, we conclude that $\Sigma_1 \cap \Sigma_2$ is equal to the union of the points A thus obtained, i.e., to a circle centered at a point of $\overleftrightarrow{O_1 O_2}$ and contained in a plane perpendicular to $\overleftrightarrow{O_1 O_2}$. But this is precisely what we wished to show.

6. Adapt, to the present case, the proof of Proposition 10.16.

7. Let O be the center and r the medial line of Γ. If B is any point of Γ, the fact that A does not lie in the plane of Γ assures that $\overleftrightarrow{AB} \not\perp r$. Hence, the bisecting plane of AB intersects r at a single point O', so that $\overline{O'A} = \overline{O'B} = \overline{O'C}$ for every $C \in \Gamma$ (note that we have used the result of the previous problem in the

last equality). Letting Σ be the sphere of center O' and radius equal to such a common distance, we conclude that Σ contains A and Γ.

8. According to the previous problem, we can let Σ be the sphere containing A and Γ. Let O be the center of Σ and $A' \in \Sigma$ be the antipodal of A. In plane (AOA_i), if $A'_i \in \overrightarrow{AO}$ is such that $A'_i \widehat{A_i} A = 90°$, then A'_i lies in α_i and in the circle of center O and radius OA, which is an equator of Σ. However, since $A'_i \in \overrightarrow{AO}$ and $A'_i \neq A$, we have $A'_i = A'$.

9. Let Γ_1 and Γ_2 denote the circles, and A and B their points of intersection. Prove that the bisecting plane of AB contains the medial lines of Γ_1 and Γ_2. Then, show that such medial lines are concurrent, say at O, and that O is the center of the desired sphere.

10. Adapt, to the present case, the hint given to the previous problem.

11. Since $(ACP) = (BDP)$, the Theorem of Intersecting Chords assures that A, B, C and D all lie in a circle Γ_1. Analogously, A, B, E and F are in a circle Γ_2. However, since $\overleftrightarrow{AB}$, $\overleftrightarrow{CD}$ and $\overleftrightarrow{EF}$ are noncoplanar, circles Γ_1 and Γ_2 do not lie in a single plane. It thus suffices to apply the result of Problem 9.

12. Firstly, suppose that $\overleftrightarrow{AB}$ is parallel to α, and let Σ be a sphere passing through A and B and tangent to α at T. Let $\beta = (ABT)$, let Γ be the circle of intersection of Σ and β, and t the parallel to $\overleftrightarrow{AB}$ passing through T. Show that t lies in β and is tangent to Γ; then, conclude that $\overline{AT} = \overline{BT}$, so that T belongs to the intersection of the bisector plane of AB with α. Finally, letting r be such a line of intersection and T be any point of it, show that there exists a single sphere tangent to α at T and passing through A and B. Now, assume that $\overleftrightarrow{AB}$ is not parallel to α and let P denote its point of intersection with α. If Σ is a sphere passing through A and B and tangent to α at T, apply the Theorem of Intersecting Chords in the plane (ABT) to get $\overline{PT}^2 = \overline{PA} \cdot \overline{PB}$, so that T lies in the circle of center P and radius $\sqrt{PA \cdot PB}$. Conversely, if $T \in \alpha$ is such that $\overline{PT}^2 = \overline{PA} \cdot \overline{PB}$, then, letting Γ be the circle passing through A, B and T, it follows again from the Theorem of Intersecting Chords that $\overleftrightarrow{PT}$ is tangent to Γ at T. If r is the medial line of circle Γ and s is the line perpendicular to α through T, then r and s are not parallel, yet both lie in the plane that passes through T and is perpendicular to $\overleftrightarrow{PT}$. Hence, r and s are concurrent, and letting O denote its point of intersection, it follows that the sphere of center O and radius OT passes through A and B and is tangent to α at T.

13. Let P be such a point and D be the midpoint of AB. Since $\overleftrightarrow{DP} \subset (ABP)$ and $(ABP) \perp \overleftrightarrow{CP}$, we have $C\widehat{P}D = 90°$; analogously, $C\widehat{O}D = 90°$. On the other hand, since AOB and APB are right triangles of hypotenuse AB, we have $\overline{PD} = \frac{1}{2}\overline{AB} = \overline{OD}$. Hence, Pythagoras' theorem gives

$$\overline{CO}^2 = \overline{CD}^2 - \overline{OD}^2 = \overline{CD}^2 - \overline{PD}^2 = \overline{CP}^2,$$

so that $\overline{CO} = \overline{CP}$. Thus, P belongs to the sphere Σ, of center C and radius $\overline{CO}$. Nevertheless, note that plane (APB) is tangent to Σ and intersects lines r and s; hence, (APB) is not parallel to either r or s, which excludes from Σ the six endpoints of its diameters parallel to r, s or t. Conversely, letting $P \in \Sigma$ be different from these six points, it is immediate to see that the plane tangent to Σ at P intersects r in a point A and s in a point B. Moreover, triangle AOB continues to be right at O, as do triangles CPD and COD. Hence, a reasoning quite similar to that of the first part furnishes $\overline{PD} = \overline{OD} = \frac{1}{2}\overline{AB}$, so that $A\widehat{P}B = 90°$.

15. For item (a), suppose that α is parallel to the generatrices g_1 and g_2. If g is a generatrix of $\mathcal{C}$ distinct from g_1 and g_2, conclude that g intersects α. Draw a plane β, perpendicular to e and not passing through V, and for $j = 1, 2$ let Q_j denote its point of intersection with g_j. If $\Gamma = \beta \cap \mathcal{C}$, we have $Q_1, Q_2 \in \Gamma$. Take points $A_1, A_2 \in \Gamma \setminus \{Q_1, Q_2\}$, with $A_1 A_2 \cap Q_1 Q_2 \neq \emptyset$, and show that the generatrices $\overleftrightarrow{VA_1}$ and $\overleftrightarrow{VA_2}$ intersect α in distinct leaves of $\mathcal{C}$. For item (b), let β be the plane which contains e and is perpendicular to α, and γ be the plane which contains g and is parallel to α. Then $\beta \perp \gamma$ and, if β does not contain g, show that line g', symmetric (in γ) to g with respect to $\beta \cap \gamma$, is a generatrix of the cone, distinct from g and also parallel to α. Finally, apply the result of item (a) to reach a contradiction.

16. For the first part of (a), it suffices to note that if A' is the inverse of $A \neq O$, then $A' \neq O$ and A is the inverse of A'. For the second, $I(\mathcal{F}_2) \subset \mathcal{F}_1$ implies $\mathcal{F}_2 = I(I(\mathcal{F}_1)) = (I \circ I)(\mathcal{F}_1) = \mathcal{F}_1$. For (b), argue exactly as in the planar case.
    For (d), let α be any plane containing $\overleftrightarrow{OX}$ and Γ be the circle of intersection of α and Σ, so that Γ is a great circle of Σ. In α, the inverse of Γ (with respect to the restriction of I to α) is the line r, perpendicular to $\overleftrightarrow{OX}$ and passing through X'. As α turns around $\overleftrightarrow{OX}$, the circle Γ spans Σ and r spans σ, which is the inverse of Σ. For (e) argue as in (d), or apply the result of d together with (a).
    For (f), as in (d) let α be any plane containing $\overleftrightarrow{OX}$, and AB be the diameter of Σ contained in $\overleftrightarrow{OX}$. If Γ is the circle of intersection of α and Σ, then Γ is a great circle of Σ, of diameter AB. In α, the inverse of Γ (with respect to the restriction of I to α) is a circle Γ' of diameter $A'B'$; also, as α turns around $\overleftrightarrow{OX}$, the circles Γ and Γ' span Σ and Σ'. Finally, for (g) argue as in the previous items, i.e., reduce the problem to the plane. For instance, given spheres Σ_1 and Σ_2 tangent at $T \neq O$ and not passing through O, let O_1 and O_2 be their centers and Γ_1 and Γ_2 their equators in the plane $\alpha = (OO_1O_2)$. Note that Γ_1 and Γ_2 are tangent at T, and so are Γ'_1 and Γ'_2 at T'. Thus, show that this implies the tangency of Σ'_1 and Σ'_2 at T'.

17. For (a), note that the points of Γ remain fixed under I, and apply the result of items (d) and (e) of the previous problem. For item (b), use the similarity

of the right triangles ONA'' and OAS. Finally, for (c) we can assume that $\Gamma_1 \neq \Gamma$. Then, let Σ_1 be the sphere of equator Γ_1, so that $S \notin \Sigma_1$. Since Γ_1 is the intersection of Σ_1 and α, we conclude that $I(\Gamma_1)$ is the intersection of $I(\Sigma_1)$ and $I(\alpha)$; however, $I(\Sigma_1)$ is a sphere and $I(\alpha) = \Sigma$, so that $I(\Gamma_1)$ is the intersection of two spheres, hence, a circle.

18. Applying the inversion of center T and ratio $k = 4$, show that Γ_1' and Γ_2' are parallel planes, at a distance of 4 units from each other. If Σ_1, Σ_2, Σ_3 are the original spheres, show that Σ_1', Σ_2', Σ_3' are pairwise tangent spheres of radius 2, lying in the strip of the space bounded by Γ_1' and Γ_2' and tangent to them. Now, let $\Sigma_1(U; r)$ and XY be the diameter of Σ_1 lying along $\overleftrightarrow{TU}$. If V is the center of Σ_1', show that $\overline{TV} = \sqrt{\frac{124}{3}}$ and hence that

$$2r = \frac{k^2 \cdot \overline{X'Y'}}{TX' \cdot TY'} = \frac{4^2 \cdot 4}{\left(\sqrt{\frac{124}{3}} - 2\right)\left(\sqrt{\frac{124}{3}} + 2\right)} = \frac{12}{7}.$$

Section 10.4

1. Adapt the proof of Theorem 10.25 to this case, thus considering a pair of Dandelin spheres. More precisely, draw spheres Σ_1 and Σ_2, both tangent to the plane α and to the leaf of the cone that α intersects; for $j = 1, 2$, let F_j be the point of tangency of Σ_j with α, and let Q_j be the point of intersection of Σ_j with a fixed generatrix g. Show that F_1 and F_2 are the foci, and $\overline{Q_1 Q_2}$ is the major axis of the desired ellipse.

2. We shall sketch the proof in the case of a cone of revolution, the proof for a cylinder of revolution being much easier. The proof parallels the one presented in the text for hyperbolas. More precisely, given an ellipse $\mathcal{E}$ of focal distance $2c$ and major axis $2a$, show that one only needs to assure the existence of spheres $\Sigma_1(O_1; R_1)$, $\Sigma_1(O_2; R_2)$ (the Dandelin spheres for $\mathcal{E}$) such that $\overline{O_1 O_2} > R_1 + R_2$,

$$\sqrt{\overline{O_1 O_2}^2 + (R_2 - R_1)^2} = 2a, \quad \sqrt{\overline{O_1 O_2}^2 - (R_2 + R_1)^2} = 2c.$$

Take any real ℓl satisfying $2a > \ell l > 2c$, set $\overline{O_1 O_2} = \ell l$ and solve the system of equations above for R_1, R_2.

3. In the notations of Fig. 10.36 and of the proof of Corollary 10.28, let α' be the plane parallel to α and passing through O_1. If $\alpha' \cap \Gamma = \{X_1, X_2\}$, let s_1, s_2 be the tangents to Γ at X_1, X_2, respectively. We claim that π maps the asymptotes of $\mathcal{H}$ to s_1, s_2. Indeed, since $X_i \in s_i \cap r_\beta$, if $t_i = \pi^{-1}(t_i)$, then $t_i \cap \mathcal{H} = \emptyset$. On the other hand, if $Z_i \in \Gamma \setminus \{X_i\}$ and $Y_i = \pi^{-1}(Z_i)$, then $Y_i \in \mathcal{H}$ and Y_i approaches

t_i as Z_i approaches s_i. Therefore, there is no other option left for t_i than being an asymptote of $\mathcal{H}$.

4. Adapt, to the present case, the hint given to Problem 3, page 317.
5. The first part follows from item (c) of Proposition 10.29. The second part follows the second part of item (b) of the proposition, together with Example 9.18.
6. Take a second secant s to $\mathcal{C}$ passing through P, with $\mathcal{C} \cap s = \{C, D\}$ and $\overleftrightarrow{AC}$ $\cap \overleftrightarrow{BD} = \{R\}$, $\overleftrightarrow{AD} \cap \overleftrightarrow{BC} = \{Q\}$. Then, $\left(\overleftrightarrow{QA}, \overleftrightarrow{QB}, \overleftrightarrow{QP}, \overleftrightarrow{QU} \right)$ is a harmonic pencil, and the result follows from Theorem 9.16.
7. The given condition is equivalent to Q_i lying in the polar of P with respect to $\mathcal{C}_i$, for $i = 1, 2$. Hence, Proposition 10.32 assures that P must lie in the polar of Q_i with respect to $\mathcal{C}_i$, for $i = 1, 2$. There is at most one such point P.
8. As in other proofs of results of this section, take a central projection that maps the conic to a circle, apply Brianchon's theorem to that circle (and to the corresponding hexagram) and then project backwards.
9. Apply Brianchon's theorem to the hexagrams $AMBCPD$ and $ABNCDQ$.

Section 11.1

1. If Q denotes the point in which AP intersects face BCD, then the triangles ABQ, ACQ and ADQ are congruent by SAS, so that $\overline{BQ} = \overline{CQ} = \overline{DQ}$. Hence, $\overleftrightarrow{AQ}$ is the medial line of BCD and, since $P \in \overleftrightarrow{AQ}$, we conclude that $\overline{PB} = \overline{PC} = \overline{PD}$.
2. Let α stand for the plane of the basis and β for the plane of one of the lateral faces of the pyramid. Show that the center I of an inscribed sphere must necessarily lie in the bisector plane of the dihedral angle formed by α and β and containing the pyramid. Use this to conclude that I is the only point of intersection of such bisector planes, so that it is uniquely determined.
3. If Σ denotes a sphere passing through $V, A_1, A_2, \ldots, A_n$, then the plane of the basis of the pyramid intersects Σ along a circle, which passes through $A_1, A_2, \ldots, A_n$. Conversely, if $A_1 A_2 \ldots A_n$ is cyclic, invoke the result of Problem 7, page 359.
4. The lateral area of the frustum is clearly equal to $\pi(Rg - R'g')$, where g and g' denote the generatrices of the given cone and the cone of revolution of basis Γ' and vertex V. Show that $\frac{g}{R} = \frac{g'}{R'} = \frac{\sqrt{d^2 + (R - R')^2}}{R - R'}$; then, substitute the expressions for g and g' obtained from these equalities to arrive at the desired result.
5. Firstly, note that the medial plane of BC is $\alpha = (ADP)$, where P stands for the midpoint of BC. Hence, α leaves M and N in distinct half-spaces, so that $\overline{BM} < \overline{CM}$. Then, we have $\overline{MN} < \overline{BM} + \overline{BN} < \overline{CM} + \overline{BN}$. Now, if M' denotes the symmetric of M with respect to (ACD), we have $\overline{MN} = \overline{M'N}$ and $\overline{CM} = \overline{CM'}$. Hence,

$$\overline{CM} = \overline{CM'} < \overline{BM'} < \overline{BN} + \overline{M'N} = \overline{BN} + \overline{MN}.$$

Analogously, $\overline{BN} < \overline{CM} + \overline{MN}$.

6. Let P be the foot of the altitude of $ABCD$ relative to A, and M be the intersection of $\overleftrightarrow{BP}$ and $\overleftrightarrow{CD}$. Then, $\overleftrightarrow{AP} \perp (BCD)$ implies $\overleftrightarrow{AP} \perp \overleftrightarrow{CD}$ and, since $\overleftrightarrow{BP} \perp \overleftrightarrow{CD}$, we have $(ABP) \perp \overleftrightarrow{CD}$. If Q is the foot of the altitude of triangle ABM dropped from B, then $\overleftrightarrow{BQ} \subset (ABP)$ implies $\overleftrightarrow{BQ} \perp \overleftrightarrow{CD}$. However, since $\overleftrightarrow{BQ} \perp \overleftrightarrow{AM}$, we have $\overleftrightarrow{BQ} \perp (\overleftrightarrow{AM}, \overleftrightarrow{CD}) = (ACD)$. Hence, BQ is an altitude of the tetrahedron and, clearly, $\overleftrightarrow{AP}$ and $\overleftrightarrow{BQ}$ do intersect. Analogously, the other two altitudes of the tetrahedron are concurrent with AP. Now, suppose that Q is the orthocenter of ACD, and let H denote the point of intersection of $\overleftrightarrow{AP}$ and $\overleftrightarrow{BQ}$, and R the foot of the altitude of the tetrahedron dropped from C. From the first part, $\overleftrightarrow{CR}$ intersects $\overleftrightarrow{AP}$ and $\overleftrightarrow{BQ}$. Also if $\overleftrightarrow{CR}$ intersects $\overleftrightarrow{AP}$ at H', with $H' \neq H$, then H' is the only point of intersection of $\overleftrightarrow{CR}$ with the plane (ABP), so that $\overleftrightarrow{CR} \cap \overleftrightarrow{BQ} = \emptyset$, which is an absurd. Hence, $\overleftrightarrow{CR}$ passes through H and, by the same token, we show that the altitude of $ABCD$ dropped from D also passes through H. It is now relatively easy to show that the feet of the altitudes of the tetrahedron dropped from the vertices C and D are the orthocenters of the corresponding faces.

7. Let α be such a plane, and M, N, P and Q, respectively, be the points at which the edges BC, AC, AD and BD meet α. Then, $\overleftrightarrow{MQ} \| \overleftrightarrow{CD}$, so that triangle BMQ is equilateral, of side length x, say. Analogously, triangles CMN and DPQ are equilateral of side length $a - x$, while ADN is equilateral of side length x. Now, letting R denote the midpoint of CD, we have $\overleftrightarrow{BR}$, $\overleftrightarrow{AR} \perp \overleftrightarrow{CD}$, so that $(ABR) \perp \overleftrightarrow{CD}$; in particular, $\overleftrightarrow{AB} \perp \overleftrightarrow{CD}$. However, since $\overleftrightarrow{MN} \| \overleftrightarrow{AB}$ and $\overleftrightarrow{MQ} \| \overleftrightarrow{CD}$, we have $Q\widehat{M}N = 90°$. It follows that $MNPQ$ is a rectangle of side lengths x and $a - x$, thus of area $-x^2 + ax$. By applying the theory of maxima and minima of quadratic functions, we conclude that the maximum value for the area of $MNPQ$ is found when points M, N, P and Q are the midpoints of the edges they belong to.

8. Let $\Sigma(O; R)$ be the sphere circumscribed to $ABCD$, and k be the common value of the equal products in the statement of the problem. The section of Σ through plane (AOB) is an equator Γ of Σ; by applying the theorem of intersecting chords to Γ, we obtain $k = \overline{AE} \cdot \overline{BE} = R^2 - \overline{OE}^2$, so that $\overline{OE} = \sqrt{R^2 - k}$. Since the same reasoning is valid with respect to the other five edges of the tetrahedron, we conclude that $\overline{OE} = \overline{OF} = \overline{OG} = \overline{OH} = \overline{OI} = \overline{OJ} = \sqrt{R^2 - k}$. Hence, points E, F, G, H, I and J are all situated on the sphere of center O and radius $\sqrt{R^2 - k}$.

9. If R is the midpoint of CD, we have $M \in BR$ and $N \in AR$; thus, AM and BN are cevians of triangle ABR, so that they intersect at a point G. Since $\frac{\overline{RN}}{\overline{RA}} = \frac{1}{3} = \frac{\overline{RM}}{\overline{RB}}$, we have $RMN \sim RBA$ by SAS, so that $\overleftrightarrow{MN} \parallel \overleftrightarrow{AB}$ and $\overline{MN} = \frac{1}{3}\overline{AB}$. In turn, the parallelism of $\overleftrightarrow{MN}$ and $\overleftrightarrow{AB}$ guarantees that $GMN \sim GAB$, and hence that $\frac{\overline{MG}}{\overline{AG}} = \frac{\overline{MN}}{\overline{AB}} = \frac{1}{3}$. We likewise conclude that AM and CP intersect at a point G', whence $\frac{\overline{MG'}}{\overline{AG'}} = \frac{1}{3}$; then, Problem 2, page 108 gives $G = G'$. Analogously, we show that DQ also passes through G.

10. Let V be the apex of the pyramid, $ABCD$ its basis, O the foot of its altitude, M the midpoint of VO and $\alpha = (ADM)$. Then, O is the point of intersection of AC and BD. Moreover, since $A, M \in (AVC)$, the half-line $\overrightarrow{AM}$ intersects VC at a point P; analogously, $\overrightarrow{DM}$ intersects VB at a point Q. Hence, AQ, DP and PQ are all contained in α, so that the desired section is the quadrilateral $ADPQ$. We claim that it is an isosceles trapezoid of bases a and $\frac{a}{3}$. In order to prove this, let l be the length of the lateral edges of the pyramid. By applying Menelaus' theorem to triangle COV, with respect to the collinear points A, M and P, we obtain $\overline{VP} = \frac{1}{2}\overline{CP}$, and hence $\overline{VP} = \frac{1}{3}\overline{VC} = \frac{l}{3}$; analogously, $\overline{VP} = \frac{l}{3}$. Therefore: (i) $\overline{BQ} = \overline{CP}$, so that $ABQ \equiv DCP$ by SAS and then $\overline{AQ} = \overline{DP}$; (ii) $VPQ \sim VCB$ by SAS, whence $\overline{PQ} = \frac{1}{3}\overline{CB} = \frac{a}{3}$ and $\overleftrightarrow{PQ} \parallel \overleftrightarrow{BC}$. However, since $\overleftrightarrow{BC} \parallel \overleftrightarrow{AD}$, we conclude that $\overleftrightarrow{PQ} \parallel \overleftrightarrow{AD}$. Now, if d stands for the length of the altitude of $ADPQ$, Pythagoras' theorem furnishes $d^2 + \left(\frac{\overline{AD}-\overline{PQ}}{2}\right)^2 = \overline{AQ}^2$. It follows from Stewart's relation applied to triangle ABV that $\overline{AQ}^2 = \frac{4l^2+3a^2}{9}$, and the result of Proposition 11.1 gives

$$d^2 = \frac{4l^2 + 3a^2}{9} - \frac{a^2}{9} = \frac{4l^2 + 2a^2}{9} = \frac{4}{9}(a^2 + h^2).$$

Finally, $A(ADPQ) = \frac{d}{2}(\overline{AD} + \overline{PQ}) = \frac{4}{9}a\sqrt{a^2+h^2}$.

11. The feet of the perpendiculars dropped from A_1 and A_3 to $\overleftrightarrow{VA_2}$ coincide, say at P, so that $\theta = A_1\widehat{P}A_3$. If $\overline{A_1P} = \overline{A_3P} = h$, then the cosine law applied to triangle A_1PA_3 gives

$$2h^2(1 - \cos\theta) = \overline{A_1A_3}^2 = \left(2R\sin\frac{2\pi}{n}\right)^2,$$

where R stands for the circumradius of $A_1A_2\ldots A_n$. However, if O denotes the center of $A_1A_2\ldots A_n$, then triangle OA_1A_2 gives $y = 2R\sin\frac{\pi}{n}$, so that

$$4R^2\sin^2\frac{2\pi}{n} = y^2 \cdot \frac{\sin^2\frac{2\pi}{n}}{\sin^2\frac{\pi}{n}} = 4y^2\cos^2\frac{\pi}{n};$$

hence, $1 - \cos\theta = \frac{2y^2 \cos^2 \frac{\pi}{n}}{h^2}$. Finally, if M be the midpoint of $A_1 A_2$, then $\overline{VM} = \sqrt{x^2 - \frac{y^2}{4}}$ and $\overline{VM} \cdot \overline{A_1 A_2} = \overline{A_1 P} \cdot \overline{V A_2}$, so that $h = \sqrt{x^2 - \frac{y^2}{4}} \cdot \frac{y}{x}$. It now suffices to substitute this expression for h into the formula above for $1 - \cos\theta$.

12. If θ is the plane dihedral angle formed by the pair of planes (ABC), (ACD), then the result of the previous problem gives $1 - \cos\theta = \frac{2x^2}{4x^2 - y^2}$. Now, let δ be the other plane dihedral angle, M the midpoint of AB and $\overline{CM} = \overline{GM} = l$. The cosine law applied to triangle CMG gives $2l^2(1 - \cos\delta) = \overline{CE}^2$. In order to compute $\overline{CE}$, let O be the center of $ABDF$ and X and Y be the centers of ABD and ABF. From $CGO \simeq XYO$, $\overline{XY} = \frac{y}{3}$ and $\overline{OX} = \frac{y\sqrt{6}}{12}$ we get $\overline{CE}\sqrt{3} = 2\overline{CO}\sqrt{2}$, so that

$$1 - \cos\delta = \frac{4\overline{CO}^2}{3l^2} = \frac{16\overline{CO}^2}{3(4x^2 - y^2)}.$$

Then

$$\theta = \delta \Leftrightarrow \frac{2x^2}{4x^2 - y^2} = \frac{16\overline{CO}^2}{3(4x^2 - y^2)} \Leftrightarrow x\sqrt{3} = 2\overline{CO}\sqrt{2}.$$

Finally, since $\overline{CO} = \overline{CX} + \overline{XO} = \sqrt{x^2 - \frac{y^2}{3}} + \frac{y\sqrt{6}}{12}$, we conclude that $\theta = \delta$ is equivalent to the equation

$$x\sqrt{3} = 2\left(\sqrt{x^2 - \frac{y^2}{3}} + \frac{y\sqrt{6}}{12}\right)\sqrt{2},$$

which is in turn equivalent to $3x - y = 2\sqrt{6x^2 - 2y^2}$. By squaring both sides and setting $u = \frac{x}{y}$, we easily arrive at $5u^2 + 2u - 3 = 0$, whence $u = \frac{3}{5}$.

13. Let C_i be the foot of the perpendicular dropped from to the line $\overleftrightarrow{A_i A_{i+1}}$. If α stands for the plane angle of the dihedral angle formed by the planes of a lateral face and of the basis of the pyramid, it is immediate to see that $\overline{PB_i} = \overline{PC_i} \tan\alpha$. Hence,

$$\overline{PB_1} + \overline{PB_2} + \cdots + \overline{PB_n} = (\overline{PC_1} + \overline{PC_2} + \cdots + \overline{PC_n})\tan\alpha.$$

Now, Problem 7, page 156, shows that $\overline{PC_1} + \overline{PC_2} + \cdots + \overline{PC_n} = na$, where a denotes the apothem of $A_1 A_2 \ldots A_n$. Therefore, $\overline{PB_1} + \overline{PB_2} + \cdots + \overline{PB_n} = na\tan\alpha$.

14. Apply Weitzenböck's inequality (cf. Problem 7, page 261) to each of the faces of the tetrahedron and, then, add the results.

15. Apply the formula for the squared length of a median to each one of the triangles ABN, BCD and ACD.

16. Draw, through B and C, lines r and s, respectively parallel to $\overleftrightarrow{CD}$ and $\overleftrightarrow{AB}$. If $\alpha = (\overleftrightarrow{AB}, r)$ and $\beta = (\overleftrightarrow{CD}, s)$, then $\alpha \parallel \beta$ and $ABCD$ lies in the region of space situated between α and β. Since the distance $d(\alpha, \beta)$ between α and β is less than or equal to the length of each line segment joining a point of α to a point of β, we have $d(\alpha, \beta) \leq \overline{MN}$, where M and N stand for the midpoints of AB and CD, respectively. Analogously, if P, Q, R and S respectively denote the midpoints of the edges AC, BD, AD and BC, we conclude that $ABCD$ can be placed between two parallel planes situated at a distance from each other less than or equal to $\overline{PQ}$ and $\overline{RS}$. Hence, letting $d = \min\{\overline{MN}, \overline{PQ}, \overline{RS}\}$, it suffices to show that d is less than or equal to the given expression. Finally, note that $3d^2 \leq \overline{MN}^2 + \overline{PQ}^2 + \overline{RS}^2$ and apply the formula of the previous problem to compute $\overline{MN}^2$, $\overline{PQ}^2$ and $\overline{RS}^2$ in terms of the lengths of the edges of the tetrahedron.

17. Let $ABCD$ be a regular tetrahedron of edge length l. Let also M, N, P and Q be points respectively situated in the edges AD, BD, BC and AC, with $\overleftrightarrow{MN} \parallel \overleftrightarrow{AB}$, $\overleftrightarrow{MQ} \parallel \overleftrightarrow{CD}$ and $MNPQ$ being a parallelogram. Since $\overline{MN} = \overline{MD}$ and $\overline{MQ} = \overline{AM}$, we have $\overline{MN} + \overline{MQ} = \overline{MD} + \overline{AM} = l$, so that the perimeter of $MNPQ$ is equal to $2l$. Hence, we can pass $ABCD$ through a rope loop of length $2l$ in the following way: firstly, we start by passing the edge AB through the loop; then, we slide the loop through the surface of the tetrahedron, adjusting it in order to always form a parallelogram (as parallelogram $MNPQ$ above). The tetrahedron will finish traversing the loop through edge CD. In order to see that $2l$ is the smallest possible length, note that there are two essentially distinct ways of beginning the traversal: (i) two of the vertices of $ABCD$ pass simultaneously through the loop: in this case, the length of the loop is, obviously, at least $2l$. (ii) One of the vertices of $ABCD$, say A, is the first to pass through the loop: suppose, without loss of generality, that B is the second vertex to pass through the loop. At this moment, the loop of minimal length will be tightly adhered to the surface of $ABCD$, thus forming a triangle BPQ, with $P \in AC$ and $Q \in AD$. Since $\overline{BP} = \overline{DP}$ and $\overline{BQ} = \overline{CQ}$, the length of the loop is at least $\overline{DP} + \overline{CQ} + \overline{PQ}$. However, the result of Problem 11, page 44, shows that such a sum is at least $2l$.

18. Let α be a plane parallel to $\overleftrightarrow{AC}$ and $\overleftrightarrow{BD}$, and A', B', C', D' denote the orthogonal projections of A, B, C, D into α. Apply Ptolemy's inequality (cf. Theorem 7.37) to the convex quadrilateral $A'B'C'D'$, noticing that $\overline{A'C'} = \overline{AC}$, $\overline{B'D'} = \overline{BD}$, $\overline{A'B'} < \overline{AB}$ etc.

19. For $1 \leq i \leq 4$, let r_i denote the radius of $\mathcal{S}_i$. For $1 \leq i < j \leq 4$, since $\overleftrightarrow{A_iA_j}$ joins the centers of $\mathcal{S}_i$ and $\mathcal{S}_j$, such spheres are tangent at a point $P_{ij} \in \overleftrightarrow{A_iA_j}$. Thus, we have $\overline{A_iP_{ij}} = r_i$, $\overline{A_jP_{ij}} = r_j$ and $\overline{A_iA_j} = r_i + r_j$. It then follows that, for $\{i_1, i_2, i_3, i_4\} = \{1, 2, 3, 4\}$,

$$r_{i_1} = \overline{A_{i_1}A_{i_2}} + \overline{A_{i_1}A_{i_3}} + \overline{A_{i_1}A_{i_4}} - (\overline{A_{i_2}A_{i_3}} + \overline{A_{i_2}A_{i_4}} + \overline{A_{i_3}A_{i_4}}).$$

For $1 \le i < j \le 4$, let $Q_{ij} = Q_{ji}$ denote the point of tangency of A_iA_j with Σ. Since the tangents drawn from A_i to Σ all have the same length, we have

$$\overline{A_1Q_{12}} = \overline{A_1Q_{13}} = \overline{A_1Q_{14}}, \quad \overline{A_2Q_{12}} = \overline{A_2Q_{23}} = \overline{A_2Q_{24}},$$

$$\overline{A_3Q_{13}} = \overline{A_3Q_{23}} = \overline{A_3Q_{34}}, \quad \overline{A_4Q_{14}} = \overline{A_4Q_{24}} = \overline{A_4Q_{34}},$$

$$\overline{A_1Q_{12}} + \overline{A_2Q_{12}} = \overline{A_1A_2}, \quad \overline{A_1Q_{13}} + \overline{A_3Q_{13}} = \overline{A_1A_3},$$

$$\overline{A_1Q_{14}} + \overline{A_4Q_{14}} = \overline{A_1A_4}, \quad \overline{A_2Q_{23}} + \overline{A_3Q_{23}} = \overline{A_2A_3},$$

$$\overline{A_2Q_{24}} + \overline{A_4Q_{24}} = \overline{A_2A_4}, \quad \overline{A_3Q_{34}} + \overline{A_4Q_{34}} = \overline{A_3A_4}.$$

From this, it is easy to conclude that $\overline{A_1Q_{12}} = \overline{A_1P_{12}}$, and hence that P_{12} and Q_{12} coincide; analogously, P_{ij} and Q_{ij} coincide, for $1 \le i < j \le 4$. Suppose that Σ' is internally tangent to S_1, S_2, S_3, S_4 (the case in which Σ' is externally tangent to S_1, S_2, S_3, S_4 can be dealt with in pretty much the same way). Let R and R' denote the radii of Σ and Σ', respectively. Since triangle $OQ_{ij}A_j$ is right at Q_{ij} and $\overline{OA_j} = R' + r_j$ and $\overline{A_jQ_{ij}} = r_j$, Pythagoras' theorem gives

$$(R' + r_j)^2 = R^2 + r_j^2.$$

Therefore, $r_j = \frac{R^2 - R'^2}{2R'}$, so that $r_1 = r_2 = r_3 = r_4$. Letting r stand for this common value, it follows that $\overline{A_iA_j} = r_i + r_j = 2r$ for $1 \le i < j \le 4$, and the tetrahedron is regular.

Section 11.2

1. Firstly, suppose that the give prism has a circumscribed sphere Σ, centered at O. Since Σ circumscribes the pyramid of apex A_1' and basis $A_1A_2 \ldots A_n$, it follows from Problem 3, page 386, that Σ is unique. Now, the bases of the prism are inscribed in the circles defined by the intersections of Σ with the planes containing them. On the other hand, $A_1A_2A_2'A_1'$ is a parallelogram inscribed in a circle (the one defined by the intersection of Σ with the plane containing it); thus, it is a rectangle. Analogously, $A_1A_nA_n'A_1'$ is a rectangle, so that $\overleftrightarrow{A_1A_1'} \perp \overleftrightarrow{A_1A_2}, \overleftrightarrow{A_1A_n}$, and hence $\overleftrightarrow{A_1A_1'} \perp (A_1A_2 \ldots A_n)$. Therefore, the prism is a right one. Conversely, if the prism is right and has cyclic bases, inscribed in circles centered at P and P', respectively, show that the whole

prism is inscribed in the sphere centered at the midpoint O of PP' and radius equal to $\frac{1}{2}\,\overline{PP'}$.

2. $D\widehat{H}C = 45°$ implies that $CDHG$ is a square. Now, setting $\overline{DH} = a$, compute $\overline{FH} = a\sqrt{3}$ and $\overline{BH} = 2a$. Finally, apply the cosine law to BHG to find $\cos G\widehat{B}H = \frac{\sqrt{3}}{2}$.

3. Since the diagonals of the cube form equal angles with all of the edges at which they are incident, each of the 8 triangles is an equilateral triangle, say of edge length ℓ. Then, show that the area of each octagon is equal to $4 - \ell^2$, so that $4 - \ell^2 = \frac{\ell^2\sqrt{3}}{4}$.

4. Firstly, show that $X \in AC$. Then, use Pythagoras' theorem to conclude that X coincides with the center of $ABCD$. Finally, apply the cosine law to compute $A\widehat{X}E$.

5. Let O be the center of the cube and S and T be the midpoints of AA' and XY, respectively. Note that S, O, Z and S, T, Z are triples of collinear points, so that O, S, T and Z are all collinear. In particular, the plane (XYZ) passes through the midpoint S of AA'. Analogously, (XYZ) passes through the midpoints U of $C'D'$ and V of BC, so that the planar section is the hexagon $SXUZVY$, of center O. Now, the midsegment theorem gives $\overline{UX} = \frac{1}{2}\overline{A'C'} = \frac{a\sqrt{2}}{2}$, and likewise for the other edges of the hexagon. Also, $\overline{OU} = \frac{1}{2}\overline{UY} = \frac{1}{2}\overline{BC'} = \frac{a\sqrt{2}}{2}$ an, by the same token, the distances from O to the vertices of the hexagon are all equal to $\frac{a\sqrt{2}}{2}$. Hence, $SXUZVY$ is regular of edge length $\frac{a\sqrt{2}}{2}$, and its area equals $\frac{3a^2\sqrt{3}}{4}$.

6. Let the parallelepiped have parallel faces $ABCD$ and $A'B'C'D'$, with $\overleftrightarrow{AA'} \parallel \overleftrightarrow{BB'} \parallel \overleftrightarrow{CC'} \parallel \overleftrightarrow{DD'}$ and $\overline{AB} = a$, $\overline{AD} = b$ and $\overline{AA'} = c$. Also, let α be the plane that intersects it. Let the regular hexagon be $KLMNOP$, such that (with no loss of generality) $K \in AB$, $L \in BC$, $M \in CC'$, $N \in C'D'$, $O \in A'D'$, $P \in AA'$. Since $\overline{KL} = \overline{NO}$, $L\widehat{K}B = O\widehat{N}D'$, $K\widehat{L}B = N\widehat{O}D'$, we have $BKL \equiv D'NO$, and hence $\overline{BK} = \overline{D'N}$. However, since $\overleftrightarrow{KN} \parallel \overleftrightarrow{OP}$, we have $\overline{AK} = \overline{D'N}$ too, and thus $\overline{AK} = \overline{BK} = \frac{a}{2}$. Analogously, L, M, N, O, P are the midpoints of the edges of the parallelepiped to which they belong. Now, if ℓ stands for the length of the edges of the hexagon, Pythagoras' theorem gives $a^2 + b^2 = 4\ell^2$, $a^2 + c^2 = 4\ell^2$, $b^2 + c^2 = 4\ell^2$. By looking at these equalities as a system of equations in a, b and c, we get $a = b = c = \ell\sqrt{2}$.

7. If $\mathcal{B}$ is a basis of the prism and γ is the plane of $\mathcal{B}$, then $\mathcal{B}$ can be seen as the orthogonal projections of $\mathcal{P}$ and $\mathcal{Q}$ onto γ. Apply item (b) of Problem 4, page 359, to obtain

$$A(\mathcal{P}) + A(\mathcal{Q}) = \frac{A(\mathcal{B})}{\cos\theta} + \frac{A(\mathcal{B})}{\sin\theta}.$$

Then, apply elementary Calculus (cf. [5], for instance) to minimize $f(\theta) = \frac{1}{\cos\theta} + \frac{1}{\sin\theta}$ when $0 < \theta < \frac{\pi}{2}$.

8. Item (i) essentially follows from $\sigma, \theta \in \left(0, \frac{\pi}{2}\right)$, together with the fact that $ADCB$, $EHGF$ are congruent and parallel squares. Item (ii) is similar; one just has to notice that $\overleftrightarrow{DH}$, $\overleftrightarrow{BF} \perp r$, so that $\overleftrightarrow{D'H'}$, $\overleftrightarrow{B'F'} \perp r$ too. For item (b), start by showing that

$$A(C'B'F'E'H'D') = A(A'D'C'B') + A(H'D'A'E') + A(A'B'F'E')$$
$$= A(A'D'C'B') + A(B'F'H'D')$$
$$= A(ADCB)\cos\sigma + \overline{B'F'} \cdot PQ,$$

where we have used item (b) of Problem 4, page 359, in the last equality and P and Q are the feet of the perpendiculars dropped from B' and D' to r, respectively. Subsequently, show that $\overline{B'F'} = \sin\sigma$ and $\overleftrightarrow{BP}$, $\overleftrightarrow{DQ} \perp r$ too; then, look at the triangular pyramids $ABB'P$ and $ADD'Q$ to compute $\overline{PQ} = \overline{AP} + \overline{AQ} = \cos\theta + \sin\theta$. Finally, for item (c), apply Proposition 7.21 twice to get

$$A(C'B'F'E'H'D') = \cos\sigma + (\sin\theta + \cos\theta)\sin\sigma$$
$$\leq \cos\sigma + \sqrt{2}\sin\sigma \leq \sqrt{3},$$

with equality if and only if $\theta = \frac{\pi}{4}$, $\cos\sigma = \frac{1}{\sqrt{3}}$ and $\sin\sigma = \frac{\sqrt{2}}{\sqrt{3}}$ (whence σ is the angle between AB and the diagonal AG).

Section 11.3

1. Let a stand for the length of the edges of $ABCD$. Take a cube $WXYZW'X'Y'Z'$, of bases $WXYZ$ and $W'X'Y'Z'$ (labeled in the usual way) and edge length $\frac{a}{\sqrt{2}}$. Tetrahedron $WYX'Z'$ is regular of edge length a, and we may assume that $A = W$, $B = Y$, $C = X'$ and $D = Z'$. This being the case, we have $C' = Z$ and $D' = X$, so that we wish to compute the angle between the planes (WYZ) and $(WX'X)$. However, since they are planes of adjacent faces of the cube, the desired angle is $90°$.
2. For each of these tetrahedra $SABC$, the proof of Proposition 11.9 shows that O is the center of the rectangular parallelepiped that has the trirectangular tetrahedron $SABC$ as one of its *corners*. If M is the midpoint of BC, then $\overleftrightarrow{OM} \parallel \overleftrightarrow{AS}$ and $\overline{OM} = \frac{1}{2}\overline{AS}$. If P is the point in which OS intersects the face ABC, then $OMP \sim ASP$ by AA. Hence, $\frac{\overline{OP}}{\overline{PS}} = \frac{\overline{OM}}{\overline{AS}} = \frac{1}{2}$, so that $P \in OS$ is the point such that $\overline{OP} = \frac{1}{3}\overline{OS}$. However, since O and S do not depend on the choices of A, B and C, the same happens with P.

3. If $ABCD$ is isosceles, Example 10.13 shows that the common perpendicular to $\overleftrightarrow{AD}$ and $\overleftrightarrow{BC}$ joins the midpoints of AD and BC, and likewise for the other two pairs of reverse edges. Conversely, let $\overline{AD} = a'$, $\overline{BC} = a$, $\overline{AB} = b'$, $\overline{CD} = b$, $\overline{AC} = c'$, $\overline{BD} = c$, and suppose that the common perpendiculars to the pairs of reverse edges of $ABCD$ join the midpoints of such edges. Let M be the midpoint of AC and N that of BD. Since $\overleftrightarrow{MN} \perp \overleftrightarrow{AC}$, the line segment MN is at the same time median and altitude of ANC, so that $\overline{AN} = \overline{CN}$. However, since AN and CN are medians of ABD and BCD, the formula for the length of a median applied to both sides of $\overline{AN} = \overline{CN}$ gives $(a')^2 + (b')^2 = a^2 + b^2$; analogously, $(a')^2 + (c')^2 = a^2 + c^2$ and $(b')^2 + (c')^2 = b^2 + c^2$. In view of such equalities, it is immediate to conclude that $a = a'$, $b = b'$ and $c = c'$.

4. If $ABCD$ is isosceles, we saw in the previous problem that the common perpendiculars to the pairs of reverse edges join the midpoints of such edges. Then, let M, N, P and Q denote the midpoints of the edges AD, BC, AC and BD. By applying the midsegment theorem to triangles ACD and BCD, we obtain $\overline{NQ} = \frac{1}{2}\overline{CD} = \overline{MP}$ and $\overline{NQ} \parallel \overline{CD} \parallel \overline{MP}$. Hence, M, N, P and Q are coplanar and the quadrilateral they form has equal and parallel opposite sides, so that it is a parallelogram. It follows that MN and PQ intersect at their respective midpoints, and a similar argument holds for the third common perpendicular. For an alternative proof, it suffices to identify the common perpendiculars under consideration in Fig. 11.21, noticing that they concur at the center of the rectangular parallelepiped associated to the isosceles tetrahedron.

5. If the tetrahedron $ABCD$ is isosceles, the two previous problems assure that the common perpendiculars to its pairs of reverse edges join the midpoints of such edges and intersect at their respective midpoints. Now, a quick glance at the proof of Theorem 11.13 allows us to conclude that such a common point O is the center of the rectangular parallelepiped associated to the tetrahedron, which coincides with the circumcenter of $ABCD$. Since $\overline{OA} = \overline{OB} = \overline{OC} = \overline{OD}$ and the faces of $ABCD$ are congruent, it is immediate to show that, in the tetrahedra $OABC$, $OABD$, $OACD$ and $OBCD$, the altitudes dropped from the vertex O have equal lengths. In this, this means that O lies at equal distances from the faces of $ABCD$, so that it is the incenter of $ABCD$. Conversely, let $ABCD$ be a tetrahedron in which the incenter I and circumcenter O coincide. Let also r and R denote the radii of the spheres inscribed and circumscribed to $ABCD$, respectively. We know that the intersection of (ABC) with the circumscribed sphere is a circle whose center (which is also the circumcenter of ABC) coincides with the projection of O on (ABC). However, since O and I coincide and the projection of I onto (ABC) lies in the interior of triangle ABC, we conclude that the circumcenter of ABC lies in its interior, so that ABC is acute. Analogously, the remaining faces of $ABCD$ are also acute. Now, the coincidence of I and O guarantees that the circumradii of the faces of $ABCD$ are all equal to $\sqrt{R^2 - r^2}$. By applying the sine law to the faces ABC and ACD, we obtain

$$\frac{\overline{AC}}{\sin A\widehat{D}C} = 2\sqrt{R^2 - r^2} = \frac{\overline{AC}}{\sin A\widehat{B}C},$$

so that $\sin A\widehat{D}C = \sin A\widehat{B}C$. However, since $A\widehat{D}C, A\widehat{B}C < 90°$, it follows that $A\widehat{D}C = A\widehat{B}C$. Likewise, $A\widehat{D}B = A\widehat{C}B$ and $B\widehat{D}C = B\widehat{A}C$, which in turn gives us

$$A\widehat{D}C + A\widehat{D}B + B\widehat{D}C = A\widehat{B}C + A\widehat{C}B = B\widehat{A}C = 180°.$$

Since the same reasoning holds true for the other vertices of $ABCD$, Proposition 11.10 assures that $ABCD$ is isosceles.

6. By the previous problem, it suffices to show that the barycenter and circumcenter coincide. To this end, and in the notations of Fig. 11.21, we can assume that the tetrahedron is $WX'YZ'$. However, since WY', $X'Z$, YW' and $Z'X$ intersect at the center O of the rectangular parallelepiped $WXYZW'X'Y'Z'$, it suffices to prove that WY', $X'Z$, YW' and $Z'X$ intersect the faces $X'YZ'$, WYZ', $WX'Z'$ and $WX'Y$ at their respective barycenters. In order to show that WY' intersects face $X'YZ'$ at its barycenter (the other cases are entirely analogous), apply, to the trirectangular tetrahedron $X'YY'Z'$, the hint given to Problem 2.

7. Review the proof of Theorem 11.13.

Section 12.1

1. In the notations of the statement of the corollary, split $A_1A_2\ldots A_k$ into the spherical triangles $A_1A_2A_i$, for $3 \le i \le k$. Then, apply Girard's theorem to each one of these triangles.

2. Apply formula (12.1) to this situation; the computations in Example 12.1 may help. You must find $2\pi R(R - d)$ as the answer.

3. Let O denote the center of the Earth, let P be the pilot's position and T be a point in the horizon, with respect to P (i.e., T is the point of contact of a tangent drawn from P to the Earth). If d stands for the distance from O to the plane of the horizon points, use metric relations in the right triangle OPT to obtain $(h + R)d = R^2$. Then, notice that the portion of the surface that can be seen by the pilot is a spherical cap, and apply the result of the previous problem.

4. Note that $A(ABC) = \frac{1}{4}A(\Sigma) = \pi R^2$. Now, apply Girard's theorem.

5. The given condition assures that, in every triangle P_iPP_j, with $i \ne j$, we have $P_i\widehat{P}P_j > 60°$. Therefore, if C_i is the (infinite) cone with vertex P, axis $\overrightarrow{PP_i}$ and opening $30°$, then each pair of $C_1, \ldots, C_n$ do not have common interior points. Hence, if Σ is the sphere of radius 1 centered at P, then

$$A(\Sigma) > \sum_{i=1}^{n} A(\Sigma \cap C_i).$$

Now, in the notations of Problem 2, $\Sigma \cap C_i$ is a spherical cap of height $1 - \frac{\sqrt{3}}{2}$, whence $A(\Sigma \cap C_i) = 2\pi \left(1 - \frac{\sqrt{3}}{2}\right)$. Thus, the area inequality above gives $4\pi > 2\pi \left(1 - \frac{\sqrt{3}}{2}\right)n$, so that $n < 4(2 + \sqrt{3}) \cong 14.9$. This shows that $n \leq 14$. For the example, we start by trying to achieve $n = 14$, and we do this by placing P_1, P_2, P_3, P_4 as the vertices of a square inscribed in an equator Γ of Σ, and P_{13} and P_{14} as the North and South poles of Σ with respect to Γ. Then, we put each of P_5 to P_{12} at the center of each of the eight spherical triangles having as their vertices one of the points P_{13} or P_{14}, together with two non antipodal points chosen from P_1, P_2, P_3, P_4. Thus, if we place P_5 at the center of $P_1 P_2 P_{14}$, then $P_5 \widehat{P} P_1 = P_5 \widehat{P} P_2 = P_5 \widehat{P} P_{14}$ so that $\overrightarrow{PP_5}$ lies along the diagonal of the cube with edges PP_1, PP_2, PP_{14}. However, a simple computation shows that the common value θ of those angles satisfies $\cos\theta = \frac{1}{\sqrt{3}}$, so that (with the aid of a calculator) $\theta \cong 54°$. Then, the described arrangement does not provide enough room for 14 cones for 6°. Nevertheless, we can achieve $n = 13$ by the following procedure: we erase P_{14} and move P_1, P_2, P_3, P_4 downwards until new positions P_1', P_2', P_3', P_4', such that $P_1' P_2' P_3' P_4'$ is a square parallel to Γ and with $P_1 \widehat{P} P_2 = P_2 \widehat{P} P_3 = P_3 \widehat{P} P_4 = P_4 \widehat{P} P_1 = 60°$. Since we needed to free enough space for just $2 \cdot 6 = 12°$ more and $P_1 \widehat{P} P_1' = 45°$, such a rearrangement of P_1, P_2, P_3, P_4 allows us to move P_5 to P_{12} slightly downwards too, in order to have 13 pairwise nonintersecting cones.

Section 12.2

1. Let $\mathcal{B}(A; R)$ be an open ball contained in $\mathcal{C}$. If $C \in AB \setminus \{B\}$ and $r = \frac{\overline{BC}}{\overline{AB}} \cdot R$, show that $\mathcal{B}(C; r) \subset \mathcal{C}$.

2. Use the definition of boundary to show that if $P \notin \partial \mathcal{A}$, then there exists $R > 0$ such that $\mathcal{B}(P; R)$ is contained either in $\mathcal{A}$ or in $\mathcal{A}^c$. Then, conclude that $\mathcal{B}(P; R) \cap \partial \mathcal{A} = \emptyset$, whence $(\partial \mathcal{A})^c$ is open.

3. Firstly, note that if a face of a convex polyhedron has k edges, then the polyhedron has at least $2k$ edges. From this, conclude that if there existed a convex polyhedron with exactly 7 edges, then all of its faces would be triangles. However, this being the case and letting F denote the number of faces of the polyhedron, show that we should have do poliedro $3F = 2 \cdot 7$, which is impossible.

4. Apply the result of Lemma 12.8.

5. Argue as in Lemma 12.8.

6. Let $V > 3$ be the number of vertices and P be one of the vertices of the polyhedron. Since P is adjacent to the other $V - 1$ vertices, which in turn are pairwise adjacent, we conclude that the polyhedron has exactly $(V - 1) + (V - 3) = 2V - 4$ faces and $(V - 1) + \binom{V-1}{2} = \frac{1}{2}(V^2 - V)$ edges. Then, it follows from Euler's relation that $V^2 - 7V + 12 = 0$, whence $V = 4$. Therefore, $F = 4$ and $E = 6$. It is now straightforward to show that we have got a tetrahedron.

7. By the sake of contradiction, suppose that there existed a convex polyhedron $\mathcal{P}$ for which all of its plane sections are triangles. If $\mathcal{P}$ has a vertex incident to more than three edges, then cutting $\mathcal{P}$ with a plane sufficiently close to such a vertex would give us a non triangular plane section; therefore, exactly three edges are incident to each vertex of $\mathcal{P}$. If a face of $\mathcal{P}$ has more than three edges, then cutting $\mathcal{P}$ with a plane parallel and sufficiently close to such a face would also give us a non triangular section; hence, all faces of $\mathcal{P}$ are triangular. Now, let V, E and F denote the numbers of vertices, edges and faces of $\mathcal{P}$, respectively. The remarks above furnish $3V = 2E = 3F$, so that $V = F$ and $E = \frac{3F}{2}$. Euler's relation thus gives $V = F = 4$ and $E = 6$. Now, arguing as in the solution to the previous problem, we conclude that $\mathcal{P}$ is a tetrahedron. However, it is immediate to show that every tetrahedron has a plane section which is a quadrilateral, and we have arrived at a contradiction.

8. We know that $V - E + F = 2$, with $F = F_3$. Hence, Lemma 12.8 gives $3F = 2E$. Now, let A be any vertex of the polyhedron. Since each face is an equilateral triangle and the sum of the angles at A of the faces incident to such a vertex is less than $360°$, we conclude that A is incident to at most 5 faces. However, since each face has three vertices, it follows that $3F \leq 5V$. Hence, $V \geq \frac{3F}{5} = \frac{2E}{5}$, whence

$$2 = V - E + F \geq \frac{2E}{5} - E + \frac{2E}{3} = \frac{E}{15}.$$

9. Use (12.5) and adapt the construction delineated in item i. of Remark 12.10.

10. Choose a face $\mathcal{F}$ of the polyhedron such that the distance from P to such a face is as small as possible. Let Q be the foot of the perpendicular dropped from P to the plane of $\mathcal{F}$, and assume that $Q \notin \mathcal{F}$. Then, PQ intersects the surface of the polyhedral at a point Q', with $\overline{PQ'} < \overline{PQ}$. If $\mathcal{F}'$ is a face of the polyhedron to which Q' belongs, then $d(P; \mathcal{F}') \leq \overline{PQ'} < \overline{PQ}$, which is an absurd.

11. Since each face is triangular, we have $2E = 3F = 36$, hence $E = 18$. Now, Euler's theorem gives $V = 2 - F + E = 8$; since $V = V_3 + V_6$ and $3V_3 + 6V_6 = 2E = 36$, we immediately get $V_3 = V_6 = 4$. Let A be a vertex incident to 6 edges (hence faces), say ABC, ACD, ADE, AEF, AFG, AGB. The eight vertex H cannot be incident to 6 others, for otherwise there would be no vertex of degree 3; hence, H is incident to 3 other vertices, which up to this point will be incident to at least 4 other vertices. Since there are only 4 vertices incident to 6 others, it is not difficult to see that H is either incident to B, D, F or C, E, G. Suppose, without loss of generality, that H is incident to B, D, F. Then, B, D, F are pairwise incident, and it is now pretty clear that the polyhedron

can be seen as the result of gluing the triangular pyramids $ABDC$, $ABFG$, $ADFE$, $BDFH$ to the tetrahedron $ABDF$. Take one of these pyramids, say $ABDC$. If not all of $\overline{CA}$, $\overline{CB}$, $\overline{CD}$ are equal, say $\overline{CA} = \overline{CB} = x$, $\overline{CD} = y$ (the other cases are entirely analogous), then $\overline{AB} = \overline{BD} = y$, $\overline{AD} = x$ and it is not difficult to show that the three dihedral angles between two of the planes (ABC), (ACD), (BCD) cannot be all equal. Therefore, $\overline{CA} = \overline{CB} = \overline{CD}$, and hence $\overline{AB} = \overline{AD} = \overline{BD}$. Since a similar reasoning holds for all of the other pyramids, we conclude that the polyhedron is the result of gluing the *regular* triangular pyramids $ABDC$, $ABFG$, $ADFE$, $BDFH$ to the *regular* tetrahedron $ABDF$. If $x < y$, then all of the lateral edges of the four triangular pyramids have length x, whereas the edges of $ABDF$ have length y. The plane dihedral angle α between two adjacent faces of each of the triangular pyramids is easy to compute and is such that $\cos \alpha = \frac{\cos \theta}{1 + \cos \theta}$, where $\theta = A\widehat{C}D$. In turn, by applying the cosine law to triangle ACD, we get $\cos \theta = 1 - \frac{y^2}{2x^2}$, so that $\cos \alpha = \frac{2x^2 - y^2}{4x^2 - y^2}$. Now, if M is the midpoint of AD and $\overline{CM} = \overline{EM} = l = \sqrt{x^2 - \frac{y^2}{4}}$, then the plane dihedral angle β between faces ACD and ADE is such that (from the cosine law applied to triangle CME) $2l^2 - 2l^2 \cos \beta = \overline{CE}^2$. Therefore, one just needs to compute $\overline{CE}$ in terms of x and y to find $\cos \beta$ in terms of x and y, and then to find $\frac{x}{y}$ by equating $\cos \alpha = \cos \beta$. For what is left to do, let X and Y be the centers of ABD and ADF, respectively, so that $\overline{AX} = \overline{AY} = \frac{y}{\sqrt{3}}$ and $\overline{XY} = \frac{y}{3}$. Since $\overline{CX} = \overline{EY} = \sqrt{x^2 - \frac{y^2}{3}}$, some Plane Euclidean Geometry in the trapezoid $CXYE$ gives the desired expression for $\overline{CE}$. The reader is to find $\frac{x}{y} = \frac{3}{5}$.

Section 12.3

1. Start by tiling the space with equal cubes in the usual way. Then, split each cube into one regular tetrahedron and four trirectangular tetrahedra (as in Fig. 11.21 when the parallelepiped is a cube), in such a way that, in the common face of any two neighboring cubes, the faces of the two trirectangular tetrahedra of one cube match with those of the two trirectangular tetrahedra of the other. If V is any vertex of the right angles of one such trirectangular tetrahedron, then the eight trirectangular tetrahedra having V as a common vertex of right angles can be assembled to form a regular octahedron.
2. In the notations of the proof of Example 12.13, the desired edge length is equal to the edge length of the regular pentagon inscribed in a circle of radius $r = \frac{2R}{\sqrt{5}}$. In order to compute it, apply the result of Problem 2, page 254. Alternatively, let O be the center of the sphere and ABC be one of the faces of the icosahedron. Use Girard's theorem to show that the angles of the spherical triangle ABC are all equal to $\frac{2\pi}{5}$ radians; then, notice that such an angle is equal to the plane angle

of the dihedral angle formed by planes (OAB) and (OAC) and which contains ABC. Now, let $P \in OB$ be the foot of the perpendicular dropped from A to OB, which coincides with the foot of the perpendicular dropped from C to OB (for $OABC$ is a regular triangular pyramid). Then, $\angle APC$ is the plane angle of the dihedral angle formed by planes (OAB) and (OAC) and which contains ABC, so that $A\widehat{P}C = \frac{2\pi}{5}$. Letting $\overline{AB} = \overline{AC} = \overline{BC} = l$ and $\overline{AP} = \overline{CP} = h$, it follows from the cosine law applied to ACP that $l = 2h \sin \frac{\pi}{5}$. Compute the area of OAB in two distinct ways to arrive at $Rh = l\sqrt{R^2 - \left(\frac{l}{2}\right)^2}$. Finally, obtain

$$l = R\sqrt{4 - \frac{1}{\sin^2 \frac{\pi}{5}}}.$$

Section 13.1

1. Split the octahedron into two quadrangular pyramids satisfying the following conditions: their common basis a parallelogram of area equal to half the area of the face of the parallelepiped with edge lengths a and b; their heights with respect to this common basis are both equal to $\frac{c}{2}$.

2. Let $\mathcal{P}$ be a parallelepiped of the same height h as $\mathcal{C}$, and whose basis $\mathcal{B}$ is a parallelogram of area equal to that of the closed disk $\mathcal{D}$ that forms the basis of the cylinder. Further, suppose that $\mathcal{B}$ and $\mathcal{D}$ both lie in the same plane α, with $\mathcal{C}$ and $\mathcal{P}$ contained in the same half-spaces with respect to α. If plane α' is parallel to α, the equality of the heights of the cylinder and the parallelepiped guarantee that $\mathcal{C} \cap \alpha' \neq \emptyset$ if and only if $\mathcal{P} \cap \alpha' \neq \emptyset$. Moreover, when this does happen, such intersections are respectively congruent to $\mathcal{D}$ and $\mathcal{B}$, hence have equal areas. Therefore, it follows from Proposition 13.3 that

$$\mathcal{V}(\mathcal{C}) = \mathcal{V}(\mathcal{P}) = A(\mathcal{B})h = A(\mathcal{D})h = \pi R^2 h.$$

3. For the second part, compute the volume of $ABCD$ as $\frac{1}{3}A(ABC)h$, where h stands for the height of $ABCD$ with respect to ABC.

4. Use Theorem 11.13 to compute the volume of $ABCD$ as the difference between the volume of the rectangular parallelepiped associated to it and the sum of the volumes of four trirectangular tetrahedra. Then, compute each of these volumes in terms of the edges of the parallelepiped, and the edges of the parallelepiped in terms of a, b and c.

5. Let $\mathcal{T} = ABCD$ and I denote its incenter. Decompose $\mathcal{T}$ into the four tetrahedra $ABCI$, $ABDI$, $ACDI$ and $BCDI$. Then, apply the formula of Proposition 13.7 to each such tetrahedra and add the results thus obtained. For items (a) and (b) use the first part, together with the results of the two previous problems.

6. Decompose $ABCD$ into the four tetrahedra $ABCP$, $ABDP$, $ACDP$, $BCDP$. Then, apply the formula of Proposition 13.7 to each one of them, and add the results.

7. If S denotes the total area of the tetrahedron and ABC is its face of greatest area, show that $A(ABC) \geq \frac{S}{4}$. Then, letting h denote the height of $ABCD$ relative to ABC, apply the formula for the volume of a tetrahedron, together with the result of Problem 5.

8. Use the properties of the barycenter of a tetrahedron to show that the height of $BCDG$ with respect to BCD is $\frac{1}{4}$ of the height of $ABCD$ with respect to this same basis.

9. Let V be the vertex of the cone, O the center of its basis, AB the chord along which the plane intersects such basis and M the midpoint of AB. Then, $\overline{AB} = 2\sqrt{3}$, $A\widehat{O}B = 60°$ and $V\widehat{M}O = 45°$. With these data at your disposal, successively compute the radius of the basis, $\overline{OM}$ and $\overline{VO}$.

10. Observe that two of the solids are solid cones of revolution, whereas the third one is the union of two solid cones of revolution with a common basis and disjoint interiors.

11. Item (a) follows from the fact (easily proved with the aid of the midsegment theorem for triangles) that if we translate F_1 and F_2 along the planes containing them, then the corresponding planar sections S are all congruent polygons. For (b), start by observing that (a) allows us to assume that the straightline joining the centers of F_1 and F_2 is perpendicular to the planes of them. Therefore, the midsegment theorem for triangles shows that S is a $2n$-gon whose edges are alternately equal to $\frac{a}{2}$ and $\frac{b}{2}$ and whose angles are all equal to $\frac{(n-1)\pi}{n}$. In order to compute its area in terms of a, b and n, proceed as in the hint given to Problem 7, 254.

12. Decompose the icosahedron into twenty congruent regular triangular pyramids. Then, letting l be the length of the edges of the icosahedron, computed with the aid of Problem 2, page 425), show that the height of each of the twenty pyramids is equal to $\sqrt{R^2 - \left(\frac{l}{\sqrt{3}}\right)^2}$. Proceed in a likewise manner for the dodecahedron, this time using the result of Problem 4, page 425.

13. For the sake of simplicity, we shall work the proof for a frustum of right triangular prism, the proof for a general one being totally analogous. Let ABC be the basis of the frustum, and a, b and c be the lengths of its lateral edges, respectively incident at A, B and C, with $a < b, c$ (cf. Fig. 13.7, page 439). Through the endpoint of the edge of length a draw (as shown in the figure), a plane parallel to the basis of the frustum. It falls decomposed into a right triangular prism of lateral edge length a, and a quadrangular pyramid whose basis is a right trapezoid of bases $b - a$ and $c - a$ and height $\overline{BC}$. If h denotes the height of such a quadrangular pyramid, then h is equal to the altitude of ABC relative to the side BC. Thus,

$$\mathcal{V}(\text{frustum}) = A(ABC) \cdot a + \mathcal{V}(\text{piramide})$$

$$= A(ABC) \cdot a + \frac{1}{3} \cdot \frac{[(b-a)+(c-a)]\overline{BC}}{2} \cdot h$$

$$= \frac{(b+c-2a)}{3} \cdot \frac{\overline{BC}h}{2}$$

$$= A(ABC)a + \frac{(b+c-2a)}{3} A(ABC)$$

$$= A(ABC) \cdot \frac{(a+b+c)}{3}.$$

14. Use Cavalieri's principle to show that the desired volume is equal to that of the solid obtained by the rotation, around the axis of abscissas, of the trapezoid of vertices $(0, 0)$, $(3, 0)$, $(3, 22)$ and $(0, 4)$.

15. For item (a), assume without loss of generality that the midpoint of LM lies in the line segment MH. Then, by applying Pythagoras' theorem in NHM, we get $\overline{NH}^2 = \overline{MN}^2 - \overline{MH}^2 \le 1 - \left(\frac{x}{2}\right)^2$. For (b), suppose (also without loss of generality) that AD is the only edge of $ABCD$ whose length is not necessarily less than or equal to 1. If $\overline{BC} = a$ and h is the altitude of triangle ABC with respect to BC, then (a) gives $h \le \sqrt{1 - \frac{a^2}{4}}$. If d is the altitude of $ABCD$ with respect to (ABC), then $d \le k$, where k is the altitude of BCD with respect to BC; since $k \le \sqrt{1 - \frac{a^2}{4}}$ (again from (a)), we conclude that $d \le \sqrt{1 - \frac{a^2}{4}}$. Thus,

$$\mathcal{V}(ABCD) = \frac{1}{3} A(ABC)d = \frac{1}{3} \cdot \frac{ah}{2} \cdot k \le \frac{a}{6}\left(1 - \frac{a^2}{4}\right),$$

whence $\mathcal{V}(ABCD) \le \frac{1}{8} \Leftrightarrow a(4 - a^2) \le 3$. Now, since $a \le 1$, we have $a(4 - a^2) = a(2 - a)(2 + a) \le 3a(2 - a) \le 3$.

16. Letting P denote the point where α intersects the edge CD, the planar section of $ABCD$ through α is triangle ABP. Hence, α divides $ABCD$ into the tetrahedra $ABDP$ and $ABCP$. Since $A(ABC) = A(ABD)$, in order to conclude they have equal volumes it suffices to show that their altitudes dropped from P have equal lengths. However, since $P \in \alpha$, we know that P is at equal distances from the planes (ABC) and (ABD).

17. Construct the rectangular parallelepiped associated to $ABCD$, with bases $AC'BD'$ and $A'CB'D$ such that $\overleftrightarrow{AA'}$, $\overleftrightarrow{BB'}$, $\overleftrightarrow{CC'}$ and $\overleftrightarrow{DD'}$ are parallel. Then, draw a plane α containing $\overleftrightarrow{MN}$, and let P and P' (resp. Q and Q') denote its intersections with lines $\overleftrightarrow{BC'}$ and $\overleftrightarrow{B'C}$ (resp. $\overleftrightarrow{AD'}$ and $\overleftrightarrow{A'D}$), respectively. If R is the point of intersection of $\overleftrightarrow{QQ'}$ and $\overleftrightarrow{AD}$, and S that of $\overleftrightarrow{PP'}$ and $\overleftrightarrow{BC}$, then α cuts $ABCD$ along the quadrilateral $NRMS$. In order to show that α

divides $ABCD$ into two solids of equal volumes, prove that the portions of the rectangular parallelepiped lying in each of the half-spaces determined by α and exterior to $ABCD$ have equal volumes; for example, start by showing that $\mathcal{V}(AMQR) = \mathcal{V}(BMPS)$.

18. For item (a), let ℓ denote the length of the edges of the tetrahedron. If the parallels to A_1A_2, A_1A_3 and A_1A_4 through P intersect $A_2A_3A_4$ in A_2', A_3', A_4', respectively, then one of the 14 pieces is the (regular) tetrahedron $PA_2'A_3'A_4'$. Show

$$\mathcal{V}(PA_2'A_3'A_4') = \left(\frac{d_1}{h}\right)^3 \mathcal{V}(A_1A_2A_3A_4) = \frac{d_1^3}{h^3}.$$

Now, note that 4 of the 14 pieces are parallelepipeds, with each one of them having each of A_1, A_2, A_3, A_4 as one of its vertices; we shall compute the volume of the parallelepiped having A_2 as a vertex. To this end, let B_2B_3 (with $B_2 \in A_1A_2$ and $B_3 \in A_1A_3$) be the intersection of the plane through P and parallel to $A_2A_3A_4$ with $A_1A_2A_3$. If $\overline{B_2B_3} = \ell_1$, show that $\frac{\ell_1}{h-d_1} = \frac{\ell}{h}$ then, note that $A_1B_2B_3$ is equilateral, whence $\overline{A_2B_2} = \overline{A_1A_2} - \overline{A_1B_2} = \ell - \ell_1 = \frac{\ell d_1}{h}$. Analogously, if C_1C_2 ($C_1 \in A_1A_3$ and $C_2 \in A_2A_3$) is the intersection of the plane through P and parallel to $A_1A_2A_4$ with $A_1A_2A_3$, compute $\overline{A_2C_2} = \frac{\ell d_3}{h}$. If $B_2B_3 \cap C_1C_2 = \{D\}$, then the volume of the parallelepiped with vertex A_2 is

$$A(A_2C_2DB_2)d_4 = \frac{\ell d_1}{h} \cdot \frac{\ell d_3}{h} \cdot \frac{\sqrt{3}}{2} \cdot d_4 = \frac{1}{3} \cdot \frac{\ell^2\sqrt{3}}{4} \cdot h \cdot \frac{6d_1d_3d_4}{h^3}$$

$$= \mathcal{V}(A_1A_2A_3A_4)\frac{6d_1d_3d_4}{h^3} = \frac{6d_1d_3d_4}{h^3}.$$

For item (b), we only need to show that $f(P) < \frac{3}{4}$. To this end, let $x_i = \frac{d_i}{h}$ and note that (from the result of Problem 6) $x_1 + x_2 + x_3 + x_4 = 1$. Hence, $f(P) = 1 - \sum_i x_i^3 - 6\sum_{i<j<k} x_ix_jx_k$. Now, some easy algebra gives

$$1 = (x_1 + x_2 + x_3 + x_4)^3 = \sum_i x_i^3 + 6\sum_{i<j<k} x_ix_jx_k + 3\sum_{i\neq j} x_i^2x_j$$

$$= 1 - f(P) + 3\sum_{i<j} x_ix_j(x_i + x_j),$$

so that $f(P) = 3\sum_{i<j} x_ix_j(x_i + x_j)$. Hence,

$$f(P) < \frac{3}{4} \Leftrightarrow \sum_{i<j} x_ix_j(x_i + x_j) < \frac{1}{4}.$$

This last inequality can be established in the following way: firstly, assume (without loss of generality) that $x_1 \geq x_2 \geq x_3 \geq x_4$ and show that $\sum_{i<j} x_i x_j (x_i + x_j) < \sum_{i<j} u_i u_j (u_i + u_j)$, where $u_1 = x_1$, $u_2 = x_2$, $u_3 = x_3 + x_4$, $u_4 = 0$; then, assume (also without loss of generality) that $u_1 \geq u_2 \geq u_3$ and show that $\sum_{i<j} u_i u_j (u_i + u_j) < \sum_{i<j} v_i v_j (v_i + v_j)$, where $v_1 = u_1$ and $v_2 = u_2 + u_3$. Further details can be found in Example 5.25 of [5].

Section 13.2

1. Adapt, to the present case, the proof of Theorem 13.11.
2. Firstly, note that the desired volume is equal to $|V_1 + V_2 - V_3|$, where V_1, V_2 and V_3 stand (in some order) for the volumes of the frustums of right cones generated by the rotations of sides AB, AC and BC around r; then, compute each of V_1, V_2 and V_3 with the aid of Theorem 13.13. Finally, use some Plane Geometry to show that d is equal to the arithmetic mean of the distances of vertices A, B and C to r.
3. Adapt, to the present case, the proof of Theorem 13.13.
4. Apply the result of the previous problem to the functions $f, g : [-R, R] \to \mathbb{R}$, given by $f(x) = d + \sqrt{R^2 - x^2}$ and $g(x) = d - \sqrt{R^2 - x^2}$.
5. Given a partition $P = \{a = x_0 < x_1 < \cdots < x_k = b\}$ of $[a, b]$, let $\mathcal{R}_j$ denote the rectangle of basis $[x_{j-1}, x_j]$ and height $f(\xi_j)$, with $\xi_j \in (x_{j-1}, x_j)$; approximate the barycenter of $\mathcal{R}_j$ by $G_j(\xi_j, \frac{1}{2}f(\xi_j))$. Let ρ be the mass density of $\mathcal{R}$ (which we are assuming to be constant) and m_j and m be the masses of $\mathcal{R}_j$ and $\mathcal{R}$, respectively, so that $m_j = \rho f(\xi_j)(x_j - x_{j-1})$ and $m = \rho \int_a^b f(x)dx$. It follows from (13.1) that $\sum_{j=1}^k m_j(\xi_j - x_G, \frac{1}{2}f(\xi_j) - y_G) = (0, 0)$, or also $\sum_{j=1}^k m_j \xi_j - \sum_{j=1}^k m_j x_G = 0$ and $\frac{1}{2}\sum_{j=1}^k m_j f(\xi_j) - \sum_{j=1}^k m_j y_G = 0$. In these equalities, substitute the expressions for m_j and m, and use that $m = \sum_{j=1}^k m_j$; then, let $|P| \to 0$ to conclude the problem with the aid of Riemann's theorem (cf. Section 10.3 of [5], for instance).
6. Firstly, recall that the point of intersection of the medians of ABC is $(\frac{a+b}{3}, \frac{h}{3})$. Then, show that $\int_0^a f(x)dx = \frac{ah}{2}$, $\int_0^a xf(x)dx = \frac{ah}{2}\left(\frac{a+h}{3}\right)$ and $\frac{1}{2}\int_0^a f(x)^2 dx = \frac{ah}{2} \cdot \frac{h}{3}$.
7. Adapt, to the present case, the hint given to that problem. You must arrive at the formulas

$$x_G = \frac{\int_a^b x(f(x) - g(x))dx}{\int_a^b (f(x) - g(x))dx} \quad \text{and} \quad y_G = \frac{\frac{1}{2}\int_a^b (f(x)^2 - g(x)^2)dx}{\int_a^b (f(x) - g(x))dx}.$$

8. Use that $A = \int_a^b (f(x) - g(x))dx$, $d = y_G = \frac{\frac{1}{2}\int_a^b (f(x)^2 - g(x)^2)dx}{\int_a^b (f(x) - g(x))dx}$ and
   (according to Problem 3) $V = \pi \int_a^b (f(x)^2 - g(x)^2)dx$.

9. Pappus' theorem gives $\frac{4}{3}\pi R^3 = 2\pi Ad$, with $A = \frac{1}{2}\pi R^2$.

10. According to Pappus, the volume of such a torus equals $2\pi \cdot \pi R^2 \cdot d = 2\pi^2 Rd$.

Glossary

Problems tagged with a country's name refer to any round of the corresponding national mathematical olympiad. For example, a problem tagged "Brazil" means that it appeared in some round of some edition of the Brazilian Mathematical Olympiad. Problems proposed in other mathematical competitions, or which appeared in mathematical journals, are tagged with a specific set of initials, as listed below:

1. **APMO**: Asian-Pacific Mathematical Olympiad.
2. **Austrian-Polish**: Austrian-Polish Mathematical Olympiad.
3. **BMO**: Balkan Mathematical Olympiad.
4. **Baltic Way**: Baltic Way Mathematical Contest.
5. **Crux**: Crux Mathematicorum, a mathematical journal of the Canadian Mathematical Society.
6. **EKMC**: Eötvös-Kürschák Mathematics Competition (Hungary).
7. **IMO**: International Mathematical Olympiad.
8. **IMO shortlist**: Problem proposed to the IMO, though not used.
9. **Israel-Hungary**: Binational Mathematical Competition Israel-Hungary.
10. **Miklós-Schweitzer**: The Miklós-Schweitzer Mathematics Competition (Hungary).
11. **NMC**: Nordic Mathematical Contest.
12. **OCM**: State of Ceará Mathematical Olympiad.
13. **OCS**: South Cone Mathematical Olympiad.
14. **OBMU**: Brazilian Mathematical Olympiad for University Students.
15. **OIM**: Iberoamerican Mathematical Olympiad.
16. **OIMU**: Iberoamerican Mathematical Olympiad for University Students.
17. **ORM**: Rioplatense Mathematical Olympiad.
18. **Putnam**: The William Lowell Mathematics Competition.
19. **TT**: The Tournament of the Towns.

© Springer International Publishing AG, part of Springer Nature 2018

535

A. Caminha Muniz Neto, *An Excursion through Elementary Mathematics, Volume II*,
Problem Books in Mathematics, https://doi.org/10.1007/978-3-319-77974-4

Bibliography

1. A.V. Akopyan, A.A. Zaslavsky, *Geometry of Conics* (Providence, AMS, 2007)
2. T. Apostol, *Calculus*, vol. 1 (Wiley, New York, 1967)
3. T. Apostol, *Calculus*, Vol. 2 (Wiley, New York, 1967)
4. V.G. Boltianski, *Figuras Equivalentes y Equicompuestas*. Lecciones Populares de Matemáticas. (In Spanish.) (MIR, Moscow, 1981)
5. A. Caminha, *An Excursion Through Elementary Mathematics I - Real Numbers and Functions* (Springer, New York, 2017)
6. A. Caminha, *An Excursion Through Elementary Mathematics III - Discrete Mathematics and Polynomial Algebra* (Springer, New York, 2018)
7. H.S.M. Coxeter, S.L. Greitzer, *Geometry Revisited* (MAA, Washington, 1967)
8. D.G. de Figueiredo, *Números Irracionais e Transcendentes*. (In Portuguese.) (SBM, Rio de Janeiro, 2002)
9. G.B. Folland, *Real Analysis: Modern Techniques and Their Applications* (Wiley, New York, 1999)
10. S.L. Greitzer, *International Mathematical Olympiads, 1959–1977* (MAA, Washington, 1978)
11. T.L. Heath, *The Thirteen Books of Euclid's Elements* (Dover, Mineola, 1956)
12. D. Hilbert, *Foundations of Geometry* (Open Court Publ. Co., Peru, 1999)
13. R. Honsberger, *Mathematical Gems II* (MAA, Washington, 1976)
14. R. Honsberger, *Episodes in Nineteenth and Twentieth Century Euclidean Geometry* (MAA, Washington, 1995)
15. R. Honsberger, *Ingenuity in Mathematics* (MAA, Washington, 1970)
16. R. Johnson, *Advanced Euclidean Geometry* (Dover, Mineola, 2007)
17. S. Lang, *Algebra* (Springer, New York, 2002)
18. E.L. LIMA, *Medida e Forma em Geometria* (Sociedade Brasileira de Matemática, 1997)
19. E.E. Moise, *Elementary Geometry from an Advanced Standpoint* (Addison-Wesley, Boston, 1963)
20. D. Pedoe, *Geometry, a Comprehensive Course* (Dover, Mineola, 1988)
21. J. Roberts, *Elementary Number Theory: a Problem Oriented Approach* (The MIT Press, Boston, 1977)
22. W. Rudin, *Principles of Mathematical Analysis* (McGraw-Hill, New York, 1976)
23. K. Ueno, K. Shiga, S. Morita, *A Mathematical Gift, II: The Interplay Between Topology, Functions, Geometry and Algebra* (AMS, Providence, 1995)
24. I.M. Yaglom, *Geometric Transformations I* (MAA, Washington, 1962)
25. I.M. Yaglom, *Geometric Transformations II* (MAA, Washington, 1968)
26. I.M. Yaglom, *Geometric Transformations III* (MAA, Washington, 1973)
27. I.M. Yaglom, A. Shenitzer, *Geometric Transformations IV* (MAA, Washington, 2009)

© Springer International Publishing AG, part of Springer Nature 2018
A. Caminha Muniz Neto, *An Excursion through Elementary Mathematics, Volume II*,
Problem Books in Mathematics, https://doi.org/10.1007/978-3-319-77974-4

Index

© Springer International Publishing AG, part of Springer Nature 2018
A. Caminha Muniz Neto, *An Excursion through Elementary Mathematics, Volume II*,
Problem Books in Mathematics, https://doi.org/10.1007/978-3-319-77974-4